Applied Technology Integration in Governmental Organizations:
New E-Government Research

Vishanth Weerakkody
Brunel University, UK

INFORMATION SCIENCE REFERENCE

Hershey · New York

Director of Editorial Content:	Kristin Klinger
Director of Book Publications:	Julia Mosemann
Acquisitions Editor:	Lindsay Johnston
Development Editor:	Myla Harty
Publishing Assistant:	Casey Conapitski
Typesetter:	Casey Conapitski
Production Editor:	Jamie Snavely
Cover Design:	Lisa Tosheff

Published in the United States of America by
Information Science Reference (an imprint of IGI Global)
701 E. Chocolate Avenue
Hershey PA 17033
Tel: 717-533-8845
Fax: 717-533-8661
E-mail: cust@igi-global.com
Web site: http://www.igi-global.com

Library of Congress Cataloging-in-Publication Data

Applied technology integration in governmental organizations : new e-government research / Vishanth Weerakkody, editor.
       p. cm.
   Includes bibliographical references and index.
   Summary: "This book provides organizational and managerial directions to support the greater use and management of electronic or digital government technologies in organizations, while epitomizing the current e-government research available"--Provided by publisher.
   ISBN 978-1-60960-162-1 (hardcover) -- ISBN 978-1-60960-164-5 (ebook) 1. Public administration--Data processing. 2. Public administration--Information resources management. 3. Internet in public administration. I. Weerakkody, Vishanth.
   JF1525.A8A67 2011
   352.3'80285--dc22
                                    2010046636

British Cataloguing in Publication Data
A Cataloguing in Publication record for this book is available from the British Library.

All work contributed to this book is new, previously-unpublished material. The views expressed in this book are those of the authors, but not necessarily of the publisher.

Table of Contents

Detailed Table of Contents

Chapter 1

Jayoti Das, Elon University, USA

Cassandra DiRienzo, Elon University, USA

John Burbridge Jr., Elon University, USA

Using cross-country data from 140 countries, this empirical study extends past research by examining the impact of trust on the level of e-government. The major empirical finding of this research shows that, after controlling for the level of economic development and other socio-economic factors, trust as measured by ethnic and religious diversity, is a significant factor affecting e-government usage.

Chapter 2

Sivaporn Wangpipatwong, Bangkok University, Thailand

Wichian Chutimaskul, King Mongkut's University of Technology Thonburi, Thailand

Borworn Papasratorn, King Mongkut's University of Technology Thonburi, Thailand

This study empirically examines website quality toward the enhancement of the continued use of e-Government websites by citizens. The website quality under examination includes three main aspects, which are information quality, system quality, and service quality. The participants were 614 country-wide e-Citizens of Thailand. The data were collected by means of a web-based survey and analyzed by using multiple regression analysis. The findings revealed that the three quality aspects enhanced the continued use of e-Government websites, with system quality providing the greatest enhancement, followed by service quality and information quality.

Chapter 3

Roy H. Segovia, San Diego State University, USA

Murray E. Jennex, San Diego State University, USA

James Beatty, San Diego State University, USA

Can web design improve the way governments serve their constituencies through the use of information technology [i.e., electronic government (e-government)]? This paper proposes that the use of paralingual web design can overcome possible trust issues in e-government with bilingual populations. An experiment was conducted where active e-government web pages were converted to paralingual format and then site visitors were surveyed regarding their trust in the content and readability. Readability was surveyed to ensure that the paralingual format did not reduce site ease of use and usefulness (in this case ease of use and usefulness can only be viewed as readability, as no other functions are possible using the web sites.) The results of the experiment show that trust was improved for the minority language speakers, while the majority language speakers remained neutral with neither group indicating significant decrease in readability. These findings have important implications for societies with large bilingual or multilingual populations, where issues of trust among minority speakers and majority speakers may exist, as they indicate that paralingual web design can help reduce these trust issues.

Chapter 4

Hana Abdullah Al-Nuaim, King Abdulaziz University, Saudi Arabia

High speed wireless networks and mobile and web-based services are changing the way we, as consumers of information, communicate, learn, do business and receive services. Successful e-commerce models have raised the expectations of citizens to have government agencies and organizations provide public services that are timely and efficient. With the growth and development of arab cities, especially in the capitals, life becomes a little bit harder for citizens dealing with highly bureaucratic government agencies as their demands for basic services increase. Although e-readiness in the region has grown considerably with impressive progress, arab cities have been clearly absent from studies on worldwide e-municipal websites. In this study, an evaluation checklist for was used to evaluate official municipal websites of arab capitals. The study found that these websites were not citizen centered, suffered from fundamental problems, had some features that were inoperable and did not follow basic guidelines for any municipal website. These sites were dominated by aesthetics and technical novelties alone, providing inactive information rather than the inclusion of interactive e-services with immediate feedback and easy to use, navigable interfaces.

Chapter 5

Ferne Friedman-Berg, FAA Human Factors Team - Atlantic City, USA
Kenneth Allendoerfer, FAA Human Factors Team - Atlantic City, USA
Shantanu Pai, Engility Corporation, USA

The Federal Aviation Administration (FAA) Human Factors Team, Atlantic City conducted a usability assessment of the www.fly.faa.gov website to examine user satisfaction and identify site usability issues. The FAA Air Traffic Control System Command Center uses this website to provide information about airport conditions, such as arrival and departure delays, to the public and the aviation industry. The most important aspect of this assessment was its use of quantitative metrics to evaluate how successfully users with different levels of aviation-related expertise could complete common tasks, such as determining the amount of delay at an airport. The researchers used the findings from this assessment to make design recommendations for future system enhancements that would benefit all users. They discuss why usability assessments are an important part of the process of evaluating e-government websites and why their usability evaluation process should be applied to the development of other e-government websites.

Chapter 6

This chapter examines the role e-government has over citizens' when they initiate contact with their government. It also compares the influence that other contact channels have on citizens' contacts with government. A public opinion survey is analyzed to determine what factors explain the different methods of contacting government, namely through the phone, e-government, visiting a government office, or a combination of approaches. This chapter also analyzes citizens' preferred method of contacting government, examining different types of information or assistance that citizens' can get from government. The results of this study indicate that e-government is just one of many possible service channels that citizens use, with the phone being the most common. The overall importance of the survey results indicate that e-government is just one contact channel for citizens, and resources should also be devoted towards other contact channels given their importance as well to citizens.

E-government strategies empower citizens through online access to services and information. Consequently, governments – including in developing countries – are implementing e-government. In this study, a survey examined available services and targeted users in Tanzania. Ninety-six government agencies responded: 46% had implemented e-government using websites. Most services (60-90%) relate to disseminating information; online transactions were the least available services. Government-affiliated staff constituted the majority (60-85%) of users. This implies that emerging e-government services mostly address internal needs (government-to-government), and one-way dissemination of information (government-to-citizen). While agencies exhibited a gradual extension to businesses (government-to-business), citizen-to-government and business-to-government relationships were minimal. Finally, the study compares Tanzania's web-presence with select countries, draws its wider implications, and advocates further research on the nature and needs of users.

The study investigates the user acceptance of automated teller machine (ATM) and transit applications (Touch 'n Go) which are embedded in Malaysian multipurpose smart identity card named as MyKad. A research framework was developed based on a well known user acceptance model i.e. Unified Theory of Acceptance and Use of Technology (UTAUT) model. Five hundred questionnaires were randomly distributed in the Multimedia Super Corridor, Malaysia. The data were analyzed using descriptive and inferential statistics. The results show that Malaysians do not have strong intentions of using the two applications. This can be explained by factors shown in the research framework: performance expectancy, effort expectancy, social influence, facilitating conditions, perceived credibility, and anxiety. Malaysians have little understanding of their benefits and the efforts needed to use them. In addition, they have the misconception that there are insufficient facilities to support the usage of the applications. Consequently, there is no social support to use the applications. Moreover, they perceive that the applications do not have credibility. Besides, they are unsure if they use of the applications would cause anxieties. As a result, few Malaysians have intentions of using the applications. Recommendations were given to increase the acceptance and to resolve the discovered issues. The present research can be replicated to study user acceptance of other applications in MyKad/smart identity cards in other countries (e.g. Hong Kong, India and the sultanate of Oman).

Chapter 9

    Kenneth L. Hacker, New Mexico State University, USA
    Shana M. Mason, New Mexico State University, USA
    Eric L. Morgan, New Mexico State University, USA

The objective of this article is to examine how the inequalities of participation in network society governmental systems affect the extent that individuals are empowered or disempowered within those systems. By using published data in conjunction with theories of communication, a critical secondary data analysis was conducted. This critical analysis argues that the Digital Divide involves issues concerning how democracy and democratization are related to computer-mediated communication (CMC) and its role in political communication. As the roles of CMC/ICT systems expand in political communication, existing Digital Divide gaps are likely to contribute to structural inequalities in political participation. These inequalities work against democracy and political empowerment for some people, while at the same time producing expanded opportunities of political participation for others. This raises concerns about who benefits the most from electronic government in emerging network societies.

Chapter 10

    Dionysis Kefallinos, National Technical University of Athens, Greece
    Maria A. Lambrou, University of the Aegean Business School, Greece
    Efstathios D. Sykas, National Technical University of Athens, Greece

In this paper we propose a model for a risk assessment tool directed towards and tailored specifically for e-government projects. Our goal is to cover the particular threats pertinent to the e-government project context and provide an interface between the broader philosophy of IT governance frameworks and the technical risk assessment methodologies, thus aiding in the successful and secure implementation and operation of e-government infrastructures. The model incorporates a wide range of applicable risk areas, grouped into eleven levels, as well as seven accompanying dimensions, assembled into a checklist-like matrix, along with an application algorithm and associated indices, which an evaluator can use to calculate risk for one or for multiple interacting projects.

Chapter 11

Local Government Authorities (LGAs) are complex organisations whose heterogeneous operational structures can be greatly enhanced by effectively using of Information Technology (IT) to support improvements in the quality of services offered to citizens. While the benefits of IT cannot be disputed, there are several concerns about its success as LGAs are confronted with the challenges of synchronising their cross-departmental business processes and integrating autonomous IS. This paper examines a potentially important area of IT infrastructure integration in LGAs through Enterprise Application Integration (EAI). The adoption of EAI solutions is a burgeoning phenomenon across several private and public organisations. Nevertheless, where EAI has added efficacy to the IT infrastructures in the private domain, LGAs have also been slow in adopting cost-effective EAI solutions. The shortage of research studies on EAI adoption in LGAs presents a knowledge gap that needs to be plugged. The research methodology followed consisted of an in-depth analysis of two case studies by using the research tools of interviews, observation and referring to archival documents. This research is timely as the demand for integrated service delivery increases, the issues of harmonising business processes and integrating IS becomes pertinent. The conclusion and lessons that can be learnt from this research is that integrating IT infrastructures through EAI achieves significant efficiency in delivering end-to-end integrated electronic Government (e-Government) services.

Chapter 12

This paper examines the adoption of free wireless Internet parks (iPark) by Qatari citizens as a means of accessing electronic services from public parks. The Qatari government has launched the free wireless Internet parks concept under their national electronic government (e-government) initiative with a view to providing free Internet access for all citizens whilst enjoying the outdoors. By offering free wireless internet access, the Qatari government hopes to increase the accessibility of e-government services and encourage their citizens to actively participate in the global information society with a view to bridging the digital divide. The adoption and diffusion of iPark services will depend on user acceptance and the availability of wireless technology. This paper examines an extended technology acceptance model (TAM) that proposes individual differences and technology complexity in order to determine perceived usefulness and perceived ease of the iPark initiative by using a survey-based study. The paper provides a discussion on the key findings, research implications, limitations, and future directions for the iPark initiative in Qatar.

The advent and widespread use of the Internet enables governments to connect directly to citizens and businesses. This results in a decrease of the transaction costs and a reduction of the administrative burden for governments, citizens and businesses. One consequence of this is the bypassing of existing parties that are interacting as intermediaries between the clients and provider. Based on intermediation theory, two case studies are analyzed which counter the argument of the bypassing of intermediaries. It is possible to adopt a re-intermediation strategy, in which intermediaries are used as a value-adding service delivery channel. The empirical evidence shows that private sector intermediaries can be employed for service innovation. The intermediaries provide service delivery channels that are closer to the natural interaction patterns of the users of public services than direct service delivery channels. Furthermore, the intermediaries can bundle public services with their own. For governments, this implies that only adopting a direct interaction strategy, which is often motivated by a desire to reduce transaction costs, is too narrow an approach from the point of view of demand-driven government. As such, direct interaction strategies need to be complimented by a re-intermediation strategy employing private sectors intermediaries in order to advance towards a truly demand-driven and client centered government.

An increasing level of cooperation between public administrations nowadays on national, regional and local level requires methods to develop interoperable eGovernment solutions and leads to the necessity of an efficient evaluation and requirements engineering process that guides the establishment of systems and services used by public administrations in the European Union. In this paper, we propose a framework to systematically gather and evaluate requirements for eGovernment in the large. The evaluation framework is designed to support requirements engineers to develop a suitable evaluation and requirements engineering process with respect to interoperable eGovernment solutions. The methodology is motivated and explained on the basis of a European research project.

Electronic tax filing is an emerging area of e-government. This research proposes a model of e-filing adoption that identifies adoption factors and personal factors that impact citizen acceptance of electronic filing systems. A survey administered to 260 participants assesses their perceptions of adoption factors, trust and self-efficacy as they relate to e-file utilization. Multiple linear regression analysis is used to evaluate the relationships between adoption concepts and intention to use e-filing systems. Implications for practice and research are discussed.

Chapter 16

Abdullah Al-Shehry, Prince Nayef College, Saudi Arabia
Simon Rogerson, De Montfort University, UK
N. Ben Fairweather, De Montfort University, UK
Mary Prior, De Montfort University, UK

The e-government paradigm refers to utilizing the potential of Information and Communication Technology (ICT) in the whole government body to meet citizens' expectations via multiple channels. It is, therefore, a radical change within the public sector and in the relationship between a government and its stakeholders. In the light of that, the Kingdom of Saudi Arabia has a keen interest in this issue and thus it has developed a national project to implement e-government systems. However, many technological, managerial, and organisational issues must be considered and treated carefully before and after going online. Based on an empirical study, this paper highlights the key organisational issues that affect e-government adoption in the Kingdom of Saudi Arabia at both national and agency levels.

Chapter 17

APetter Gottschalk, Norwegian School of Management, Norway

A stage model for knowledge management systems in policing financial crime is developed in this chapter. Stages of growth models enable identification of organizational maturity and direction. Information technology to support knowledge work of police officers is improving. For example, new information systems supporting police investigations are evolving. Police investigation is an information-rich and knowledge-intensive practice. Its success depends on turning information into evidence. This paper presents an organizing framework for knowledge management systems in policing financial crime. Future case studies will empirically have to illustrate and validate the stage hypothesis developed in this paper.

Chapter 18

Katarina Giritli-Nygren, Midsweden University, Sweden
Katarina Lindblad-Gidlund, Midsweden University, Sweden

The idea of eGovernment is moving rapidly within supra-national and national and local institutions. At every level leaders are interpreting the idea, attempting to grasp either the next step or indeed the very essence of the idea itself. This paper outlines a diagnostic framework, resting on three different dimensions; translation, interpretative frames and sensemaking, to create knowledge about the translation processes and by doing so, emphasize enactment rather than vision. The diagnostic framework is then empirically examined to explore its possible contribution to the understanding of the complexity of leader's translating and mediating the idea of eGovernment in their local context. In conclusion it is noted that the diagnostic framework reveals a logic of appropriateness between local mediators, eGovernment, different areas of interest and appropriate organisational practices.

Chapter 19

Karl Löfgren, Roskilde University, Denmark
Eva Sørensen, Roskilde University, Denmark

Since the late 1990s, an explicit goal of most industrialized states has been to integrate electronic access to government information and service delivery, examples being 'the 24/7 agency' or 'Joined-up governance'. This aim, which goes beyond the establishment of 'single' governmental websites, calls for both horizontal, as well as vertical integration of otherwise separate public agencies and authorities who are supposed to collaborate towards 'joint' and 'needs-based' electronic solutions to the benefit of citizens. While many authors have described this implementation of a policy aim in purely technical interoperability terms, we frame this development as a policy process of metagoverning self-regulating networks. This paper is primarily a theoretical think piece in which we will present a systematic framework for the analysis of meta-governing the policy process of electronic government. In addition to the value of framing the process as a metagovernance process, we wish to discuss how the metagovernance approach also sheds light on whether or not the on-going process of vertical and horizontal integration leads to more or less centralization, and whether it may contribute to a more democratic process. Our arguments will be supported by empirical illustrations mainly adopted from Scandinavian research.

Chapter 20

Fabio Perez Marzullo, Federal University of Rio de Janeiro, Brazil
Jano Moreira de Souza, Federal University of Rio de Janeiro, Brazil

This chapter presents an IT Governance Framework and a Competency Model that was developed to identify the intellectual capital and the strategic actions needed to implement an efficient IT Governance programme in Brazilian Government Offices. This work is driven by the premise that the human assets of an organization should adhere to a set of core competencies in order to prioritize and achieve business goals that, when seen from a government perspective are related to public resources management. IT Governance may help the organization to succeed in its business domain; consequently, through effective investment policies and strategic decisions on IT assets, the organization can come up with a business-IT alignment proposal, capable of enabling and achieving highly integrated business services.

Preface

TEN YEARS OF ELECTRONIC GOVERNMENT IMPLEMENTATION: THE CHALLENGES ENCOUNTERED AND LESSONS FOR FUTURE PROGRESS

ABSTRACT

Electronic government was formally introduced to the public sector in late 1990s in developed countries. Ever since its inception proponents of electronic government have been presenting it as a formula for improving the public sector through making the public sector more e-business like. In this perspective, the main objectives of public sector transformation promoted by electronic government involves improving efficiency, effectiveness and productivity of services for citizens while maintaining a customer centric focus and cost orientation. As a result, governments around the world were encouraged to embrace electronic government as a way of reaching out to more citizens through improved availability of services. However, research has proved that the implementation of the concept is highly complex and faced with many challenges while the adoption of services delivered through electronic government has been sparse. This chapter examines some of the challenges and complexities faced by the public sector when implementing electronic government, key indicators for its adoption, current trends and future directions.

INTRODUCTION

With the introduction of electronic service delivery in the public sector in the late 1990s, the concept of electronic government (e-government) has established itself as one of the most influential enablers of transformation in public administrations around the world. E-government has forced public sector organisations to realise the importance of making their services more efficient, transparent and available (Irani et al., 2007; Weerakkody et al., 2007). With citizens becoming more Internet savvy after experiencing good electronic business (e-business) services from the private sector, many begin to expect the same high standards from public sector organisations. In this respect, e-government promises to emulate the private sector by offering services that are more citizen centric and accessible (Al-Shafi, 2008). When analysing the existing literature on e-government, different researchers have identified various factors that entice governments to implement electronic service (Weerakkody et al., 2007; Irani et al., 2007; Chen et al., 2006; Gichoya, 2005; Chircu and Lee, 2005; Reffat, 2003; Moon, 2002). For instance, Moon (2002) argues that e-government can help restore public trust by coping with corruption, inefficiency, ineffectiveness and policy alienation. Conversely, lack of access to e-Services (Chircu and Lee, 2005), and digital divide (InfoDev, 2002; Reffat, 2003; John and Jin, 2005; Carter and Bélanger, 2005; Ifinedo and Davidrajuh, 2005; Chen et al., 2006; Carter and Weerakkody, 2008; Thomas and Streib, 2003; and Dwivedi and Irani, 2009) are factors that may influence trust and thereby impede the further take-up of electronic services. These findings indicate that various researchers and practitioners have attempted to offer insights into the implementation and acceptance of e-government services in different national contexts (Al-Shafi and Weerakkody, 2008).

Given the aforementioned discussion, a common argument that surfaces from the literature is that e-government offers many benefits and can potentially avail opportunities to developing countries (Ndou 2004; Karunananda and Weerakkody 2006; Irani et al. 2007). Yet, although, the benefits of electronic services are well documented, the implementation of e-government services in the public sector has faced many challenges in both developed and developing countries. The reason for such challenges are often explained by researchers as influenced by the complexity of the radical changes that are introduced to the public sector which are primarily driven by the Internet and its array of associated Information and Communication Technologies (ICTs) (Weerakkody et al., 2009). Given this context, it is widely accepted in the literature that e-government is much more complex and radical than any previous efforts of IT induced change experienced in the public sector (such as for instance in the era of new public management) (Weerakkody et al., 2007; Pollitt, 2000). In this respect, further research is needed to explore the impact of e-government implementation and diffusion particularly in countries where large sums of money and resources have been invested to this effect.

This chapter will examine some of the complexities and challenges of e-government implementation and adoption through a review of existing literature and offer discussions on current trends and future directions for e-government. The author's own experience of researching the subject and handling articles submitted to the International Journal of Electronic Government Research (as it's Editor-In-Chief) is used to draw ideas for the chapter. The next section in the chapter reflects on some of the past experiences of e-government. This is followed by examining some of the challenges faced by the public sector when implementing e-government. Next, key indicators for successful e-government adoption are discussed. Finally, the chapter concludes by visiting the current trends in e-government implementation and diffusion and outlining future directions.

EVOLUTION OF ICT ENABLED TRANSFORMATION IN THE PUBLIC SECTOR: A REFLECTION OF THE PAST

The public sector has experienced many ICT influenced change initiatives in the last three decades. By the late 1970s the New Right movement which advocated economically liberal and increased socially conservative policies proposed to deregulate the market and reduce the size of the government. Their ideas raised doubts about the ability of governments in the developed countries to meet citizens' needs arguing that the traditional public administration was very bureaucratic and inefficient (Massey and Pyper 2005). According to the New Right it is only through market competition that efficiency of governments can be achieved. New Right theorists strongly suggested that governments not only need reforms but also need to adopt private sector management techniques and practices to deal with the inefficiencies and ineffectiveness of the old public administration. These ideas found an audience in the political leadership of Western countries looking for alternatives to resolve the crisis in the welfare societies (see Levitas 2005; Delton 2007). For example these New Right ideas found an audience in the Conservative government that came to power in the United Kingdom in 1979, and Reagan's election in 1978 in the US provided some momentum for market-oriented reforms in the public sector (Richard and Smith 2005). However these ideas had remained on the periphery of the debate about the role of government and outside the mainstream of policy making until the mid 1970s. By the 1980s they moved to the centre stage of policy making of western governments (see Pollitt 2000). Consequently New Public Management (NPM) was evolved in mid 1980s promising to improve public services by making public sector organisations much more 'business-like' and 'market oriented' with emphasis on accountability and responsiveness to citizens' needs (Diefenbach 2009).

Although the above objectives are included in the concept of New Public Managemnt (NPM), some argue that 'NPM is a slippery label' as different theorists and practitioners stress different things in the conceptualisation of NPM (Manning 2000). For instance, public sector agencies have tried to replicate management concepts influenced by NPM ideas such as Business Process Re-engineering, Lean Management and Total Quality Management with a view of improving key public services such as healthcare, transportation and social services and local government (Pollitt, 2000). In the early 1990s in particular large sums of money were invested in transforming public institutions both the North America and Western Europe. However, research has shown that these initiatives have not met the expectations of citizens and delivered policy outcomes which have resulted in services that improve the day to day lives of ordinary citizens. Rather, the NPM influenced transformations have resulted in the tax payers' money being wasted and creating social exclusions. An example would be that of e-business ideas being borrowed from the private sector and applied in the context of e-government without due consideration of the wider context nor stakeholders using the services offered by e-government. Critics have attributed lack of adopted and digital divide, which are two major barriers to e-government diffusion, as a direct cause of attempting to mimic private sector concepts without adequate forethought to how such concepts will impact society. Yet, research has shown that these management concepts have all contributed positively to the private sector and have various strengths. Therefore, it is wrong to criticise the ideas, rather a more considerate approach to applying these ideas and policies by creating the appropriate environment that accounts for the various stakeholder and social needs may results in better outcomes in delivering suitable services that help reduce the current digital divide and lack of adoption of e-government services.

E-GOVERNMENT IMPLEMENTATION CHALLENGES AND COMPLEXITIES AND THEIR IMPACT ON PUBLIC SECTOR TRANSFORMATION EFFORTS

Prior research has proved that e-government implementation is surrounded by organisational, techno-logical, political, economic and social issues which have to be considered and treated carefully in order to facilitate the transformation of traditional public services to an e-services context (Irani et al., 2007; Weerakkody et al., 2007; Carter and Belanger, 2005; Beynon-Davies and Williams, 2003; Gil-García and Pardo, 2005; Heeks, 2006;). On the other hand, from an adoption perspective, e-government services are yet to be universally accepted as a medium for accessing online public services. E-government can be broadly viewed as the adoption of ICT in government organisations to improve public services. For many countries, e-government implementation efforts began in the late 1990s. As said before, the e-government led implementation of ICT in public administration during the last ten years has offered better, faster and more transparent means for citizens and businesses to interact with government.

Equally, it has also created a platform for better collaboration and information sharing between various government agencies. Implementation efforts in most countries have now evolved from basic information provisioning to more integrated service offerings that involve cross agency process and information systems (IS) transformation to enable more joined-up and citizen-centric e-government services. However, public sector process transformation is a complex undertaking involving distributed decision-making that requires a good understanding of the political context, business processes and technology as well as design and engineering methods capable of breaking through the traditional boundaries that exist between public organisation units. Conversely, from a demand perspective extensive efforts are required to increase citizens' awareness about the transformation of the delivery of government services and their online availability. As discussed before, in order to prevent digital divide in terms of using e-government services, it is also necessary that citizens from all facets of society are equipped with basic ICT skills as well as private and or public access to high-speed internet connections (Dwivedi and Irani, 2009; Irani et al. 2009). Furthermore, from an organisational perspective e-government has introduced an environment where most public institutions such as healthcare, social services, education and employment have struggled with the need to balance issues such as transparency and opaqueness, or social inclusion and professionalism. Consequently, there has been increasing pressure on the academic and practitioner communities for research that focuses on bridging the gap between e-government theory and practice.

In the aforementioned backdrop, various researchers and practitioners have attempted to offer insights into the implementation, acceptance and diffusion of e-government services. The last few years has seen e-government being regarded with the same level of importance that e-business was treated with in the mid 1990s. Consequently, in the last two years in particular, transformational government (or t-government for short) has emerged as the parallel of business process reengineering (BPR) that the private sector witnessed in the early 1990s. While early e-government efforts focused on e-enabling customer facing, front-office processes, t-government entails the same principles as BPR and focuses on ICT enabled transformation of both front- and back-office processes in public sector organisations.

While e-government services have been in place for over ten years now, the attention has gradually shifted towards more comprehensive and citizen centric services and the concept of e-participation is widely discussed in Western Europe in particular. These comprehensive services require more advanced and intelligent mechanisms for information exchange and management. When coping with such a shift, public organisations which are often characterised as bureaucratic and inefficient are bound to undergo major process and IS related problems. A number of studies have focused on exploring these issues and the noteworthy studies are by Heeks (1999; 2006), Layne and Lee (2001), Tan and Pan (2003), Jansssen and Cresswell (2005) and Kamal et al. (2009). Consequent to these issues being addressed, as mentioned before, in the last three years or so the concept of transformational government has emerged, which encompasses a broader perspective of public administration, since t-government is seen as the final stage of fully functional electronic service delivery for the public sector. This concept is now gaining momentum particularly in Europe (Dhillon et al. 2008; Irani et al. 2008; Weerakkody and Dhillon 2008). Yet, many sceptics may argue that this is simply a fashionable trend where buzzwords are used to renew efforts and gain publicity. In essence, transformational government implies reengineering back office processes and IT systems and mirrors the same principles practiced in the private sector during the BPR movement (Beynon-Davies and Martin 2004; Irani et al. 2007; Tan and Pan 2003; Weerakkody et al. 2007). This can be a daunting task since BPR has proved to be largely unsuccessful despite the wide publicity and hype associated with the concept (Avison et al. 2001; Willcocks 1995)

Challenges Facing E-Government Implementation

While it is evident that e-government is an effective driver for economic growth and significant cost reductions, conversely there remain many challenges which impede the exploration and utilisation of its opportunities (Al-Sebie and Irani, 2005; Gilbert *et al.*, 2004; Ndou, 2004; Jaeger and Thompson, 2003). The multidimensionality and complexity of e-government initiatives implies the existence of an extensive multiplicity of challenges that impede implementation and management (Ndou, 2004). For example, as reported by Stoltzfus (2005), e-government is costly, involves tremendous risks, requires a skilled technical pool of resources, and a stable technical infrastructure. Implementing e-government necessitates the evaluation of the following risk factors: political stability, an adequate legal framework, trust in government, importance of government identity, the economic structure, the government structure (centralised or not), levels of maturity within the government and citizen demand (Basu, 2004). Furthermore, inherent issues of e-government include: security and privacy, homeland security, diverse educational levels of users, accessibility issues, and prioritisation of e-government over basic functions of government, building citizen confidence in e-government, and whether certain forms of government do better with e-government than others (Jaeger, 2003).

Ke and Wei, (2004) also assert that many e-governments efforts – to turn vision into reality - have been obstructed by various challenges. Several researchers from academia and industry have argued that the emergence of e-government is a fundamental transformation of government, which entails profound changes in its structure, process, culture and behaviour of the individual in the public sector (Irani *et al.*, 2005; Prins, 2001; Howard, 2001). This is because the e-government paradigm includes changing the operational activities of government agencies to carry out its work. Thus, public sector are bound to face challenges such as overcoming resistance to change, privacy, security and possibly a lack of top management support in implementing e-government, which need to be addressed (Al-Shehry *et al.*, 2006; West, 2004; Ndou, 2004). Literature indicates that there is no single list of challenges to e-government initiatives (Gil-Garcìa and Pardo, 2005; Aldrich *et al.*, 2002; Layne and Lee, 2001). These are merely a handful of challenges to e-government initiatives reported in the normative literature. In addition to this, numerous other researchers have put forward their empirical findings on challenges to e-government initiatives in different disciplines. However, although there are very few notable consistencies across the different disciplines and research findings, the common themes that keep emerging can be categorised into the following seven groupings: (a) organisational, (b) technological, (c) social, (d) managerial, (e) operational, (f) strategic and (g) financial (Al-Shehry *et al.*, 2006; Al-Sebie and Irani, 2005; Gilbert *et al.*, 2004; West, 2004; Ndou, 2004; Jaeger and Thompson, 2003; Prins, 2001). For example, in a report by Government Accountability Office, (2001), these challenges are identified as: (a) sustaining committed executive leadership, (b) building effective e-government business cases, (c) maintaining a citizen focus, (d) protecting personal privacy, (e) implementing appropriate security controls, (f) maintaining electronic records, (g) maintaining a robust technical infrastructure, (h) addressing IT human capital concerns, and (i) ensuring uniform service to the public. Kushchu and Kuscu, (2003) highlight technological challenges to e-government, e.g. (a) infrastructure development, (b) payment infrastructure, (c) privacy and security, (d) accessibility, (e) legal issues and (f) compatibility. Gil-Garcìa and Pardo, (2005) put forward their work on e-government challenges and group them into five categories: (a) information and data, (b) information technology, (c) organisational and managerial, (d) legal and regulatory and (e) institutional and environmental.

Drawing from the aforementioned theoretical arguments which are drawn from the literature on e-government research, a conceptual taxonomy that maps the key factors influencing the implementation of e-government under the four broad themes of organisational, technology, social, and political is proposed in Table 1.

KEY INDICATORS FOR ENTICING E-GOVERNMENT ADOPTION

While the abovementioned factors impact Public Sector agencies' efforts to implement e-government services that may affect the outcomes of implementation success, a number of other factors need to be addressed to ensure the satisfactory adoption of implemented e-services. These can broadly be synthesised into six key indicators as explained next.

Trust

In the context of e-government, trust is defined as the observation of confidence in using various transactions of an e-government website and believing that the government body has implemented a reliable and secure system (Carter and Belanger, 2005). Trust gives indicators of the user's belief in security, privacy, and confident, which can arise due to involvement of financial transactions and/or personal information (ibid). Additionally, trust of individuals and institutions is an important factor of e-government adoption; lack of user's trust can result in a major challenge to the acceptance of e-government services (Warkentin et al., 2002). Practically, Gefen et al. (2003) posit that trust in the agency has a strong impact on the adoption of a technology diffused by that institution. Before endorsing e-government initiatives, citizens must believe government agencies demonstrate the competence and technical savvy necessary to implement and secure e-government systems. Transparent, accurate, reliable interactions with e-government service providers will enhance citizen trust and acceptance of e-government services (Al Shafi and Weerakkody, 2008). On the contrary, broken promises and fraudulent behaviour from government officials and employees will decrease trust and increase opposition to these initiatives (Carter and Weerakkody, 2008). Oxendine et al. (2003) compare citizen adoption of electronic networks in different regions of the US (Oxendine et al., 2003). They found that system adoption was more prominent in localities where citizens are more trusting. Due to the impersonal nature of the Internet, citizens must believe that the agency providing the service is reliable. Wang and Emurian (2005) posit lack of trust as one of the most formidable barriers to e-service adoption, especially when financial or personal information is involved.

Carter and Belanger (2005) found that perceived ease of use, compatibility and trust were significant indicators of users' intentions to use e-government services. Based on studies conducted during 2003, 2004 and 2005, Carter and Belanger proved that, trust can determine the adoption of e-government services. These studies suggested that any increase in the level of trust will be positively related to higher level of intention to use e-government services. Carter and Belanger (2005) found that there are few users who never used electronic payments because of others' bad experiences with fraud and privacy issues in an e-business context. Nevertheless, the authors also argue that this may not necessarily lead to negative feedback for individuals and institutions in terms of use of the Internet for e-government.

Table 1. Factors Influencing E-Government Implementation

A Taxonomy of Factors Influencing E-government Implementation		
Themes	**Factors Impacting E-government Implementation**	**Description**
Organisational	Organisational structure	The relatively enduring allocation of work roles and administrative mechanisms that creates a pattern of interrelated work activities and allows the organisation to conduct, coordinate, and control its work activities.
	Power distribution	User (government employee) resistance as well as employees losing their authority and power over traditional business processes.
	Information system strategy alignment	Alignment of strategies between different Information systems
	Prioritisation of deliverables	Prioritisation of deliverables which will ensure the most strategically significant services are managed and delivered appropriately in time.
	Future needs of the organisation	Project is a long-term initiative, and the adoption and implementation of e-government systems need time and appropriate models to support that implementation for the future needs of the organisation.
	Organisational culture	Groups of programming in the brain which differentiate members of different organisations.
	Employees training	Employees and managers need to get familiar with work under new circumstances, and to be prepared for changes.
Technological	Information Technology (IT) standards	IT assets are to be acquired, managed, and utilized within the organisation and act as the glue that links the use of physical and intellectual IT assets.
	Security and privacy issue	Security issues that consist of computer security, privacy and confidentiality of personal data.
	System integration	Integrates a system across different roles that provide a full and real 'one stop shop'. This integration assumes that all participant agency efforts are joined together.
	E-government portal and access	Portal access and availability of a payment gateway service 24/7 to process transactions.
Political	Government support	The commitment, involvement and support of the government's top authorities that would enable e-government officials to implement the project with more confidence.
	Funding	E-government initiatives are long term projects; therefore, these projects need long term financial support from the government.
	Leadership	Government officials or politicians to whom others turn when missions need to be upheld, breakthroughs made, and performance goals reached on time and within budget.
	Legislation and legal	regulations and legislation that acts to cope with the changes that are caused by e-government systems and include e-signatures, archiving data protection, preventing computer crimes and hackers, and the freedom of information.
Social	Citizen centric	E-government activities that focus on citizens needs and deliver services that add value to the citizen.
	Awareness	Awareness campaigns that promote e-government services to achieve more citizen participation and to achieve successful implementation.
	Digital divide	Digital divide that includes access to information; transaction services; and citizen participation.

Privacy

In an e-government context, public privacy means protecting the data of individuals and institutions during any interactions with online public administration services. Privacy gives the individuals and institutions universal access and enables them to use e-government services in a healthy environment (ibid). Gefen et al. (2003) has argued that privacy can cause a major concern for e-government adoption, since the government databases hold vast amounts of personal information. Moreover, today, all internet user's are concerns with misuse and frauds such as unauthorised tracking, sharing of information with third parties and specially financial information (e.g. credit card number); therefore, public administration has to ensure that the system used in data collection is kept private and cannot be used without user's authorisation, thus, the users will trust the e-government interactions which will ensure the enhancement of e-government adoption and privacy is generated (Gefen, et al., 2003).

Confidence

Many studies on e-government implementation highlighted the importance of including trust and privacy when examining adoption to achieve better understanding of user acceptance of electronic services (Gefen *et al.*, 2003; Carter and Belanger 2005). Online interactions and transactions between various individuals and governments involve exchange of sensitive information e.g. credit card details, therefore, it is important for the adoption of such services that individuals have a high level of trust and privacy in e-government (Carter and Weerakkody, 2008). Therefore, governments need to employ trust and privacy building strategies to increase citizens' confidence in e-enabled services (ibid). Carter and Weerakkody (2008) argued that, in terms of e-government interactions, there are many dimensions of confidence such as transparency and accuracy of the services provided. Those dimensions contribute to boosting individuals' and institutions' confidence in the adoption of e-government services. Conversely, any unaccomplished processes or misleading behaviour from government officials and employees will decrease citizens' confidence in e-government services.

Gefan et al., (2002) and Carter and Weerakkody (2008) emphasised that e-government is much less known than e-commerce but the technologies employed in e-government implementation to achieve confidence are generally considered to require a higher level of security to safeguard citizens data; they also need to demonstrate this to citizens in a convincing manner than required for e-commerce. This is particularly important as the users of e-government services are the general public who represent various different demographic contexts; the levels of Internet and ICT exposure and experience they possess may differ vastly when compared to e-business users. Therefore, their concerns with privacy and trust towards e-government services are likely to range from being highly confident to not-confident at all. Given these literature findings, it is proposed that the evaluation of usage and adoption of e-government services by individuals and institutions needs to take into account the issue of confidence as adoption is determined by the level of confidence users have on the e-services.

Efficiency

One of the key indicators that need to be considered when evaluating the success of online government services is the level of *efficiency* they offer in comparison to traditional services. In terms of IT systems or e-government, efficiency can be defined as the ratio of useful output to total input in any system. In the e-government context improved efficiency can only be achieved by reengineering and redesigning the front and back office processes and the IT systems that support the online services. Often governments succeed in e-enabling their services without paying much attention to the level of optimisation they achieve in a new service compared to the traditional way of offering the same service prior to e-government. This is one of the most difficult indicators to measure as there is bound to be differences in expectations and opinions between the service provider (government agency) and the user (citizen or business) regarding the optimum level of *efficiency* in a service and/or system. E-service optimisation can only be achieved by redesigning the underlying system, evaluating the performance and reengineering again in a continuous cycle until optimal performance efficiency is reached, particularly in the eyes of the end-user or citizen. Certainly, most of the features used to improve the *efficiency* are of a technical nature and will involve systems development and programming activity to optimise search and access, data processing and decision making as well as developing user friendly systems that take into account good principles of human computer interaction.

Availability

The core advantage of implementing e-government in any country is to make available a large number of services on a 24/7/365 basis. Therefore, the digital services act as a key motivating factor for various stakeholders benefiting from uninterrupted availability of different levels of e-government services ranging from the cataloguing of basic information to more complex transactions or dealings with government. *Availability* encapsulates the different types, styles, quantities, and levels of electronic services that are offered by any government to its citizens and businesses / institutions within the state. Empirical studies on e-government by different researchers have identified *availability* as an important factor in the adoption and diffusion of e-government. Studies have confirmed that when key services are not visible in government websites citizens often lose interests in e-government interactions (Carter and Weerakkody, 2008). This obviously translates to lower levels of e-government adoption resulting ultimately in e-government failure.

Accessibility

While research on e-government adoption has established availability as a key performance indicator for e-government implementations, *accessibility* allows the digital services provided by government to be offers to a wide range of users (citizens and businesses). *Accessibility* refers to unhindered and convenient access to online services by citizens of all demographic variations and institutions of all types irrespective of economic status, gender, age, physical ability or level of ICT literacy to minimize any digital divide (or social exclusion) in the respective society or country. *Accessibility* also encapsulates various online services that are offered by government websites and the medium by which they are offered (such as on a PC, mobile phone, PDA or through digital TV). Various e-government studies have time and again highlighted the need to offer online services using a multi-channel system (or services that are compatible with other media devices) to encourage adoption. Similarly, it is imperative that e-government services are offered in national languages to overcome any digital divide.

While efficiency, availability and accessibility encapsulates the key indicators that directly influence e-government adoption and success, other factors such as offering e-government services that maintain high levels of security and privacy will help build citizens trust and thereby confidence in using the services. Collectively these key indicators will then contribute to improving adoption.

CONCLUSION AND FUTURE DIRECTIONS FOR E-GOVERNMENT

The above discussions have shown that e-government implementation efforts have faced a plethora of challenges due to the complex nature of implementation changes and the diversity of stakeholders involved in the implementation as well as the adoption process. Some countries have managed to overcome these challenges with exemplary strategies while others have stagnated for many years overwhelmed by these challenges. Sadly though, many countries that have successfully implemented e-government have struggled to improve adoption of the electronic services that are provided resulting in digital divide and exclusion of certain segments of society such as the elderly, disabled, less computer literature and poorer citizens. This chapter has outlined some of these challenges and complexities faced by the public sector during their efforts to implement electronic government in the last decade. It has also looked at some of the key indicators for e-government adoption success by reviewing the extant literature.

Different countries around the world have tried to deal with the different challenges they face using different strategies. In particular various strategies have been adopted to encourage adoption of e-government across the world. To deal with digital divide and social exclusion, North American countries such as the US and Canada are currently considering alternative and multi-channel methods of service delivery that complement and run in parallel with 'standard internet based' e-government services. In particular, the US and Canada have now given the 'choice of the method of access' to public services back to the citizens where citizens have more control of their data and their interactions with the government. This can often involve reverting back to face to face meetings with public agencies or officials to complete a transaction or find answers to questions that citizens may have. Such interactions may be facilitated using a citizen service centre or by having the citizen directly visiting a public sector agency or contacting through telephone. Other methods of service delivery may include the use of digital television or mobile phones to deliver services and facilitate interaction with citizens. When examining such methods, one cannot avoid noticing the fact that these pioneering countries of e-government in North America who have led e-government rankings since the United Nations began their league tables and benchmarking initiatives, are now adopting multi channel service delivery to reduce the digital divide caused by e-services.

In contrast, many European countries such as the UK, Germany, Belgium, France, Netherlands, Italy etc., are placing more emphasis on ICT although with similar objectives of delivering services using multi-channels. The key difference between Europe and North America is that the latter is emphasising more on alternative channels such as mobile phones (which has technological focus), whereas the former is considering face to face contact in order to ensure that no citizens are left out of essential public services. In this context, although faced with digital divide and adoption related problems in terms of e-services, policy makers in Europe still consider using ICT to improve citizen participation albeit with a different approach. For instance, the most recent discussion amongst European policy makers has been the use of social networking sites such as face book or twitter to encourage citizens' participation in e-government services and public policy making processes. These new initiatives are commonly referred to as e-participation. While such initiatives will encourage a certain segment of society to participate in public debate and engagement, they will certainly not resolve issues relating to lack of e-government service adoption or digital divide in terms of using e-services.

In recent years, because of the need to create a culture of engaging stakeholders in the public policy making processes, many European governments have also turned to transformations that are influenced by a mix of NPM, change management and ideas of social innovation. As explained above, some of these new transformation initiatives have aimed to create an environment for seeking suggestions and encouraging participation and inclusion in public services (for an example see the EC's e-participation initiatives: http://www.european-eparticipation.eu/). These initiatives encourage citizens' participation in policy making processes by engaging with their governments using different digital media or channels (such as e-government using the Internet, mobile phones, digital television, social media etc.) as well as emphasis on using 'crowd sourcing' where citizens are given the responsibility to resolve complex societal problems by coming up with their own solutions. However, such initiatives will only be successful if suggestions are acted upon by relevant public sector agencies. In this context, the public sector can adopt good practices from the private sector, such as the ability to excel in customer relationship management. For instance, the competitive environment in the ICT sector (e.g. telecoms), retail (e.g. supermarket chains) and travel and hospitality (airlines) provide excellent examples of how companies react to their customer suggestions and expectations. These customer relationship management principles offer many lessons for public sector agencies who have often engaged in passive engagement with their stakeholders (citizens, businesses and employees). A common practice used when dealing with public sector agencies may involve completing *service satisfaction surveys* or passively dropping a *suggestion leaflet* into a box. Comparatively, such methods of passive stakeholders' involvement fails to account for the various health, transportation, social services, education, etc., needs of the diverse citizens that make up various societies in different countries across the world.

In some western countries public organisations in key service sectors such as health, transport and local government that have experienced financial and/or performance problems have adopted transformational government initiatives to go into partnerships with the private sector resulting in the increased number of Private-Public-Partnership (PPPs). For instance, in countries such as the UK, parts of the National Health Service (NHS), Transport and local government services have been privatised. Conversely, the recent global economic downturn in particular has forced many governments to step in and bail out private sector organisations and Non-governmental organisations (NGOs) at the expense of the tax payers' money. This environment has attracted many European and North American governments to consider the nationalisation of critical private institutions such as banks, energy and water suppliers. Consequently, this has now placed the public sector at the heart of managing efficient delivery of critical services. Inevitably this requires public sector managers to assume similar roles played by private sector managers as well as acquiring the right skills and leadership qualities required to deliver these services. For example, the history of PPPs in the UK goes back to the 1980s, when the privatization of public services started and also spread to many other countries (Yescombe, 2007). The main drivers for this were the beliefs that there should be a 'roll-back of the state' with the private sector providing services where this is more efficient, and that the introduction of competition leads to a better service and lower cost for the citizen, as well as less waste of economic resources, especially if services are supplied free or below cost by the state. This was in sharp contrast with the 20th-century trend for public utilities to be provided by the state. For example, the British private finance initiatives program was aimed at extending these benefits of privatisation to core public services which could not be privatised. However, it is important to point out the fact that there are important differences between privatisation and PPPs, some of which make it difficult for a PPP to achieve the same results as a privatisation (Grimsey & Lewis, 2007; Yescombe, 2007).

Given the aforementioned discussion, it is fair to state that the emphasis on e-government diffusion and adoption is now somewhat different between leading countries that have pioneered the concept in the last decade. Yet, the focus remains the same, 'citizens at the heart of public sector service delivery'. Which part of the developed world will be more successful in terms of reducing the digital divide will be an interesting observation? In this respect, international agencies such as the UN and OECD will need to take responsibility to evaluate progress with much more sophisticated matrix and methods than those currently used.

REFERENCES

Aldrich, D., Bertot, J. C. & Mcclure, C. R. (2002) E-government: Initiatives, developments, and issues. *Government Information Quarterly,* 19, 349-355.

Al-Shafi, S., and Weerakkody, V. (2008), 'The Use Of Wireless Internet Parks To Facilitate Adoption And Diffusion Of E-Government Services: An Empirical Study In Qatar,' Proceedings of the 14th Americas Conference on Information Systems (AMCIS 2008), Toronto, Ontario, 2008.

Al-Sebie, M. & Irani, Z. (2005) Technical and organisational challenges facing transactional e-government systems: an empirical study. *Electronic Government, an International Journal* 2, 247-276.

Al-Shafi, S. (2008) Free Wireless Internet park Services: An Investigation of Technology Adoption in Qatar from a Citizens' Perspective. *Journal of Cases on Information Technology*, 10, 21-34.

Al-Shafi, S. & Weerakkody, V. (2008b) The Use Of Wireless Internet Parks To Facilitate Adoption And Diffusion Of E-Government Services: An Empirical Study In Qatar. *Paper presented at the Proceedings of the 14th Americas Conference on Information Systems (AMCIS 2008),* Toronto, Canada

Al-Shehry, A., Rogerson, S., Fairweather, N. B. & Prior, M. (2006) The Motivations For Change Towards E-Government Adoption: Case Studies From Saudi Arabia. *eGovernment Workshop '06 (eGOV06), Brunel University, West London, UB8 3PH*

Avison, D.E., Dwivedi, Y.K., Fitzgerald, G., and Powell. P. (2008). The Beginnings of a New Era: Time to Reflect on 17 Years of the *ISJ, Information Systems Journal*, 18(1), 5-21.

Avison, D., Fitzgerald, G. & Powell, P. (2001) Reflections on information systems practice, education and research: 10 years of the Information Systems Journal. *Information Systems Journal*, 11(1), 3-22

Basu, S. (2004) E-government and developing countries: an overview. *International Review of Law, Computers & Technology,* 18, 109-132.

Beynon-Davies, P. and Williams, M. D. (2003). Evaluating Electronic Local Government in the UK. *Journal of Information Technology*, 18(2), 137-149.

Beynon-Davies, P. and Martin, S. (2004). Electronic Local Government and the Modernisation Agenda: Progress and Prospects for Public Service Improvement. *Local Government Studies,* 30(2), 214-229.

Carter. L. and Weerakkody. V. (2008) 'E-government adoption: A cultural comparison'. *Information Systems Frontiers* 10:4, 473-482

Belanger, F. and Carter, L. (2008). Trust and Risk in E-government Adoption. *Journal of Strategic Information Systems,* 17(2): 165-176.

Butler, T. (2003). An Institutional Perspective on Developing and Implementing, Intranet- and Internet-Based Information Systems, *Information Systems Journal,* 13(3): 209–231.

Carter, L. and Belanger, F. (2005). The Utilization of E-government Services: Citizen Trust, Innovation and Acceptance Factors. *Information Systems Journal,* 15(1), 5-25.

Currie, W.L. and Guah, M.W. (2007). Conflicting institutional logics: a national programme for IT in the organisational field of healthcare. *Journal of Information Technology,* 22(3), 235-247.

Chen, C., Tseng, S. & Huang, H. (2006) A comprehensive study of the digital divide phenomenon in Taiwanese government agencies. *International Journal Of Internet And Enterprise Management,* 4, 244-256.

Chircu, A. M. & Lee, D. H. D. (2005) E-government: key success factors for value discovery and realisation. *Electronic Government, an International Journal,* 2, 11-25.

Delton L (2007) Public Management, Democracy and Politics in The Oxford Hand Book of Public Management (Eds) E. Ferlie, L. E. Lynn and C. Pollitt. Oxford, Oxford University Press

Diefenbach T (2009) "New Public Management in Public Sector Organisations: The Dark Side of Managerialistic 'Enlightment'", *Public Administration* 87(4): 892–909

Dhillon, G.S., Weerakkody, V. and Dwivedi, Y.K. (2008). Realising transformational stage e-government: a UK local authority perspective. *Electronic Government, An International Journal,* 5(2), 162–180.

Dwivedi, Y.K. and Irani, Z. (2009). Understanding the Adopters and Non-adopters of Broadband. *Communications of the ACM,* 52(1), 122-125.

Dwivedi, Y.K. and Williams, M.D. (2008). The Influence of Demographic Variables on Citizens' Adoption of E-Government. *Electronic Government: An International Journal,* 5(3), 261–274.

Dwivedi, Y.K., Papazafeiropoulou, A., Gharavi, H. and Khoumbati, K. (2006). Examining the Socioeconomic Determinants of Adoption of an E-Government Initiative 'Government Gateway'. *Electronic Government - An International Journal,* 3(4), 404–419.

Gefen D, Karahanna E, Straub D. (2003), Trust and TAM in Online Shopping: An Integrated Model. *MIS Quarterly,* 27(1); 51-90

Gil-García, J. R. and Pardo, T. A. (2005). E-Government Success Factors: Mapping Practical Tools to Theoretical Foundations. *Government Information Quarterly,* 22 (2): 187–216.

Gilbert, D., Balestrini, P. & Littleboy, D. (2004) Barriers and benefits in the adoption of e-government. *International Journal of Public Sector Management,* 17, 286-301.

Grimsey, D., & Lewis, M. K. (2007). Public Private Partnerships: The Worldwide Revolution in Infrastructure Provision and Project Finance. Cheltenham, UK: Edward Elgar Publishing Limited.

Heeks, R. (1999). Reinventing Government in the Information Age. International Practice in IT-enabled Public Sector Reforem. Routledge, NY.

Heeks, R. (2006). Implementing and Managing e-Government: An International Text. Sage Publication, London, England.

Heeks, R. and Stanforth, S. (2007). Understanding e-Government project trajectories from an actor-network perspective, *European Journal of Information Systems*,16 (2), 165-178.

Howard, M. (2001) e-Government Across the Globe: How Will" e" Change Government? *Government finance review,* 17, 6-9.

Infodev (2002) The e-Government Handbook for Developing Countries. Center of Democracy and Technology.

Ifinedo, P. & Davidrajuh, R. (2005) Digital divide in Europe: assessing and comparing the e-readiness of a developed and an emerging economy in the Nordic region. *Electronic Government, an International Journal,* 2, 111-133.

Irani, Z., Love, P. E. D., Elliman, T., Jones, S. & Themistocleous, M. (2005) Evaluating e-government: learning from the experiences of two UK local authorities. *Information Systems Journal,* 15, 61-82.

Irani, Z., T. Elliman, and Jackson, P. (2007). Electronic transformation of government in the U.K: A research agenda, *European Journal of Information Systems* 16(4) 327-335.

Irani, Z., Dwivedi, YK and Williams, M.D. (2009). Understanding Consumer Adoption of Broadband: An Extension of Technology Acceptance Model. *Journal of Operational Research Society*, 60(10), 1322-1334.

Irani, Z, Love, P.E.D. and Jones, S. (2008), Learning lessons from evaluating eGovernment: Reflective case experiences that support transformational government, *The Journal of Strategic Information Systems*, 17 (2), pp.155-164.

Janssen, M. and Cresswell, A. (2005). An Enterprise Application Integration Methodology for E-Government. *Journal of Enterprise Information Management*, 18, 5, 531-547.

Jaeger, P. T. & Thompson, K. M. (2003) E-government around the world: lessons, challenges, and future directions. *Government Information Quarterly,* 20, 389-394.

Kamal, M. M., Weerakkody, V. and Jones, S. (2009). The Case of Enterprise Application Integration in Facilitating E-Government Services in a Welsh Authority. *International Journal of Information Management*, 29(2), 161-165.

Karunananda, A. & Weerakkody, V. (2006) E-government Implementation in Sri Lanka: Lessons from the UK. *Proceedings of the 8th International Information Technology Conference.* Colombo, Sri Lanka.

Ke, W., & Wei, K. K. (2004) Successful e-government in Singapore. *Communications of the ACM,* 47, 95-99.

Layne, K., and Lee, J. (2001). Developing Fully Functional E-government: A four-stage model. *Government information quarterly,* 18(2),122-136.

Levitas R. (2005) The Inclusive Society? Social Exclusion and New Labour. Basingstoke, Palgrave Macmillan.

Massey A and R Pyper (2005). Public Management and Modernisation in Britain. Basingstoke, Palgrave Macmillan

Moon, M. J. (2002) The evolution of e-government among municipalities: rhetoric or reality? *Public administration review*, 424-433.

Ndou, V. (2004) E–Government FOR Developing Countries: Opportunities & Challenges. *EJISDC*, 18, 1-24.

#Norris, D. F. & Moon, M. J. (2005). Advancing E-Government at the grass roots: Tortoise or hare? *Public Administration Review*, 65(1), 64-75.

Oxendine, A., Borgida, E., Sullivan, J. L., & Jackson, M. S. (2003). The importance of trust and community in developing and maintaining a community electronic network. *International Journal of Human-Computer Studies*, 58(6), 671–196.

Pollitt, C. (2000) "Is the Emperor in his Underwear? An analysis of the impacts of public management reform." *Public Mnagement, 2(2)*

Prins, C. (2001) Electronic government. Variations on a concept. Designing E-Government. On the Crossroads of Technological Innovation and Institutional Change, 1-5.

Reffat, R. (2003) Developing A Successful E-Government. *School Of Architecture, Design Science And Planning*. University Of Sydney, Australia.

Richard, D. and M. J. Smith, (2005) Governance and Public Policy in the United Kingdom Oxford, Oxford University Press

Scholl, H. J. J. (2001). Applying stakeholder theory to e-government: benefits and limits, *presented at 1st IFIP Conference on e-Commerce, e-Business, and e-Government*, Zurich, Switzerland.

Shareef, M.A., Kumar, U., Kumar, V., and Dwivedi, Y.K. (2009). Identifying Critical Factors for Adoption of E-Government. *Electronic Government: An International Journal*, 6(1), 70-96

Tan, C., Pan, S. and Lim, E. (2005). Managing stakeholder interests in e-government implementation: Lessons learned from a Singapore e-government project, *Journal of Global Information Management*, 13(1), 31-53.

Tan, C. W. and Pan, S. L. (2003). Managing e-Transformation in the Public Sector: An e-Government Study of the Inland Revenue Authority of Singapore (IRAS), *European Journal of Information Systems*, 12(4), pp. 269-281.

Warkentin, M., Gefen, D., Pavlou, P. and Rose, M. (2002). Encouraging Citizen Adoption of e-Government by Building Trust. *Electronic Markets*, 12(3), 157-162

Weerakkody, V. and Dhillon, G. (2008). Moving from E-Government to T-Government: A Study of Process Reengineering Challenges in a UK Local Authority Context, *International Journal of Electronic Government Research*, 4(4), 1-16.

Weerakkody, V., Dwivedi, Y.K, and Kurunananda, A. (2009) Implementing E-Government in Sri Lanka: Lessons from the UK. *Information Technology for Development,* 15(3), 171-192

Weerakkody V, Janssen M and Hjort-Madsen K (2007). Realising Integrated E-government Services: A European Perspective. *Journal of Cases in Electronic Commerce,* 3(2), 14–38.

Welch E.W, Hinnant CC and Moon MJ (2005). Linking Citizen Satisfaction with E-Government and Trust in Government. *Journal of Public Administration Research and Theory* 15(3), 371-391.

West, D. (2004). E-Government and the transformation of service delivery and citizen attitudes, *Public Administration Review, 64*(1), 15-27.

Willcocks, L. (1995). IT-enabled Business Process Re-engineering: Organisational and Human Resource Dimensions, *Journal of Strategic Information Systems,* vol.4 (3), pp.279-301

Yescombe, E. R. (2007). Public-Private Partnerships: Principles of Policy and Finance. Oxford: Butterworth-Heinemann (Elsevier).

Chapter 1
Global E-Government and the Role of Trust:
A Cross Country Analysis

Jayoti Das
Elon University, USA

Cassandra DiRienzo
Elon University, USA

John Burbridge Jr.
Elon University, USA

ABSTRACT

Using cross-country data from 140 countries, this empirical study extends past research by examining the impact of trust on the level of e-government. The major empirical finding of this research shows that, after controlling for the level of economic development and other socio-economic factors, trust as measured by ethnic and religious diversity, is a significant factor affecting e-government usage.

INTRODUCTION

In today's global economy, a country's level of e-government, or the use of the Internet and other communication technologies to provide government services, has emerged as an important policy tool for government. Recent studies indicate that the use of e-government is growing throughout the world (Mossberger, Tolbert, and Stansbury, 2003 and Larsen & Rainie, 2002). Formally defined as "the delivery of government information and services online via the Internet or other digital means" (West 2000, paragraph 7), e-government

offers many benefits to its major stakeholders. Implemented properly, it can be a cost effective method to deliver public services which can result in significant gains for the national economy. Through the Internet a government can provide information and allow citizens, businesses, and other governments to make a wide variety of transactions and to participate in various forums. E-government initiatives foster competitiveness by ensuring the integration of countries into the global community. A country that engages in e-government signals to the international community that it is open, transparent, and efficient, and creates an environment conducive to its users

DOI: 10.4018/978-1-60960-162-1.ch001

by streamlining procedures and providing easy access to a variety of public services (Thomas & Streib, 2003 and Peterson & Seifert, 2002).

Given its value to a country both in regard to domestic and international relations, it is important to understand the factors that encourage the use of e-government. To date, the majority of e-government research studies are narrowly defined case studies which are qualitative in nature (Devadoss et al., 2002 and Ke & Wei, 2004). In regard to quantitative research, the primary focus has been on the availability of technical infrastructure and its usage at a country-level, while others have considered how digital technology and online services allow citizens to accomplish tasks easily (West, 2000 and Steyaert, 2004). Broader studies by West (2003) and United Nations Department of Economic and Social Affairs (UNDESA) (2003) have explored how the use of the Internet and other digital communication technologies transforms the role of the government in regard to the level of economic and human development. However, only a few recent studies such as Kovačić (2005) and Rose (2005), have explored which factors drive the level of e-government within a country. These studies have found that several "hard", socio-economic and institutional factors such as freedom of expression, civil rights, level of development, and infrastructure significantly affect a country's e-government usage. However, little to no research has considered "soft" factors such as trust which can also impact the successful implementation and adoption of e-government.

Many researchers have considered trust as a dimension of "social capital" or "civil society". In particular, Putnam (1995, pg. 67) defines social capital as the "...features of social organization such as networks, norms, and social trust that facilitate coordination and cooperation for mutual benefit". Persell et al. (2001, pg. 206) describe the qualitative dimension of civil society as the "... social attitudes such as loyalty and trust, social practices such as civility and cooperation, and the health and safety of its members." Further, Leana

and van Buren III (1999, pg. 538) define the term "organizational social capital" as "...a resource reflecting the character of social relations within the organization, realized through members' levels of collective goal orientation and shared trust."

Regardless of how this social dimension is defined, a substantial number of studies have found evidence that higher levels social capital, which includes trust, is a catalyst for greater economic growth, development, and prosperity at the institutional, firm, and country-level. Specifically, at the country level, Knack and Keefer (1997) find that higher levels of social capital, as defined by trust and civic norms, lead to greater economic performance. Further, Coleman (1990), Putnam (1993), and Fukuyama (1995) all theorize that social capital, largely defined as trust between social groups, can significantly affect growth and economic success at the institutional level and Tsai and Ghoshal (1998) find that higher levels of social capital and trust significantly and positively affect product innovation at the firm level. Finally, Alesina and La Ferrara (2002, pg. 212) state "...a small but growing literature stresses that measures of 'trust' strictly defined or broader measures of 'social capital' are associated with effective public policies and more successful economic outcomes."

In regard to trust and communication, Granovetter (1973) states that based on social network theory, trust can be transferred to channels of communication which now include e-government. Trust can play an important role in facilitating the development of e-government within a country as any digital medium is a social platform through which individuals interact or transact with other citizens, businesses, or governments. Moore (1999) and Maskell (2000) have cited trust as an important factor in the development of new ideas and a prerequisite for any successful technological change as trust ultimately influences whether innovative ideas and new forms of communication are able to flourish. Further, Volken (2002) states that trust can be viewed as a cultural resource that facilitates innovative actions. Since e-government is considered an innovative technology, its success should be dependent on the level of trust between parties. Society must believe in the integrity of the applications and the controlling governmental entities. As discussed in Bélanger et. al (2002) and Lee and Turban (2001), there are privacy and security issues as well as reliability and integrity issues related to electronic commerce which also apply to e-government (Warkentin et al., 2002; Carter & Bélanger, 2003; 2005). As noted in the E-government Handbook (2007), the success of e-government projects and ultimately faith in e-governance depends on building trust between agencies, governments, businesses, and citizens. In this light, the success of e-government is dependent on the perceptions of trustworthiness among the stakeholders as users need to have confidence both in the government and the technology in order to successfully engage in e-government.

This study contributes to the advancements of e-government research by empirically exploring the relationship between trust and e-government. Using cross-country data from 140 countries, the relationship between trust and e-government is examined using a regression analysis. In addition, a cluster analysis is used to explore emerging global patterns in the use of e-government. If trust emerges as a significant factor as to why countries differ in the level of e-government, policy makers working to enhance e-government usage also need to consider policies and methods to promote trust.

LITERATURE REVIEW

As previously discussed, trust is generally considered a dimension of social capital. In regard to trust itself, the literature provides definitions which generally differ based on the academic discipline (Bhattacharya et al., 1998). Personality psychologists define trust as an individual characteristic while social psychologists view trust from the perspective of behavioral expectations of others (Kim and Prabhakar, 2000). From an organizational behavior perspective, Currall and Judge (1995) define trust as an individual's reliance on another party under conditions of dependence and risk. Marketing research has defined trust as the willingness of one party to rely upon and have confidence in an exchange partner (Morgan and Hunt, 1994 and Mooreman et al., 1993). In regard to ethics, Baier (1986) defines trust as the reliance on others ability to properly care for things that have been assigned to their custody or supervision which can include such tangible things such as money or children or intangible things such as democracy and societal norms (Tschannen-Moran and Hoy, 2000). Finally, in reference to information technologies (IT), Gefen et al. (2005, pg. 54) states that to trust means to have expectations about others' behavior and that "One of the central effects of this trust in the context of IT adoption is to increase the perceived usefulness (PU) of Information Technology (IT) associated with the trustee's agency."

Given its intangible nature, trust is challenging to quantify and measure. In regard to social trust, Fukuyama (2001) discusses the difficulty of measuring it and Collier (2002) stated that trust is essentially impossible to measure directly and that for empirical evaluations, proxy indicators are necessary. Researchers have generally created measures of trust derived from surveys that were developed for specific studies or have used responses from the World Values Survey (WVS) (Knack and Keefer, 1997, La Porta et al., 1997, Glaeser et al., 2000, Collier (2002), and Johansson-Stenman et al., 2005). The studies which have developed their own surveys have generally focused on particular regions and are limited in scope. Further, as Collier (2002) notes, the WVS trust data is only available for 45 countries and, of the country data available, the majority of the countries are industrialized and the transitioning countries are under-represented.

However, there is a growing literature which suggests that diversity, as measured by ethnic, linguistic, and / or religious heterogeneity, can be an appropriate proxy for the level of trust within a society. In a United States (U.S.) study, Putnam (1995) finds that the greater the diversity within a community, the less likely neighbors will trust each other and, in general, greater diversity yields lower levels of trust. Similarly, Fukuyama (1995) argues that within a society there is a boundary of trust such that people in relationships within that boundary are trusted more than people outside that boundary and Glaeser et al. (2000) find that trust is easier to facilitate among people that are homogeneous. Warkentin et al. (2002) establishes the connection between group diversity and the level of trust by stating that differences among groups will increase the likelihood of distrust between the groups. Newton and Delhey (2005) and Anderson and Paskeviciute (2006) find that greater levels of ethnic diversity are associated with lower levels of trust at the country level and Van Parijs (2004) suggests that the welfare of a nation or government is negatively affected by diversity. Further, in an Australian study, Leigh (2006) finds that trust is lower in ethnically and linguistically diverse communities and Johansson-Stenman et al. (2005) find that different religious affiliations significantly affect trust. In summary, this research suggests that the greater the diversity, the lower the level of trust.

Given this literature, this study uses diversity in regard to ethnicity, linguistics, and religiosity to proxy trust. In particular, the Fractionalization Index created by Alesina et al. (2003) which separately measures ethnic (E), linguistic (L), and religious (R) diversity for 190 countries, is used in this analysis. Alesina et al. (2003) employs the Herfindahl index methodology and the index represents the probability that two randomly selected individuals from a population belong to different groups. The Fractionalization Index is computed in the following manner:

$$FRACT_j = 1 - \sum_{i=1}^{N} s_{ij}^2,$$

Where s_{ij} is the share of the group i $(i=1,\ldots..N)$ in country j. A measure close to 0 would imply a less diverse or more homogenized society, while a measure closer to 1 suggests the opposite. Thus, the larger the fractionalization index the more diverse the society.

Using this measure, countries in Sub-Saharan Africa show the highest degree of fractionalization for all measures; ethnicity 0.66, linguistic 0.63, and religious diversity 0.50. The least ethnically fractionalized countries are South Korea and Japan. Countries in western and southern Europe reflects low levels of ethnic diversity (0.18), linguistic diversity (0.20), and religious diversity (0.31) on average, while the U.S. was found to be more ethnically, linguistically, and religiously diverse compared to most of Europe. Furthermore, South Africa, the U.S., and Australia were found to be the most religiously diverse countries. Low levels of religious diversity were predominantly Catholic (Italy and Ireland), Protestant (Scandinavia), and of course, Muslim countries.

METHODOLOGY

In this section, the regression model used to test the relationship between trust, as proxied by diversity, and e-government is described. First, the hypothesized relationships between the three different measures of diversity and e-government are presented. This discussion is followed by a presentation of the control variables necessary for the regression analysis as well as the hypothesized relationships between these variables and e-government. The proxies used for each the control variables as well as for e-government are then discussed. Finally, a graphical representation of the model is presented which is followed by a discussion of the descriptive statistics for all variables used in the analysis as well as a correlation matrix and discussion of the relationships.

Ethnic Diversity

Much research (Alesina & La Ferrara 2002; Delhey & Newton, 2004; Leigh, 2006; and Zak & Knack, 2001) has found that ethnic homogeneity raises "social capital" and trust which makes it easier to develop social networks that disseminate information and knowledge. In contrast, Collier (1998) finds that cultural and ethnic heterogeneity tend to hamper nation-building as such societies are more prone to polarization and social conflict. Further, Mauro (1995) finds a negative correlation between ethno-linguistic fractionalization and political stability, bureaucratic efficiency, and institutional efficiency, while Shleifer and Vishny (1993) find that ethnically diverse societies are likely to have higher rates of corruption and less trust. Annett (2001) and Ritzen et al. (2001) also find that ethnic diversity results in poor quality of government services and institutions. Several others such as Rodrick (1999) and Svensson (1998) note that ethnically polarized societies cause governments to marginalize investment in legal infrastructure. Since ethnically diverse countries tend to experience more bureaucratic, institutional, and political inefficiency as well as inferior government services and infrastructure, it can be hypothesized that:

H1: Countries that are more ethnically fractionalized than others experience lower levels of e-government.

Linguistic Diversity

Linguistic heterogeneity reduces trust both at the native and immigrant level, as it often fails to promote equal treatment of the diverse languages and reduces the quality of information transmitted across different groups (Cook & Cooper, 2003). Leigh (2006) finds that trust is lower in linguistically heterogeneous communities; however, Delhey and Newton (2004) find no relation between linguistic heterogeneity and

trust. Empirical studies by Mauro (1995), Annett (2001), and Barro (1999) find that linguistic diversity causes greater political instability and inefficient public policies due to increased transaction costs. However linguistically homogenous societies have historically been more effective in communicating new ideas among themselves making the adoption of a *lingua franca* important. Since e-government represents a platform for communication and networking across agencies, and linguistic heterogeneity decreases the quality of information and often trust amongst its citizens and agencies, it can be hypothesized that:

H2: Countries that are more linguistically fractionalized experience lower levels of e-government.

Religious Diversity

Alesina and Ferrara (2002) state that trust is influenced by individual's religious beliefs as different religions have different attitudes toward social interactions. Some believe that religious fundamentalists are less trusting of diverse religious groups (Uslaner, 2002), while others contend that certain religious beliefs like the protestant ethics promote higher levels of trust (Inglehart & Baker, 2000). Since most religious doctrines contain some elements that promote trust and some elements set boundaries or limitations on trust, classifying some religious beliefs as promoting trust and not others is speculative. Further, while a religiously homogenous society may be more trusting of each other, intolerance towards a minority religion can develop and a more religiously diverse society can be more accepting of other religious affiliations and beliefs.

Studies examining religious allegiance and trust have not found a significant relationship between the two (Johansson-Stenman et al., 2006; Alesina & Ferrara, 2002; and Delhey & Newton, 2004). Although religious diversity has been linked with inefficiencies in the distribution of public goods and government, drawing any clear conclusion regarding the presence of different religions within a country and its effect on civic engagement and e-government is not readily apparent (Banerjee et al., 2005). Therefore, the hypothesis regarding religious diversity and e-government is stated:

H3: A country's degree of religious fractionalization will have no impact on e-government.

Control Factors

Before a statistical analysis examining the relationship between trust, as proxied by diversity, and e-government can be performed, the other socio-economic and institutional factors which are known to affect e-government need to be controlled in order to prevent model misspecification. As noted by Easterly and Levine (2001) and Alesina et. al (1999) institutional factors such as education, role of governance, and political systems often mitigate the effect of diversity. For example, ethnic and linguistic diversity can have a greater adverse effect on a country's level of trust and e-government if there is a lack of institutional stability and democratic processes hampered. In an analysis exploring the relationship between trust and e-government, all of these other factors must be controlled for in order to prevent model misspecification. A brief discussion of the control variables, the hypothesized relationships, and their proxies follows.

Political Rights and Civil Liberties

A democratic society usually reflects greater openness, freedom of expression, civil rights, and political stability and are generally more successful in regard to the development of e-government. Moon et al. (2005) state that countries with less democratic governments tend to marginally employ e-government as such governments generally do not want to engage in transparent and interactive relations with citizens or other agencies. Further, West (2003) suggests that a country's level of political development is critical in determining its level of commitment to e-government. Advocates of democracy and freedom of communication consider political and civil liberties as a necessity for greater access to government and public services (Zinnbauer, 2001). As stated by Rose (2005), the greater the acceptance of civil liberties and freedom of expression the greater the provision of any digital e-facility. Therefore, it can be hypothesized that:

H4: Countries with higher levels of democratic and political freedom experience higher levels of e-government.

The 2004 Political Rights (*PR*) and Civil Liberties (*CL*) indices by Freedom House are used to proxy the democracy afforded to the citizens. Freedom House conducts an annual survey concerning civil liberties and political rights in nations, emphasizing the importance of democracy and freedom. The indices are created using a survey method to judge all countries by a single standard which emphasizes the importance of democracy and freedom. In a free society, political rights allow people to participate freely in the political process, including the right to vote, compete for public office, and to elect representatives. Civil liberties include the freedom of expression and media independence, the right to associate and organize the rule of law, and personal autonomy without interference from the state. Rates are assigned separately to political rights and civil liberties on a scale of one to seven, with one representing the most free and seven the least. Following Klitgaard et al. (2005), the average of a country's *PR* and *CL* rates (noted *PCR*) is used in this analysis to represent a country's level of democratic freedom.

Economic Freedom

According to Ciborra and Navarra (2003), the pre-conditions for e-government are stability and an open economy and Watson and Mundy (2001) suggest that institutions favoring poor governance practices might actually resist the growth of e-government. Singh et al. (2004) state that an economically free nation with a stable legal and monetary system, efficient labor and product markets, and open trade and investment opportunities, provides a more competitive and dynamic environment in which e-government could flourish. Further, Rose (2005) states that if governments are more open with higher levels of institutional freedom, they are more receptive to e-government compared to countries with higher levels of economic and social instability and inefficient bureaucracies. Therefore, it can be hypothesized that:

H5: Countries with higher levels of economic and institutional freedom experience higher levels of e-government.

The 2004 Index of Economic Freedom (*EFI*) by the Heritage Foundation is used to capture distortions across institutional factors such as economic restrictions and barriers within a country. The EFI index considers 50 economic freedom variables which are divided into ten broad categories; trade policy, fiscal burden of government, government intervention in the economy, monetary policy, capital flows and foreign investment, banking and finance, wages and prices, property rights, regulation, and informal market activity. Each of these categories is assigned a score and then these 10 categories are averaged and an overall economic freedom score between one (most economically free) and five (least economically free) is assigned to each country. The Heritage Foundation classifies the economic freedoms afforded to a country in four broad categories: "Free", countries with an average overall score of 1.99 or less; "Mostly Free", countries with an average overall score of 2.00 to 2.99; "Mostly Un-free", countries with an average overall score of 3.00 to 3.99; and "Repressed", countries with an average overall score of 4.00 or higher.

Economic Development

Finally, it is necessary to control for the existing level of economic development within a country. Norris (2001) finds that more affluent industrialized economies are more conducive to online access to governments and the digital divide often creates a poor-information society. Hargittai (1999) finds economic wealth as measured by Gross Domestic Product (GDP) per capita to be an essential ingredient in promoting digital platforms. Further, Kaufmann (2004) states that countries with higher per capita GDP are better able to afford and support e-government and Rose (2005) cites GDP per capita, urbanization, personal computers, and telephone lines per thousand people as being important in the development of e-government. Finally, Singh et al. (2004) emphasize the role of human capital and existing Information and Communication Technology (ICT) infrastructure in the growth and development of e-government. Therefore, it can be hypothesized that:

H6: Countries that enjoy higher levels of economic development will have higher levels of e-government.

GDP per capita (measured in constant 2000 U.S. dollars) is used to represent a country's level of economic development and was collected from the World Bank (2004). GDP per capita is selected to proxy economic development as other measures of economic development, such as the Human Development Index, include sub-indices that incorporate human capital measures. As discussed in the following section, e-government is proxied by an index that includes human capital. Thus, GDP per capita is employed in this study to proxy economic development in order to prevent model misspecification.

Measuring E-Government

Several measures of e-government focus on either a country's potential to engage in e-government (Economist Intelligence Units, 2004), or a country's actual level of e-government related activities (World Economic Forum, 2007). The measure of e-government used in this analysis, the E-government Index 2005 (*EGOV*) developed by the United Nations (U.N.), incorporates both the potential for and the implementation of e-government and is the most complete assessment of e-government across countries (Singh, et al. 2004).

The *EGOV* or e-government readiness index is an average of the (1) telecommunication infrastructure index, (2) the human capital index and (3) the web measure index as created and defined by the U.N. This composite index captures the overall ability of a country to engage and implement e-government. It is important to note that this index includes the two major factors important for presence and diffusion of e-government, infrastructure and education, while the web measure is intended to capture the attitude of governments to inform, transact, and network. The index ranges

from zero (low levels of e-government) to one (high levels of e-government). For example, the U.S. has the highest value of *EGOV* (0.9062) and is followed by northern European countries such as Denmark (0.9058) and Sweden (0.8983). Among the developing countries, the Republic of Korea (0.8727) and Singapore (0.8503) lead with Estonia (0.7347), Malta (0.7012), and Chile (0.6963) following. Regionally, Europe and North America generally have greater values of *EGOV* while South-Central Asia and Africa lag behind.

The above discussion defines all of the variables and their proxies used in this analysis as well as their hypothesized relationships with e-government. In an effort to provide a visual synopsis of the above discussion, Figure 1 provides a graphical representation of the research model.

Descriptive Statistics

A sample of 140 countries is used to test the previously stated hypotheses. *EGOV* represents 2005 data, while the data for the control variables represents 2004 data, with the exception the *E*, *L*, and *R* which are only available for 2003. The control variables are appropriately lagged as their affect on *EGOV* cannot be expected to occur immediately. Table 1 provides a summary

Table 1. Variable summary and descriptive statistics

Variable	Proxy	Mean*	St. Deviation*
E-government	E-government Index (*EGOV*)	0.465	0.204
Ethnic Diversity	Ethnic Fractionalization Index (*E*)	0.452	0.254
Linguistic Diversity	Linguistic Fractionalization Index (*L*)	0.397	0.284
Religious Diversity	Religious Fractionalization Index (*R*)	0.440	0.238
Political Rights / Civil Liberties	Freedom House (*PCR*)	3.275	1.858
Economic Freedom**	Economic Freedom Index (*EFI*)	1.076	0.235
Economic Development**	GDP per capita (*GDPPC*)	7.742	1.602

*n = 140

** A series of scatter plots and tests indicated that the relationships between EFI and EGOV and GDPPC and EGOV are log-linear and that the relationship between natural logarithm of each of these controls with EGOV was linear. Therefore, the natural log of both EFI and GDPPC were taken before the regression analysis was performed and the statistics above are based on the natural log of each of these variables.

Table 2. Spearman correlation matrix

	EGOV	E	L	R	PCR	LnEFI	LnGDPPC
EGOV	1						
E	-0.595**	1					
L	-0.494**	0.686**	1				
R	-0.023	0.267**	0.315**	1			
PCR	-0.679**	0.363**	0.280**	-0.014	1		
LnEFI	-0.708**	0.405**	0.326**	0.015	0.703**	1	
LnGDPPC	0.869**	-0.555**	-0.525**	-0.098	-0.597**	-0.724**	1

*p< 0.05 **p< 0.01

and descriptive statistics of the variables used in the analysis.

Table 2 provides the correlation matrix for all of the variables. Each of the variables was tested for normality using the Jacque-Bera test for normality. At 95% confidence, each of the variables, with the exception of *EGOV* was found to be non-normal. Given that one of the assumptions for the Pearson measure of correlation is normality, the Spearman rank correlation was used to measure the correlation between all of the variables and these results are presented in Table 2.

As seen in Table 2, *E* is positively and significantly correlated with *PCR* and *LnEFI* and negatively and significantly correlated with *LnG-DPPC*. These correlation measures suggest that countries with greater ethnic diversity tend to have lower levels of political rights and civil liberties (higher values of the *PCR* index indicate less freedoms), economic freedoms (higher values of the *EFI* index indicate less freedom) and economic development. Similar to ethnic diversity, *L* is positively and significantly correlated with *PCR* and *LnEFI* and negatively and significantly correlated with *LnGDPPC* which suggest that countries with greater linguistic diversity tend to have lower levels of political rights and civil liberties, economic freedoms, and economic development. Finally, there are no significant cor-

Figure 1. Composite model illustration

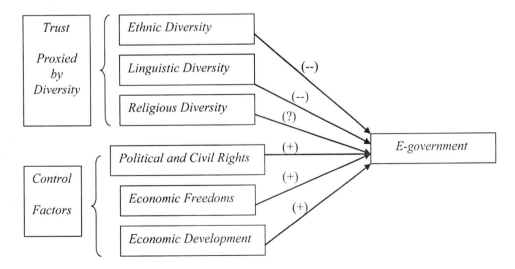

relation measures between religious diversity and the control variables.

In regard to e-government, *EGOV* is significantly and negatively correlated with *E*, *L*, *PCR*, and *LnEFI*. These relationships suggest that countries with higher *EGOV* values tend to have less ethnic and linguistic diversity and more political rights, civil liberties, and economic freedom. *EGOV* is significantly and positively correlated with *LnGDPPC* which suggests that countries that are more economically developed tend to have higher levels of e-government. Finally, the correlation between religious diversity, *R* and *EGOV* is not significant. These correlations suggest existing interrelationships amongst the variables, without establishing any definite causal relationship. In order to explore the directional role of trust and e-government across countries, a regression analysis is performed.

Findings

To test the previously stated hypotheses, an ordinary least squares regression analysis was performed. The regression equation is defined as:

$$EGOV = \beta_0 + \beta_1 E_1 + \beta_2 L_2 + \beta_3 R_3 + \beta_4 PCR + \beta_5 LnEFI + \beta_6 LnGDPPC + \varepsilon$$

The Ordinary Least Squares (OLS) regression results are presented in Table 3. The significance levels of the coefficient estimates, the Adjusted R^2 value, and the F statistic indicate that the regression model provides a good fit to the data.

Tests for multicollinearity and heteroscedasticity validate the results shown in Table 3. The test of multicollinearity for each of the independent variables using the variance-inflation factor (VIF) shows that the VIF values range from 1.139 to 2.90. Given that VIF values greater than 5.3 have been suggested as cutoffs (Hair et al., 1992), multicollinearity does not appear to be a problem in this analysis. Furthermore, White's (1980) general test for heteroscedasticity indicates that the residuals

are homoscedastic suggesting that the regression results were not influenced by heteroscedasticity.

The regression allows the coefficients associated with *E*, *L*, and *R*, to be interpreted as the marginal effect of an increase in each of the factors on *EGOV*, holding all other independent variables constant. The results show that the coefficient on ethnic fractionalization is significant and negative suggesting that the marginal impact of becoming heterogeneous with regard to ethnic diversity (going from zero to one) results in an average decline of approximately 0.14 of a country's level of *EGOV*. This suggests that, while controlling for other factors known to affect e-government, countries with greater levels of ethnic diversity have lower levels of e-government on average, which supports H1.

The coefficient on linguistic fractionalization is negative, but not significant. The negative sign supports H2; however, the coefficient is insignificant which suggests that linguistic diversity does not significantly affect e-government which is contrary to H2. This result is not surprising as Alesina et al. (2003) state that there is a high correlation between ethnic and linguistic fractionalization measures and in regression analyses often linguistic diversity is insignificant when ethic diversity is included in the analysis. The implication is that when variables such as political and economic freedom and level of development are accounted for, linguistic diversity does not significantly affect e-government.

Finally, the coefficient on religious fractionalization was found to be positive and significant which does not support H3. As previously discussed, religious diversity can potentially have positive and negative effects on e-government and making general statements or hypotheses regarding the impact of religious diversity on e-government would be speculative. However, in this analysis, the positive effects, such as the possibility of religiously diverse societies to be more tolerant and trusting, outweighed the possible negative effects of religious diversity and countries with more religious diversity or religious freedom engage in higher levels of e-government on average. The marginal impact of becoming heterogeneous with regard to religion (going from zero to one) results in an average increase of approximately 0.09 of a country's level of *EGOV*.

All three of the control variables in this study prove to be relevant. The coefficient on *PCR* is statistically significant and negatively related to *EGOV* which confirms H4. Thus, countries with higher levels of democratic and political freedom experience higher levels of e-government on average. The coefficient on *LnEFI* was found to be negative and significant suggesting that countries with higher levels of economic and institutional freedom experience higher levels of e-government on average, thus supporting H5. Finally, the coefficient on *LnGDPPC* is positive and significant, supporting H6 that an economically prosperous country is more likely to advance e-government efforts.

In summary, the results of the cross country regression analysis suggest that both ethnic and religious diversity significantly affect a country's level of e-government even after controlling for such factors as democratic and economic freedom and economic development. The findings indicate that, on average, countries that are less diversified with respect to ethnicity, more diversified in terms of religion, have greater levels of political, civil, and economic freedoms, and higher levels of economic development enjoy greater levels of e-government.

The cross country regression results are general and show the broad relationships between the variables across all countries. A country-based cluster analysis is performed in the next section in order to highlight any emerging global trends in e-government. The cluster analysis is designed to divide the countries into groups based on their similarity in regard to all significant variables from the regression analysis.

Cluster Analysis

A cluster analysis is used to group the 140 countries into distinct clusters using trust, socio-economic, and institutional variables as the criteria. The independent variables, with the exception of linguistic diversity, as it is not significant at 5% level, are used as the country characteristics upon which the clusters are formed. A cluster analysis groups objects, in this case countries, into groups such that the objects within a group are most similar to each other with respect to specified characteristics and least similar to objects in the other groups. In other words, the cluster analysis groups the individual countries based on how similar or "close" they are in regard to their ethnic and religious diversity, economic freedom, political and civil liberties and level of development.

A non-hierarchical cluster analysis is performed using the squared Euclidean distance as the measure of how close two countries are in regard to the five variables; *E, R, PCR, LnEFI,* and *LnGDPPC*. This particular type of cluster analysis requires that the number of clusters to be created in the analysis be set prior to performing the analysis. This is an exploratory process, and no hard guidelines have been suggested for determining the number of clusters to be created in a non-hierarchical analysis.

After considering different cluster analysis results, a grouping of two countries clusters was chosen and is presented in Tables 4 and 5. The results can be interpreted as follows. Country Cluster 1 represents the group of countries that are most homogenous in regard to their ethnic and religious fractionalization, political, civil, and economic freedoms, and economic development and most heterogeneous to the Country Cluster 2 with respect to these variables.

Given these country groupings, the differences in the level of e-government across the clusters should be detectable. To explore this theory further, Table 6 provides the mean *EGOV* value for each of the clusters. Cluster 1 has a

Table 3. Regression results

	Coefficient Estimates	**Std Err**	**t Stat**	**VIF**
Intercept	0.07507	0.10703	0.70	0
E	-0.13975**	0.04199	-3.33	2.204
L	-0.00351	0.03771	-0.09	2.218
R	0.08976**	0.03224	2.78	1.139
PCR	-0.01832**	0.00519	-3.53	1.806
LnEFI	-0.11249*	0.05224	-2.15	2.900
LnGDPPC	0.07692**	0.00761	10.10	2.880

Adj. R2 = 0.8282 F stat = 112.68** *p <0.05; **p<0.01

Table 4. Country cluster 1

Albania	Ghana	Nicaragua
Algeria	Guatemala	Niger
Armenia	Guinea	Nigeria
Azerbaijan	Guinea Bissau	Oman
Bahrain	Honduras	Pakistan
Bangladesh	India	Paraguay
Belarus	Indonesia	Philippines
Benin	Iran	Russia
Bolivia	Jordan	Saudi Arabia
Bosnia and Herzegovina	Kazakhstan	Senegal
Burkina Faso	Kenya	Sierra Leone
Cambodia	Kuwait	Sri Lanka
Cameroon	Kyrgyzstan	Swaziland
Central African Republic	Laos	Syria
Chad	Lebanon	Tajikistan
China	Lesotho	Tanzania
Colombia	Libya	Togo
Congo, Republic	Macedonia	Tunisia
Cote d´Ivoire	Madagascar	Turkey
Ecuador	Malawi	Uganda
Egypt	Malaysia	Ukraine
Equatorial Guinea	Mali	United Arab Emirates
Ethiopia	Mauritania	Uzbekistan
Fiji	Moldova	Venezuela
Gabon	Mongolia	Vietnam
Gambia	Morocco	Zambia
Georgia	Mozambique	Zimbabwe
	Nepal	

Table 5. Country cluster 2

Argentina	Greece	Peru
Australia	Guyana	Poland
Austria	Hungary	Portugal
Belgium	Iceland	Romania
Belize	Ireland	Singapore
Botswana	Israel	Slovakia
Brazil	Italy	Slovenia
Bulgaria	Jamaica	South Africa
Canada	Japan	South Korea
Chile	Latvia	Spain
Costa Rica	Lithuania	Suriname
Croatia	Luxembourg	Sweden
Cyprus	Malta	Switzerland
Czech Republic	Mauritius	Thailand
Denmark	Mexico	Trinidad and Tobago
Dominican Republic	Namibia	United Kingdom
Estonia	Netherlands	Uruguay
Finland	New Zealand	USA
France	Norway	
Germany	Panama	

lower level of *EGOV* compared to Cluster 2. Further, at 99% confidence, a *t*-test indicated that Cluster 1 has a significantly lower mean *EGOV* level than Cluster 2. Therefore, the cluster analysis demonstrates that if countries are grouped by their similarities in regard to their ethnic and religious fractionalization, political, civil, economic freedoms, and level of economic development, significant differences in their level of e-government can be detected.

Cluster 1 generally includes countries that are less developed and are in the process of enhancing their infrastructure, which will hopefully lead to increased e-government usage. Interestingly, Cluster 1 includes countries like China, India, and Russia that are clearly developing nations with a lot of internet infrastructure in place, but perhaps are not as engaged in e-government as other similar countries. The urban centers of such countries may be heavy users of commercial internets, but it appears that they do not use the Internet for public service which would reach the greater majority of the population, especially across the rural parts of these countries. Alterna-

Table 6. EGOV means by cluster

Clusters	Mean (*EGOV*)	Sample Size
Cluster 1 *Low E-government*	0.336	82
Cluster 2 *High E-government*	0.646	58

tively, Cluster 2 represents countries with high e-government usage and includes developed countries such as the U.S., Sweden, and the United Kingdom as well as emerging nations such as Chile, Slovenia, and Romania that are actively engaged in e-government. In this light, the cluster analysis helps us understand the global patterns in regard to country usage of e-government given the diversity and control variables previously discussed.

DISCUSSION AND CONCLUSION

Trust is a fundamental element to the development process associated with computerization. The Internet age has seen a significant number of successful business applications built on trust using Web-based technologies. For example, eBay established the credibility of online auctions while Amazon became the pacesetter for purchasing via the Internet. Without trust, neither buyers nor sellers would use eBay, Amazon, or other online providers for products and services. In regard to e-government, trust can be considered an even more important issue as it involves most of the citizenry as users. Thus, for e-government to be successful the citizenry must trust their government.

The relationship between e-government and trust was empirically tested in this study. Trust, proxied by ethnic, linguistic, and religious diversity, was found to significantly affect the level of e-government within a country. In particular, countries with less ethnic diversity and more religious diversity, higher levels of economic development, political and civil rights, and economic freedom were found to have greater levels of e-government. Furthermore, a cluster analysis was performed to illustrate the global patterns in e-government. Policy implications can be drawn from this analysis. For example, countries belonging to Cluster 1 that experience lower levels of e-government need to address public and social issues influencing trust among groups, while developing

e-government applications. One way to address this issue is to ensure that the viewpoints of different groups have been considered. For example, many U.S. governmental Web sites now have a Spanish language version allowing Hispanics to feel comfortable performing transactions. National leaders must also realize that they often belong to a dominant ethnic or religious group and others are often suspicious of the party in power. Since the success of e-government services relies heavily on trust, those in power have the responsibility of developing applications that are representative of all groups. Countries that are unable to pursue such strategies may find themselves falling further behind in their ability to compete with the nations of the world.

Regarding the importance of successful e-government practices across countries, it has been argued that e-government promotes overall governance and enhances a country's competitiveness in the global community (IDABC 2005, Taylor et al. 2003). Countries with high levels of e-government are more open and transparent in their policies and practices. Such countries often engage in higher levels of corporate governance, are better connected with the regional and global community, and can be good investment opportunities as they generally experience lower financial risk. Further, e-government can potentially impact the development of open source software. Since open source software lies in the public domain, in order for society to adopt it there needs to be a higher level of trust. Hopefully, as countries address issues related to trust and e-government, they will also benefit from open source as a viable technological platform. Given these positive ramifications of e-government, leaders and policy makers should place a higher priority on the promotion of e-government.

While the above discussion clearly points to the importance of e-government and its relationship to trust, further research is still needed in this area. For example, as better measures of trust become available, future research can test the robustness of these results with larger cross-sectional data sets. In conclusion, though this study has established the importance of trust and the other factors such as level of development and freedom, there are other issues that need to be considered in order to comprehensively analyze the differences across countries with regard to e-government. For example, some societies put a premium on interpersonal contact between governmental representatives and the citizenry and in such societies, a wider acceptance of governmental transactions via the Internet maybe necessary. Thus, one can infer that a more holistic approach to the acceptance of e-government across countries should be adopted by policy makers.

REFERENCES

Alesina, A., Baqir, W., & Easerly, W. (1999). Public Goods and Ethnic Divisions. *The Quarterly Journal of Economics*, *114*(4), 1243–1284. doi:10.1162/003355399556269

Alesina, A., Devleeschauwer, A., Easterly, W., Kurlat, S., & Wacziarg, R. (2003). Fractionalization. *Journal of Economic Growth*, *8*, 155–194. doi:10.1023/A:1024471506938

Alesina, A., & La Ferrara, E. (2002). Who Trusts Others? *Journal of Public Economics*, *85*, 207–234. doi:10.1016/S0047-2727(01)00084-6

Anderson, C., & Paskeviciute, A. (2006). How Ethnic and Linguistic Heterogeneity Influence the Prospects for Civil Society: A Comparative Study of Citizenship Behavior. *The Journal of Politics*, *68*(4), 783–802. doi:10.1111/j.1468-2508.2006.00470.x

Annett, A. (2001). Social Fractionalization, Political Instability, and the Size of the Government. *IMF Staff Papers*, *48*(3), 561–592.

Baier, A. C. (1986). Trust and AntiTrust. *Ethics*, *96*, 231–260. doi:10.1086/292745

Banerjee, A., Iyer, L., & Somanathan, R. (2005). History, Social Divisions, and Public Goods in Rural India. *Journal of the European Economic Association*, *3*, 639–647.

Barro, R. J. (1999). Determinants of Democracy. *The Journal of Political Economy*, *107*(2), 158–183. doi:10.1086/250107

Bélanger, F., Hiller, J., & Smith, W. (2002). Trustworthiness in electronic commerce: the role of privacy, security, and site attributes. *The Journal of Strategic Information Systems*, *11*, 245–270. doi:10.1016/S0963-8687(02)00018-5

Carter, L., & Bélanger, F. (2003). *Diffusion of Innovation and Citizen Adoption of E-government Services*. The Proceedings of the 1st International E-Services Workshop, 57–63.

Carter, L., & Bélanger, F. (2005). The Utilization of e-government Services: Citizen Trust, Innovation and Acceptance Factors. *Information Systems Journal*, *15*(1), 5–25. doi:10.1111/j.1365-2575.2005.00183.x

Ciborra, C., & Navarra, D. D. (2003). "Good Governance and Development Aid: Risks and Challenges of E-government in Jordan". Paper presented at the IFIP WG 8.2 - WG 9.4, Athens, Greece.

Collier, P. (1998). *The Political Economy of Ethnicity*. Washington, DC: World Bank.

Collier, P. (2002). Social Capital and Poverty: A Microeconomic Perspective. In Van Bastelaer, T. (Ed.), *The Role of Social Capital in Development* (pp. 19–41). Melbourne: Cambridge University Press. doi:10.1017/CBO9780511492600.003

Cook, K., & Cooper, R. (2003). Experimental Studies of Cooperation, Trust, and Social Exchange. In Ostrom, E., & Walker, J. (Eds.), *Trust and Reciprocity, Interdisciplinary Lessons from Experimental Research* (pp. 209–244). New York: Russell Sage Foundation.

Currall, S., & Judge, T. (1995). Measuring Trust Between Organizational Boundary Role Persons. *Organizational Behavior and Human Decision Processes, 1995*(64), 151–170. doi:10.1006/obhd.1995.1097

Delhey, J., & Newton, K. (2004). "Social Trust: Global Pattern or Nordic Exceptionalism?" *Wissenschauftszentrum Berlin für Sozialforschung (WZB) Discussion Paper* SP I 2004-202.

Devadoss, P. R., Pan, S. L., & Huang, J. C. (2002). Structurational Analysis of e-government Initiatives: A Case Study of SCO. *Decision Support Systems, 34*(3), 253–269. doi:10.1016/S0167-9236(02)00120-3

E-government Handbook. (2007). Available online http://www.cdt.org/egov/ handbook/trust.shtml, Accessed March 15, 2007.

Easterly, W., & Levine, R. (2001). What have we learned from a decade of empirical research on growth? It's not Factor accumulation: Stylized facts and growth models. *The World Bank Economic Review, 15*(2), 177–219. doi:10.1093/wber/15.2.177

Economist Intelligence Units's. *E readiness Rankings* (EIU). Available online http://globaltechforum.eiu.com/in dex.asp?layout=rich_story&doc_id=6427 Accessed January 10, 2007.

Freedom House. (2006). *Freedom of the World.* Available online http://www.freedomhouse.org/ accessed January 10, 2007.

Fukuyama, F. (1995). *Trust: The Social Virtues and the Creation of Prosperity.* New York: Free Press.

Fukuyama, F. (2001). Social Capital, Civil Society and Development. *Third World Quarterly, 22*(1), 7–20. doi:10.1080/713701144

Gefen, D., Rose, G., Warkentin, M., & Pavlou, P. A. (2005). Cultural Diversity and Trust in IT Adoption: A Comparison of Potential e-Voters in the USA and South Africa. *Journal of Global Information Management, 13*(1), 54–78.

Glaeser, E. L., Laibson, D., Scheinkman, J., & Soutter, C. (2000). Measuring Trust. *The Quarterly Journal of Economics, 115*(3), 811–846. doi:10.1162/003355300554926

Granovetter, M. (1973). The Strength of Weak Ties. *American Journal of Sociology, 78*(6), 1360–1380. doi:10.1086/225469

Hair, J. F. Jr, Anderson, R. E., Tatham, R. L., & Black, W. C. (1992). *Multivariate Data Analysis.* New York: Macmillian.

Hargittai, E. (1999). Weaving the Western Web: Explaining Differences in Internet Connectivity Among OECD Countries. *Telecommunications Policy, 23*(10/11).

Inglehart, R., & Baker, W. (2000). Modernization, Cultural Change, and the Persistence of Traditional Values. *American Sociological Review, 65*(1), 19–51. doi:10.2307/2657288

Johansson-Stenman, O., Mahmud, M., & Martinsson, P. (2006). *Trust and Religion: Experimental evidence from Bangladesh.* Department of Economics, Göteborg University, Mimeo.

Kaufmann, D. (2004). "*Corruption, Government and Security: Challenges for the Rich country and the World*" Chapter in the Global Competitiveness Report 2004/2005. Available online http://siteresources.worldbank.org/INTWBIGOVANTCOR/Resources/ETHICS.xls Accessed January 11, 2007.

Ke, W., & Wei, K. K. (2004). Successful e-government in Singapore. *Communications of the ACM, 47*(6), 95–99. doi:10.1145/990680.990687

Kim, K., & Prabhakar, B. (200) "Initial Trust, Perceived Risk, and the Adoption of Internet Banking," *Proceedings of ICIS 2000*, Brisbane, Australia, Dec. 10-13, 2000.

Klitgaard, R., Justesen, M.K., & Klemmensen R. (2005). "The Political Economy of Freedom, Democracy and Terrorism", Dept. of Political Science and Public Management University of Southern Denmark, mimeo.

Kovačić, Z. J. (2005). The Impact of National cultures on Worldwide E government Readiness. *Informing Science Journal, 8*, 143–159.

La Porta, R., Lopez-de-Silanes, F., Shleifer, A., & Vishny, R. (1997). Trust in Large Organizations. *American Economic Review Papers and Proceedings, LXXXVII*, 333–338.

Larsen, E., & Rainie, L. (2002). "The rise of the e-citizen: How people use government agencies' web sites". *Pew Internet and American Life Project*. Available Online http://www.pewinternet. org/ reports/toc.asp?Report=57 accessed January 13, 2007.

Lee, M., & Turban, E. (2001). A Trust Model for Internet Shopping. *International Journal of Electronic Commerce, 6*, 75–91.

Leigh, A. (2006). Trust Inequality and Ethnic Heterogeneity. *The Economic Record, 82*(258), 268–280. doi:10.1111/j.1475-4932.2006.00339.x

Maskell, P. (2000). Social Capital, Innovation, and Competitiveness. In Baron, S., Field, J., & Schuller, T. (Eds.), *Critical Perspectives*. New York: Oxford University Press.

Mauro, P. (1995). Corruption and Growth. *The Quarterly Journal of Economics, 110*(2), 681–712. doi:10.2307/2946696

Moon, J. M., Welch, W. E., and Wong, W. (2005). "What Drives Global E-governance? An Exploratory Study at a Macro Level" Proceedings of the 38th Hawaii International Conference on System Sciences, pp 1-10.

Moore, M. (1999). Truth, Trust and Market Transactions: What Do We Know? *The Journal of Development Studies, 36*(1), 74–88. doi:10.1080/00220389908422612

Moorman, C., Deshpand, R., & Zaltman, G. (1993). Factors Affecting Trust in Market Research Relationships. *Journal of Marketing, 57*, 81–101. doi:10.2307/1252059

Morgan, R. M., & Hunt, S. (1994). The Commitment-Trust Theory of Relationship Marketing. *Journal of Marketing, 58*(3), 20–38. doi:10.2307/1252308

Mossberger, K., Tolbert, C., & Stansbury, M. (2003). *Virtual inequality: Beyond the digital divide*. Washington, DC: Georgetown University Press.

Newton, K., & Delhey, J. (2005). Predicting Cross-National Levels of Social Trust: Global Pattern or Nordic Exceptionalism? *European Sociological Review, 21*(4), 311–327. doi:10.1093/esr/jci022

Norris, P. (2001). *Digital divide: Civic engagement, information poverty, and the Internet worldwide*. New York: Cambridge University Press.

Persell, C., Green, A., & Gurevich, L. (2001). Civil Society, Economic Distress, and Social Tolerance. *Sociological Forum, 16*(2), 203–230. doi:10.1023/A:1011048600902

Peterson, E., & Seifert, J. (2002). Expectation and challenges of emergent electronic government: The promise of all things E? *Perspectives on Global Development and Technology, 1*(2), 193–212. doi:10.1163/156915002100419808

Putnam, R. D. (1995). Bowling Alone: America's Declining Social Capital. *Journal of Democracy, 6*(1), 65–78. doi:10.1353/jod.1995.0002

Ritzen, J., Easterly, W., & Woolcock, M. (2001). "On 'Good' Politicians and 'Bad' Policies: Social Cohesion, Institutions and Growth", *Working Paper, Washington DC. World Bank*

Rodrick, D. (1999). Where did all the growth go?' External Shocks, Social Conflict, and Growth Collapses. *Journal of Economic Growth, 1*, 149–187.

Rose, R. (2005). A Global diffusion model of e- Governance. *Journal of Public Policy, 25*(1), 5–27. doi:10.1017/S0143814X05000279

Shleifer, A., & Vishny, R. (1993). Corruption. *The Quarterly Journal of Economics, 108*(3), 599–618. doi:10.2307/2118402

Singh, H., Das, A., & Joseph, D. (2004). "Country-level determinants of e-government maturity". Nanyang Technological University, Working Paper.

Steyaert, J. C. (2004). Measuring the Performance of Electronic Government Services. *Information & Management, 41*, 369–375. doi:10.1016/S0378-7206(03)00025-9

Svensson, J. (1998). Investment, Property rights and Political Instability: Theory and Evidence. *European Economic Review, 42*(7), 1317–1342. doi:10.1016/S0014-2921(97)00081-0

The Heritage Foundation. (2005). *Index of Economic Freedom.* Retrieved Dec 20, 2006 from http://www.heritage.org/research/features/index/.

The Heritage Foundation. *Index of Economic Freedom* (2005), Available online http://www.heritage.org/index/, accessed January 15, 2007.

The World Bank Group. (*GDPPC* 2004). Available online http://www.worldbank.org/data, accessed January 15, 2007.

Thomas, J. C., & Streib, G. (2003). The New Face of Government: Citizen-Initiated Contacts in the Era of E-government. *Journal of Public Administration: Research and Theory, 13*(1), 83–102. doi:10.1093/jpart/mug010

Tschannen-Moran, M., & Hoy, W. K. (2000). A Multidisciplinary Analysis of the Nature, Meaning, and Measurement of Trust. *Review of Educational Research, 70*(4), 547–593.

United Nations Department of Economic and Social Affairs (UNDESA). (2003). World Public Sector Report 2003: "e-government at the Crossroads".

Uslaner, E. M. (2002). *The Moral Foundations of Trust.* Cambridge: Cambridge University Press. doi:10.1017/CBO9780511614934

Van Parijs, P. (2004). *Cultural Diversity versus Economic Solidarity.* Brussels: De Boeck Universite.

Volken, T. (2002). "Elements of Trust: The cultural dimensions of internet diffusion revisited", *Electronic Journal of Sociology*, Available online http://epe.lac-bac.gc.ca/100/201/300/ejofsociology/2005/01/volken.html, accessed March 10, 2007.

Warkentin, M., Gefen, D., Pavlou, P., & Rose, G. (2002). Encouraging Citizen Adoption of E-government by Building Trust. *Electronic Markets, 12*, 157–162. doi:10.1080/101967802320245929

Watson, R. T., & Mundy, B. (2001). A Strategic Perspective of Electronic Democracy. *Communications of the ACM, 44*(1), 27–30. doi:10.1145/357489.357499

West, D. M. (2000). "Assessing E-government: The Internet, Democracy, and Service Delivery by State and Federal Governments", Available online,http://www.insidepolitics.org/ egovtreport00.html, accessed March 2, 2007.

West, D. M. (2003). "Global E-government, 2003". Available onlinehttp://www.insidepolitics.org/egovt03int.pdf, accessed March 3, 2007. White, H., "A Heteroskedasticity—Consistent Covariance Matrix Estimator and a Direct Test for Heteroskedasticity," *Econometrica,* 48(4): 817–838.

World Economic Forum. (WEF). (2007). *The Networked Readiness Index,* available online, http://www.weforum.org/pdf/gitr/rankings2007.pdf, accessed March 2, 2007.

World Values Study Group. (1994). *World Values Survey, 1984-93* [Computer File]. ICPSR version. Ann Arbor MI: Institute for social Research [producer], 1994. Ann Arbor, MI: Inter-university Consortium for Political and Social Research [distributor].

Zak, P. J., & Knack, S. (2001). Trust and Growth. *The Economic Journal, 111*(April), 295–321. doi:10.1111/1468-0297.00609

Zinnbauer, D. (2001). Internet, Civil Society and Global Governance: The Neglected Political Dimension of the Digital Divide. *Information & Security., 7*, 45–64.

This work was previously published in International Journal of Electronic Government Research (IJEGR), edited by Vishanth Weerakkody, pp. 1-18, copyright 2009 by Information Science Reference (an imprint of IGI Global)

Chapter 2
Quality Enhancing the Continued Use of E-Government Websites:
Evidence from E-Citizens of Thailand

Sivaporn Wangpipatwong
Bangkok University, Thailand

Wichian Chutimaskul
King Mongkut's University of Technology Thonburi, Thailand

Borworn Papasratorn
King Mongkut's University of Technology Thonburi, Thailand

ABSTRACT

This study empirically examines Web site quality toward the enhancement of the continued use of e-government Web sites by citizens. The web site quality under examination includes three main aspects, which are information quality, system quality, and service quality. The participants were 614 country-wide e-citizens of Thailand. The data were collected by means of a web-based survey and analyzed by using multiple regression analysis. The findings revealed that the three quality aspects enhanced the continued use of e-government Web sites, with system quality providing the greatest enhancement, followed by service quality and information quality.

INTRODUCTION

Electronic government, so called e-government, has been broadly defined as the use of information and communication technology (ICT) to transform government by making it more accessible, effective, and accountable (infoDev & CDT, 2002). The Internet is indeed the most powerful and popular means of delivering e-government. Hence, Web sites have been employed as a platform for delivering a wide range of government services electronically.

By using e-government Web sites, citizens can conveniently access government information and services and gain greater opportunities to participate in the democratic process (Fang, 2002). Citizens can access government information and

DOI: 10.4018/978-1-60960-162-1.ch002

services anywhere and anytime. Thus, the time spent in traveling and waiting is reduced. From the government's point of view, the more citizens that use e-government Web sites, the more operation and management costs are reduced.

To obtain these benefits, the initial adoption and subsequent continued use of e-government Web sites by citizens are required. In general, an information system indicated that its eventual success depends on its continued use rather than first-time use (Bhattacherjee, 2001; Limayem, Hirt, & Cheung, 2003). Likewise, initial use of e-government Web sites is an important indicator of e-government success. However, it does not necessarily lead to the desired outcome unless a significant number of citizens move beyond the initial adoption and use e-government Web sites on a continual basis. To enhance the continued use, this study proposes that quality of e-government Web sites is one significant factor.

According to DeLone and McLean (2002), the three quality aspects, information quality, system quality, and service quality, are the determinants that effect user's intention to use an information system. In practice, these three aspects have been employed to study the initial intention to use the information system and to evaluate the quality of information system (e.g., Lee & Kozar, 2006; Negash, Ryan, & Igbaria, 2003; Wilkin & Castleman, 2002). However, there is a lack of prior research that uses information quality, system quality, and service quality to examine the continued use in the context of e-government Web sites.

This study therefore aims to examine the Web site quality toward enhancement of the continued use of e-government Web sites by citizens. The population of interest for this study is e-citizens of Thailand, a group of citizens who has experienced Thailand's e-government Web sites. The reason that makes Thailand an ideal place to study is that e-government is considered a new innovation to Thai citizens and is conceived as a fundamental element to encourage the country development.

In the next section, the background of this study is briefly reviewed. Thereafter, the research model and hypotheses development, research methodology, and data analysis are presented. Finally, the discussion, limitations, and suggestions for future research are given.

BACKGROUND OF STUDY

DeLone and McLean's Information System Success Model

In order to ascertain the success of an information system, DeLone and McLean (1992) proposed the Information System Success Model (referred hereafter as the 'D&M IS Success Model') as shown in Figure 1. The model asserts that system quality and information quality are the determinants of system use and user satisfaction which effect individual and organizational impact respectively.

However, Pitt, Watson, and Kavan (1995) noticed that commonly used measures of information system effectiveness focus on the products, rather than the services. They then proposed that service quality needs to be considered as an additional measure of the D&M IS Success Model. DeLone and McLean (2002) therefore reformulated the D&M IS Success Model by including service quality as an additional determinant that effects the use and user satisfaction as shown in Figure 2.

D&M IS Success Model has become popular for the specification and justification of the measurement of the dependent variable in an information system research. In the summer of 2002, a citation search yielded 285 refereed papers in journals and proceedings that referenced D&M IS Success Model during the period 1993 to mid-2002 (DeLone & McLean, 2003). In practice, a number of empirical studies (e.g., Iivari, 2005; Molla & Licker, 2001; Seddon & Kiew, 1994) gave support for the associations among the measures identified in the D&M IS Success Model.

Figure 1. Original DeLone and McLean's information system success model (DeLone & McLean, 1992, p.87)

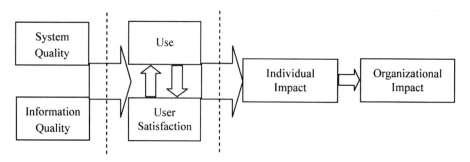

E-Government in Thailand

Like many countries, Thailand has been fully cognizant of both the potentials and the benefits of e-government. Since 1996, the e-government development has been driven by the National Information Technology Committee (NITC). Several programs, such as computer training for mid-level officers, specifying minimum requirements of information technology equipment for government agencies, and the establishment of a Chief Information Officer (CIO) in the public sector, have been imposed to support and promote this initiative (Chamlertwat, 2001). The School-Net Thailand (a national school informatization program to empower all schools to access a large pool of information resources using the Internet), Government Information Network or GINet (a government backbone network to facilitate intra- and inter- agencies communication and information exchanges), and the development of legal infrastructure to support the application of information technology have also been initiated under the first National IT Policy Framework for the year 1996–2000 (IT 2000) (Ateetanan, 2001).

In March 2001, the two-year period e-government project was established by NITC to establish a framework for building up e-government and to implement some pilot projects. In March 2002, the National IT Policy Framework for the year 2001–2010 (IT 2010) was approved by the cabinet and e-government was a manifest flagship, in addition to e-industry, e-commerce, e-society, and e-education. Subsequently, in September 2002, the cabinet further endorsed the first National ICT Master Plan for the year 2002–2006. The master plan devises seven key strategies. One of which is e-government (NECTEC, 2003).

Figure 2. Reformulated DeLone and McLean's information system success model (DeLone & McLean, 2002, p.2974)

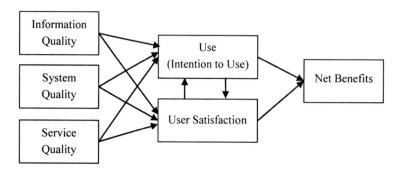

As a result, e-government has been developing rapidly in Thailand. In early 2004, NECTEC initiated the first on-line survey of government e-services. The survey revealed that all 267 government agencies have Web sites to provide information to the public (NECTEC, 2005a). With accordance to the global ranking, the E-government Readiness Survey of United Nations (United Nations, 2003; United Nations, 2004; United Nations, 2005) reported that Thailand owned an E-government Readiness Ranking at 56 from 191 global countries in 2003, moved up to the 50th rank in 2004, and finally at the 46th rank in 2005.

In terms of overall usage, there is not much usage of e-government Web sites compared to other Web site categories. According to the Truehits 2005 Award (Truehits, 2006) government Web sites occupied the 15th rank from 19 Web site categories and had proportion of usage only 1.64%, which went down 0.13% from the previous year.

At this stage it is not clear if Thai citizens will continue using Thailand's e-government Web sites. Hence, better understanding of the factors that enhance citizens to continue using e-government Web sites can create greater value for Thailand's government and also other governments all over the world.

RESEARCH MODEL AND HYPOTHESES DEVELOPMENT

Based on the review of the aforementioned literature, the conceptual research model used to guide this study is proposed as shown in Figure 3. The model is based on the three quality aspects of the D&M IS Success Model, adapting to the e-government Web site context. In the following, the meaning of all constructs and the theories supporting the relationships are presented.

Information Quality

According to DeLone and McLean (1992), information quality is concerned with the measure of the information that the system produces and delivers. When applied to this study, the information

Figure 3. Conceptual research model

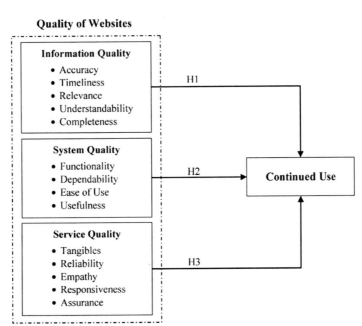

quality focuses on characteristics of information produced by e-government Web sites.

Quality of information is believed to be the most salient factor for predicting customer decision-making behavior (Jeong & Lambert, 2001) and user intention to use a particular system (DeLone & McLean, 1992, 2002; Molla & Licker, 2001). Furthermore, information quality has long been found associated with customer or user satisfaction in previous empirical studies (Seddon & Kiew, 1994; Spreng, MacKenzie, & Olshavsky, 1996; Szymanski & Hise, 2000; Negash, Ryan, & Igbaria, 2003; Iivari, 2005).

Concerning to the case of e-government, the quality of information on the Web sites is very significant since most citizens use e-government Web sites for informational purposes (Accenture, 2004), and the first phase of e-government implementation is to publish government information (infoDev & CDT, 2002). If e-government Web sites contain low information quality, they are useless. Furthermore, high quality information encourages citizens to use the Web sites (Cullen & Hernon, 2004). Hence, the following hypothesis is proposed:

H1: Information quality of e-government Web sites enhances the continued use of e-government Web sites.

According to a review of related literature (Bailey & Pearson, 1983; Doll & Torkzadeh, 1988; Wang & Strong, 1996), the fundamental

dimensions of information quality is composed of five dimensions: accuracy, timeliness, relevance, understandability, and completeness. This study thereby uses these five dimensions to measure citizens' perceptions toward information quality of e-government Web sites. Table 1 shows a brief definition of each dimension.

System Quality

According to DeLone and McLean (1992), system quality is concerned with the measure of the actual system which produces the output. The system quality in this study therefore focuses on features and performance characteristics of e-government Web sites regarding the quality in use or the citizen's view of quality.

System quality, in the sense of quality in use, has been found as a significant determinant of overall user satisfaction (DeLone & McLean, 1992, 2002; Seddon & Kiew, 1994; Negash, Ryan, & Igbaria, 2003; Iivari, 2005), user acceptance (Bevan, 1999), and system use (DeLone & McLean, 1992, 2002). The more satisfied the user is with the system the more he or she will be inclined to use it. Conversely, if system use does not meet the user's needs, satisfaction will not increase and further use will be avoided (Baroudi, Olson, & Ives, 1986). Therefore, this study postulates that:

H2: System quality of e-government Web sites enhances the continued use of e-government Web sites.

Table 1. Information quality dimensions

Dimension	Definition	Contributing Authors
Accuracy	The information is correct and reliable	Bailey and Pearson (1983) Doll and Torkzadeh (1988) Wang and Strong (1996)
Timeliness	The information is current and timely	
Relevance	The information corresponds to the need and is applicable for the task at hand	
Understandability	The information is clear and easy to comprehend	
Completeness	The information has sufficient breadth and depth for the task at hand	

Table 2. System quality dimensions

Dimension	Definition	Contributing Authors
Functionality	The required functions are available in the system	Bailey and Pearson (1983) Doll and Torkzadeh (1988) Davis (1989)
Dependability	The system is accurate and dependable over time	
Ease of Use	The system can be accessed or used with relatively low effort	
Usefulness	The benefits that the user believes to derive from the system, including convenience, saving time, and saving cost	

Based on a review of related literature (Bailey & Pearson, 1983; Doll & Torkzadeh, 1988), this study identifies and categorizes the characteristics related to the quality in use and user satisfaction into four core dimensions: functionality, dependability, ease of use, and usefulness. Ease of use and usefulness are also excerpted from Davis's (1989) Technology Acceptance Model (TAM). This study therefore uses theses four dimensions to measure citizens' perception toward system quality of e-government Web sites. The definition of each dimension is summarized in Table 2.

Service Quality

Service quality refers to the quality of personal support services provided to citizens through e-government Web sites, such as answering questions, taking requests, and providing sophisticated solutions to citizen's problems. This definition is consistent to service quality of DeLone and McLean (2002, 2004) that concerns the measure of the user support services delivered by the service provider.

Prior literature on marketing has indicated that service quality is an important determinant of customer satisfaction (Cronin & Taylor, 1992; Bitner, Booms, & Mohr, 1994; DeLone & McLean, 2002, 2004) and repeat patronage (Zeithaml, Berry, & Parasuraman, 1996), especially in pure service situations where no tangible object is exchanged (Parasuraman, Zeithaml, & Berry, 1985; Solomon, Surprenant, Czepiel, & Gutman, 1985). With regards to e-government, service quality is needed since citizens differ in education, knowledge, and experience. The service quality therefore acts as an enabler of the citizen's capability to use e-government Web sites. Hence, this leads to the hypothesis:

H3: Service quality of e-government Web sites enhances the continued use of e-government Web sites.

Table 3. Service quality dimensions

Dimension	Definition	Contributing Authors
Tangibles	Physical facilities, equipment, and appearance of personnel	Parasuraman, Zeithaml, and Berry (1988)
Reliability	Ability to perform the promised service dependably and accurately	
Empathy	Caring, individualized attention the service provider gives its customers	
Responsiveness	Willingness to help customers and provide prompt service	
Assurance	Knowledge and courtesy of employees and their ability to inspire trust and confidence	

Based on the SERQUAL developed by Parasuraman, Zeithaml, and Berry (1988), the quality of service is composed of five dimensions: tangibles, empathy, reliability, responsiveness, and assurance as defined in Table 3. The SERVQUAL is both a reliable and a valid measure of service quality and is also applicable to a wide variety of service contexts (Parasuraman, Zeithaml, & Berry, 1988). Thus, the SERVQUAL dimensions are used to measure citizens' perception toward system quality of e-government Web sites.

RESEARCH METHODOLOGY

Participants

The participants were 614 e-citizens from five regions of Thailand. The majority of the participants are living in the capital of Thailand (Bangkok) and vicinity (77.36%), followed by the central region (9.77%), the northern region (5.70%), the northeast region (3.58%), and the southern region (3.58%). The dispersion of participants in this study was comparable to the Internet user profile of Thailand, wherein Internet users are concentrated in Bangkok and vicinity, and the rest are distributed in other regions with nearly equivalent proportion (NECTEC, 2005b). Figure 4 shows the participant dispersion in this study compared to the Internet user profile of Thailand.

Demographic characteristics of the overall participants are summarized in Table 4. The proportion of the gender of participants is equal. Most of them are between 21–30 years of age (64.01%), have a bachelor's degree (54.89%), work in private sectors (58.79%), and have monthly income between 10,001–20,000 Baht (36.64%). About half of participants (62.21%) have experienced Internet for 6-10 years.

In terms of experience with e-government Web sites, the most frequently mentioned experience is searching, inquiry, or complaint (79.32%), followed by online transactions (68.24%) and downloading forms (60.75%). The five most frequently mentioned topics are tax (75.90%), tourism (56.84%), education (55.37%), citizen registration (35.67%) and communication (31.60%). The participants' experience with e-government Web sites is illustrated in Table 5.

Instrument Development

The questionnaire was used as an instrument to gather data from participants. The measurement items for information quality, system quality, and service quality were rated on a 5-point Likert scale (1=Strongly Disagree; 2=Disagree; 3=Neutral;

Figure 4. Dispersion of participants compared to Internet user profile of Thailand

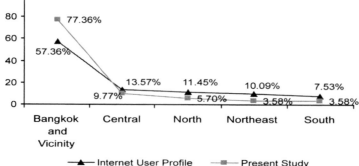

Table 4. Participants' demographic characteristics

Characteristics		Frequency	Percent [a]
Gender	Female	307	50.00
	Male	307	50.00
Age	21−30 years	393	64.01
	31−40 years	179	29.15
	41−50 years	35	5.70
	Older than 50 years	7	1.14
Level of Education	Bachelor's Degree	337	54.89
	Master's Degree	268	43.65
	Doctorate Degree	9	1.47
Occupation	Student	51	8.31
	Government Employee	113	18.40
	State Enterprises Employee	30	4.89
	Private Sector Employee	361	58.79
	Self Employment	54	8.79
	Unemployed	3	0.49
	Retiree	2	0.33
Monthly Income	Less than or equal to 5,000 Baht	12	1.95
	5,001−10,000 Baht	75	12.21
	10,001−20,000 Baht	225	36.64
	20,001−30,000 Baht	130	21.17
	More than 30,000 Baht	172	28.01
Years on Internet	Less than 1 year	2	0.33
	1−5 years	92	14.98
	6−10 years	382	62.21
	More than 10 years	138	22.48

[a] Percentages subject to rounding

4=Agree; 5=Strongly Agree). Table 6 lists the measurement items.

The validity of the questionnaire was strengthened through an extensive review of the literature and an agreement among professionals. In addition, the pretest through 25 convenience samples was employed to determine if the intended audiences had any difficulty understanding the questionnaire and whether there were any ambiguous or biased questions. Based on the feedback of the pretest, one reverse word item was dropped since it caused confusion.

To ensure the measurement items are measuring the same construct, the most widely used measure named Cronbach's alpha was employed for each construct to measure the internal consistency among all items. As observed from Table 7, the reliability analysis gave alpha coefficients exceeding .70 which are typically regarded as an acceptable reliability coefficient (Nunnaly, 1978); one exception was for the scale of completeness which is a little bit lower than .70. However, the lower limit of acceptability may decrease to .60 in exploratory research (Hair, Anderson, Tatham, &

Table 5. Participants' experience with e-government Web sites

Experience [a]		Frequency	Percent [b]
Type of Use	Search / Inquiry / Complaint	487	79.32
	Conduct Online Transaction	419	68.24
	Download Form	373	60.75
Topic of Use	Tax	466	75.90
	Tourism	349	56.84
	Education	340	55.37
	Citizen Registration	219	35.67
	Communication	194	31.60
	Research	155	25.24
	Transportation	140	22.80
	Employment	138	22.48
	Recreation	137	22.31
	Health	130	21.17
	Commerce	110	17.92
	Foreign Affairs	97	15.80
	Housing	92	14.98
	Safety and Regulation	88	14.33
	Government Welfare	82	13.36
	Agricultural	71	11.56
	Public Utility	66	10.75
	Industry	65	10.59
	Politics	59	9.61
	Family and Community	40	6.51

[a] Participants could tick all that apply
[b] Percentages subject to rounding

Black, 1998). Therefore, the items measuring the constructs were acceptable and reliable.

Data Collection

In order to eliminate costs, data coding time, and human-error, and to easily reach citizens in different geographic areas across the country, a web-based survey with a probability list-based method, which samples participants based on a list, was employed to collect data. The bundled script program was used to check and advise participants, thereby ensuring that all items in the questionnaire were filled in completely and appropriately. To take steps toward ensuring the integrity of the data, the IP address of each participant and the time used for completing the survey were recorded.

After the survey was uploaded to the server, the 3,600 invitations for participation, including a link to the Web site, were randomly emailed to an alumni mailing list in a variety of faculties (e.g., science, agriculture, engineering, pharmacology, liberal arts, business administration, information technology, and humanities) of five universities across five regions of Thailand. The selection of these five universities resulted from a three-stage sampling. First, a stratified sampling was

Table 6. List of measurement items

Construct	Dimension	Description
Information Quality	Accuracy	Using e-government Web sites enables me to have accurate information.
		I can trust the information on e-government Web sites.
	Timeliness	Using e-government Web sites enables me to access the newest information.
		Using e-government Web sites enables me to access up-to-date information when compare to deal with other sources.
	Relevance	Using e-government Web site enables me to have information that is relevant to the site.
		Using e-government Web site enables me to have the information that I need.
	Understandability	Information on e-government Web sites is easy for me to comprehend.
		Information on e-government Web sites is clear for me.
	Completeness	Using e-government Web sites enables me to access adequate information.
		I find information on e-government Web sites is sufficient for the task at hand.
System Quality	Functionality	E-government Web sites provide necessary information and forms for downloading.
		E-government Web sites provide necessary online transactions.
		E-government Web sites provide service functions that I need.
		E-government Web sites present a variety of services
	Dependability	E-government Web sites perform right at the first time.
		Every time I request e-government Web sites, the Web sites are available.
		The government will not misuse my personal information.
		I feel safe in my online transaction with e-government Web sites.
	Ease of Use	I can easily login to e-government Web sites.
		Getting the information that I want from e-government Web sites is easy.
		It is easy for me to complete transactions through e-government Web sites.
		The organization and structure of e-government Web sites is easy to follow.
	Usefulness	Using e-government Web sites enable me to accomplish tasks more quickly.
		The results of using e-government Web sites are apparent to me.
		Using e-government Web sites can cut traveling expense.
		Using e-government Web sites can lower traveling and queuing time.
		Using e-government Web sites enable me to do business with the government anytime, not limited to regular business hours.

continued on following page

Table 6. Continued

Service Quality	Tangibles	If I need help, I can find a way to reach a government staff such as email or webboard on e-government Web sites.
		There is staff who will respond to my request indicated on e-government Web sites.
	Reliability	If I send a request via email or webboard to the government, I will receive the right solution from the government staff.
		If I send a request via email or webboard to the government, I will receive the solution that matches to my needs from the government staff.
	Empathy	If I send a request via email or webboard to the government, I will receive the response that shows the willingness to help from the government staff.
		If I send a request via email or webboard to the government, I will receive the response that shows the friendliness of the government staff.
	Responsiveness	If I send a request via email or webboard to the government, I will receive prompt response from the government staff.
		If I have a problem with e-government Web sites, the government staff will quickly resolve my problem.
	Assurance	The government staff seem to have sufficient knowledge to answer my questions.
		The government staff seem to have an ability to solve my problem.
Continued Use		In the future, I would not hesitate to use e-government Web sites.
		In the future, I will consider e-government Web sites to be my first choice to do business with the government.
		In the future, I intend to increase my use of e-government Web sites.

Table 7. Reliability analysis results

Construct	No. of Items	Mean	SD	Cronbach's Alpha
Information Quality	10	3.067	.604	.899
Accuracy	2	3.394	.748	.839
Timeliness	2	2.715	.795	.783
Relevance	2	3.148	.725	.737
Understandability	2	2.997	.766	.840
Completeness	2	3.079	.741	.688
System Quality	17	3.100	.617	.925
Functionality	4	3.068	.722	.837
Dependability	4	3.019	.699	.749
Ease of Use	4	2.875	.716	.872
Usefulness	5	3.372	.811	.878
Service Quality	10	2.587	.778	.951
Tangibles	2	2.649	.961	.819
Reliability	2	2.582	.899	.937
Empathy	2	2.625	.868	.925
Responsiveness	2	2.406	.870	.905
Assurance	2	2.674	.843	.907
Continued Use	3	3.232	.865	.873

performed to cluster Thailand into five regions. Second, a simple random sampling was done to select a university corresponding to each of the five regions. Third, a simple random sampling was employed to select some email addresses corresponding to each of the five selected universities. Sending the invitation emails to the alumni mailing list can guarantee that the participants have experience with the Internet and hence probably enabled us to reach e-citizens. In addition, previous studies indicated that e-government Web sites are particularly popular among those who have at least a college education (Larsen & Rainie, 2002; Wangpipatwong, Chutimaskul, & Papasratorn, 2005). Finally, 1,159 e-mail addresses turned out to be invalid and the invitation emails could not be delivered to the recipients. However, there were 2,441 valid e-mails that did reach the recipients.

Responses to the survey were collected for a two-month period (February 1, 2006 to March 31, 2006). Respondents were screened according to whether they had experience with e-government Web sites. Only those who had previous experience continued with the survey. Out of 799 responses, 614 responses indicated experience with e-government Web sites. All these 614 responses were then used in the analysis after they were verified to be valid and complete without any unusual data or multiple responses. The number of valid responses conforms to finite population sampling formula (Yamane, 1973), along with a 95% confidence level and a 5% precision level.

Data Analysis

A multiple regression was chosen as the appropriate method to examine whether information quality, system quality, and service quality of e-government Web sites will enhance the continued use of e-government Web sites. Together with the analysis, assumptions of multivariate normal distribution, linearity, and homogeneity of variance

Table 8. Regression analysis results of information quality, system quality, and service quality on continued use

Construct	Unstandardized Coefficients		Standardized Coefficients	t	p	Collinearity Statistics
	B	Std. Error	Beta			VIF
(Constant)	.293	.144		2.031	.043	
Information Quality	.218	.067	.153	3.251	.001	2.371
System Quality	.547	.073	.390	7.515	.000	2.901
Service Quality	.221	.044	.199	5.072	.000	1.660

$R^2 = .434$; F = 155.793; p = .000

were tested. There were no violations of these assumptions. The Variance Inflation Factor (VIF) less than 5 confirms the lack of multicollinearity (Studenmund, 1992). Finally, the number of cases is very well above the minimum requirement of 50+8k for testing the multiple correlation and 104+k for testing individual predictors, where k is the number of independent variables (Green, 1991).

ANALYSIS AND RESULTS

To examine the Web site quality toward enhancement of the continued use of e-government Web sites, information quality, system quality, and service quality were simultaneously regressed on the continued use of e-government Web sites. The results revealed that these three quality aspects significantly accounted for 43.4% of the variance in the continued use of e-government Web sites (R^2 = .434, F = 155.793, p < .001). As shown in Table 8, system quality (β = .390, p < .001) yielded the greatest enhancement on the continued use, followed by service quality (β = .199, p < .001), and information quality (β = .153, p < .01). Therefore, all proposed hypotheses were supported.

Afterward, all fourteen dimensions corresponding to information quality, system quality, and service quality were regressed, using stepwise method to investigate the finest model of enhancement. As shown in Table 7, there were five dimensions, usefulness (β = .413, p < .001), empathy (β = .179, p < .001), accuracy (β = .106, p < .01), assurance (β = .102, p < .05), and relevance (β = .079, p < .05) which formed the finest model of enhancement.

Table 9. Regression analysis results with the finest model

Dimension	Unstandardized Coefficients		Standardized Coefficients	t	p	Collinearity Statistics
	B	Std. Error	Beta			VIF
(Constant)	.283	.137		2.067	.039	
Usefulness	.441	.041	.413	10.826	.000	1.713
Empathy	.179	.042	.179	4.298	.000	2.043
Accuracy	.123	.043	.106	2.846	.005	1.635
Assurance	.105	.043	.102	2.406	.016	2.111
Relevance	.094	.046	.079	2.066	.039	1.715

$R^2 = .483$; F = 113.514; p = .000

CONCLUSION AND DISCUSSION

The aim of this study was to examine the Web site quality toward enhancement of the continued use of e-government Web sites by citizens. The study was motivated by the lack of empirical studies that uses information quality, system quality, and service quality to examine the continued use in the context of e-government Web sites.

As predicted, the results revealed that Web site quality corresponding to information quality, system quality, and service quality enhanced the continued use of e-government Web sites. The higher the level of information quality, system quality, and service quality, the higher the citizens' intention to continue using e-government Web sites. The results thereby corroborate that information quality, system quality, and service quality enhance not only initial intention as DeLone and McLean (2002) asserted, but also the continued use in the context of e-government Web sites.

Further, the results also revealed that system quality provided the greatest enhancement on the continued use of e-government Web sites, followed by service quality and information quality. This outcome resembles e-business study (Lee & Kozar, 2006) that found online customers considered system quality as the greatest significant factor in selecting the most preferred e-business Web sites.

When considering the dimensional perspective, the results showed that there were five dimensions which formed the finest model of enhancement. These five dimensions were ordered in significance as usefulness (of system), empathy (of service), accuracy (of information), assurance (of service), and relevance (of information). This outcome thereby suggests that government should ensure that these five dimensions are well integrated in the e-government Web sites. The following are some of suggestions.

- **Usefulness:** E-government Web site should provide useful services compared to the traditional way, such as convenience, saving time, and saving cost.
- **Empathy:** Responsible staff should give caring and individualized attention to citizen, such as providing individualized attention to individual concerns and requests, through email communication rather than a generic auto-reply message.
- **Accuracy:** Information on e-government Web site should be correct and reliable.
- **Assurance:** Responsible staff should have the knowledge and ability to inspire trust and confidence. The staff should provide impeccable response to convey trust and confidence to citizens.
- **Relevance:** Information on e-government Web site should be relevant to the site and corresponds to the need.

Furthermore, government should obviously disclosure the usefulness of e-government Web sites. The government may highlight the unique features of the Web sites compared to dealing with government staff for the same services and promote the idea that the Web sites facilitate the access to services anywhere and anytime with saving time and cost.

To conclude, it is a necessity for the government to recognize the quality of e-government Web sites, since it enhances the continued use of the Web sites. Government should ensure that the significant dimensions corresponding to information quality, system quality, and service quality are well established. Finally, the next challenge for government involves changing the citizens' perception and the means in which the information, system, and service are presented and delivered to the citizens corresponding to their needs.

LIMITATIONS AND RESEARCH DIRECTIONS

Although the study provides meaningful implications, it has two limitations. First, the dimensions used to measure information quality, system quality, and service quality are equally weighted. Future research may try using dimensions that are unequally weighted. Second, this study intends to elicit data from e-citizens who are ready for e-government. To regard the digital divide, future research should elicit the data from citizens who have lower level of education, lower income, and also citizens who lack access to the Internet.

REFERENCES

Accenture (2004). *e-government Leadership: High Performance, Maximum Value*. Retrieved May 7, 2005 from: http://www.accenture.com/xdoc/en/ind ustries/government/ gove_egov_value.pdf

Ateetanan, P. (2001). *Country Report Thailand*. Retrieved November 14, 2006 from: http:// unpan1.un.org/intradoc/groups/public/ documents/ APCITY/UNPAN012806.pdf

Bailey, J. E., & Pearson, S. W. (1983). Developing a tool for measuring and analyzing computer user satisfaction. *Management Science*, *29*(5), 530–545. doi:10.1287/mnsc.29.5.530

Baroudi, J. J., Olson, M. H., & Ives, B. (1986). An empirical study of the impact of user involvement on system usage and information satisfaction. *Communications of the ACM*, *29*(3), 232–238. doi:10.1145/5666.5669

Bevan, N. (1999). Quality in use: Meeting user needs for quality. *Journal of Systems and Software*, *49*(1), 89–96. doi:10.1016/S0164-1212(99)00070-9

Bhattacherjee, A. (2001). Understanding information systems continuance: An expectation-confirmation model. *Management Information Systems Quarterly*, *25*(3), 351–370. doi:10.2307/3250921

Bitner, M. J., Booms, B. H., & Mohr, L. A. (1994). Critical service encounters: The employee's viewpoint. *Journal of Marketing*, *58*(4), 95–106. doi:10.2307/1251919

Chamlertwat, K. (2001). Current status and issues of e-government in Thailand. *15th Asian Forum for the Standardization of Information Technology*, Kathmandu, Nepal.

Cronin, J. J. Jr, & Taylor, S. A. (1992). Measuring service quality: A reexamination and extension. *Journal of Marketing*, *56*(3), 55–68. doi:10.2307/1252296

Cullen, R., & Hernon, P. (2004). *Wired for Well-being Citizens' Response to E-government*. Retrieved March 15, 2005 from: http://www.e-government.govt.n z/docs/vuw-report-200406/

Davis, F. D. (1989). Perceived usefulness, perceived ease of use, and user acceptance of information technology. *Management Information Systems Quarterly*, *13*(3), 319–340. doi:10.2307/249008

DeLone, W. H., & McLean, E. R. (1992). Information systems success: The quest for the dependent variable. *Information Systems Research*, *3*(1), 60–95. doi:10.1287/isre.3.1.60

DeLone, W. H., & McLean, E. R. (2002). Information systems success revisited. *Proceedings of the 35th Hawaii International Conference on System Science, 3*(1), 2966–2976.

DeLone, W. H., & McLean, E. R. (2003). The DeLone and McLean model of information systems success: A ten year update. *Journal of Management Information Systems*, *19*(4), 9–30.

DeLone, W. H., & McLean, E. R. (2004). Measuring eCommerce success: Applying the DeLone and McLean information system success model. *International Journal of Electronic Commerce, 9*(1), 31–47.

Doll, W. J., & Torkzadeh, G. (1988). The measurement of end-user computing satisfaction. *Management Information Systems Quarterly, 12*(2), 259–274. doi:10.2307/248851

Fang, Z. (2002). E-government in digital era: Concept, practice, and development. *International Journal of the Computer . The Internet and Management, 10*(2), 1–22.

Green, S. B. (1991). How many subjects does it take to do a regression analysis? *Multivariate Behavioral Research, 26*(3), 499–510. doi:10.1207/s15327906mbr2603_7

Hair, J. F. Jr, Anderson, R. E., Tatham, R. L., & Black, W. C. (1998). *Multivariate Data Analysis* (5th ed.). Upper Saddle River, NJ: Prentice-Hall.

Iivari, J. (2005). An empirical test of the DeLone-McLean model of information system success. *The Data Base for Advances in Information Systems, 36*(2), 8–27.

infoDev and CDT (Center for Democracy and Technology) (2002). *The E-government Handbook for Developing Countries.* Retrieved November 14, 2006 from: http://www.cdt.org/egov/handbook/2002-11-14egovhandbook.pdf

Jeong, M., & Lambert, C. U. (2001). Adaptation of an information quality framework to measure customers' behavioral intentions to use lodging web sites. *Hospital Management, 20*(2), 129–146. doi:10.1016/S0278-4319(00)00041-4

Larsen, E., & Rainie, L. (2002). *The Rise of the E-Citizen: How People Use Government Agencies' Web Site.* Retrieved April 12, 2006 from: Available: http:// www.pewinternet.org/pdfs/PIP_Govt_Web site_Rpt.pdf

Lee, Y., & Kozar, K. A. (2006). Investigating the effect of Web site quality on E-Business success: An analytic hierarchy process (AHP) approach. *Decision Support Systems, 42*(3), 1383–1401. doi:10.1016/j.dss.2005.11.005

Limayem, M., Hirt, S. G., & Cheung, C. M. K. (2003). Habit in the context of IS continuance: Theory extension and scale development. *Proceedings of the 11th European Conference on Information Systems (ECIS 2003).* Retrieved April 12, 2006 from: http://is2.lse.ac.uk/asp/aspecis/20030087.pdf

Molla, A., & Licker, P. S. (2001). E-Commerce systems success: An attempt to extend and re-specify the Delone and Maclean model of IS success. *Journal of Electronic Commerce Research, 2*(4), 131–141.

NECTEC (National Electronics and Computer Technology Center). (2003). *Thailand Information and Communications Technology (ICT) Master Plan (2002-2006).* Retrieved November 6, 2006 from: http://www.nectec.or.th/pld/masterplan/ document /ICT_Masterplan_Eng.pdf

NECTEC (National Electronics and Computer Technology Center). (2005a). *Thailand ICT Indicators 2005.* Retrieved November 6, 2006 from: http://www.nectec.or.th/ pub/book/ ICTIndicators.pdf

NECTEC (National Electronics and Computer Technology Center). (2005b). *Internet User Profile of Thailand 2005.* Retrieved November 6, 2006 from: http://www.nectec.or.th/ pld/internetuser/Internet %20User%20Profile%202005.pdf

Negash, S., Ryan, T., & Igbaria, M. (2003). Quality and effectiveness in web-based customer support systems. *Information & Management, 40*(8), 757–768. doi:10.1016/S0378-7206(02)00101-5

Nunnaly, J. C. (1978). *Psychometric Theory* (2nd ed.). New York, NY: McGraw-Hill.

Parasuraman, A., Zeithaml, V. A., & Berry, L. L. (1985). A conceptual model of service quality and its implications for future research. *Journal of Marketing, 49*(4), 41–50. doi:10.2307/1251430

Parasuraman, A., Zeithaml, V. A., & Berry, L. L. (1988). SERVQUAL: A multiple-item scale for measuring consumer perceptions of service quality. *Journal of Retailing, 64*(1), 12–40.

Pitt, L. F., Watson, R. T., & Kavan, C. B. (1995). Service quality: A measure of information systems effectiveness. *Management Information Systems Quarterly, 19*(2), 173–185. doi:10.2307/249687

Seddon, P. B., & Kiew, M.-Y. (1994). A partial test and development of the DeLone and McLean's model of IS success. *Proceedings of the International Conference on Information Systems (ICIS 94)*, 99–110.

Solomon, M. R., Surprenant, C. F., Czepiel, J. A., & Gutman, E. G. (1985). A role theory perspective on dyadic interactions: the service encounter. *Journal of Marketing, 49*(1), 99–111. doi:10.2307/1251180

Spreng, R. A., MacKenzie, S. B., & Olshavsky, R. W. (1996). A reexamination of the determinants of consumer satisfaction. *Journal of Marketing, 60*(3), 15–32. doi:10.2307/1251839

Studenmund, A. H. (1992). *Using Econometrics: A Practical Guide* (2nd ed.). New York, NY: HarperCollins.

Szymanski, D. M., & Hise, R. T. (2000). E-Satisfaction: An initial examination. *Journal of Retailing, 76*(3), 309–322. doi:10.1016/S0022-4359(00)00035-X

Truehits (2006). *Truehits 2005 Awards*. Retrieved November 16, 2006 from: http://truehits.net/awards2005

United Nations. (2003). *UN Global E-government Survey 2003*. Retrieved November 16, 2006 from: http://unpan1.un.org/int radoc/groups/public/documents/un/ unpan019207.pdf

United Nations. (2004). *UN Global E-government Readiness Report 2004: Towards Access for Opportunity*. Retrieved November 16, 2006 from: http://unpan1.un.org/intradoc/ groups/public/documents/un/unpan019207.pdf

United Nations. (2005). *Global E-government Readiness Report 2005: From E-government to E-Inclusion*. Retrieved November 16, 2006 from: http://unpan1.un.org/intradoc/ groups/ public/documents/un/unpan021888.pdf

Wang, R. Y., & Strong, D. M. (1996). Beyond accuracy: What data quality means to data consumers. *Journal of Management Information Systems, 12*(4), 5–34.

Wangpipatwong, S., Chutimaskul, W., & Papasratorn, B. (2005). Factors influencing the use of eGovernment Web sites: Information quality and system quality approach. *International Journal of the Computer, the Internet and Management, 13*(SP3), 14.1–14.7.

Wilkin, C., & Castleman, T. (2003). Development of an instrument to evaluate the quality of delivered information systems. *Proceedings of the 36th Hawaii International Conference on System Sciences*. Retrieved December 11, 2007 from: http:// csdl2.computer.org/comp/proce edings/hicss/2003/1874/08/187480244b.pdf

Yamane, T. (1973). *Statistics: An Introductory Analysis* (3rd ed.). New York, NY: Harper & Row.

Zeithaml, V. A., Berry, L. L., & Parasuraman, A. (1996). The behavioral consequences of service quality. *Journal of Marketing, 60*(2), 31–36. doi:10.2307/1251929

This work was previously pulished in International Journal of Electronic Government Research (IJEGR), edited by Vishanth Weerakkody, pp. 19-35, copyright 2009 by Information Science Reference (an imprint of IGI Global)

Chapter 3
Paralingual Web Design and Trust in E-Government

Roy H. Segovia
San Diego State University, USA

Murray E. Jennex
San Diego State University, USA

James Beatty
San Diego State University, USA

INTRODUCTION

Electronic government (e-government) is the use of Information and Communication Technologies (ICT), including the Internet, by government organizations to facilitate providing information and services to their constituents. E-government Web sites provide everything from basic information about governmental bodies and issues, to online services such as registering vehicles and applying for employment and for permits. More recent e-government services include e-consultation, which is citizen participation and response to forthcoming consultations and decisions on matters of public interest (Jadu, 2005). The impetus to implement e-government can be attributed to cost control and improved service to citizens. Another driver is government's growing awareness of the need to attain more democratic governance (Coleman and Gotze, 2001; OECD, 2001), coupled with a widespread public interest in the potential of ICT to empower citizens and to increase government accountability and transparency (Hart-Teeter, 2003). An example is the United States E-Government initiative targets use of improved Internet-based technology to make it easier for citizens and businesses to interact with the government, save taxpayer dollars, and streamline citizen-to-government communications (USOMB, 2005). These many drivers make it likely that e-government will be a lasting ICT application leading e-government system designers to look for tools and methodolo-

DOI: 10.4018/978-1-60960-162-1.ch003

gies that will ensure their acceptance and use by the intended users.

This article introduces one such potential design methodology, paralingual Web design, and uses an experiment to test this design methodology to see if it has potential for improving system acceptance and success. Paralingual is a Web design methodology for presenting information in more than one language. Paralingual Web design involves placing content in the desired languages but instead of having separate pages for each language as is common in a bi or multilingual Web design, the bi or multilingual content is placed side by side on the same page. The inspiration for this article is the trend towards localization in ecommerce systems and a concern that there may be a localization issue for e-government when the target population is bi or multilingual. Localization is defined by the Localization Industry Standards Association (LISA, 2008) as the process of modifying specific products or services for specific markets. In the case of e-government this involves tailoring e-government Web sites to fit the constituent market and in the case of a constituent market that speaks more than one language, allowing for these multiple languages. The concern driving this experiment is that there may be a trust issue affecting the success/adoption of e-government should these systems fail to take into account the bi or multi lingual aspects of their constituents.

A premise of information systems, IS, is that for an IS to be successful the intended system users must "use" the system where Rai et al. (2001) consider "use" to be the consumption of the outputs of the IS by the users as measured in terms such as frequency of use, amount of time of use, numbers of access to the IS, usage pattern, etc. General thinking is that the more an IS is used, the more successful the IS. Two of the more widely accepted IS success/acceptance models, the DeLone and McLean (1992 and 2003) IS Success Model and the Davis (1989) Technology Acceptance Model, TAM, incorporate "use" as

a measure of success (DeLone and McLean) or successful adoption (TAM) through constructs such as Intent to Use, Perceived Usefulness, and Perceived Ease of Use.

Several authors (Gefen, et al., 2002; Tan, et al., 2005, Tan, et al., 2008, Warkentin, et al., 2002) suggest that use of e-government is influenced by the trust that potential users have with e-government. This article hypothesizes that this trust in e-government, and thus subsequent use, can be increased in bi and multi lingual societies by using paralingual Web design. This allows readers who are bilingual to easily see both versions and readily determine if the same information is being said in each version. It is expected that trust will be increased through this citizen validation process.

The contribution of this research is showing designers of e-government Web pages how the Web pages can be designed to improve trust in a bi-lingual constituency. While this research did not test this design approach in a multi-lingual environment it is expected this design can also be applied to e-government Web pages for these constituencies.

LITERATURE REVIEW

This article draws from three main bodies of literature, the trust, paralingual, and IS acceptance/success literatures. These literatures are summarized below and provide the theoretical foundation for the article. The trust literature is presented first as it provides the issue of concern for the article. The paralingual literature is second to provide the background for why the proposed design methodology is a good solution for the trust issue. The IS acceptance/success literature is provided third as it helps provide the framing for the experiment.

Trust can be defined as "the subjective assessment of one party [trustor] that another party [trustee] will perform a particular transaction according to his or her confident expectations, in an environment characterized by uncertainty" (Ba and Pavlou, 2002, p. 245.) For e-government this means users trusting that the e-government service is providing correct information, that data will be protected, and that transactions will be conducted in a secure manner and recorded appropriately.

Trust in government has historically been problematic in the United States of America as constituent citizens are known to have a high level of distrust in their governing bodies. Trust in government has been declining for more than three decades and has been the topic of a substantial amount of research in political science (Levi and Stoker, 2000; Hibbing and Theiss-Morse, 2002). In the state of California, a recent study exposed an unexpectedly high level of distrust in government by California citizens. During 2004, a series of dialog-oriented seminars were held by Viewpoint Learning in various locations in California. One of the seven major findings of the study was that an underlying issue was profound mistrust of government and elected officials. Furthermore, this mistrust was more intense and persistent than expected, outstripping the levels that have been measured by polls and focus groups (Rosell, Gantwerk, and Furth, 2005).

In addition to the trust issues above, there are known issues with trust in e-government Web sites. This is clearly the effect of the general mistrust by citizens in their government bodies, as mentioned previously. The principal reason given for mistrust of the Web is an artifact of the internet itself. Namely, the internet is now perceived to be beyond the control of the hosts and providers in terms of security and trust. Despite the use of lock icons, digital signatures, passwords, privacy policy statements, and other security techniques, internet users feel that hosts and providers have lost control of the digital data transport medium as well as the software infrastructure that supports it, impeding the growth of e-government (Mercuri, 2005.) To counter this, the International Telecommunication Union (ITU) is providing support for national e-government projects including enhancing security and trust in the use of public networks (Khalil-babnet, 2005).

Improving trust in government and e-government is a critical issue. In the study by Viewpoint Learning, citizens voiced a strong desire to find constructive solutions to problems facing the state (Rosell, Gantwerk, and Furth, 2005). In a geographical area with a high proportion of bilingual speakers, usage of e-government Web sites may be improved in the same way as has been shown effective in electronic commerce (ecommerce). That is, with regard to language issues, researchers have found that customers are far more likely to buy products and services from Web sites in their own language, even if they can read English well. Furthermore, attention to site visitors' needs should be an important consideration in Web design because such attention can help a site build trust with customers (Schneider, 2003). Gassert (2004) suggests building trust through knowledge by using ICT for better education and information. Additionally, LaVoy (2001) supports the use of e-government as a way of improving trust by improving accountability. Finally, Gefen, et al. (2002) view trust in government as the main driver for e-government adoption. Their analysis show data privacy concerns create the biggest barrier to adoption of e-government. While this form of e-government service, online tax service, consists of real transactions, the trust issue dealt with in this research comes even before citizens attempt such transactions. Namely, the citizens must be given a reason to simply trust in the information that is on a government Web site.

One of the earliest known examples of written multi-language information is the Rosetta Stone. According to Wikipedia (2006a), this archaeological artifact is a granite stone with writing in three different written scripts dated to about 200 BCE (Before Common Era, essentially the same as BC). It contains Greek, Demotic Egyptian, and Egyptian hieroglyphics. The message in the three scripts is the same and is a decree by the Egyptian ruler Ptolemy V regarding taxes and temple construction. The purpose of having the message written in three languages adjacent to each other was to solve a difficult linguistic problem..

A contemporary example of multi-language dissemination of important information is a requirement in the California Labor Code about employers posting worker information in English and Spanish. According to the Labor Code, several important documents, such as the Notice of Workers Compensation Coverage, must be posted in Spanish and English whenever there are employees of Spanish descent (California Labor Code, 2006.) There are similar requirements regarding Minimum Wage, Pregnancy Disability Leave and the California Family Rights Act (OSHA4LESS. COM, 2006)

The term "paralingual" is used to define the layout of information using two sets of text in different languages on the same page, such as in a Web page. The term was coined as an extension of the word "bilingual." Para is a Greek prefix that means beside, near, or alongside (Wikipedia, 2006). Therefore, paralingual refers to two languages adjacent to each other on the same page. Paralingual Web pages are almost non-existent on the Web. Although many Web sites are now multi-language Web sites, the common layout for these Web sites is to separate the languages to separate pages. This commonality is reflected in the standards on localization (LISA, 2008.). This localization can be found in e-government. Cunliffe et al (2002) reports a case study of developing a bilingual Web site in English and Welsh for users in Wales. This study focused on Web site design for just two languages and recognized that there are many bilingual areas in the world (Cunliffe et al, 2002). One of the most important aspects of designing for bilingual Web site content is to provide rich interconnectivity between materials in the two languages (Cunliffe et al, 2002). While not the main focus of the study, Cunliffe, et al. (2002) does discuss the options for placement of two languages on the same page.

Paralingual Web design is expected to affect e-government use two ways: increased use due to increased trust and possible decrease in use due to impacts on ease of use and usefulness. As discussed above it is expected to help improve trust. The two models mentioned in the Introduction help to predict the probable impact of paralingual Web design. TAM was developed by Davis (1989) as an explanation of the general case determinants of computer acceptance that are capable of explaining user behavior across a broad range of systems, technologies, and user populations. The model includes use as a determinant but indicates that use is determined by ease of use and perceived usefulness, attitude, and intention to use (see Figure 1). TAM is a derivative of Fishbein and Ajzen's (1975) Theory of Reasoned Action (TRA) model. TRA focuses on situation specific personal beliefs and attitudes, and the effects of the beliefs of others who can influence the individual. The fundamental premise of TRA is that individuals will adopt a specific behavior if they perceive it will lead to positive outcomes (Compeau and Higgins, 2001). However, adoption is also influenced by two factors, Perceived Usefulness and Perceived Ease of Use. Perceived Usefulness reflects that an individual's perception of usefulness influences their intention to use the technology primarily through the creation of a positive attitude. This is consistent with the TRA, which holds that attitude (an individual's positive or negative feelings about performing a behavior) influence behavioral intention. Geffen, et al. (2002) found that trust impacts Perceived Usefulness with increased trust improving Perceived Usefulness. Perceived Ease of Use reflects the user's assessment of how easy a system is to learn and use. TAM includes ease of use as a separate belief construct based on the concept of self-efficacy (an individual's judgment of his/her ability to organize and execute tasks necessary to perform a behavior). Ultimately, TAM predicts that if paralingual Web design can improve trust with the users while not reducing perceptions of ease

Figure 1. Technology acceptance model (Davis, 1989)

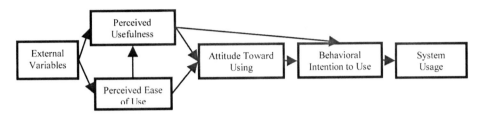

of use or usefulness then it can be expected that paralingual Web design will be accepted by users.

DeLone and McLean (2003) revision of the IS Success Model also helps predict the impact of paralingual Web design. This is a causation model that implies that system quality, information quality, and service quality will lead to increased use or increased intent to use which will lead to benefits and success of the system (see Figure 2.) The intent to use construct is important for this article as it is similar to the ease of use and usefulness constructs from TAM, especially when intent to use is operationalized using the Perceived Benefit Model (Thompson, Higgins, and Howell, 1991). Additionally, trust is reflected in the information quality dimension as users must be able to trust the information in the system for there to be quality. Paralingual Web design provides the trust in the information quality and aids in perceived usefulness in intent to use. Ultimately, the IS Success Model predicts that im-

proved trust will help improve information quality which will increase intent to use/actual use leading to benefits and system success.

The conclusion from the literature review is that Paralingual Web design may be a design tool that will support building trust in content and process by e-government users. As mentioned earlier, Geffen, et al. (2002) consider this essential for e-government adoption as it support perceived usefulness. The only concern is an impact to perceived ease of use resulting in the experiment design discussed in the next section.

RESEARCH METHODOLOGY

This study sought to determine whether paralingual Web site design can improve trust in e-government Web sites while maintaining ease of use and usefulness for both the reader and the provider government. In a bi or multilingual en-

Figure 2. DeLone and McLean's (2003) Revisited IS success model

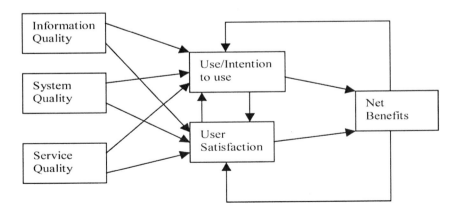

vironment, the ability to communicate concepts across diverse cultures and languages has become increasingly important, especially when issues of trust are involved. Further, e-government has rapidly increased in usage, making it even more essential for Web designers to conscientiously strive to ensure that concepts have the same meaning across cultures. This study focuses on paralingaul issues in a highly bilingual populated location in the United States, that of San Diego and Tijuana, with a total combined population of over 5 million as of 2004 and over 64 million annual border crossings. The subject municipality is National City, a city of 54,260 (2000 census) located approximately 10 miles from the United States border with Mexico. The municipality population is 60% Hispanic/Latino (all or part), 39% white (all or part), and 30% other (all or part where other includes Asian, Hawaiian/Pacific Islander, Black, and American Indian) (Note that the sum is greater than 100% due to respondents reporting belonging to more than one race) (National City, 2008). Actual percentage of English-Spanish bilingual residents is not reported but is understood to be large. The subject municipality was selected because of its expected bilingual population and its willingness to participate in the experiment. It should be noted that simply because the municipality population is heavily Hispanic or Latino does not mean that the local language is predominately Spanish. This is a US city and the citizens are US citizens. English is the majority language spoken in National City as evidenced by the National City Web site being in English only. However, due to its proximity to the border and its large Hispanic/Latino population it is reasonable to assume there is a substantial bilingual population.

To create the experiment, three informational Web pages were converted to paralingual format consisting of English and Spanish text placed horizontally adjacent to each other. Informational pages were chosen as the National City Web site is primarily informational and does not perform any financial transactions. This is considered acceptable as the majority of e-government Web sites in 2004 were informational rather than financial transaction focused (ICMA, 2004). Translation of the English content in the original selected Web pages was performed taking into account the following:

- Variations in the style and vocabulary of Spanish. A style to reflect the local style was chosen. This style is mostly Mexican in its structure and vocabulary, so these were kept in perspective at all times.
- The level of writing was kept at approximately a high school level of comprehension (the same as the English version).
- The Spanish translation was written to conform strictly to correct language structure, syntax, and spelling. This is demonstrated by the correct application of diacritical marks, such as accents, tildes, and umlauts.

The translation task was performed by a native Spanish speaker with professional training and experience as a translator. The translated content was then evaluated and modified by a Spanish language professor with a specialization in translation studies.

A survey was generated using survey monkey to gauge the opinions of visitors to the experimental Web pages. The survey consisted of eight items. Items 1-4 dealt with demographics of the respondents and were asked at the request of National City:

1. What is your age range? (18-24, 25-34, 35-44, 45-54, 55-64, 65 or over)
2. Are you a resident of National City? (Yes, No)
3. What language or languages do you use for communication (speaking, reading, writing)? (English only, Spanish only, mostly English with some Spanish, about half English and half Spanish, mostly Spanish

with some English, some other combination of languages) (Note that this question was used to group responses)

4. Have you visited the National City official Web site before now? (Yes, No)

Items 5 and 6 operationalized trust. Item 5 queries improved trust due to the information on the Web page while item 6 queries improved trust due to the paralingual format. Two items were used to ensure that improvement in trust was due to paralingual format and not to just reading the information (both use a 7 point Likert scale with 1 being Strongly Disagree and 7 being Strongly Agree):

5. Please respond to this statement: I have a greater trust now than before in my understanding of the National City Web site.

6. Please respond to this statement: I have a greater trust now than before in the information on the National City Web site because it is in English and Spanish side by side.

The seventh item was also asked at the request of National City and queried how aware the respondent was of the multilingual nature of National City:

7. Are you aware that other residents in National City speak more than one language? (Yes, No, Not sure)

The final item operationalized ease of use and usefulness by querying perceptions of site readability. Only readability is measured as no transactions are performed with the paralingual Web sites:

8. Please respond to this statement: It was easy for me to read the pages with English and Spanish next to each other: (No, it was very difficult, No, it was somewhat difficult, It was neither difficult nor easy, Yes, it was somewhat easy, Yes, it was very easy)

Municipality officials encouraged participation by constituents and residents in the vicinity through a series of bilingual public announcements encouraging individuals to visit the modified Web pages and to complete a brief survey documenting their opinions on the Web site. The respondents to the survey had the choice of filling out the survey in English or Spanish, presumably selecting their primary language of communication. The respondents who answered the online surveys represent a portion of Web users who visited the National City Web site and chose to view at least one of the paralingual pages. This constitutes a self-selected sample. The visitors to the Web site may have been responding to one of the various methods of advertisement of the research, or may have been incidental visitors to the Web site who then decided to follow a link from the home page to one of the paralingual pages. While there are problems with self selected samples, it is acceptable for this experiment as the goal was to get real users to respond to the Web pages and the inherent difficulties in recruiting a random sample on a public Web page. Data was collected over a three month period with 133 responses being collected. Data were grouped based on the respondent's answer to the question asked about the respondent's primary language for communication. The range of response choices was from "English only" to "Spanish only"; an additional choice was "some other combination of languages." In order to generate analytical results that are more representative of the language component of the sample data, the English sample data included only those who indicated "English only," "Mostly English," or "some other combination of languages." The Spanish sample data included those that indicated "Spanish only," "mostly Spanish," or "half English and half Spanish." This resulted in 97 English responses and 36 Spanish responses.

Statistical data analysis was used to analyze the collected data. Beatty (2000), Siegel (1956), McClave, Benson, and Sincich (2008), Jaccard and Becker (1990), and others have stressed the importance of selecting appropriate statistical procedures that correspond to the methods used to assign numerical values, as well as the type and level of data being analyzed. Since the questionnaires employed in the current study are based on a traditional Likert format, with anchor points ranging from "strongly disagree" to "strongly agree," nonparametric statistics were deemed appropriate. The nonparametric tools used include the Mann-Whitney U test, the Wilcoxon T test, and Spearman's rho. The Mann-Whitney U test is an appropriate tool for comparing central tendencies between *independent* responses to Question 5 and 6 *across* the two language groups, while the Wilcoxon matched pairs T test is an appropriate tool for comparing central tendencies across the *matched* pairs of data. With regard to tests of association, Spearman's rho is an appropriate measure for examining the correlation between matched responses to Questions 5 and 6 *within* each language group. When deemed appropriate and meaningful, means and standard deviations have been calculated for various data sets as well. All statistical tests were performed via SPSS and validated using other software. All results are reported in the format recommended by Jaccard and Becker (1990).

All collected surveys were used for the analysis. There was no exclusion or limitation of surveys based on any criteria. However, it should be noted that some respondents did not fill out all survey questions. For example, a small portion of both English and Spanish respondents did not fill out the second half of the surveys. Thus, the data used in different components of the analysis consist of different sample sizes, depending on the questions being examined. The software used for these analytical tests accounted for the missing data and thus calculated test values based only on the number of questions actually answered.

RESULTS AND DISCUSSION

For all tests, the alpha value was chosen to be $\alpha = 0.05$. For ease and uniformity of comparison between all tests, the z value will be used. For the chosen $\alpha = 0.05$ and 2-tailed tests, the critical z value is $z = \pm 1.96$.

Mann-Whitney U on Grouping by Language Choice

The Mann-Whitney U test was used to compare the central tendencies of responses to item 5: I have a greater trust now than before in my understanding of the National City Web site, based on the 97 English-based responses and the 36 Spanish-based responses. This item was designed to measure differences between respondents' improvement of trust in their *understanding* of the information on the paralingual page only. The Mann-Whitney U and its corresponding *z*-value (-1.907) resulted in no statistical difference in perceptions between the English and Spanish respondents.

A second Mann-Whitney U test was used to compare the central tendencies of responses to item 6: I have a greater trust now than before in the information on the National City Web site because it is in English and Spanish side by side, again based on the 97 English-based responses and the 36 Spanish-based responses. This item was designed to measure differences between the respondents' improvement of trust in regard to the *information* on the paralingual page, based on having the information in the two languages on the same page. This test yielded a *z*-value of -4.406 (p <.01), indicating that there were significant differences between the responses of the English and Spanish respondents with regard to this question.

These results partially support the hypothesis that use of paralingual Web pages will increase trust in the page content and government sponsor. Since trust levels are different only when the paralingual information is considered, and the

Spanish respondents showed higher medians than the English respondents in both items 5 and 6, the Spanish respondents show an increased level of trust based on the paralingual content.

Wilcoxon T Test on Grouping by Language Choice

The Wilcoxon matched-pairs T test was used to compare the central tendencies on data sets similar to those tested with the Mann-Whitney U test, items 5 and 6. The calculated z value for English items 5 and 6 is z = -3.188. This indicates that there is a statistical difference in the means of the answers to each item. For Spanish items 5 and 6, the calculated z value is z = -1.034 and we fail to reject the null hypothesis, therefore we conclude that there is no statistical difference in the means of the answers to each item.

This result seems to indicate that Spanish speakers improved their trust because the text was in Spanish, this does not necessarily support that a paralingual Web design is necessary to improve trust and that simply being bi-lingual in some format will increase trust of the minority speakers.

Spearman's rho on Grouping by Language Choice

Spearman's rho was used to examine the relationship between responses on English item 5 (n=97) and 6 (n=97) to determine how well the English respondents' answers correlated between the two items regarding trust. The test measures the level of correlation between these two items. The two-tailed test yielded ρ = 0.504 (p <.01), which was statistically significant.

Spearman's rho was also calculated for Spanish items 5 (n=36) and 6 (n=36). The calculated correlation, ρ = 0.563 (p<.01), is significant for a 2-tailed test as calculated by SPSS.

Spearman's rho was performed to see the degree of correlation between items 5 and 6 in each group. The rho value for English items 5 and 6

is 0.504, so the correlation is "significant". For Spanish items 5 and 6 the rho value is 0.563, also significant. Thus each of the two groups answered consistently in the trust questions.

Readability Reflecting Ease of Use and Usefulness of the Paralingual Web Pages

Item 8 of the survey is a measure of the readability as reflecting ease of use and usefulness of the paralingual pages. The choices for responses had a range of five from "it was very difficult" to "it was very easy." The percentages of the English respondents (n=97) who answered "it was somewhat easy" or "it was very easy" is 61.3% and the percentage of Spanish respondents (n=36) answering similarly is 85.7%. This is interpreted to imply that having the page in paralingual format did not diminish readability and thus ease of use and usefulness, important to predicting acceptance of paralingual Web pages by users. However, it is also implied that English respondents were more likely to find an impact to ease of use and usefulness. This may suggest that these respondents will be less accepting of a change to paralingual format as they perceive paralingual format to be less useful.

Alternative Analytical Calculations

Performing nonparametric tests is the appropriate method for analysis of ordinal data. These results have been shown in the previous section. However, means and standard deviations are more commonly understood and therefore used more commonly to describe data. Table 1 is a summary of the means and standard deviations for items 5, 6, and 8.

CONCLUSION

This article is primarily intended to provide evidence to support government decision makers,

Table 1. Means and standard deviations of survey data

Item	Value Range	English: Mean/StdDev	Spanish: Mean/StdDev
5	1 - 7	4.2209/1.785	4.8182/1.286
6	1 - 7	3.5930/1.811	5.0606/1.540
8	1 - 5	3.88/1.259	4.366/0.994

e-government researchers, and e-government Web designers in applying paralingual Web page design for improving trust in government in regions where there is a high proportion of bilingual residents. An experiment was performed to test the hypothesis that paralingual Web design will improve trust in the content of the e-government Web page without significantly affecting ease of use and usefulness. This hypothesis was confirmed, but not quite as expected. It was found that the paralingual format improved trust for the minority speaker but not the majority speaker. Upon reflection this is an expected finding.

An additional finding with respect to ease of use and usefulness was noted. It was found that in general respondents did not find the paralingual format hard to read, however, it was noted that the majority speakers (English) were less enthusiastic than the minority (Spanish) speakers about the paralingual Web design. This finding has implications for the adoption of paralingual Web design in that it may show that there will be resistance to adoption of a paralingual Web approach by the majority speakers. The implication to policy makers is that there needs to be additional research done with respect of citizen attitudes towards bilingual government prior to implementing a paralingual Web strategy. This is particularly important in different regions of the United States where it could be expected that the generally favorable bilingual acceptance attitudes found in California may not exist.

The conclusion of this article is that paralingual Web design is useful for e-government in areas with significant bilingual populations. However, there are limitations to this approach as it appears

that there is a risk of backlash and rejection from the majority speaking population. The implication for policy makers is that paralingual Web design should be used when there are known trust issues between majority and minority speakers that translate into trust issues with government and e-government initiatives.

Limitations

This experiment has a small sample with limited items testing improved trust and ease of use and usefulness. The conclusion that paralingual Web design improved trust for the minority Spanish speakers is supported by the statistical analysis but it cannot be totally discounted that trust may have been increased simply because the content was in Spanish. Additionally, this experiment only looked at informational Web pages and the conclusions may not apply to financial or other transactional Web pages. Finally, only one city was looked at, one minority language used (Spanish,) and the sample population was self selected meaning that the results may not be reflective of all populations, cultures, and languages.

Areas for Future Research

There are several areas for future research of which the first are those areas that address the limitations to this research. This includes further studies using other languages and locations; obtaining a large sample size; and using transaction based Web pages in addition to informational pages. It is expected that there may be ease of use and usefulness issues associated with paralingual Web design used for transactional Web pages which could affect adoption of the pages and requiring that the trust improvement from paralingual Web design be balanced against ease of use and usefulness.

An additional area for future research is in using paralingual Web design in a multilingual environment. In this case multilingual implies more than two languages and is a topic of relevance in many countries in Europe and Asia (Finland and Switzerland are two examples). While there are historical examples of multilingual designs, none are in Web page layouts (the Rosetta Stone is an example of a three language representation). Given the limitations on screen areas, especially for mobile and/or handheld devices, a paralingual Web design involving three or more languages may not be practical or may induce substantial ease of use and usefulness issues. Research needs to be done to see if this is a practical approach or if more traditional approaches of having multiple versions of the same Web page each in a different language is a preferable approach.

A final area for future research is in e-government policy. The conclusion that majority speakers may resist paralingual implementations is very important to policy makers. Research into what factors may influence majority speakers to accept and adopt paralingual implementations or which factors may influence majority speakers to reject paralingual implementations is critical to policy makers for crafting appropriate e-government strategies and policies. It is expected that paralingual Web design will only be appropriate in regions where there is significant lack of trust by the minority speakers but this needs to be confirmed.

REFERENCES

Ba, S. and Pavlou, P.A., (2002). Evidence of the Effect of Trust Building Technology in Electronic Markets: Price Premiums and Buyer Behavior, MIS Quarterly, (26:3), 2002, pp. 243-268.

Beatty, J. R. (2000). *Statistical Methods* (6th ed.). New York: McGraw-Hill.

Becerra, H. (2006). Welcome to Maywood, Where Roads Open Up for Immigrants. Los Angeles Times, March 21, 2006, p. A-1.

Bharat, K. and Broder, A. (1998). A technique for measuring the relative size and overlap of public Web search engines. In 7th WWW,.

California Labor Code. Section 3550(d). Accessed on March 24, 2006 from http://www.leginfo.ca.gov/cgi-bin/displaycode?section=lab&group=03001-04000&file=3550-3553.

Castelli, W. A., Huelke, D. F., & Celis, A. (1969). Some basic anatomic features in paralingual space. *Oral Surgery, Oral Medicine, and Oral Pathology*, *27*(5), 613–621. doi:10.1016/0030-4220(69)90093-0

Coleman, S., & Gøtze, J. (2001). Bowling Together: Online Public Engagement in Policy Deliberation. Hansard Society, London. Retrieved October 4, 2005 from http://bowlingtogether.net.

Compeau, D., & Higgins, C. (1995). Computer Self-Efficacy: Development Of A Measure And Initial Test. *Management Information Systems Quarterly*, *19*(2), 189–211. doi:10.2307/249688

Cunliffe, D., Jones, H., Jarvis, M., Egan, K., Huws, R., & Munro, S. (2002). Information Architecture for Bilingual Web Sites. *Journal of the American Society for Information Science and Technology*, *53*(10), 866. doi:10.1002/asi.10091

Davis, F. (1989). Perceived Usefulness, Perceived Ease Of Use, And User Acceptance Of Information Technology. *Management Information Systems Quarterly*, *13*, 319–339. doi:10.2307/249008

DeLone, W. H., & McLean, E. R. (1992). Information Systems Success: The Quest for the Dependent Variable. *Information Systems Research*, *3*, 60–95. doi:10.1287/isre.3.1.60

DeLone, W. H., & McLean, E. R. (2003). The DeLone and McLean Model of Information Systems Success: A Ten-Year Update. *Journal of Management Information Systems, 19*(4), 9–30.

Fishbein, M., & Ajzen, I. (1975). *Belief, Attitude, Intention And Behavior: An Introduction To Theory And Research.* Reading, MA: Addison-Wesley.

Gassert, H. (2004). "How to Make Citizenz Trust E-Government," University of Fribourg, E-Government Seminar, Information Systems Research Group. Available from http://edu.mediagonal.ch/unifr/ egov-trust/slides/html/title.html accessed November 29, 2005.

Gefen, D., Warkentin, M., Pavlou, P. A., & Rose, G. M. (2002). EGovernment Adoption. Eighth Americas Conference on Information Systems, Association for Information Systems.

Girion, L. (2006). Language Becoming an Issue for Health Insurers. Los Angeles Times, March 20, 2006, p. C-1.

Grefenstette, G., & Nioche, J. (2000). Estimation of English and non-English Language Use on the WWW. Xerox Research Centre Europe. Available from http://arxiv.org/pdf/cs.CL/0006032 accessed March 19, 2006.

Grönlund, A., & Horan, T. A. (2004). Introducing E-Gov: History, Definitions, and Issues. *Communications of the Association for Information Systems, 15*, 713–729.

Gulli, A., & Signorini, A. (2005). The Indexable Web is More than 11.5 Billion Pages. WWW 2005, May 10–14, 2005, Chiba, Japan.

Hart-Teeter. (2003). The New E-government Equation: Ease, Engagement, Privacy and Protection. A report prepared for the Council for Excellence in Government, Retrieved November 27, 2005 from http://www.excelgov.org/usermedia/images/uploads/PDFs/egovpoll2003.pdf.

Hibbing, J. R., & Theiss-Morse, E. (2002). *Stealth Democracy: Americans' Beliefs About How Government Should Work.* Cambridge: Cambridge University Press. doi:10.1017/CBO9780511613722

International City/County Management Association (ICMA). (2004). Electronic Government Survey 2004. accessed February 22, 2008 from: http://icma.org/upload/bc/attach/%7B 9BA2A963-DDCC-40B7-836D-F1CFC17DCD98%7Degov2004Web.pdf

Jaccard, J., & Becker, M. A. (1990). *Statistics for the Behavioral Sciences.* Belmont, CA, USA: Wadsworth Publishing Company.

Jadu, (2005). Enterprise Content Management for Public and Private Sector. Available from http://www.jadu.co.uk/ego v/jadu_egov_econsultation.php accessed November 26, 2005.

Khalil-babnet, M. WSIS Prepcom-2: Cybersecurity an Issue for All. Available from http://www.babnet.net/en_det ail.asp?id=935 accessed November 29, 2005.

La Prensa-San Diego. Available from http://laprensa-sandiego. org/rates/rates.html accessed November 27, 2005.

LaVoy, D. J. (2001, Fall). Trust and Reliability. *Public Management, 30*(3), 8.

Levi, M., & Stoker, L. (2000). Political trust and trustworthiness. *Annual Review of Political Science, 3*, 475–507. doi:10.1146/annurev.polisci.3.1.475

Localization Industry Standards Association. LISA, (2008). Localization. Accessed February 23, 2008 from http://www.lisa.org/Localization.61.0.html.

Macias, E., & Temkin, E. (2005). Trends And Impact Of Broadband In The Latino Community. Tomás Rivera Policy Institute, 2005. Available from http://www.trpi.org/PDFs/broadband.pdf accessed March 21, 2006.

McClave, J. T., Benson, P. G., & Sincich, T. (2008). *Statistics for Business & Economics, 10/E*. Upper Saddle River, New Jersey, USA: Prentice Hall.

Mercuri, R. T. (2005). Trusting in Transparency. Association for Computing Machinery. *Communications of the ACM, 48*(5), 15. doi:10.1145/1060710.1060726

Moller, R. M. (2000). Profile of California Computer and Internet Users. California Research Bureau. California State Library. Available from http://www.library.ca.gov/crb/00/01/00-002.pdf accessed April 9, 2006.

National City. (2008). National City Web Site. Available from http://www.ci.national-city.ca.us/main.asp accessed February 22, 2008.

National Performance Review. (1993). *From Red Tape To Results: Creating A Government That Works Better And Costs Less*. Washington, D.C.: Government Printing Office.

Organization for Economic Co-operation and Development (OECD). (2001). *Citizens as Partners*. Information, Consultation and Public Participation in Policy-Making.

Osborne, D., & Gaebler, T. (1992). *Reinventing Government: How The Entrepreneurial Spirit Is Transforming The Public Sector*. Reading, MA: Addison-Wesley.

OSHA4LESS.COM, "Sales Book - lew 2.pdf", 2006. Received as attachment to electronic mail message, March 24, 2006.

Paralinguism in the Theatres and the International Theatre Festivals. Sasho Ognenovski. Intercultural Communication, ISSN 1404-1634, 1999, August, issue 1. Available from http://www.immi.se/intercultural/ accessed November 19, 2005.

Rai, A., Lang, S., & Welker, R. (2002). Assessing the Validity of IS Success Models: An Empirical Test and Theoretical Analysis. *Information Systems Research, 13*(1), 50–69. doi:10.1287/isre.13.1.50.96

Rosell, S., Gantwerk, H., & Furth, I. (2005). Listening To Californians: Bridging The Disconnect. Viewpoint Learning, Inc. Retrieved January 15, 2006 from http://www.viewpointlearni ng.com/pdf/HI_Report_FINAL.pdf.

Sandoval, V. A., & Adams, S. H. (2001). Subtle Skills for Building Rapport: Using Neuro-Linguistic Programming in the Interview Room. Available from http://www.fbi.gov/publications/leb/2001/august2001/aug01p1.htm accessed November 19, 2005.

Schneider, G. P. (2003). *Electronic Commerce, Fourth Annual Edition*. Boston: Thomson Course Technology.

Secretariat, Treasury Board of Canada, (2006). Communications Policy of the Government of Canada. Available from http://www.tbs-sct.gc.ca/pubs_ pol/sipubs/comm/comm1_e.asp#04 accessed March 21, 2006.

Siegel, S. (1956). *Nonparametric Statistics for the Behavioral Sciences*. New York: McGraw-Hill Book Co.

Tan, C.-W., Benbasat, I., & Cenfetelli, R. T. (2008). Building Citizen Trust towards e-Government Services: Do High Quality Web sites Matter? Proceedings of the 41st Hawaii International Conference on System Sciences, IEEE Computer Society Press.

Tan, C. W., Pan, S. L., & Lim, E. T. K. (2005). Towards the Restoration of Public Trust in Electronic Governments: A Case Study of the E-Filing System in Singapore", Proceedings of the 38th Annual Hawaii International Conference on System Sciences, IEEE Computer Society Press

2005The Year of Languages. (2005). Special project of the American Council on the Teaching of Foreign Languages (ACTFL). Available from http://www.yearoflanguages.org/i4 a/pages/index.cfm?pageid=3419 accessed March 21, 2006.

Thompson, R. L., Higgins, C. A., & Howell, J. M. (1991). Personal Computing: Toward a Conceptual Model of Utilization. *Management Information Systems Quarterly*, (March): 125–143. doi:10.2307/249443

United States Census Bureau. Census 2000. "QT-PL. Race, Hispanic or Latino, and Age: 2000. Data Set: Census 2000 Redistricting Data (Public Law 94-171) Summary File Geographic Area: National City city, California." Available from http://factfinder.census.gov/servlet/Q TTable?_bm=y&-geo_id=16000US0650398&-qr_name=DEC_2000_PL_U_QTPL&-ds_name= D&-_lang=en&-redoLog=false, accessed March 2, 2006.

United States Department of Labor. The Family and Medical Leave Act of 1993, 29 CFR 825.300, 2006. Available from http://www.dol.gov/dol/allcfr/ESA/T itle_29/Part_825/29CFR825.300. htm accessed March 24, 2006.

United States Office of Management and Budget (USOMB). (2005), E-Gov: Powering America's Future With Technology. Retrieved October 5, 2005 from http://www.whitehouse.gov/omb / egov/index.html.

Warkentin, M., Gefen, D., Pavlou, P. A., & Rose, M. (2002). Encouraging Citizen Adoption of e-Government by Building Trust. *Electronic Markets*, *12*(3), 157–162. doi:10.1080/101967802320245929

Wikipedia, (2006). Paralanguage. Available from http://en.wikipedia.org/wiki/P aralanguage#Linguistics accessed September 19, 2006.

Wikipedia, (2006a). Rosetta Stone. Available from http://en.wikipedia.org/wi ki/Rosetta_stone accessed September 20, 2006.

Chapter 4
Evaluation of Arab Municipal Websites

Hana Abdullah Al-Nuaim
King Abdulaziz University, Saudi Arabia

ABSTRACT

High speed wireless networks and mobile and web-based services are changing the way we, as consumers of information, communicate, learn, do business and receive services. Successful e-commerce models have raised the expectations of citizens to have government agencies and organizations provide public services that are timely and efficient. With the growth and development of Arab cities, especially in the capitals, life becomes a little bit harder for citizens dealing with highly bureaucratic government agencies as their demands for basic services increase. Although e-readiness in the region has grown considerably with impressive progress, Arab cities have been clearly absent from studies on worldwide e-municipal websites. In this study, an evaluation checklist for was used to evaluate official municipal websites of Arab capitals. The study found that these websites were not citizen centered, suffered from fundamental problems, had some features that were inoperable and did not follow basic guidelines for any municipal website. These sites were dominated by aesthetics and technical novelties alone, providing inactive information rather than the inclusion of interactive e-services with immediate feedback and easy to use, navigable interfaces.

INTRODUCTION

High speed wireless networks and mobile and web-based services are changing the way we communicate, learn, do business and receive services. As consumers of information, our relationship

with education, government and commerce is changing through e-learning, e-government and e-commerce respectively. With such a fast delivery of private sector services, e-commerce has raised citizens' expectations to have government agencies and organizations provide public services that are timely and efficient. Citizens think of governments as large bodies of bureaucratic pa-

DOI: 10.4018/978-1-60960-162-1.ch004

perwork that impedes their active participation, creating legitimacy problems, which can only be solved by bridging the gap between citizens and the government through a more active and open relationship (Daemen and Schaap, 2000).

E-government initiative development should be supported by improved Information and Communication Technology (ICT) infrastructures. This would give accessibility to all citizens in an effort to bridge the digital divide and to create more informed citizens willing to participate in the development of their city. It also should be supported by the following elements (Intel, 2003): delivery of operational efficiency that is result-driven and accountable; delivery of government policy with faster implementation that extends best practices more effectively; access by citizen-centered users anytime and from anywhere; and development of strategic investment models to stimulate revenue for economic development and to encourage partnerships.

Therefore, governments need to continuously evaluate the needs and expectations of their citizens by getting them "out of line" and putting them online due to changing financial and social situations in light of rapid technological developments (Pardo, 2000). In the United States, President Obama's presidential transition team treated government websites as core government business function rather than information technology projects where they recommended the development of government-wide guidelines for disseminating content in universally accessible formats such as news feeds, videos, pod-casts, etc (Goldberg, 2009). US Government agencies have an editor-in-chief appointed for every website they maintain as do popular commercial websites, in order to develop and enforce website polices to ensure the site is effective and usable and that the content is accurate, relevant and written in a style all citizens would understand (Goldberg, 2009).

E-MUNICIPALITIES

ICT's transform cities from local and inhibited social, economic, political and cultural living spaces to global centers, carrying urban and local values into the global arena (Çukurçayir and Eroglu, 2010). The most important citizen to government relationship is that of citizens' relationship with their city, because municipalities directly affect the daily service needs of citizens; they also experience the most frequent per capita level of service contact, generating heightened demands with high-level economic prospects (Prychodko, 2001).

City municipalities should aim to increase the consciousness and co-responsibility of the residents for the quality of life in the city, to have respect for the rules of living together, and to improve the performance of the city's services (Rodriguez, 2005). E-municipality can be defined as a municipality realizing all kinds of communication, business and service offers in an electronic environment (Akinci, 2004).

The incentives for both citizens and governments to utilize and provide ICT services include the following (Town of Freetown, Ma, 2007; Center for Democracy and Technology and Infodev, 2002; Pardo, 2000; Daemen and Schaap, 2000; Goldberg, 2009):

- Provide citizens with a single point of access to information and services
- Ensure timely access to information any day and at any time
- Ease currently bureaucratic and labor-intensive procedures, saving money and increasing productivity
- Build services around citizens' choices
- Notify public agencies and citizens of emergency situations
- Stem corruption by automating procedures and processes, especially in revenue-generating areas improving returns while increasing government credibility

- Reduce delays between production of and access to information, and provide timelier updates of materials
- Facilitate social inclusion
- Increase citizen participation in decision-making and online polls
- Decrease phone calls and walk-in visits for routine information and forms
- Provide easily-accessed basic information to people who may need to move to the city
- Provide access to minutes, budgets, reports, municipal meeting times and agendas and other documents, increasing residents' civic IQ
- Advertise contracts, job notices, and land buyouts
- Reduce printing and faxing bills because residents can download forms and reports rather than request printed copies
- Promote local education of young residents and schools

Creating and maintaining a local government website for a city and having an effective web presence is extremely challenging and depends on the information technology infrastructure of the country and the e-literacy of its citizens. Government websites should be carefully crafted front-ends that hide numerous distinct agency procedures and processes, and the risk of not understanding, or underestimating the complex mixture of requirements and technological, managerial and policy related challenges is a costly failure (Pardo, 2000).

A study by Moon (2002) on the evolution of e-government surveyed 1471 e-municipalities in the USA and found that the size and type of government are significant institutional factors in the implementation of e-government. He also found that the larger the government, the more the municipality will actively pursue e-government and that the major barriers to its development are lack of technical, personal and financial capabilities (Moon, 2002). Also, in a study evaluating 500

Dutch e-municipalities, it was found that the municipalities were autonomous, lacking coordination and a shared vision of e-government (Mulder, 2004). But the quality of Dutch municipal web sites is encouraged and monitored by means of a national "hit list" that ranks all websites by their scores based on certain but limited sets of criteria (De Jong and Lentz, 2006). According to De Jong and Lentz, (2006) the criteria used for evaluation of the Dutch municipal websites have several limitations: (1) they regularly focus on manifest website characteristics that can be easily coded; (2) they consider the amount of information provided on the website and the degree of interactivity to be a matter of *the more the better*; and (3) they contain rather superficial and controversial criteria for assessing the usability of website.

Government services require queuing during office hours, long waits, confrontations with bureaucrats and the occasional bribe. These situations are more prevalent in under-privileged communities; yet, however complicated and challenging, there have been many digital government successes at local governments throughout the world.

RESEARCH OBJECTIVE

In the Arab world, citizens receive most of their basic services from government agencies, which are highly bureaucratic in nature. Employees are rarely trained in customer service or reprimanded for inefficient work and customer dissatisfaction or complaints from citizens are often ignored. With the growth and development of Arab cities, especially in the capitals, life becomes a little bit harder for citizens as their demand for basic services increase. Each Arab country has its own government structure and people are not aware who does what any more. Also, women are entering the workforce in larger numbers in all Arab countries, allowing them to own homes and businesses. With this, their need for municipal services has grown dramatically, yet their entry into local municipal agencies for services is not facilitated, especially in a country like Saudi Arabia where there is segregation of the sexes.

Most Arab cities were clearly absent from Holzer and Kim's (2005) report on the Digital Governance in Municipalities Worldwide, which ranked 81 municipal websites for cities from 98 countries with an online population over 160,000. Only five Arab cities in Africa and seven Arab cities in Asia were selected for the study. Based on the evaluation of two native speakers, only five cities with an official website were ranked: Cairo-45, Dubai-50, Riyadh-52, Amman-65 and Beirut-66.

Based on Moon's (2002) conclusion that relates the size of the local government to its pursuit of e-government, it was the objective of this research to evaluate the municipality websites of Arab capitals based on their content, architecture, layout, and website design, using an evaluation checklist for municipality websites.

E-READINESS OF ARAB COUNTRIES

E-readiness measures the quality of a country's ICT infrastructure and the ability of its consumers, businesses and governments to use ICT to their benefit. Technologies deployed and services provided still remain underutilized by people and countries around the world. This is far from maximizing utility (Economic Intelligence Unit, 2009). Although Internet population penetration is only 10 percent in the Middle East compared to 69.7 percent in North America, between the year 2000-2007, Internet usage worldwide has grown 265.6 percent while in the Middle East, the growth was 920.2 percent (Internet World Stats, 2007). A study by the Economist Intelligence Unit (2006; 2009) on e-readiness of the Middle East and northern Africa (Table 1) found that:

- The region has made major improvements in Internet connectivity in recent years, especially in the mobile market.
- In Saudi Arabia, for example, Internet usage grew by 1000 percent between 2000 and 2005 and is in the process of building the Middle East's largest IP network for Internet transport and wireless communications.
- Government encouragement, competition and economic growth have all contributed to a booming mobile market.
- The region added over 20 million new subscribers in 2005, while a market like Algeria recorded triple-digit growth.
- Jordan is growing as quickly, and is the Middle East's most competitive mobile market. It leapt from 90th place in 2005 to 15th in the 2008 UN e-participation index.
- This growth in the region may be due to good progress in terms of liberalization and privatization in these countries.

In the last few years, the region as a whole has advanced its e-government readiness, while some countries have an e-readiness index higher than the world average of 0.4514 in 2008 compared to 0.4267 in 2005 (United Nation, 2008; UN, 2010).

This could be attributed to the worldwide considerable cost reduction of ICT infrastructure and accessibility. For example, in 2006 broadband prices for DSL connections in 30 developed countries have been reduced by 19 percent with a 29 percent increase in the speed of the connection (UN, 2007). Such a reduction has led to implementation of new technologies in many developing countries, without national governments having to invest heavily in land-based infrastructures (UN, 2007). Although the ten largest mobile operators in the Middle East and Africa are now collectively adding over 12 million new subscribers every quarter, the region still has a long way to go in establishing a solid foundation of connectivity (EIU, 2009). In developed countries, the growth

Table 1. E-readiness of the Middle East and Africa ranking

Country	2008 Index	2010 Index	2005 Ranking	2008 Ranking	2010 Ranking
UA Emirates	0.6301	0.5349	42	32	49
Bahrain	0.5723	0.7363	53	42	13
Jordan	0.5480	0.5278	68	50	51
Qatar	0.5314	0.4928	62	53	62
Kuwait	0.5202	0.5290	75	57	50
Saudi Arabia	0.4935	0.5142	80	70	58
Lebanon	0.4840	0.4388	71	74	93
Oman	0.4691	0.4576	112	84	82
Syrian Arab Republic	0.3614	0.3103	132	119	133
Iraq	0.2690	0.2996	118	151	136
Yemen	0.2142	0.2154	154	164	164
Egypt	0.4767	0.4518	99	79	86
Libya	0.3546	0.3799	....	120	114
Algeria	0.3515	0.3181	123	121	131
Tunisia	0.3458	0.4826	121	124	66
Morocco	0.2944	0.3287	138	140	126
Sudan	0.2186	0.2542	150	161	154

Source: United Nations, 2008, 2010

of the Internet is fueled by the need for and ultimately dependence on e-commerce. In contrast, e-commerce is still in its infancy in the Arab world, and according to Alexa.com – the web traffic information company, the top websites accessed are entertainment information and social networking websites. Arab governments, however, clearly have begun recognizing the value of electronic services to their citizens by committing to large financial investments in e-government.

ASSESSMENT AND MEASUREMENT OF MUNICIPAL WEBSITES

Citizen-centered websites are defined as "intentions-based, goal-based, user-focused, or task-based sites. These sites take as their starting point what the user wants to do online. They reflect the point of view and accommodate the general knowledge of their users. By way of contrast, the structure of most municipal sites follows the organization chart of the local government" (Massachusetts, 2004).

Based on the above definitions of a city's homepage and citizen-centered websites, when evaluating municipality websites it's imperative to ask the most important questions (LaVigne, Simon, Dawes, Pardo, & Berlin, 2001):

1. Does the municipality know who its audience is?
2. Can people contact the municipality or ask questions easily?
3. Is the navigation of the site easy?
4. Are the services the municipality provides useful?
5. Is this site worth finding? Would people visit again? Or does it waste their time?
6. Is this site clear, easy to access, attractive, efficient, and effective in each of its components: content, text, images/video, sound,

databases, graphics, buttons, supporting labels, index, links to external sites, tools, e-mail, search, transaction processing, and downloadable forms?

There has been extensive research conducted on the importance of e-government and its rankings based on certain criteria. However, little has been done in terms of the assessment and measurement of the efficiency and effectiveness of municipal websites for citizens as the main public service provider for those who live and visit the city, services that should be delivered intuitively, without effort and with immediate feedback and results.

Kaylor, Deshazo and Van Eck's (2001) evaluation of American cities created a rubric for benchmarking the degree to which functions and services were web-enabled for each city without entering the domain of website evaluation and focusing solely on the evolution of the cities' e-government. However, in Holzer and Kim's (2005) evaluation of websites for cities worldwide, where a city's main homepage was defined as "the official website where information about city administration and online services are provided by the city," five equally weighted categories were used for their evaluation instrument and each item for every category was coded on either a scale of four-points (0, 1, 2, 3) or a dichotomy of two-points (0, 3 or 0, 1).

Researching the maturity level of e-governments, Rodríguez, Estevez, Giulianelli and Vera (2009) defined a measurement framework that included 152 metrics for assessing website design and content grouped into the following features:

1. Information about the institution or the services provided by it
2. Functionality of the services offered through the website
3. Truthfulness of the information published which should be real, relevant and up to date
4. Promotion of citizen participation by having one and two way interaction services

5. User friendliness of the website through its ease of use
6. Usability of the website
7. Accessibility of the website by people with disabilities
8. Navigability for efficient browsing of the website pages

In an effort to easily evaluate local government websites, the Maxwell School at Syracuse University formally established the Community Benchmarks Program (1999), to support local governments and nonprofits through the use of comparative measures. These would improve performance and accountability as part of a continuous improvement effort. They developed a quick evaluation checklist for municipal websites based on the following categories (Conners, Koretz, Knowle & Thibodeau, 1999):

Content:

- Meeting information: times, dates, places, agendas and minutes
- Budget information: current annual budget of the city
- Services: a list of services provided by the city and an explanation of each
- Contact information
- Feedback/e-mail capabilities
- "How-to" information: complete guide or explanation of how to obtain different permits/licenses
- Relevant links
- Search capability.

Architecture:

- Site organization: systematic navigation around site
- Thematic separation: businesses, residents and tourists can easily locate all relevant information
- Number of links to find specific information

- Persistent navigation: consistent use of links found in the same location on every page

Layout:

- Readability of graphics and text: text is clear with no spelling or grammatical errors
- Page layout tested on different browsers: site adjusts to screen
- Overall design of page

Website design:

- Proper loading and working order of all graphics, links and buttons
- Frequency of updates; all dates, times and schedules are current and correct
- Presence of a Contact Webmaster link
- Registration with search engines and browsers

METHOD

Some of the evaluation instruments for assessing municipal websites such as the 152 measurement metrics used by Rodriguez et al. (2006) and the 98 item instrument developed by Holzer and Kim's (2005) are too many and complex to be assessed by the average user. These can only be used by researchers and experts in the field who have prior instructions or training on how to use them. Also some of the categories used such as security and privacy (which examines privacy policies and issues related to authentication), and citizen participation (which involves examining how local governments engage citizens to participate online) are not policies commonly implemented by Arab city municipalities. These municipalities are not democratic in nature, nor are their websites. To be able to use a simple evaluation instrument that can be intuitive for the average

citizen, the researcher modified the Municipality Evaluation Checklist (Table 2) developed by the Maxwell School's Community Benchmarks Program by excluding the Internet browser item since Microsoft's Internet explorer was the default browser and no other browser was available for evaluation; also three essential items were added: News, a site map and the ability to find a site by guessing its URL. The latter is especially important since part of the service to citizens is to ensure that those who look for information find it. According to domain naming guidelines, it is more effective to name the site using the name of the organization, institute, or company followed by the domain type than by the country's extension. This naming technique is apparent with the highest ranking city municipalities worldwide such as: shanghai.gov.cn, seoul.gov.kr, nyc.gov and cityofsydney.nsw.gov.au.

According to the original checklist developers, for each criterion the website received a rating between 1 and 5 (Conners, Koretz, Knowle & Thibodeau, 1999):

1. A rating of one means the municipality does not have the information or that the information provided is extremely poor.
2. A rating of two means the municipality has the information to some extent, but it is not fully developed.
3. A rating of three means the municipality has the information, it is developed, but there is room for improvement.
4. A rating of four means the municipality has the information and it is solidly developed.
5. A rating of five means the municipality has provided complete information and no improvement is necessary.

After conducting a heuristic evaluation of the Arab city websites, the researcher found that many of the checklist's criteria is absent or available but not functioning. Therefore, the researcher adopted the dichotomous measures used by Holzer and

Table 2. The modified quick evaluation checklist of municipal websites

	Rating
Municipality Name:	
Website Address:	
News	
1. Content-website contains the following:	
Meeting information	
Budget information	
Services	
Contact information	
Feedback/email capability	
"How-to" information	
Relevant links	
Search capability	
Site map	
2. Architecture:	
Site well organized	
Separated thematically (i.e. Business, residents, tourists)	
Number of links to find specific information	
Persistent navigation	
3. Layout:	
Graphics and text easily readable	
Overall design of page:	
4. All graphics, links and buttons work or load properly:	
5. Frequency of updates:	
6. webmaster – someone is responsible for site:	
7. Registration with search engines and browsers:	
Ability to find site by guessing URL	
Ability to find site by searching key words	
Search Engine 1:	
Search Engine 2:	
Comments:	

Kim (2005) which correspond to ratings of "0" or "1" to assess the presence or absence of the most essential items on the checklist. For every item listed, a score of 1 was given if the item was available or done well; a fraction of that score (.5) was given if the item was available but had problems or was not functioning and 0 if it was not available. Therefore, the maximum points a site could receive was 22.

In an attempt to pilot the modified municipal evaluation checklist, it was important for the researcher to assess the performance of the top three ranked municipal websites of Holzer and Kim's (2005) Digital Governance Report: Seoul, New York and Shanghai (Figure 1,2 and 3).

The websites were searched for using two of the most popular search engines available online: Google and Yahoo. The keywords used in all three were the name of the city, municipality, city government and website. Only the sites that have been branded.gov were considered because visitors looking for official government information must be confident of the credibility of the site. All three cities were found from the first hit.

Using the checklist, the cities of Seoul, New York and Shanghai websites received a score of 22, since they have all the necessary information available. The navigation, graphics and search facilities within the sites all work properly and efficiently. The websites for all three cities are clearly citizen-centered websites, since they connect their residents and community through many services offered by the city and the private sector. Most services needed by citizens are considered: health, housing, transportation, traffic, contacts, culture, how to information, news, relevant links, etc. The websites offers separate sections for businesses, residents, and visitors and the search facility and navigation techniques are fully functional. Labels are clear; paths to relevant information are short and most need only one link. The homepage of each site is well organized and the layout of the structure is clear. These sites are easy to learn and easy to use and users would be less likely to get lost or confused and can predict where information will be displayed.

Researchers who rate e-government services in different cities have selected cities with a certain number of Internet users as a cut-off mark. This type of data is not available for most Arab cities and since the capitals of all Arab countries have the highest population and receive the most funding for government development projects (with, perhaps, the exception of Dubia in UAE), only Arab capitals were selected for this research. To be able to compare the performance of the three top websites in Holzer and Kim's longitude study

Figure 1. City of Seoul website

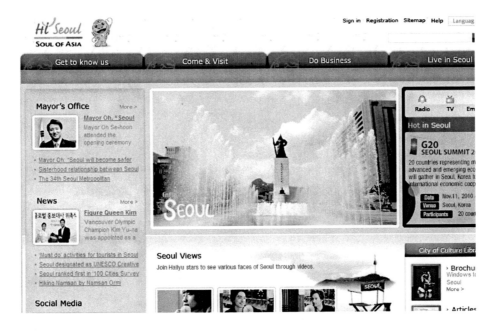

Figure 2. City of New York website

Figure 3. City of Shanghai website

with municipal websites of Arab country capitals, the following procedure was followed:

1. The capitals of all Arab countries were identified.

2. For each city, a search was conducted using the same keywords used during the pilot evaluation with the addition of Arabic and English written Arabic words including baladia /baladiya and amannah the Arabic words for municipality. The search was conducted using Google and Yahoo in English and Google Saudi Arabia for the search in Arabic.

3. Emails were sent to different organizations and government agencies in countries where no website of the capital was found to determine if in fact there was a city website of the capital but it was named differently. No response was received from any of them.

4. Some websites offer both English as well as Arabic versions, but since this study is for citizen centered websites and the citizens' mother tongue is Arabic, only the Arabic website for each city was evaluated.

5. The search resulted in only six countries in the Arab world which have a city website for their capitals: Amman, Beirut, Muscat, Riyadh, Doha, Kuwait. A government website for Cairo was not found, although the Longitude study rated Cairo's website as number 45 overall. Since that report did not reference the URL of the websites they used in the study, an email was sent to the organization that published the report – E-Governance Institute, SPAA, Rutgers-Newark- requesting the URLs they used. The websites they used for Riyadh, Beirut, and Amman all corresponded to the ones found by this research except for Cairo. The Website http://www.ipgd.idsc.gov.eg/ used for evaluation was a portal for Egypt and not Cairo. Therefore, the city of Cairo was excluded from this research.

The researcher and five other government employees conducted the evaluation. All evaluators are Computer Science graduates and are experts in Internet use; none were aware of municipality rankings within the literature. The evaluators had no other additional piece of information on the topic other than the checklist to assess the availability and functionality of each item for all six Arab municipality websites. Each evaluator was asked to go through the site thoroughly and fill in the evaluation form.

RESULTS

Results in Table 3 show the average score out of 22 for each municipality. Detailed scores on each item on the checklist can be found in Al-Nuaim (2009).

Some items were not evaluated because some evaluators were faced with long delays for pages to load and were not able to complete the evaluation. The discrepancy between evaluators was due to some evaluators finding fault with some of the criteria listed. For example, some links did not work properly and some items were found by some of the evaluators while other evaluators failed to find them. This is a problem that is persistent in Arabic websites where designers of the site fail to consider link placement and naming conventions when designing leading to user confusion. The researcher and evaluators' comments were used for further analysis of the sites.

Based on the results and comments of the evaluators, the researcher found the following:

1. The cities of Riyadh and Amman have an acceptable score (16.306, 14.93,) but much more needs to be done. This score reflects that they only have 74 percent and 67 percent of what should be available and functioning within the site.

2. Since the cities of Kuwait, Beirut, Doha and Muscat received a score less than 14, they

Table 3. The website of 6 municipalities of Arab Capitals

Doha www.Baladdiya.gov.qa	Riyadh www.alriyadh.gov.sa	Municipality Website Address
12.052	16.306	Total Score
Kuwait www.baladia.gov.kw/	Muscat www.Mctmnet.gov.om	Municipality Website Address
13.524	8.328	Total Score
Beirut cms.beirut.gov.lb	Amman www.Ammancity.gov.jo	Municipality Website Address
9.156	14.93	Total Score

need to revaluate their website, since what they offer lacks the basic requirements for a municipal website and does not meet the needs of citizens.

3. The organization of all the sites was ad hoc and lacks structure, making navigation difficult.

4. The Arab municipality websites were not citizen centered, with, perhaps, the exception of Amman.

5. Amman city website was extremely slow to navigate and hindered evaluation by most evaluators due to the amount of graphics and images used.

6. The city of Muscat had two URL's for its website. In addition to the one used in this evaluation, www.mn.gov.om was found.

7. The website for Doha was actually the website of the ministry, although the URL uses the Arabic word baladiya which means municipality.

8. With the exception of Amman, the websites do not offer separate sections for businesses, residents, and visitors.

9. Quality of content: A website's content defines a website's purpose, relevance, currency, freshness, user appeal, and ease of navigation, to determine whether visitors will return for repeat business (Muninetguide,

2006). Based on this definition, the researcher found the following:

○ **Accuracy:** Only content experts can assess accuracy. This was not done during evaluation

○ **Purpose:** It wasn't really clear what the ultimate goal of these websites was and who were the intended users. Most are there just to have a web presence and to provide information about the city, but that information is general and does not affect the daily lives of citizens. For example, the website for the city of Beirut on the surface caters to tourists more than the citizens of the city themselves.

○ **Relevance:** Many of the links provided were not relevant to citizen-centered needs. Many contained extra information about the weather, currency, etc.

○ **Currency:** The websites did not update their content regularly and do not have a "last updated date" in the footnote of the page. Some news bulletins were clearly outdated.

○ **Freshness:** Most the information is mundane and not very interesting, with the exception of Beirut and

Amman which they provide cultural and tourist information.

- ◦ **User Appeal:** The use of animation in some of the sites is against web accessibility guidelines and is extremely distracting and frustrating to view.
- ◦ **Ease of Navigation:** The depth of the sites for the information available was appropriate, so finding information normally required finding the link and the information would probably be on the next page requested. But that shows the limited depth and breadth of the site's structure.

10. These website suffer from other fundamental problems:
    - ◦ All but Amman either don't have a search capability or have it but it was not functioning.
    - ◦ Most have very limited interactive services; in fact most of these services only offer a list of required documents for face-to-face service.
    - ◦ Contact information for the municipality was absent, even though there should be different forms of contact information.

CONCLUSION

This study evaluated the only six Arab capitals with official municipal websites to assess the current state of their websites. Survey results show that these websites were not citizen centered, suffered from fundamental problems, and lacked basic requirements for any municipal website with some inoperable features and limited interactive services. This evaluation of municipalities only addresses their website, not the municipality itself; differing budgets or levels of authority of any municipality were not part of this research. If a municipality realized the importance of being online and providing e-services for citizens,

then it has already committed to the principle itself and has staff responsible for the website. Any limitations of the website are due to how this staff designs and develops the website and the interactive services management wanted to offer citizens. For these particular websites, it may be a problem of knowledge, skill and policy, not of possibilities and budgeting.

To address the limitations of Arab municipality websites, municipalities or local governments need to first identify their real audiences, their capabilities and their needs that can be serviced online. Surveying citizens and employees of a municipality would be the first step to identifying those needs. Analysis, design and development of municipalities' websites must be centered on the needs and capabilities of the customer of the municipality services, and must follow well known quality frameworks for e-government design guidelines. Periodic evaluation of the websites needs to be conducted with close monitoring of the progress of the delivery of services and users' satisfaction with the available features and resources.

Since it is the ultimate goal of most governments to move towards a digital society and Internet use is growing in the Middle East faster than any other region, Arab municipalities have no choice but to invest their resources in creating an interactive online presence. Although e-government is still in its infancy in the Arab world, the ICT infrastructure is well established in most Arab countries and clearly growing with growing budgets. However, it needs policy makers to commit to change to oversee its efficient implementation to empower people to be part of the political process. The possibility of having municipal websites as the single point of entry to most government services should be the focus of every urban Arab city due to the very nature of the culture and the benefits they will provide.

There are, however success stories since the publication of the results of this evaluation. Local government policy makers must now be fully aware of how prominent a website is for their municipality and there have been major advances in this regard. The researcher searched for Arab capital municipalities using the same keywords as in this research and found the following:

- Manamah, Sanaa, and Abu Dhabi, capitals of Bahrain, Yemen, and United Arab Emirates have municipal websites: http://www.manama-mun.gov.bh/; http://www.yemen.gov.ye; and http://www.adm.gov.ae but Sanaa's site is constantly down.
- Greater Damascus, Cairo and Khartoum, capitals of Syria, Egypt, and Sudan have websites http://www.damascus.gov.sy/, http://www.cairo.gov.eg/Pages/Default.aspx, and http://krt.gov.sd/websites.php. However, all three are not city municipal websites, but rather local state government websites.
- The municipal website of Beirut seems to no longer exist and no alternative URL was found.

With all the success stories and progress in e-government in the Arab world, other capitals not mentioned above still have no municipal websites and those capitals that do, are still being dominated by aesthetics and technical innovation alone with inactive information presentation rather than the inclusion of interactive services with immediate feedback and easy to use and navigate interfaces.

REFERENCES

Akinci, H. (2004), *Geospatial Web Services for e-Municipality*, XX ISPRS Congress. Istanbul, Turkey. Retrieved from http://www.isprs.org/proceedings /XXXV/congress/comm2/papers/210.pdf

Al-Nuaim, H. A. (2009). How "E" are Arab Municipalities? An Evaluation of Arab Capital Municipal Web Sites. [IJEGR]. *International Journal of Electronic Government Research*, 5(1), 50–63.

Center for Democracy and Technology and InfoDev. (2002). *The e-Government Handbook for Developing Countries*. Washington, D.C.: The World Bank, 2002. Retrieved from http://www.cdt.org/egov/handbook /2002-11-14egovhandbook.pdf

Connors, H., Koretz, P., Knowle, S., & Thibodeau, M. (1999). Municipal Web Sites in Onondaga County: A Study Comparing Selected Characteristics. *Community Benchmarks Program. Maxwell School of Citizenship and Public Affairs*. Syracuse University. Retrieved from http://www.maxwell.syr.edu/benchm arks/newsite/reports/web_down.html

Çukurçayır, M., & Eroğlu, H. (2010). E-Cities: A Content Analysis of the Web Pages of Heidelberg and Konya Metropolitan Municipalities. *Current Rserach Journal of Social Sciences*, 2(1), 7–12.

Daemen, H., & Schaap, L. (2000). Developments in Local Democracies: An Introduction. Daemen. In H & Schaap, L. (eds.). *Citizen and City: Developments in fifteen local democracies in Europe*. Delft: Eburon.

De Long, M., & Lentz, L. (2006). Scenario evaluation of municipal Web sites: Development and Use of an Expert-Focused Evaluation Tool. *Government Information Quarterly*, 23, 191–206. doi:10.1016/j.giq.2005.11.007

Economist Intelligence Unit. (2006). The 2006 E-Readiness Rankings. *The Economist*. Retrieved from http://graphics.eiu.com/files/ad_pdfs/2006Ereadiness_Ranking_WP.pdf

Goldberg, J. (2009). *State of Texas Municipal Web Sites: A Description of Website Attributes and Features of Municipalities with Populations Between 50,OOO-125,OOO.* Texas State University-San Marcos. Retrieved from http://ecommons.txstate.edu/arp/307

Holzer, M., & Kim, S. (2005). Digital Governance in Municipalities Worldwide: A Longitudinal Assessment of Municipal Websites Throughout the World. *The E-Governance Institute. National Center for Public Productivity.* Rutgers, The State University of New Jersey. Newark. Retrieved from http://unpan1.un.org/intradoc/groups/ public/documents/ASPA/ UNPAN022839.pdf

Internet World Stats. (2007). Internet Usage Statistics. *The Internet Big Picture World Internet Users and Population Stats. Miniwatts* Marketing Group. Retrieved from http://www.internetworldstats.com/stats.htm

LaVigne, M., Simon, S., Dawes, S., Pardo, T., & Berlin, D. (2001). *Untangle the Web: Delivering Municipal Services Through the Internet. Center for Technology in Government. University at Albany.* SUNY.

Massachusetts. (2004) *A Recipe for Success Building a Citizen-Centric Website: Commonwealth of Massachusetts.* December. Retrieved from www.mass.gov/Aitd/docs/ Cookbook_ver2.pdf

Moon, M. (2002). The Evolution of E-Government among Municipalities: Rhetoric or Reality? *Public Administration Review, 62*(4), 424–434. doi:10.1111/0033-3352.00196

Mulder, E. (2004). A Strategy for E-government in Dutch Municipalities. *Proceedings. 2004 International Conference on Information and Communication Technologies: From Theory to Application, 19*(23), 5 – 6.

Muninetguide (2006). *Striving for Online Excellence.* Retrieved from http://www.muninetguide.com/article s/Striving-for-Online-Excellence---161.php

Pardo, T. A. (2000). *Realizing the Promise of e-Government. Center for Technology in Government. University at Albany.* SUNY.

Prychodko, N. (2001). *Municipalities and Citizen-Centred Service.* Report to the Public Sector Service Delivery Council December 4, 2001. Retrieved from http://www.iccs-isac.org/eng /pubs/iccs_muni.pdf

Rodriguez, J. R (2005). IRIS project: Promoting civic attitudes in Barcelona through a customer service request platform. *The 2005 Ministerial e-Government Conference on transforming Public Services.* Retrieved from http://archive.cabinetoffice.gov. uk/egov2005conference/

Rodriguez, R., Estevez, E., Giulianelli, D., & Vera, P. (2009). Assessing E-Governance Maturity through Municipal Websites Measurement Framework and Survey Results. *Proceedings of the 6th Workshop on Software Engineering, Argentinean Computer Science Conference.* San Salvador de Jujuy, Argentina, October 5-9, 2009. Retrieved from http://egov.iist.unu.edu/cegov/ OUTPUTS/PUBLICATIONS

Town of Freetown. MA. (2007). *Top Ten Reasons to Have a Municipal Website.* Retrieved from http://town.freetown.ma.us/dept/faq _detail.asp?DeptID=WEB&FAQID=76

United Nations. (2005). *Global E-Government Readiness Report 2005: from E-Government to inclusion.* United Nations, New York, NY. Retrieved from http://unpan1.un.org/intradoc/ groups/ public/documents/un/unpan021888.pdf

United Nations. (2008). *E-Government Survey 2008: from e-Government to Connected Governance.* United Nations, NY. Retrieved from http://unpan1.un.org/intradoc/groups/ public/documents/un/unpan028607.pdf

United Nations. (2010). *E-Government Survey 2010: Leveraging E-Government at a time of Financial and Economic Crisis.* United Nations, NY. Retrieved from http://unpan1.un.org/intradoc/groups/ public/documents/un/unpan038851.pdf

Chapter 5

Aviation–Related Expertise and Usability:
Implications for the Design of an FAA E–Government Website

Ferne Friedman-Berg
FAA Human Factors Team - Atlantic City, USA

Kenneth Allendoerfer
FAA Human Factors Team - Atlantic City, USA

Shantanu Pai
Engility Corporation, USA

ABSTRACT

The Federal Aviation Administration (FAA) Human Factors Team – Atlantic City conducted a usability assessment of the www.fly.faa.gov Web site to examine user satisfaction and identify site usability issues. The FAA Air Traffic Control System Command Center uses this Web site to provide information about airport conditions, such as arrival and departure delays, to the public and the aviation industry. The most important aspect of this assessment was its use of quantitative metrics to evaluate how successfully users with different levels of aviation-related expertise could complete common tasks, such as determining the amount of delay at an airport. The researchers used the findings from this assessment to make design recommendations for future system enhancements that would benefit all users. They discuss why usability assessments are an important part of the process of evaluating e-government Web sites and why their usability evaluation process should be applied to the development of other e-government Web sites.

INTRODUCTION

On November 15, 2007, President Bush announced actions to address aviation delays during the Thanksgiving holidays. As part of this announce-

DOI: 10.4018/978-1-60960-162-1.ch005

ment, he directed people to visit the Web site fly. faa.gov, which is a Federal Aviation Administration (FAA) e-government Web site that provides real time information about airport delays.

Fourth, the federal government is using the Internet to provide real-time updates on flight delays. People in America have got to know there's a Web site called Fly.FAA.Gov; that's where the FAA transmits information on airport backups directly to passengers and their families. If you're interested in making sure that your plans can -- aren't going to be disrupted, you can get on the Web site of Fly.FAA.Gov. As well, if you want to, you can sign up to receive delay notices on your mobile phones. In other words, part of making sure people are not inconvenienced is there to be -- get transmission of sound, real-time information. (Bush, 2007)

There has also been a concerted effort by the FAA to publicize its Web site by placing advertisements in airports across the United States. Many news outlets now provide airport delay information as part of their weather forecasts, and this delay information comes, most often, directly from the fly.faa.gov Web site.

Because this Web site is the public face of a large federal agency, it is important that it presents the agency in the best light possible. An agency Web site should be a positive public relations vehicle and should not, in itself, create any public relations problems. Although use of e-government Web sites is increasing annually, low user acceptance of e-government Web sites is a recognized problem (Hung, Chang, & Yu, 2006). Many factors affect whether or not someone will use or accept an e-government Web site, including past positive experience with e-government Web sites (Carter & Bélanger, 2005; Reddick, 2005); the ease of use of the Web site (Carter & Bélanger, 2005; Horst, Kuttschreutter, & Gutteling, 2007); the perceived trustworthiness of the information presented on the Web site (Carter & Bélanger; Horst, et al., 2007); the perceived usefulness of the Web site (Hung et al., 2006); and personal factors such as education level, race, level of current internet use, and income level (Reddick, 2005). If a Web site has many functional barriers, such as having a poor layout or producing incomplete search results, customers of the site may not use it (Bertot & Jaeger, 2006).

Early work in e-government has consistently ignored studying the needs of end users, and there has been little research focusing on the demand side of e-government (Reddick, 2005). That is, *what are customers looking for when coming to an e-government Web site?* Although there have been many benchmarking surveys conducted on e-government Web sites, benchmarking surveys often do not describe the benefits provided by a Web site and only enumerate the number of services offered by that site (Foley, 2005; Yildiz, 2007). Benchmarks do not evaluate the user's perception of sites and do not measure real progress in the government's delivery of e-services. However, governments often chase these benchmarks to the exclusion of all other forms of evaluation (Bannister, 2007).

E-government academics emphasize the importance of usability testing and highlight the need to focus on Web site functionality, usability, and accessibility testing (Barnes & Vigden, 2006; Bertot & Jaeger, 2006). However, despite its importance, many organizations still are not performing usability testing on e-government Web sites. Current work often does not address the needs of different user communities, employ user-centered design, or use rigorous methods to test the services being delivered (Bertot & Jaeger; Heeks & Bailur, 2007).

Governments around the world are working to review best practices for e-government evaluation methods (Foley, 2005). Because of the social and economic benefits of providing information online, it is important that e-government Web site designs meet the needs of its targeted users. In addition, it is important to document the benefits provided by the Web site to increase public support (Foley). Carter and Bélanger (2005) point out that e-government Web sites should be easy to navigate. They note that the organization of information on the site should be congruent with citizens' needs. When consumers visit an e-government Web site, they are most frequently looking for information (Thomas & Streib, 2003), which they need to be able to find quickly and easily. If users encounter problems while using a Web site, they may become frustrated and be less likely to adopt or utilize e-government services in the future. A positive experience with an e-government Web site will be communicated to others (Carter & Bélanger), and a usable Web site can play a significant role in engendering trust in the agency itself.

Most Web usability research focuses on e-commerce sites and privately run Web sites (Hung et al., 2006), and people expect e-government Web sites to be as good or as usable as private sector sites (Irani, Love, & Montazemi, 2007). People are more likely to use an e-government Web site if the transactions with that site are compatible with previously conducted transactions on similar, non-government Web sites (Carter & Bélanger, 2005).

However, there are clear differences between e-government and e-commerce Web sites. For instance, e-government sites must provide universal accessibility so that all citizens have access to information. Additionally, e-government Web sites are accountable to the public, whereas commercial Web sites are only accountable to people who have a financial stake in the Web site. It is not always clear, however, where the boundary between these two types of sites lies (Salem, 2003). Additionally, there are often challenges faced in producing e-government Web sites that are not faced by commercial sites (Gil-Garcia & Pardo, 2005). For example, when creating e-government Web sites, designers need to consider whether the project goals align with the goals or mission of the government agency (Yildiz, 2007). They also must make sure that all project stakeholders are involved, determine whether they are in compliance with all relevant government regulations, and work within government budget cycles and changing government contractors.

The FAA and fly.faa.gov

The FAA Air Traffic Control System Command Center provides information about airport condi-

tions, such as arrival and departure delays, to the public and the aviation community via their Web site, www.fly.faa.gov. This Web site allows users to view airport conditions for specific airports.

The Web site has many different functions that help the user to search for delay information (see Figure 1). Using the Search by Region function, users are able to look up airports in different geographic regions, such as the Northeastern states and the Southeastern states. When using the Search by Airport function, users are able to search for airport delay information by typing in the name of a city, airport, or a three-letter airport code. The View by Major Airport function allows users to search for delay information using a drop down list of 40 major airports.

The site is also a repository of information for use by airlines, pilots, passengers, government personnel, academics, individual aircraft operators, and other stakeholders in the aviation community. It provides access to real-time and historical advisory information, real-time airport arrival demand information, current reroutes, and reroute restrictions. It also provides access to information related to air traffic management tools, a glossary of aviation terms, a national routes database, pilot tools for making arrival and departure reservations, a collection of National Airspace System documents, and many other air traffic tools.

The focus of this assessment was on the evaluation of site elements that the general public would access the most, such as the airport delay information and the glossary of aviation terms. From the user's point of view, the Web site needs to provide accurate information quickly, with minimal effort, while minimizing potential mistakes. The site should be easy for users to learn and provide an appealing and satisfying experience.

We faced some unique issues and challenges when evaluating the fly.faa.gov Web site. First, the fly.faa.gov Web site presents real-time, up-to-the-minute data, whereas most e-government Web sites often present static information or informa-

Figure 1. The www.fly.faa.gov home page, illustrating the View by Region, Search by Airport, View by Major Airport, and Site Map search methods

tion that changes infrequently. It was also clear that the expectations of site users were likely to be influenced by the information found on more commercial aviation sites. Because people have preconceived notions about the airlines and the reliability of information provided by airlines, it was possible that this perception could transfer to their perception of this Web site.

The Web site was also originally designed for use by people associated with the aviation industry, such as pilots and local airport authorities, who have at least a working knowledge of various aviation concepts. Because it is accessible on the internet and other travel sites have links to it, members of the traveling public (who may have little, if any, understanding of aviation or its associated jargon) also frequently use the site. The Web site is also being touted (Bush, 2007) as the first place the public should visit on the Web when looking for travel-related delays in the aviation system. Therefore, it was important to evaluate whether this site is usable by people who do not have a background in aviation. In this usability assessment, we examined how effectively people

with different levels of domain knowledge were able to use the site.

It was difficult to identify a single typology that described the Web site. Although the site often looks like a Government to Consumer (G2C) site (Hiller & Bélanger, 2001), its original purpose was to function as a Government to Business (G2B) site or a Government to Employee (G2E) site. The site allows people to perform basic transactions (Hiller & Bélanger, Stage 3), but it also attempts to be a full-service, one-stop site for many types of aviation related information (Hiller & Bélanger, Stage 4). For instance, although this evaluation did not focus on the G2B information, airlines often use the site to find delay information, and general aviation pilots use the site to make route reservations. Although the site tries to organize its content to meet the different needs of these different categories of users (Ho, 2002; Schelin, 2003), it is not clear how the organizational structure was determined or whether it is the most optimal organization for all types of users.

We conducted this formal usability assessment to determine how successfully the Web site meets these usability goals and the needs of its users, including both expert and novice users. The assessment employed techniques commonly used in usability evaluations (Ahlstrom & Longo, 2003; Nielsen, 2003). The participants completed a set of representative tasks using the Web site, while researchers observed and recorded their actions and comments. Users also answered a series of questions rating the usability of the site. The data collected through these activities helped us identify a number of problems. After identifying the final list of usability issues, we used a part of the heuristic evaluation technique (Nielsen) to determine the most critical issues. This article discusses the technique used in this evaluation, highlights some of the most critical issues, and provides suggestions to designers on how to fix them. We also discuss the benefits of applying this formal process to the development of other e-government Web sites.

METHODOLOGY

Participants

We recruited 32 adult volunteers from the FAA William J. Hughes Technical Center to serve as participants. Because the participants were FAA employees, many had greater aviation-related knowledge than the general public. However, many FAA employees, such as administrative assistants and facility support workers, do not have significant knowledge of aviation or air traffic control. We included participants of both categories.

Equipment

The laptops used in the experiment contained fully interactive offline versions of the fly.faa.gov Web site. A User Script asked the participants to use the Web site to find information to answer 17 questions: 12 asked users to search for delay information, 3 asked users to find the definitions for aviation-related terms, and 2 asked users to identify the authority to be contacted when trying to obtain specific information. The script also asked users to use the Search by Region, Search by Airport, and View by Major Airport methods for specific questions. This allowed us to evaluate the usability of each function.

Procedure

Each session lasted 30 to 45 minutes. After signing an informed consent form, the participants completed a Background Questionnaire that collected information about the participants' knowledge of computers, Web sites, and aviation terminology.

After completing the Background Questionnaire, the participants next completed the User Script. We observed each participant during the experiment and recorded pertinent actions or comments. At the end of the experiment, the participants completed a Post-Session Questionnaire, where they rated their experience and identified usability issues.

Because using participants who all had a high level of aviation-related knowledge could have biased the results, we used the data to categorize the participants into three groups (novices, moderate knowledge users, and experts), based on their aviation-related knowledge. We analyzed the data by level of expertise to determine whether aviation-related knowledge had an impact on user performance. By analyzing the results in this way, we could make recommendations targeted toward making the site usable for the different user populations. When even individuals with a high level of aviation-related expertise had trouble using certain features, this provided strong evidence that those features needed to be redesigned. Even if novices were the only ones who had a problem with a feature, we rated that problem as severe if the impact for those users was severe.

RESULTS

Background Questionnaires

The Background Questionnaire asked the participants questions regarding their familiarity with aviation-related terms and acronyms. For example, participants were asked to list three-letter abbreviations for airports (e.g., Philadelphia International Airport = PHL), or were given the three-letter abbreviations and asked to list the airports associated with those abbreviations (e.g., MIA = Miami International Airport). Using the correct responses to these and other aviation-related questions, we categorized the participants as novices (n = 8), moderate knowledge users (n = 15), and experts (n = 9). The novices were slightly younger than both the experts and those with moderate-knowledge (M_{novice} = 41.6 years, $M_{moderate}$ = 49.9 years, M_{expert} = 49.1 years). More than 70% of novices and those with moderate knowledge reported never using the fly.faa.gov Web site. In contrast, 75% of the experts reported using the Web site a few times a year.

All the participants had extensive experience using computers and the Web. Because we found no discernable differences in reported Web and computer use among the participants, we were unable to stratify the participants based on these factors.

User Script Data: Overall Analysis

Of the 12 questions that asked users to find specific delay information, the participants answered 79.4% correctly. For the subset of five delay questions that allowed the participants to use their preferred search method, the participants answered 71.2% correctly. For the subset of four Search by Airport questions, 84.5% of the participants answered the questions correctly. For the View by Major Airport question, 90.6% of participants found the correct answer; for the View by Region question, 81.3% found the correct answer; and for the **Site Map** question, 87.5% found the correct answer.

Three questions asked the participants to use the site to provide the definition of three aviation related terms and abbreviations. Although 84.4% of participants answered all three questions correctly, 6.3% answered one incorrectly, 3.1% answered two incorrectly, and 6.3% were not able to answer any of the questions. By comparing the percentage of participants who answered a question correctly, we determined that all three questions were equally difficult.

Two questions asked the participants to find whom to contact to obtain information about the status of an individual flight or why an airport was closed. For these questions, only 28.1% of the participants answered both questions correctly, 56.2% answered one incorrectly, and 15.6% answered both incorrectly.

User Script Data: Analysis by Level of Expertise

We analyzed the data by level of expertise to determine whether aviation-related knowledge had an impact on user performance. Analyzing all 17 questions, we found an effect of expertise on overall task performance, $F (2, 29) = 3.54$, $p =.04$. Post hoc pairwise contrasts indicated expert participants were able to answer significantly more questions than novices (85.6% vs. 69.1%, $p =.01$), and there was a trend suggesting moderate-level users answered more questions than novices (79.6% vs. 69.1%, $p =.07$).

We performed ordinal (linear) chi-square tests on individual questions to determine whether the percentage correct increased or decreased across the user categories (Howell, 2007). Although only three of the questions were significant, 7 of the 12 delay questions showed the expected pattern of results (see Table 1). Therefore, we also tested the binomial probability that 7 of the 12 delay questions would show the expected ordering of expert > moderate > novice. We found that it was unlikely that this pattern would occur by chance 7 out of 12 times, $p < .001$. This suggests that experts were better able to find information on the fly.faa.gov Web site than moderate users, who in turn were better than the novices. We did not find the same pattern for the aviation term or contact information questions.

We grouped the questions to analyze performance on the different subsets of questions. For the 12 questions that asked users to find specific delay information, novices, moderate-level users, and experts answered 65.6%, 81.7%, and 88% of the questions correctly, $F (2, 29) = 5.04$, $p = .01$. Post hoc pairwise contrasts indicated experts and moderate-level users were better able to find

delay information than novices ($p = .005$ and $p = .021$, respectively).

We further divided the 12 delay questions into subcategories based on search method. For the subset of questions that allowed people to find information using their preferred search method, we found an effect of expertise on user performance, $F (2, 29) = 9.93$, $p = .001$. Experts and moderate users performed better than novices when searching for delay information using their preferred search method, answering an average of 80% and 77.3% of the questions correctly, while novices only answered an average of 50% correctly ($p < .001$ for both post hoc pairwise comparisons).

Table 1. Percentage correct by level of aviation-related expertise

Questions	% Correct		
	Novices	**Moderate Users**	**Experts**
1. Los Angeles to Salt Lake City.**	75.0	100.0	100.0
2. Portland to Memphis.	25.0	53.3	33.3
3. Denver to Philadelphia. **Search by Airport.**	87.5	93.3	88.9
4. Houston to Chicago. **Search by Airport.****	62.5	73.3	100.0
5. Newark to Burlington.	50.0	73.3	88.9
6. Las Vegas to New York. **View by Major Airport.**	75.0	93.3	100
7. Phoenix to Dallas.*	12.5	73.3	77.8
8. Cincinnati/Northern Kentucky to Detroit. **View by Region.**	75.0	80.0	88.9
9. Pittsburgh to Washington DC. **Site Map.**	75.0	86.7	100.0
10. New York to San Jose. **Search by Airport.**	75.0	80.0	100.0
11. Orlando to St. Louis. **Search by Airport.**	87.5	86.7	77.8
12. Houston to Tulsa.	87.5	86.7	100.0
Using information available on the site, provide the definitions of the following aviation-related terms or abbreviations:			
13. CIGS	87.5	93.3	88.9
14. MULTI-TAXI	87.5	86.7	88.9
15. VOL	75.0	93.3	100.0
Using information available on the site, who should a visitor contact to obtain information about the following:			
16. Status of an individual flight	100.0	78.6	87.5
17. Why an individual airport was closed	50.0	26.7	44.4

* p < .10, two-tailed. * * p < .05, two-tailed.

For the four delay questions that asked users to specifically use the **Search by Airport** method, novices, moderate users, and experts answered 78.1%, 83.3%, and 91.7% of them correctly. Although these results were not statistically significant, they demonstrated the same trend as the other sets.

Post-Session Questionnaire: Overall Analysis

The Post-Session Questionnaire asked the participants to rate their subjective experience with the fly.faa.gov Web site using 6-point scales. Except for the question asking about the level of detail, higher ratings indicated positive responses and lower ratings indicated negative responses. For the question that asked the users how detailed the information on the site was, a rating of 1 indicated too little detail and a 6 indicated too much detail. For these summaries, we omitted responses from the participants who chose more than one number on the rating scale. The ratings indicated that the participants thought it was fairly easy to find information on the site ($M = 4.4$, $SD = .8$) and that they understood information once they found it ($M = 4.8$, $SD = 1.0$). The participants also found it fairly easy to navigate between pages on the site ($M = 4.9$, $SD = 1.2$) and found the design of the site to be consistent ($M = 4.9$, $SD = 1.0$). They indicated that there was somewhat too much detail ($M = 3.9$, $SD = 0.8$), but that information on the site was fairly readable ($M = 4.8$, $SD = 1.1$). Finally, they indicated that, overall, they were mostly satisfied with the site ($M = 4.7$, $SD = 0.8$). When we compared satisfaction ratings to actual performance, it was apparent that participants were not able to accurately estimate performance, given that they answered an average of 20.1% questions incorrectly. However, despite their performance, the participants still reported high satisfaction with the site. Given this dissociation between performance and satisfaction, it is important that usability experts evaluate not just user satisfaction, but actual user performance, when evaluating a Web site.

Post-Session Questionnaire: Analysis by Level of Expertise

We found no significant differences in the ratings between experts, moderate-level users, and novices. There were, however, some interesting trends in the data. The ratings on information comprehensibility indicated that experts found the information to be somewhat more comprehensible than moderate-level users, who, in turn, found the information to be more comprehensible than novices. In evaluating design and layout consistency, the experts were the least satisfied with the design consistency, with novices being the most satisfied, and moderate users falling somewhere in the middle. For the ratings on the level of detail, experts gave the highest ratings (i.e., slightly too much detail), with novices giving the lowest ratings (i.e., slightly too little detail), and moderate users falling in the middle (i.e., an appropriate level of detail).

Rating of Usability Issues

Using comments and questionnaire ratings made by the participants, along with our observations of the participants while they completed the User Script, we compiled a consolidated list of usability issues and rated the severity of each issue (for a comprehensive list, see Friedman-Berg, Allendoerfer, & Pai, 2007). When rating the severity of each problem, we considered the following factors (Nielsen, 2003).

1. **Frequency:** Is the problem very common or very rare?
2. **Impact:** How easy is it for the users to overcome the problem when navigating through the Web site?

3. **Persistence:** Can users overcome the problem once they know about it, or will the problem bother users repeatedly?

The researchers rated each issue as having high, medium, or low frequency, impact, and persistence, and then used these three ratings to determine a severity rating from 0 to 5. The severity rating scale was adapted from Nielsen (2003).

0 = I don't agree that this is a usability problem at all

1 = minor/ cosmetic problem only: not necessary to fix, should be given lowest priority

2 = usability problem: small benefit from fixing, should be given low priority

3 = moderate usability problem: moderate benefit from fixing, should be given medium priority

4 = major usability problem: important to fix, should be given high priority

5 = usability catastrophe: extremely important to fix, should be given highest priority

After each researcher independently assigned a severity rating for each issue, we averaged them to compute a consolidated severity rating (Nielsen, 2003). These consolidated severity ratings provide a good estimate of additional usability efforts needed when developers establish priorities for future enhancements. We rank ordered the usability issues from those having the highest severity rating to those having the lowest.

The following section discusses the eight usability issues that had the highest severity rating and provides suggestions and design recommendations regarding how these issues could be resolved. User interface design standards and best practices drive these suggestions (Ahlstrom & Longo, 2003). In some cases, we developed simple prototypes to demonstrate potential design concepts that designers could use to remediate some of these issues.

Issue 1: User Confusion Regarding Delay Types

The primary purpose of fly.faa.gov is to provide travelers with airport delay information. For example, a traveler going from Philadelphia to Miami might want to find out about departure delays at PHL and arrival delays at MIA. The traveler also might have some interest in the causes of delays, which can include factors like weather, airport construction, and traffic flow programs. However, the difference between delay types was not readily apparent to many participants. For example, one question asked users to find information about delays at their arrival destination. The arrival airport had no arrival delays, but did have general departure delays. Because the instructions indicated that they were arriving at that airport, the participants should have focused on the lack of an arrival delay, but only 40.6% of the participants answered this question correctly. Those who answered incorrectly seemed to be looking at the departure delay, which indicated that they did not understand which delays were relevant for them. This issue received a mean severity rating of 4.3, $SD = 0.5$.

It is important that the site provide users with the information they want without requiring them to understand difficult air traffic concepts. We also found that novices had greater difficulty in finding delay information than both moderate level users and experts. This was likely due to novice users not understanding more technical concepts. We recommend that the site not try to present difficult concepts to the lay public, but instead present information in a less technical manner. For instance, instead of referencing ground delay programs as the cause of a delay, the site could indicate that a delay was due to congestion. For users seeking more detailed information, the Web site could provide additional information about ground delay programs using links to additional pages.

Because the participants were not always able to identify relevant delays, we recommend that the site provide users with a capability that gives them easy access to pertinent delay information. For example, the site might provide an interactive tool that allows users to input departure and arrival airports or click on city pairs to generate a single report on relevant delays for air traffic traveling between a pair of airports.

Issue 2: Information Presentation: Clutter and Redundant Information

The participants' comments and researchers' observations suggested that there was too much information on the typical search results page (see Figure 2). This issue received a mean severity rating of 4.3, $SD = 0.5$. The site sometimes presented information for a single airport in multiple places on the same page. The information was dense, used too much text, and was not well organized. In many instances, the participants had difficulty finding the delays that were relevant for them. Displaying so much information can be especially problematic when users are in a hurry to find information. Users may scan too quickly and get lost. They may read the wrong line, overlook in-

formation they are looking for, or see a big block of text and give up.

We recommend simplifying and reorganizing these pages to make it easier for users to find and understand information on the page. The page could use a tabular layout arranged in columns and organized by arrivals and departures (see Figure 3). Much of the text information is not useful, creates clutter, and should therefore be removed. Because the distinction between general departure delays and destination-specific delays is not clear to users, it should be deemphasized or eliminated. Finally, all delay information related to an individual airport should be consolidated.

Presenting two sets of delay information for one airport, especially if the data are inconsistent, is confusing. The Web site should avoid going into too much technical detail regarding the causes of delays. It might instead use icons or graphics (e.g., clouds with snow, clouds with rain) to depict weather or other causes of delays. The Web site could offer links to additional information for advanced users.

Figure 2. Crowded Airport Status Information page

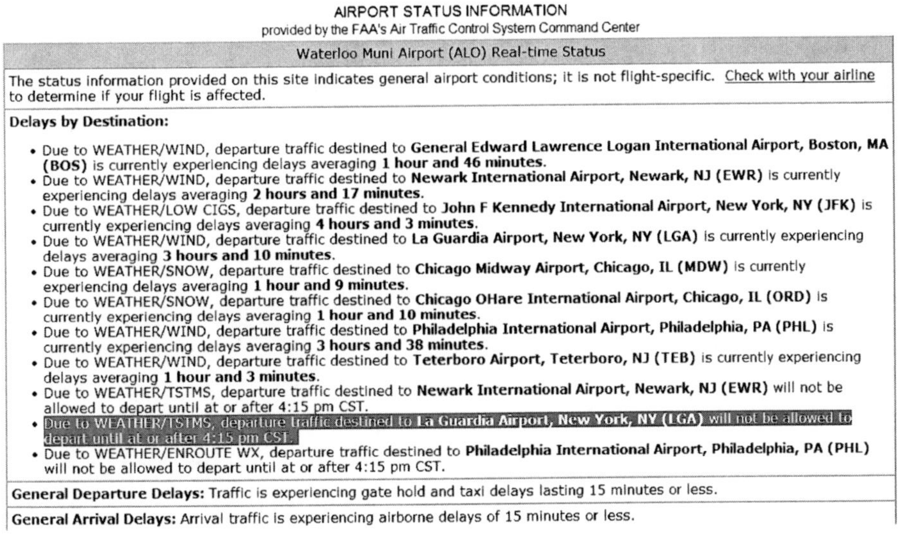

AIRPORT STATUS INFORMATION
provided by the FAA's Air Traffic Control System Command Center

Waterloo Muni Airport (ALO) Real-time Status

The status information provided on this site indicates general airport conditions; it is not flight-specific. Check with your airline to determine if your flight is affected.

Delays by Destination:

- Due to WEATHER/WIND, departure traffic destined to **General Edward Lawrence Logan International Airport, Boston, MA (BOS)** is currently experiencing delays averaging **1 hour and 46 minutes.**
- Due to WEATHER/WIND, departure traffic destined to **Newark International Airport, Newark, NJ (EWR)** is currently experiencing delays averaging **2 hours and 17 minutes.**
- Due to WEATHER/LOW CIGS, departure traffic destined to **John F Kennedy International Airport, New York, NY (JFK)** is currently experiencing delays averaging **4 hours and 3 minutes.**
- Due to WEATHER/WIND, departure traffic destined to **La Guardia Airport, New York, NY (LGA)** is currently experiencing delays averaging **3 hours and 10 minutes.**
- Due to WEATHER/SNOW, departure traffic destined to **Chicago Midway Airport, Chicago, IL (MDW)** is currently experiencing delays averaging **1 hour and 9 minutes.**
- Due to WEATHER/SNOW, departure traffic destined to **Chicago OHare International Airport, Chicago, IL (ORD)** is currently experiencing delays averaging **1 hour and 10 minutes.**
- Due to WEATHER/WIND, departure traffic destined to **Philadelphia International Airport, Philadelphia, PA (PHL)** is currently experiencing delays averaging **3 hours and 38 minutes.**
- Due to WEATHER/WIND, departure traffic destined to **Teterboro Airport, Teterboro, NJ (TEB)** is currently experiencing delays averaging **1 hour and 3 minutes.**
- Due to WEATHER/TSTMS, departure traffic destined to **Newark International Airport, Newark, NJ (EWR)** will not be allowed to depart until at or after 4:15 pm CST.
- Due to WEATHER/TSTMS, departure traffic destined to **La Guardia Airport, New York, NY (LGA)** will not be allowed to depart until at or after 4:15 pm CST.
- Due to WEATHER/ENROUTE WX, departure traffic destined to **Philadelphia International Airport, Philadelphia, PA (PHL)** will not be allowed to depart until at or after 4:15 pm CST.

General Departure Delays: Traffic is experiencing gate hold and taxi delays lasting 15 minutes or less.

General Arrival Delays: Arrival traffic is experiencing airborne delays of 15 minutes or less.

Figure 3. Airport Status Information in a redesigned format

AIRPORT STATUS INFORMATION
provided by the FAA's Air Traffic Control System Command Center

Waterloo Muni Airport (ALO) Real-time Status			
The status information provided on this site indicates general airport conditions; it is not flight-specific. <u>Check with your airline</u> to determine if your flight is affected.			
Departure Delays			**Arrival Delays**
Are you flying to:			Arrival traffic is experiencing airborne delays of 15 minutes or less.
Airport	**Delay**		
Logan International Airport, Boston, MA (BOS)	<u>1 hour 46 minutes</u>	Snow	
Newark International Airport, Newark, NJ (EWR)	<u>2 hours 17 minutes</u>	Thunderstorms	
John F. Kennedy International Airport, New York, NY (JFK)	<u>4 hours 3 minutes</u>	Congestion	
LaGuardia Airport, New York, NY (LGA)	<u>3 hours 10 minutes</u>	Emergency	
Chicago Midway Airport, Chicago, IL (MDW)	<u>1 hour 9 minutes</u>	Equipment Outage	
Chicago O'Hare International Airport, Chicago, IL (ORD)	<u>1 hour 10 minutes</u>	Construction	
All other airports	<u>Delays of 15 minutes or less</u>		

Issue 3: Overuse of Aviation-Related Acronyms and Jargon

The site uses too many aviation-specific acronyms and jargon when providing specific information about the causes of delays. This issue received a mean severity rating of 4.0, $SD = 0.0$. Aviation-specific acronyms, abbreviations, and jargon are difficult for the general public to understand, and the glossary is difficult to find. The average user of the Web site may never be aware that it exists. When the participants had to find the definition of three aviation-related terms, 16% were unable to find the definition for at least one of them. Therefore, we recommend eliminating the use of these terms when they are not essential. This would eliminate unnecessary detail, simplify the site, and make it easier to use and understand.

Issue 4: User Confusion with Using the View by Region Maps

The fly.faa.gov Web site provides users with a View by Region search function that allows users to look up airports by searching in different geographic regions. These regions include the Northeast, North Central, Northwest, Southeast, South Central, and Southwest regions, along with Alaska and Hawaii. When a user uses the View by Region function, they are taken to a map that contains only states that are part of a region. However, it is not easy for someone with little knowledge of geography to determine the region for a particular state. The participants got lost when looking for airports that were not on the main U.S. map because they were unable to determine the relationship of regional maps to the main U.S. map. This was especially difficult for states such as Ohio that lie at the edge of a region. These issues make the View by Region method difficult for the general public to use and the participants found the View by Region maps to be confusing. This issue received a mean severity rating of 4.0, $SD = 0.0$.

One question asked the participants to find delay information for an airport that was not available on the main map or on the View by Major Airport menu. Only 71.9% of the participants found the correct answer for this question, indicating that the participants had some difficulty finding information when they needed to drill down on the maps.

There are several recommendations that could alleviate some of the issues related to the use of the View by Region method. First, the site could place an outline around the different regions or use color coding to highlight the different regions on the U.S. map. This would help users identify which states belong in which region. The site could display split portions of the main U.S. map on the same page to better orient users to the different regions. To familiarize people with relevant geographic information, the site could label states, both on the main U.S. map and on the smaller regional maps. The site could also offer users a drop-down menu that listed the various airports by state.

Issue 5: Lack of User Knowledge Regarding Three-Letter Airport Identifiers

All commercial airports have three-letter identifiers, and using them is an efficient way to obtain delay information about an airport. The site provides a function that allows users to type a three-letter identifier directly into the Search by Airport text box, which will take the user to the details page for that airport. It also provides cues to site users by labeling airports on the main U.S. map with their three-letter identifiers (see Figure 1). However, many participants did not know the correct three-letter identifiers for airports and did not use the cues on the main map to determine the correct identifier. This issue received a mean severity rating of 3.3, $SD = 0.6$.

The site should emphasize that the Search by Airport text box accepts regular airport names and city names in addition to three-letter identifiers. Although the Search by Airport text box does have a label indicating that users can enter city, airport code, or airport name information in this field, we recommend that the Web site provide the user with specific examples to highlight and better explain the different search options.

Issue 6: The Search by Airport Function Returns Redundant and Irrelevant Results

City name searches using the Search by Airport function generate an intermediate results page that lists multiple airports. These listings often contain redundant and irrelevant results. This issue received a mean severity rating of 3.3, $SD = 0.6$. For example, a search for Chicago generates a search results page listing two airports: Midway and O'Hare International. The site lists each result twice, once under City Name Matches and once under Airport Name Matches (see Figure 4). This format is confusing and users may not realize that both links take them to the same information. Some participants questioned why the site listed an airport twice. We recommend that the Airport Lookup Search Results page consolidate search results and list airports only once in any search results list.

Issue 7: User Spelling and Misspellings and Their Impact on the Search by Airport Function

User spellings and misspellings can have a serious impact on the Search by Airport function. In some instances, the correct spelling does not work, but a misspelling does. For example, typing *O'Hare* does not return any results, but *Ohare* does. Typing *LaGuardia* returns no results, but *La Guardia* does. In addition, common misspellings do not produce any results at all, even when the system could provide reasonable guesses about what the user intended. For example, *Newyork* does not produce any search results at all. This issue received a mean severity rating of 3.3, $SD = 0.6$. The participants quickly became frustrated and confused when the site did not return any search results for correct spellings or reasonable misspellings. The search function should always result in a hit when the correct spelling is used, should provide "best guess" search result even

Figure 4. The www.fly.faa.gov results page for a Search by Airport search for Chicago

Airport Lookup Search Results - 'CHICAGO'

City Name Matches
* Chicago Midway Airport, Chicago, IL (MDW)
* Chicago OHare International Airport, Chicago, IL (ORD)

Airport Name Matches
* Chicago Midway Airport, Chicago, IL (MDW)
* Chicago OHare International Airport, Chicago, IL (ORD)

when users make spelling mistakes, and should ignore spacing errors.

Issue 8: Inconsistent Use of Pop-up Windows

The fly.faa.gov Web site is inconsistent in its use of pop-up windows. When users access information using the Search by Airport method or when they click on the color-coded dots on the main site map, the Web site displays the search results in a pop-up window. However, when users access information using the View by Major Airport method, the site displays the same information in the current browser window rather than in a pop-up window. This issue received a mean severity rating of 3.0, $SD = 0.0$.

During the assessment, some participants accidentally closed the browser by clicking the **Close** button when search results appeared in the main browser window. These participants had become accustomed to results appearing in a pop-up window. When search results appeared in the main browser window, they still reacted as if they were in a pop-up window and accidentally closed down the site, along with the browser.

We recommend that the site be more consistent in how it returns search results and Airport Status Information pages. Users become confused when the site responds differently to similar actions.

If the standard convention of the site is to bring up search results in pop-up windows, then the site should bring up all search results in pop-up windows.

DISCUSSION

The level of aviation-related expertise had an impact on many aspects of user performance. Experts were more likely than novices and moderate-level users to have had some prior interaction with the fly.faa.gov Web site. They were also better at finding delay information on the Web site. Experts appeared to have a better conceptual understanding of the different types of airport delays than both novices and moderate users. Finally, experts indicated that they found the information on the Web site to be slightly more comprehensible than both novices and moderate level users. Although we realize that there may be some performance decrement for people who have no affiliation with the FAA, we expect that their performance and their issues should be most similar to our novice users.

On the basis of performance differences, we recommend that the primary goal of site designers should be to make the site more usable for people who do not have an aviation background. If people in the general public visit this site without an aviation-related background, we would expect them to have substantial difficulty (a) understanding which delays were relevant for them, (b) understanding how airport delays differ from airline delays, and (c) interpreting much of the jargon used by aviation experts. Although both experts and novices use the site, simplifying the Web site should help all users, not just novices. Links to additional information can be provided for expert users.

Subjective reports indicated that the participants were generally satisfied with the fly.faa.gov Web site, and objective data revealed that they could successfully complete most tasks using the site. By evaluating user performance data in conjunction with user comments and researcher observations, we were able to identify a number of human factors issues with the Web site that we would not have identified by relying solely on subjective data.

After identifying issues, we rated each one in terms of its impact on site usability, discussed each issue in detail, identified supporting data when appropriate, and provided recommendations for improving the usability of the Web site. Many of the suggested improvements should be easy to implement and should further increase user satisfaction and site usability.

CONCLUSION

One of the primary lessons that we learned from this usability evaluation is that developers should not simply rely on subjective reports of usability when evaluating e-government Web sites. It is just as important to observe users interacting with a Web site and collect objective performance data to better identify usability issues. By having people use the Web site to find different types of information, we were better able to identify those areas of the site that caused problems for users. To encourage organizations to perform usability evaluations on e-government sites, we should

ensure that they provide value by identifying important usability issues that can be remedied through redesign. As we saw in this evaluation, subjective reports often fail to identify these issues. If research on Web site usability fails to identify significant usability issues, it is likely that such evaluations will not be used.

We also found that having researchers rate the severity of usability issues improved our evaluation. Future e-government usability assessments could reap benefits by using this technique. Many times, when a usability assessment is performed, the output of the assessment is a laundry list of issues that usability experts present to site designers. If guidance is given on issue severity or criticality, it is usually ad hoc and is not derived using any formal methodology. By requiring evaluators to explicitly rate each item on frequency, impact, and severity, they are required to think about how and in what ways the problem will affect the user. This user-centric focus is the key element of this methodology. It allows site evaluators to provide designers with a roadmap of how they can best focus their effort to provide a more optimal user experience. Additionally, we recommend that usability assessments use more than one evaluator to make severity ratings. We found that different evaluators might have different priorities, but by using combined severity ratings from three or more evaluators, you can increase the reliability of the ratings (Nielsen, 2003).

By employing an evaluation processes like the one used in this study to evaluate e-government sites, whether they are G2B sites, G2C sites, or G2E sites, designers and system developers can better allocate limited resources during the design process. In general, it is important that e-government Web site designers take into consideration the demographics of those who will use their Web site or application. If an e-government Web site or application, initially targeted to users with a specific area of expertise, is going to be redesigned for use by the general public, the site must be evaluated for usability. Based on the results of such an evaluation, changes need to be made to ensure that the site is usable by the broadest possible audience.

REFERENCES

Ahlstrom, V., & Longo, K. (Eds.). (2003). *Human factors design standard for acquisition of commercial-off-the-shelf subsystems, non-developmental items, and developmental systems (DOT/FAA/CT-03/05/HF-STD-001). Atlantic City International Airport*. NJ: FAA William J. Hughes Technical Center.

Allendoerfer, K., Friedman-Berg, F., & Pai, S. (2007). *Usability assessment of the* fly.faa.gov *Web site* (DOT/FAA/TC-07/10). Atlantic City International Airport, NJ: Federal Aviation Administration William J. Hughes Technical Center.

Bannister, F. (2007). The curse of the benchmark: An assessment of the validity and value of e-government comparisons. *International Review of Administrative Services, 73*, 171–188. doi:10.1177/0020852307077959

Barne, S. J., & Vidgen, R. T. (2006). Data triangulation and Web quality metrics: A case study in e-government source. *Information & Management, 4*, 767–777. doi:10.1016/j.im.2006.06.001

Bertot, J. C., & Jaeger, P. T. (2006). User-centered e-government: Challenges and benefits for government Web sites. *Government Information Quarterly, 23*, 163–168. doi:10.1016/j.giq.2006.02.001

Bush, G. W. (2007, November 15). President Bush discusses aviation congestion. *Office of the Press Secretary* [Press release]. Retrieved February 22, 2008, from http://www.whitehouse.gov/news/releases/2007/11/20071115-6.html

Carter, L., & Bélanger, F. (2005). The utilization of e-government services: Citizen trust, innovation and acceptance factors. *Information Systems Journal, 15*, 5–25. doi:10.1111/j.1365-2575.2005.00183.x

Foley, P. (2005). The real benefits, beneficiaries and value of e-government. *Public Money & Management, 25*, 4–6.

Gil-García, J. R., & Pardo, T. A. (2005). E-government success factors: Mapping practical tools to theoretical foundations. *Government Information Quarterly, 22*, 187–216. doi:10.1016/j.giq.2005.02.001

Heeks, R., & Bailur, S. (2007). Analysing eGovernment research. *Government Information Quarterly, 22*, 243–265. doi:10.1016/j.giq.2006.06.005

Hiller, J., & Belanger, F. (2001). *Privacy strategies for electronic government. E-Government series*. Arlington, VA: Pricewaterhouse Coopers Endowment for the Business of Government.

Ho, A. (2002). Reinventing local government and the e-government initiative. *Public Administration Review, 62*, 434–444. doi:10.1111/0033-3352.00197

Horst, M., Kuttschreutter, M., & Gutteling, J. M. (2007). Perceived usefulness, personal experiences, risk perception, and trust as determinants of adoption of e-government services in the Netherlands. *Computers in Human Behavior, 23*, 1838–1852. doi:10.1016/j.chb.2005.11.003

Howell, D. C. (2007). *Chi-square with ordinal data*. Retrieved January 22, 2007, from

http://www.uvm.edu/~dhowell/StatPages/More_Stuff/OrdinalChisq/OrdinalChiSq.html

Hung, S. Y., Chang, C. M., & Yu, T. J. (2006). Determinants of user acceptance of the e-government services: The case of online tax filing and payment system. *Government Information Quarterly, 23*, 97–122. doi:10.1016/j.giq.2005.11.005

Irani, Z., Love, P. E. D., & Montazemi, A. (2007). E-Government: Past, present, and future. *European Journal of Information Systems, 16*, 103–105. doi:10.1057/palgrave.ejis.3000678

Nielsen, J. (2003). *Severity ratings for usability problems*. Retrieved April 15, 2006, from http://www.useit.com/papers/ heuristic/severityrating.html

Reddick, C. G. (2005). Citizen interaction with e-government: From the streets to servers? *Government Information Quarterly, 22*, 38–57. doi:10.1016/j.giq.2004.10.003

Salem, J. A. (2003). Public and private sector interests in e-government: A look at the DOE's Pub-SCIENCE. *Government Information Quarterly, 20*, 13–27. doi:10.1016/S0740-624X(02)00133-8

Schelin, S. H. (2003). E-government: An overview. In Garson, G. D. (Ed.), *Public information technology: Policy and management issues* (pp. 120–137). Hershey, PA: Idea Group Publishing.

Thomas, J. C., & Streib, G. (2003). The new face of government: Citizen-initiated contacts in the era of e-government. *Journal of Public Administration: Research and Theory, 13*, 83–102. doi:10.1093/jpart/mug010

Yildiz, M. (2007). E-government research: Reviewing the literature, limitations, and ways forward. *Government Information Quarterly, 24*, 646–665. doi:10.1016/j.giq.2007.01.002

This work was previously published in International Journal of Electronic Government Research (IJEGR), edited by Vishanth Weerakkody, pp. 64-79, copyright 2009 by Information Science Reference (an imprint of IGI Global)

Chapter 6
Comparing Citizens' Use of E–Government to Alternative Service Channels

Christopher G. Reddick
The University of Texas at San Antonio, USA

ABSTRACT

This chapter examines the role e-government has over citizens' when they initiate contact with their government. It also compares the influence that other contact channels have on citizens' contacts with government. A public opinion survey is analyzed to determine what factors explain the different methods of contacting government, namely through the phone, e-government, visiting a government office, or a combination of approaches. This chapter also analyzes citizens' preferred method of contacting government, examining different types of information or assistance that citizens' can get from government. The results of this study indicate that e-government is just one of many possible service channels that citizens use, with the phone being the most common. The overall importance of the survey results indicate that e-government is just one contact channel for citizens, and resources should also be devoted towards other contact channels given their importance as well to citizens.

INTRODUCTION AND BACKGROUND

Citizen engagement is thought to be the key to the success of e-government and its development. E-government can only achieve true transformation through citizen engagement (Jones, Hackney, & Irani, 2007). Research indicates that citizens who use the Internet to contact government express higher levels of satisfaction with their contact experience (Cohen, 2006). A satisfactory contact experience by citizens improves their feelings of trust and support for government and its leaders thereby enhancing democracy (Cohen, 2006).

There is a growing body of literature that examines channel choice and e-government use (Reddick, 2005a; Pieterson, 2009). Research is starting to focus on non Internet forms of e-government such as CRM or customer relationship management systems (Schellong, 2008). There is a growing body of research that examines

DOI: 10.4018/978-1-60960-162-1.ch006

citizen-initiated contacts with government, and the approaches that governments use to handle these contacts (Reddick, 2005a; Ong & Wang, 2009).

Compared to the Internet, telephones in many developed and developing countries are a more popular means of accessing information from government. In addition, research shows that a combination of contact channels works best to increase e-government service adoption (Singh & Sahu, 2008). Citizens prefer to use a combination of contact channels; therefore, government should address this by providing multiple contact points (Chen, Huang, & Hsiao, 2006; Ebbers, Pieterson, & Noordman, 2008). This chapter presents an analysis of the most popular contact channels currently used and preferred by citizens to get information and assistance from their government.

On the supply side of e-government, there is a fairly extensive literature that indentifies the importance of e-government to public sector organizations (Coursey & Norris, 2008). However, on the demand side we know less about how and why citizens initiate contact with their government via the Internet (Thomas & Streib, 2003; Thomas & Streib, 2005; Reddick, 2005b). The purpose of this chapter is to examine citizens and their contacts with e-government through different service channels, a relatively understudied area of research.

The next section of this chapter examines the existing literature on citizens' contacts with government through e-government and factors that are predicted to explain their choice of contact channel. This section is followed by a presentation of a survey that examines what citizens look for when they go online, the service channels they use to initiate contact with government, and the preferred method of gathering government information. Towards the end of this chapter, there is a statistical analysis examining factors that explain actual and preferred contact methods. The conclusion section summarizes the most important findings of this research and provides future research opportunities.

LITERATURE ON E-GOVERNMENT AND CITIZENS

The e-government literature that examines citizens and their contact with government can be grouped into two important themes: issues and constraints and socio-economic and demographic factors. These two themes should have an impact on citizen adoption of e-government compared to other contact channels such as the phone. This section examines the literature on each of these themes with reference to its impact on citizens' use of e-government to initiate contacts with their government.

ISSUES AND CONSTRAINTS

There are several issues and constraints mentioned in the e-government literature that should affect the choice of contact channel citizens' choose to use. One measure is the level of satisfaction that the individual has with the ways things are going in their community (Cohen, 2006). User satisfaction is critical for the development of e-government (Verdegem & Verleye, 2009).Citizens that are satisfied with the ways things are going in their community, are more likely to use e-government to contact government.

A second measure of issues and constraints that individuals face is their success at finding the information that they want on the Internet (Carter & Belanger, 2005; Streib & Navarro, 2006; Teo, Srivastava, & Jiang, 2008). According to this research, individuals that are more successful at finding information online would most likely turn to this source. Individuals that have more challenges at finding the information, they may turn to other contact channels. There is research that examines the preferred method of contacting government and its relationship to satisfaction and success (Reddick, 2005a).

A third measure of issues and constraints that individuals face is privacy of personal indefinable information (PII). Individuals that are more concerned about privacy are less likely to want to go online to get information and assistance from government (Carter & Belanger, 2005; Lean, Zailani, Ramayah, & Fernando, 2009). Someone in this state of mind would most likely use an alternative contact channel, such as visiting a government office, where there may be the perception of greater security of PII.

Another measure is if someone expects to get information or services from a government website, this should translate into the individual having a greater likelihood of going online. Therefore, expectations of success should drive choice of contact channels. Research shows that increased expectation of finding government information online influences citizens' use of the Internet (Thomas & Streib, 2005). There is much research that examines e-government technology and its adoption in governments, but there is less research that examines actual user satisfaction with e-government (Verdegem & Verleye, 2009).

There is research that examines the level of trust by citizens and its impact on the use of e-government, which is another predictor of choice of service channel. One aspect of citizen interaction with e-government is the notion that it will have an impact on citizen trust in government (West, 2004; West, 2005). Research shows that e-government will not create more trust in government, but it will enhance trust in those already trusting government (Parent, Vandebeek, & Gemino, 2005).

Another issue and constraint with regards to using e-government compared to other contact channels is if the individual feels overloaded with all of the information that they receive (Goulding, 2001). The prediction is that citizens' that feel overloaded with information are not as likely to go online using e-government. These individuals believe that their ability to find the right information is impaired because they feel overwhelmed with the information they receive.

A final predictor of e-government use compared to other contact channels is control citizens' feel over how technology has influenced their lives (Bovens & Zouridis, 2002). The prediction here is that citizens that feel computers and information technology give them more control over their lives, are more likely to use e-government compared to other contact channels.

Socio-Economic and Demographic

Along with the issues and constraints on individuals when they initiate contact with government, there are several socio-economic and demographic factors that should influence whether someone will use e-government compared to other contact channels. There is existing research that examines socioeconomic factors and their impact on e-government use, with a focus on e-government enhancing trust in government (Welch, Hinnant, & Moon, 2004; Tolbert & Mossberger, 2006).

Research shows that there is a digital divide in terms of access and skills (Dugdale, Daly, Papandrea, & Maley, 2005). The digital divide refer to the difference between the information have and have-nots, which is the gap between computer literate and the computer illiterate. Lack of access to the internet by certain demographic groups is the most common form of digital divide; this is related to age, income, education, and ethnicity. There also is the skills digital divide, which is based on experience with using the Internet (Belanger & Carter, 2009).

In the digital divide, individuals that have lower access to the Internet are not as likely to use e-government compared to alternative contact channels (Helbig, Gil-Garcia, & Ferro, 2009). For instance, African-Americans, Latinos, lower income individuals, and dial up Internet users are less likely to access the Internet because of the digital divide. This should translate into these groups using other contact channels to get information or assistance from government either by the phone of visiting a government office.

Another demographic factor that should influence channel choice is gender. According to existing research, e-government content and service use is closely related to gender (Akman, Yazici, Mishra, & Arifoglu, 2005). Individuals that are younger between the ages of 18-29 should be more likely to use the Internet and e-government. In addition, individuals that are college educated should be more likely to use e-government. Citizens that work for government might have a different attitude of government, and what is can offer, and may be more likely to use e-government (Moon & Welch, 2005). There may also be differences between individuals that live in urban settings, compared to rural and suburban areas of the country. There is research that supports a rural-urban digital divide in Canada, with the odds of using the Internet in an urban area being much greater than a rural area (Noce & McKeown, 2008). These social and demographic factors are what are commonly found in the literature to explain citizen use of e-government and are used to explain service channel choice in this chapter.

Methods

This chapter used data collected from a survey produced for the Pew Internet & American Life Project. The telephone survey consisted of random selection of individuals throughout the United States. Telephone interviews were conducted by Princeton Survey Research Associates International between June 27 and September 4, 2007, among a sample of 2,796 adults 18 years of age and older. The final response rate for the survey was 27%. The methods used in this chapter to analyze the survey results were simple descriptive statistics and frequencies of responses. There was also logistic regression done because this study does a comparison of individuals that used the phone compared to other communication channels. The following section indentifies the sources of information that citizens' search for when they need information of help in a situation. This pro-vides the context of the impact of the Internet on citizens' lives.

Sources of Information and Assistance

Table 1 presents the sources of information or assistance that citizens commonly use. This information gives the impression of the role that the Internet and government play in the lives of individuals when searching for information. The most common source is to ask a professional advisor such as a doctor, lawyer, or financial expert (55.7%). The Internet was the second most commonly used source for citizens to get information or assistance (47.1%). Asking family or friends was the third most commonly used source, according to 41% of respondents. Contacting a government office or agency was done by 33.2% of respondents. Libraries were used as a source for information and help according to only 13.5% of respondents.

Overall the results in Table 1 indicate that the go to source for information and help for citizens is the Internet (Estabrook, Witt, & Rainie, 2007); it was the second most commonly used source among the ones indentified. The Internet is so important, that almost half of the population used this source when they needed information. Government was the fifth most commonly used source of information for citizens, demonstrating the important role that it plays in U.S. society.

Citizen-Initiated Contacts with Government

Table 2 examines how citizens contact a government office or agency. This question looks specifically at the responses in Table 1 from individuals that contacted a government office or agency. Recall that contacting a government office or agency was the fifth most commonly used information or assistant source. The most common method was to call the government office or agency on the phone, according to 39.2%

Table 1. When citizens need information or assistance in a situation they get help from the following sources

	Yes, I used this source %	No, I did not use that source %	Don't know refused %
Ask professional advisors, such as doctors, lawyers or financial experts	55.7	43.7	0.6
Use the internet	47.1	52.4	0.5
Ask friends and family members	41.0	58.5	0.5
Use newspapers, magazines, and books	37.4	62.2	0.4
Contact a government office or agency	33.2	66.5	0.3
Use television and radio	19.3	80.3	0.4
Go to a public library	13.5	86.1	0.3
Use another source not mentioned already	10.8	87.8	1.5
Go to a local place where you can use a computer for free	9.4	90.2	0.3

Notes: Total response is 2796

of respondents. Doing a combination of methods was the second most commonly used approach (27.4%). Visiting a government office or agency was done by 18.1% of respondents. While visiting a government office or agency website was only completed by 8.0% of respondents. Emailing a government office or agency was completed by 2.8% of respondents. Therefore, if you just look at the role of e-government in citizens' contacts, only including those that contacted government by website or email, this represents a mere 10.8% of respondents.

Overall, the results presented in Table 2 show that e-government was not a very common

method for getting information or assistance compared to the other service channels. A similar finding for low e-government use, compared to other service channels, was found in another study that examines different service channels in the Netherlands (Pieterson & Ebbers, 2008). In fact, the most common approach was to call the government office or agency on the phone. Therefore, placing too many resources on Internet development may be problematic for government because of its low usage among citizens compared to other contact channels such as the phone. The focus for government would be better served devoting resources to the phone system, or mak-

Table 2. How citizens contact a government office or agency to find information and help in their situation

	Frequency	Percent
Call on the phone	269	39.2
Do a combination of the above	188	27.4
Visit a office or agency in person	124	18.1
Visit a government website	55	8.0
Other	22	3.2
Send email	19	2.8
Write a letter	9	1.3

Notes: Total response is 686

ing sure that multiple service channels are available for citizens to contact their government (Horrigan, 2004).

Preferred Channel Choice

Table 3 provides an interesting picture of the preferred contact channels that citizens would use if they had to solve a problem or look for information. For instance, when a citizen has a personal tax issue they prefer to use the phone to contact a government office or agency (57.0%). There were 28% of respondents who would prefer to use some other method such as a combination of approaches to solve a personal tax issue. Doing research for school or work, the preferred method to contact government was the Internet, according to 52.3% of respondents. Exploring the kinds of programs that different agencies offer the preferred method is over the Internet (41.8%). The results in Table 3 indicate that for more problem-solving issues the preferred method is over the phone, while conducting information gathering or searchers the preferred method is over the Internet.

Overall, the results in Table 3 show that citizens do prefer to have choice and e-government will not solve all of the issues they need addressed. From this table, it appears that the Internet is the channel that helps citizens the most with general information, but when there is a problem citizens' will most likely turn to the phone or some other

contact channel. The following section examines more closely the issues and constraints and socio-economic and demographic predictor variables that are used to predict choice of actual and preferred contact methods.

Predictor Variables

Table 4 provides information on the various predictor variables used in this research. These variables, as previously mentioned, take into account both the issues and constraints and socio-economic and demographics of citizens and their contact with various service channels. The results in this table show that 65.8% of respondents were satisfied with the ways things were going in their community. There were 43.7% of respondents that were very successful at finding the information they wanted. Privacy was a factor in choosing a contact channel, according to 14.3% of respondents. A majority of citizens expect to get the information or services that they want from the government website (62.5%). Only 12.5% just about always and most of the time trust Washington, D.C. to do the right thing. Only 15.3% of citizens feel overloaded with the information they received from any source. In addition, only 21.6% of respondents believed that computers gave them more control over their lives.

The typical demographic makeup of those surveyed showed that males composed 38.1% of the sample, African Americans composed 15.4%

Table 3. If citizens ever needed to contact government would they prefer to use the phone, Internet, or some other way

	Over the Phone %	On the Internet %	Some other way %
A personal tax issue	57.0	11.7	28.0
Getting a license or permit for your car	12.8	22.6	59.1
Exploring government benefits for yourself or someone else	30.5	33.9	30.1
Doing research for school or work	12.5	52.3	21.1
The kinds of programs different agencies offer	26.1	41.8	24.1
Community issues such as education, crime, and traffic	30.4	24.5	38.2

Notes: Total response is 2796

of the sample, and Latinos were 8.1% of the sample. Less than 10% of those surveyed were young between 18-29 years of age. There was 39.1% of the respondents that had an average household income in 2006 of $40,000 or greater. There were 29% of respondents that were college educated. Only 34.5% of respondents lived in an urban setting. There was a majority of respondents that have low access to the Internet, meaning that they have a dial up connection to the internet and/or limited Internet access. Finally, there were only 9% of respondents that worked for federal, state, or local government. These sixteen predictor variables demonstrated issues and constraints and socio-economic and demographic characteristics that are predicted to influence choice of service channel for both actual and preferred contacts with government.

Logistic Regression Results of Actual Contacts

The following section uses these predictor variables to determine if they have an influence on the choice of citizen contact with their government for information or service requests.

Table 5 uses the subsample of citizens that have contacted government for information or service requests. This regression is composed of 686 respondents out of the total sample size of 2796. Since it is a much smaller sample, caution should be used when reviewing the results. The regressions were run for the various contact channels outlined in Table 2. The possible choices that we did regressions for were calling on the phone, visiting a government office, e-government, and a combination of methods. For the e-government logistic regression, we combined the response of visiting a government website with emailing

Table 4. Predictor variables of factors that explain choice of contact channel

Predictor Variables	Frequency	Percent
Issues and Constraints		
Satisfied with the way things are going in their community	1839	65.8
Very successful at finding the information you wanted	1223	43.7
Privacy was a factor in choosing the information source	399	14.3
Expect to get information or services from the government agency's website	1748	62.5
Just about always and most of the time trust Washington to do the right thing	350	12.5
Feel overloaded with all the information they receive	427	15.3
Computers and technology give people more control over their lives	603	21.6
Socio-Economic and Demographic		
Male	1065	38.1
African American	430	15.4
Latinos	227	8.1
18-29 age range	269	9.6
2006 Household income before taxes greater than $40,000	1094	39.1
College Educated	812	29.0
Urban	965	34.5
Low Access or infrequently use the Internet and have dial up access	1579	56.5
Employed federal, state, or local government	252	9.0

Notes: N=2796

a government office or agency. The purpose of this question is to determine the impact that these sixteen factors have on choice of channel for contacting government.

The results in Table 5 revealed that the only predictors were urban and employed in government, had an influence on whether someone contacted government by phone. With logistic regression the Odds Ratios (OR) can be used to determine the likelihood of a predictor variable explaining the dependent variable. If someone lived in an urban setting they were almost one and a half times (OR=1.41) more likely to use the phone to contact their government. If the individual was employed for the federal, state, or local government they were less likely to use the phone (OR=0.58) to contact government.

For visiting a government office, individuals with over $40,000 household income in 2006 were less likely to visit a government office for information or help with a situation (OR=0.58). An individual with a college education was less likely to visit a government office for information or help with a situation (OR=0.41). These results show that access to resources and education indicates that visiting a government office is not as likely.

Factors that explained the use of e-government show that when individuals expect to get information or services from a government agency's website they were two times more likely to use e-government (OR=2.10). E-government users were almost three times more likely to be young, between the age of 18 and 29 years (OR=2.78). E-government users were also over one and a half times more likely to have a college education (OR=1.60). The results show that e-government users are confident that they will find something useful on government websites; they tend to be younger and well educated.

Of the four modes, the most robust regression was from individuals that used a combination of methods to contact their government for information or help with a situation. Males were slightly more likely to use a combination of methods

(OR=1.34). African Americans were less likely to use a combination of methods (OR=0.56). Individuals that were college educated were more likely to be use a combination of methods (OR=1.65). Those individuals that worked for government were over one and a half times more likely to use a combination of methods (OR=1.55).

Overall, the results in Table 5 indicated that e-government users were different from those that use the phone to contact government. E-government users tended to be younger and college educated, they had higher expectations that government would deliver the information they wanted through their websites. Individuals that used the most common service channel, the phone, were more likely to live in an urban setting, but less likely to be employed by the federal government. What was most interesting is that the combination of contact methods had the greatest predictive power, compared with the other service channel choices, predicting four of the 16 independent variables.

Logistic Regression Results of Preferred Contacts

Table 6 provides information on the issues and constraints and socio-economic and demographic factors that are thought to explain the preference of citizens contacting government for information or assistance. Odds Ratios (OR) are only reported in order to conserve space in this table. Unlike the previous logistic regressions, that examined the actual use of the different service channels, the regression results for citizen preference in Table 6 explained many of the predictor variables. These logistic regressions results examined six common issues: personal tax issue, getting a license or permit for a car, exploring government benefits for yourself or someone else, doing research for school or work, the kinds of programs that different government agencies offer, and community issues such a education, crime, and traffic.

Table 5. Logistic regression examining the different methods of contacting government

Predictor Variables	Phone			Visit			Combination			E-Government		
	Beta	*Prob. Sign.*	*Odds Ratio*	*Beta*	*Prob. Sign.*	*Odds Ratio*	*Beta*	*Prob. Sign.*	*Odds Ratio*	*Beta*	*Prob. Sign.*	*Odds Ratio*
Issues and Constraints												
Satisfied with the way things are going in their community	-0.03	0.86	0.97	-0.18	0.41	0.84	-0.05	0.81	0.96	0.35	0.22	1.42
Very successful at finding the information you wanted	0.16	0.34	1.18	-0.29	0.17	0.75	0.05	0.79	1.05	0.28	0.32	1.32
Privacy was a factor in choosing the information source	-0.28	0.17	0.76	0.23	0.35	1.26	-0.01	0.96	0.99	0.06	0.84	1.06
Expect to get information or services from the government agency's website	0.00	0.99	1.00	-0.18	0.43	0.83	0.09	0.66	1.10	0.74**	0.04	2.10
Just about always and most of the time trust Washington to do the right thing	-0.09	0.74	0.91	0.00	0.99	1.00	0.34	0.30	1.41	0.19	0.68	1.21
Feel overloaded with all the information they receive	0.07	0.78	1.08	0.30	0.35	1.34	0.07	0.83	1.07	-0.80	0.16	0.45
Computers and technology give people more control over their lives	0.21	0.38	1.23	0.04	0.88	1.05	-0.27	0.36	0.77	0.05	0.92	1.05
Socio-Economic and Demographic												
Male	-0.17	0.30	0.84	0.05	0.82	1.05	0.30*	0.10	1.34	-0.27	0.31	0.76
African American	0.30	0.19	1.35	0.35	0.21	1.42	-0.58**	0.04	0.56	-0.69	0.15	0.50
Latinos	-0.02	0.96	0.98	0.28	0.46	1.32	-0.28	0.41	0.75	0.19	0.66	1.21
18-29 age range	0.01	0.98	1.01	-0.23	0.53	0.79	-0.25	0.46	0.78	1.02***	0.01	2.78
2006 Household income before taxes greater than $40,000	0.20	0.28	1.22	-0.54**	0.03	0.58	0.15	0.47	1.16	0.17	0.56	1.18
College Educated	-0.14	0.43	0.87	-0.89***	0.00	0.41	0.50***	0.01	1.65	0.47*	0.09	1.60
Urban	0.35**	0.04	1.41	-0.22	0.33	0.80	-0.03	0.89	0.97	-0.43	0.13	0.65
Low Access or infrequently use the Internet and have dial up access	0.29	0.22	1.33	-0.24	0.43	0.79	-0.38	0.17	0.69	0.00	0.99	1.00
Employed federal, state, or local government	-0.54**	0.04	0.58	-0.07	0.85	0.94	0.44*	0.09	1.55	0.26	0.46	1.29
Constant	-0.73***	0.01	0.48	-0.63*	0.07	0.53	-1.20***	0.00	0.30	-3.23***	0.00	0.04

Notes: ***Significant at the 0.01; ** significant at the 0.05; * significant at the 0.10 levels. N=686

Table 6. Logistic regressions odds ratios of preferred method of contacting government

Predictor Variables	Personal Tax Issue		Getting a license or permit for a car		Exploring government benefits for yourself of some else		Doing Research for school or work		The kinds of programs different government agencies offer		Community Issues such as education, crime, and traffic	
	Phone	Internet	Phone	Internet	Phone	Internet	Phone	Internet	Phone	Internet	Phone	Internet
Issues and Constraints												
Satisfied with the way things are going in their community	1.21**	1.43**	0.97	0.91	1.00	1.27**	1.10	1.06	1.20*	1.07	1.15	1.20
Very successful at finding the information you wanted	1.06	1.15	1.03	1.02	1.05	1.21**	0.86	1.38***	0.91	1.33***	0.90	1.13
Privacy was a factor in choosing the information source	0.82*	1.07	0.96	0.82	1.09	0.86	1.21	1.05	1.05	1.03	0.90	1.23
Expect to get information or services from the government agency's website	1.17*	1.86***	1.04	2.18***	0.73***	3.94***	0.72***	3.11***	0.77***	3.77***	0.97	2.57***
Just about always and most of the time trust Washington to do the right thing	1.10	1.36	1.20	1.16	1.06	0.82	1.34*	0.75**	1.09	1.15	0.87	1.26
Feel overloaded with all the information they receive	1.00	0.58*	1.14	0.95	1.08	0.87	1.38**	0.58***	1.15	0.66***	1.12	0.88
Computers and technology give people more control over their lives	1.06	2.34***	1.11	2.10***	1.17	1.92***	1.03	1.91***	1.22*	1.82***	1.26**	2.41***
Socio-Economic and Demographic												
Male	0.83**	1.19	1.00	1.22*	0.85*	1.18	1.17	0.88	1.11	0.97	1.01	1.01
African American	0.64***	0.87	1.06	0.65***	0.93	0.60***	1.13	0.96	0.80*	0.76***	0.76**	0.78*

continued on following page

Table 6. Continued

Latinos	0.74**	1.33	0.92	1.11	0.82	0.95	1.21	1.19	0.90	1.03	1.04	1.00
18-29 age range	0.98	1.47**	1.28	1.19	1.04	1.40**	0.63*	2.79***	0.76	1.62***	0.81	1.88***
2006 Household income before taxes greater than $40,000	1.43***	1.37***	0.70***	1.83***	0.93	1.90***	0.87	1.82***	0.80**	1.80***	0.99	1.48***
College Educated	0.94	1.13	0.89	1.89***	1.04	1.16	0.57***	1.41***	0.77***	1.45***	0.96	1.12
Urban	1.05	0.99	1.08	1.14	1.05	0.97	0.98	0.95	0.98	1.10	0.99	1.02
Low Access or infrequently use the Internet and have dial up access	0.98	0.22***	1.08	0.18***	1.53***	0.24***	2.13***	0.20***	1.45***	0.21***	1.33**	0.16***
Employed federal, state, or local government	1.06	1.09	1.18	1.38**	0.83	1.31*	0.64	2.20***	0.70**	1.47**	0.87	0.98
Constant	1.08	0.07***	0.14***	0.17***	0.42***	0.20***	0.10***	0.74***	0.35***	0.33***	0.38***	0.19***

Notes: ***Significant at the 0.01; ** significant at the 0.05; * significant at the 0.10 levels; N=2796; odds ratios are just reported to conserve space.

When examining the six issues, there were four predictors that consistently explained the preference for the Internet being used for information and help. The odds ratios indicated that citizens that expect to get information or service requests from a government agency were more likely to use the Internet for various problems they needed help with. Essentially, citizens that had high expectations of what the government provided online gave them confidence that they will be able to get things done when visiting a government website. For instance, individuals that believed they could get information on a government website were almost four times more likely to want to explore government benefits online for themselves or someone else.

The second predictor that explained the six issues citizens' dealt with when initiating contact with government was IT and its control over individual lives. Individuals that believed computers and technology gave them more control over their lives were around two times more likely to prefer to go online and visit a government website for information or assistance. For example, individuals who believed that computers gave them more control were over two times more likely to want to visit government websites for community issues such as education, crime, and traffic.

The third predictor of the six issues that citizens commonly get information or assistant from showed that household income was an important factor. If the individual's 2006 household income before taxes was greater than $40,000, they were more likely to prefer contacting the government for these six issues. For example, higher income individuals were almost two times more likely to prefer to go online to get a license or permit for their car.

A fourth predictor of using the Internet for going online to solve an issue or problem was the level of Internet access. The logistic regressions showed that individuals that infrequently used the Internet and/or had a dial up connection were less likely to go online for these six issues. This is not surprising, but it confirms much of the literature on the digital divide.

Some interesting results surface when comparing citizens' preferences for contacting government using the phone or Internet. First, individuals that expect to get information from a government website are less likely to use the phone to explore government benefits, are less likely to use the phone to do research for school, and are less likely to use the phone to examine different programs that government offers. Individuals are more likely to do all three tasks if they use the Internet. Essentially, individuals that are more comfortable with government websites tend to gravitate towards this contact channel to do research rather than use the phone.

A second finding when comparing preferences for phone and Internet for the six issues was that if someone feels overloaded with information they will use the phone to do research for school, and they are less likely to use the Internet for this purpose. Third, individuals that have a college education are less likely to do research for school over the phone, less likely to learn about programs over the phone, they are more likely to use the Internet for these purposes. Finally, individuals that have low access to the Internet are more likely to prefer the phone for research and learn about community issues.

DISCUSSION

The following section provides a discussion of the most important results of the regression analysis presented in this chapter. For individuals that contacted government through e-government, these individuals were more likely to expect to get information or services from a government website, they tended to be younger and college educated. While citizens that use the phone to contact government are more likely to live in an urban area and are less likely to be employed by government. While using a combination of methods the results indicate that citizens are more likely to be male, college educated, and employed by government, but were less likely to be African American. The results on the contacts with government through different service channels showed that the socio-economic and demographic factors partially explained all four contact channels. For the category of issues and constraints, only the expectation of getting information or services from a government agency's website was a driver of e-government use. Therefore, when citizens expected e-government, they were more likely to use this method of contact.

This chapter also examined the preferred methods of contacting government through either the phone or Internet; examining the six issues. When examining the preference for the Internet for getting information or help, the results showed that when citizens expect to get information or services from a government website they are more likely to use the Internet; individuals that believe computers give them more control over their lives are more likely to use the Internet. There also was the impact of the digital divide on those that preferred to use the Internet to contact government; individuals that have low access to the Internet and lower household income were more likely to use the phone than the Internet. When comparing phones to the internet, individuals conducting more research-oriented activities such as exploring government benefits are more likely to use the Internet than phone if they expect to find information on government websites. The college educated individual is less likely to use the phone for research purposes. Citizens making greater than $40,000 per year are more likely to use the Internet to do research than use the phone. The results show that the task at hand had an influence on service channel used. This research showed differences in Internet and phone preference was dependent upon if it was a research activity, where the internet was preferred choice controlling for other factors.

CONCLUSION AND RECOMMENDATIONS

This chapter examined citizens and their use of e-government to contact government. The purpose of this research was to examine citizens and their contacts with e-government through different service channels. The results of this study indicated that the Internet was the second most commonly used source of information and assistance for citizens; with almost half of respondents using this source. The phone was the most common service channel that citizens use to initiate contact with their government. E-government was almost the last service channel that citizens used. Other popular methods of contact was visiting the government office or agency or doing a combination of methods. The results of this study indicate that governments should be aware of three things about e-government: (1) it is only one possible service channel for citizens among many others; (2) the phone is still the most popular service channel; (3) governments should invest resources in a multiple service channels in order to adequately serve their citizens.

Another major finding of this research is that service channel use is dependent upon task at hand for the citizen. For more information gathering tasks the preferred service channel was the Internet. While for more personal issues, such as tax problems, the preferred method was to contact government over the phone. Therefore, governments should better understand demand for their service channels in order to understand how much development should be placed on each channel.

The results of this study showed that e-government use was driven by the expectation of citizens of what the government agency or department offers online. If there was a greater expectation of e-government there was more demand by citizens for e-government. Government should spend more resources promoting what they currently offer online; this would improve the perceived expectations of what is available and improve demand.

A final result shows that for information gathering the preferred method of contact government was over the Internet. For more interaction the preferred method is to call the government agency over the phone. This supports much of the qualitative literature indicating that for more complicated problems individuals turn to the Internet to initiate contact with their government. A recommendation for government would be to make sure that information placed online be regularly updated and accurate.

There are some notable limitations to this study that should be mention and future research recommendations are provided. One limitation is that the survey data used in this study was taken from a preexisting survey; therefore, some of the factors that we use are rather crude compared to more precise measures if we designed the survey ourselves. The second limitation is that this study explored service channel use and preference in one country, which can be problematic since the results would provide much different findings for developed countries like the U.S. compared to developing countries. Therefore, future research should provide a more custom built survey on citizen-initiated contacts comparing two or more countries exploring the similarities and differences in choice of contact. This research was one of the first to explore the impact of why citizens prefer one contact channel over another and the factors that lead to their choice. However, there is more research that needs to be done on this understudied area of citizens and e-government.

REFERENCES

Akman, I., Yazici, A., Mishra, A., & Arifoglu, A. (2005). E-Government: A Global View and an Empirical Evaluation of some Attributes of Citizens. *Government Information Quarterly*, *22*(2), 239–257. doi:10.1016/j.giq.2004.12.001

Belanger, F., & Carter, L. (2009). The Impact of the Digital Divide on E-Government Use. *Communications of the ACM*, *52*(4), 132–135. doi:10.1145/1498765.1498801

Bovens, M., & Zouridis, S. (2002). From Street-Level to System-Level Bureaucracies: How Information and Communication Technology is Transforming Administrative Discretion and Constitutional Control. *Public Administration Review*, *62*(2), 174–184. doi:10.1111/0033-3352.00168

Carter, L., & Belanger, F. (2005). The Utilization of E-government Services: Citizen Trust, Innovation and Acceptance Factors. *Information Systems Journal*, *15*(1), 5–25. doi:10.1111/j.1365-2575.2005.00183.x

Chen, D.-Y., Huang, T.-Y., & Hsiao, N. (2006). Reinventing Government through On-Line Citizen Involvement in the Developing World: A Case Study of Taipei City Major's E-Mail Box in Taiwan. *Public Administration and Development*, *26*(5), 409–423. doi:10.1002/pad.415

Cohen, J. E. (2006). Citizen Satisfaction with Contacting Government on the Internet. *Information Policy*, *11*(1), 51–65.

Coursey, D., & Norris, D. F. (2008). Models of E-Government: Are They Correct? An Empirical Assessment. *Public Administration Review*, *68*(3), 523–536. doi:10.1111/j.1540-6210.2008.00888.x

Dugdale, A., Daly, A., Papandrea, F., & Maley, M. (2005). Accessing E-Government: Challenges for Citizens and Organizations. *International Review of Administrative Sciences*, *71*(1), 109–118. doi:10.1177/0020852305051687

Ebbers, W., Pieterson, W., & Noordman, H. N. (2008). Electronic Government: Rethinking Channel Management Strategies. *Government Information Quarterly*, *25*(2), 181–201. doi:10.1016/j.giq.2006.11.003

Estabrook, L., Witt, E., & Rainie, L. (2007). *Information Searches that Solve Problems*. Retrieved July 31, 2009 from http://www.pewinternet.org/Reports/2007/Information-Searches-That-Solve-Problems.aspx.

Goulding, A. (2001). Information Poverty or Overload? *Journal of Librarianship and Information Science September, 33*(3), 109-111.

Helbig, N., Gil-Garcia, J. R., & Ferro, E. (2009). Understanding the complexity of electronic government: Implications from the digital divide literature. *Government Information Quarterly*, *26*(1), 89–97. doi:10.1016/j.giq.2008.05.004

Horrigan, J. B. (2004). *How Americans get in Touch with Government*. Retrieved July 31, 2009 from http://www.pewinternet.org/Reports/2004/How-Americans-Get-in-Touch-With-Government.aspx

Jones, S., Hackney, R., & Irani, Z. (2007). Towards E-government Transformation: Conceptualising "Citizen Engage". *Transforming Government: People. Process and Policy*, *1*(2), 145–152.

Lean, O. K., Zailani, S., Ramayah, T., & Fernando, Y. (2009). Factors influencing intention to use e-government services among citizens in Malaysia. *International Journal of Information Management*, *29*(6), 458–475. doi:10.1016/j.ijinfomgt.2009.03.012

Moon, M. J., & Welch, E. W. (2005). Same Bed, Different Dreams? A Comparative Analysis of Citizen and Bureaucratic Perspectives on E-Government. *Review of Public Personnel Administration*, *25*(3), 243–264. doi:10.1177/0734371X05275508

Noce, A. A., & McKeown, L. (2008). A New Benchmark for Internet Use: A Logistic Modeling of Factors Influencing Internet use in Canada, 2005. *Government Information Quarterly*, *25*(3), 462–476. doi:10.1016/j.giq.2007.04.006

Ong, C.-S., & Wang, S.-W. (2009). Managing Citizen-Initiated Email Contacts. *Government Information Quarterly*, *26*(3), 498–504. doi:10.1016/j.giq.2008.07.005

Parent, M., Vandebeek, C. A., & Gemino, A. C. (2005). Building Citizen Trust through E-Government. *Government Information Quarterly*, *22*(4), 720–736. doi:10.1016/j.giq.2005.10.001

Pieterson, W. (2009). *Channel Choice: Citizens' Channel Behavior and Public Service Channel Strategy*. Netherlands: Thesis, University of Twente.

Pieterson, W., & Ebbers, W. (2008). The Use of Service Channels by Citizens in the Netherlands: Implications for Multi-Channel Management. *International Review of Administrative Sciences*, *74*(1), 95–110. doi:10.1177/0020852307085736

Reddick, C. G. (2005a). Citizen-Initiated Contacts with Government: Comparing Phones and Websites. *Journal of E-Government*, *2*(1), 27–51. doi:10.1300/J399v02n01_03

Reddick, C. G. (2005b). Citizen Interaction with E-Government: From the Streets to Servers? *Government Information Quarterly*, *22*(1), 38–57. doi:10.1016/j.giq.2004.10.003

Schellong, A. (2008). *Citizen Relationship Management: A Study of CRM in Government*. Berlin: Peter Lang Publishing Group.

Singh, A., & Sahu, R. (2008). Integrating Internet, Telephones, and Call Centers for Delivering Better Quality E-Governance to all Citizens. *Government Information Quarterly*, *25*(3), 477–490. doi:10.1016/j.giq.2007.01.001

Streib, G., & Navarro, I. (2006). Citizen Demand for Interactive E-Government: The Case of Georgia Consumer Services. *American Review of Public Administration*, *36*(3), 288–300. doi:10.1177/0275074005283371

Teo, T., Srivastava, S. C., & Jiang, L. (2008). Trust and Electronic Government Success: An Empirical Study. *Journal of Management Information Systems*, *25*(3), 99–131. doi:10.2753/MIS0742-1222250303

Thomas, J. C., & Streib, G. (2003). The New Face of Government: Citizen-Initiated Contacts in the Era of E-Government. *Journal of Public Administration: Research and Theory*, *13*(1), 83–102. doi:10.1093/jpart/mug010

Thomas, J. C., & Streib, G. (2005). E-Democracy, E-Commerce, and E-Research: Examining the Electronic Ties between Citizens and Governments. *Administration & Society*, *37*(3), 259–280. doi:10.1177/0095399704273212

Tolbert, C. J., & Mossberger, K. (2006). The Effects of E-Government on Trust and Confidence in Government. *Public Administration Review*, *66*(3), 354–369. doi:10.1111/j.1540-6210.2006.00594.x

Verdegem, P., & Verleye, G. (2009). User-Centered E-Government in Practice: A Comprehensive Model for Measuring User Satisfaction. *Government Information Quarterly*, *26*(3), 487–497. doi:10.1016/j.giq.2009.03.005

Welch, E. W., Hinnant, C. C., & Moon, M. J. (2005). Linking Citizen Satisfaction with E-Government and Trust in Government. *Journal of Public Administration: Research and Theory*, *15*(3), 371–391. doi:10.1093/jopart/mui021

West, D. M. (2004). E-Government and the Transformation of Service Delivery and Citizen Attitudes. *Public Administration Review*, *64*(1), 15–27. doi:10.1111/j.1540-6210.2004.00343.x

West, D. M. (2005). *Digital Government: Technology and Public Sector Performance*. Princeton, NJ: Princeton University Press.

Chapter 7
Determining Types of Services and Targeted Users of Emerging E-Government Strategies:
The Case of Tanzania

Janet Kaaya
University of California-Los Angeles, USA

ABSTRACT

E-government strategies empower citizens through online access to services and information. Consequently, governments – including in developing countries – are implementing e-government. In this study, a survey examined available services and targeted users in Tanzania. Ninety-six government agencies responded: 46% had implemented e-government using websites. Most services (60-90%) relate to disseminating information; online transactions were the least available services. Government-affiliated staff constituted the majority (60-85%) of users. This implies that emerging e-government services mostly address internal needs (government-to-government), and one-way dissemination of information (government-to-citizen). While agencies exhibited a gradual extension to businesses (government-to-business), citizen-to-government and business-to-government relationships were minimal. Finally, the study compares Tanzania's web-presence with select countries, draws its wider implications, and advocates further research on the nature and needs of users.

INTRODUCTION

Many countries worldwide have recognized the importance of implementing e-government strategies as one of the key options for improving the delivery of government services to their citizens. This has been enhanced by developments in information and communication technology (ICT) coupled with tangible economic benefits that are attached to international collaborative networks brought about by various facets of globalization (UN, 2003; Rycroft & Kash, 2004). Implementing e-government strategies has created an avenue for potential citizen empowerment through direct online access to services in which the users submit their inquiries on issues affecting their development. Such an environment has further challenged participating governments to improve their gen-

DOI: 10.4018/978-1-60960-162-1.ch007

eral performance through increased efficiency, accountability and transparency (Abramson & Means, 2001; Andersen, 2004; Garson, 2004; Heeks, 2002; Ho, 2002; Holliday, 2002; Kalu, 2007; La Porte et al., 2002; Michael & Bates, 2005; UN, 2003). One of the observed major shifts associated with e-government is improved relationship between government agencies and users of government information in terms of delivery of and access to government services.

Therefore, various governments—including those from developing countries—are implementing e-government strategies at various levels of complexity, from simple to sophisticated settings. Moreover, most African countries have been motivated to adopt e-government even though they have to deal with such challenges as poor telecommunications infrastructure and a low level of awareness of the potential benefits of ICTs in development, and relatively low literacy levels (Davidrajuh, 2007; Ifinedo, 2007; Heeks, 2002; Mutula, 2002; Panagopoulos, 2004; Singh & Naidoo, 2005; UN, 2003, 2004). Some of the reasons that might have prompted these countries to implement e-government strategies are associated with the noted advantages drawn from the experiences of other regions of the world. Such advantages have been summarized by the UN (2001) as: potential for more user-centered, transparent and efficient services; cost effectiveness of service delivery; improved quality of services; and spreading all the associated benefits of implementing e-government strategies to the national economy at large. Similar or closely related benefits have been reported by Allen et al. (2001), Bertot et al. (2008), Blackstone et al. (2005), Garson (2004), Heeks (2002), La Porte, et al. (2002), Panagopoulos (2004), Riley (2000), Stowers (2004), and Whitson and Davis (2001).

Much as many governments have adopted e-government, the issue has also attracted researchers from various disciplines. They have conducted such studies as evaluating the status of e-government implementation, related services

and users, and the challenges faced by implementing governments or nations (Andersen, 2004; Ho, 2002; Holliday, 2002; Ifinedo, 2007; Lau et al., 2008; Norris, 2005; Reddick, 2004; Salem, 2003; Singh & Naidoo, 2005). For instance, Reddick (2004) takes note of the methods involved in some of e-government research programs (54):

There are essentially two streams of research on e-government growth in public administration. The first stream is the content analysis of government Web sites for specific features of e-government. The second stream of literature is the e-government surveys of local government officials. There are also studies that combine both content and survey methodologies.

In evaluating the status of e-government growth, the researchers also assess the situation of available services and associated users.

Although some of the researchers have conducted their studies on Africa (e.g. Heeks, 2002; Mutula, 2002; Singh & Naidoo, 2005; Weerakkody et al., 2007), the bulk of the literature has reported results from developed countries as a group, or from developing countries of Asia. Africa remains relatively unexplored in empirical studies mainly because many African countries are in their early stages of implementing e-government. The present exploratory study therefore, endeavors to address the gap by determining the types of available and planned e-government services in the region of Africa. Using Tanzania (well-known by the author) as a case study, it also identifies existing and future targeted users, as well as possible models of relationships or interactions between government agencies and users in e-government implementations. The study seeks to answer the following specific research questions: What types of e-government services have been adopted and what is the extent of adoption? What are the user groups for e-government services? Are there indications of users' interaction with their governments? What is the nature of interactions?

The study has addressed these questions through a survey of 96 central government agencies, some of which were already implementing e-government while others were not but intended to do so.

Until 1977, all public telecommunication services in Tanzania were operated by an East African corporation that also covered similar services in two partner East African states, Kenya and Uganda. The services in Tanzania are now operated by several providers and regulated by Tanzania Communication Regulatory Authority (for more information about communications institutions in Tanzania, visit the Ministry of Infrastructure Development homepage at http://www. infrastructure.go.tz/communications.html; also, see Table 1). Tanzania acknowledges potential contributions of ICTs to its social and economic development. Consequently, the nation formulated its national ICT policy of 2003, as a strategy for

attaining that potential (URT-MoCT (2003), with the following broad objectives (p.9):

1. To provide a national framework that will enable ICT to contribute towards achieving national development goals; and
2. To transform Tanzania into a knowledge-based society through the application of ICT.

Also, the government of Tanzania notes the potential of employing ICTs to improve the delivery of government services (Mkonya, 2007). Thus the ICT policy document includes a goal statement concerning e-government implementation (p.19):

Empower the public by building an e-Government platform that facilitates their relationship and interactions with the Government, and enhances the range and delivery of more effective public services at both central and local levels, while

Table 1. Basic characteristics of Tanzania

Size (sq. km[miles])		945,087 [364,900]
Population		34,569,232 (2002 census); 39,384,223 (2007 estimates)
Independence year		1961
Languages	National	Swahili
	Official	Swahili & English
% GDP contribution by sector (2007 est.)	Agriculture	42.8
	Industry	18.4
	Service	38.7
GDP growth rate (2007 est.)		7.3%
GDP per capita (US$; 2007 est.)		1,300
Literacy rate (2002 census)		69.4%
Life expectancy at birth (yrs) [2008 est.]		51.5
Human Development Index (2005 est.)		0.467 [world average:0.743]
E-govt Readiness Index (2005 est.)		0.3020 [world average:0.4267]
Internet users (2007 est.)		400,000
Internet Hosts (2007 est.)		20,757
Teledensity	Fixed lines (2007 est.)	0.58% [236,500]
	Mobile (2008 est.)	23.8% [9,358,000]

Sources: CIA World Factbook (2008), UN (2005), UNDP (2008), ITU (2008), Internet World Stats (2008)

also generating accurate and timely information to better shape policies, strategic plans and tactical decisions for developing and enhancing the delivery of affordable public services.

Tanzania participates in regional and international initiatives aimed to foster ICT infrastructure for development. For instance, in partnership with Kenya and Uganda (plus recent admission of neighboring Rwanda and Burundi) through the East African Community, a development strategy for East Africa was developed that also recognizes the role of information and associated ICTs in facilitating a regional integration process of these countries given their common cultural heritage (Okello, 1999).

It is generally accepted that the creation of a government website is a first step toward implementing full e-government strategies (Layne & Lee 2001; UN, 2001, 2002, 2003, 2004; Yong & Koon, 2005), and the present study adopts Relyea and Nunno's (2000) simple definition of e-government as "online government services, that is any interaction one might have with any government body or agency, using the Internet or World Wide Web" (p. 5). This research is also part of a major study employing logistic regression to examine the forces that have influenced the adoption of e-government strategies in the same study area. The next section provides a brief review of the dimensions of e-government services followed by an overview of possible government-user relationship models in e-government implementation, in order to match them to the study's findings.

E-GOVERNMENT SERVICES

Authors view the types of e-government from various facets such as those that reflect key government functions (e.g. Cook et al., 2002), those that take into consideration the users and the entire e-government-stakeholder-spectrum (e.g. UN, 2003, 2004), or even a broader dimension that incorporates government functions, associated policies,

regulations, and the society (e-society) as a whole (Heeks, 2001; Martin & Byrne, 2004). Cook et al. (2002) present four e-government dimensions: e-services, e-management, e-democracy, and e-commerce; they offer a clear-cut description of each of the dimensions and thus their categorization of e-government services constitutes a major share of the present study.

E-services involve the electronic delivery of government information and other services, in most cases over the Internet. This is the most common type of e-government. Cook et al. (2002) provide examples of e-services via government websites:

... descriptions of government departments and officials, contact information, economic development data, a calendar of events, meeting minutes, the local government law and code book, public safety information, special announcements, tourism information, polling locations, and local historical information. Sites that offer dynamic querying allow citizens to enter key words to search through board meeting agendas and minutes, park and recreation reservation calendars, and real property tax information. (p.6)

E-management entails the use of ICTs as a means of improving the management of government especially the flow of information within government structures. Heeks (2001) refers to this type of e-government as e-administration, while Riley (2000) emphasizes its importance in streamlining government administration. One common application is the use of e-mail for internal communication, but e-management also covers data and electronic record management systems and other tools that support management and delivery of government services, such as budgeting and geographical information system (GIS) The latter is elaborated by Blackstone et al. (2005) in the context of state and local governments in the United States. Cook et al. (2002) underline the importance of e-management even though it is usually not as obvious as the other dimensions of e-government services:

E-management is often slighted because it is mostly invisible to the public. But it is essential to every aspect of e-government. Without it, the services, public engagement, and high-quality, low-cost operations that e-government promises cannot be realized. (p.7)

Drawing on the experiences shown in their study of some localities in New York State, Cook et al. (2002) provide more examples of applications of e-management apart from interdepartmental e-mail communication. These include intranet-based meeting and scheduling systems, budgeting and accounting systems, as well as the geographical information system. Such systems were found to have streamlined internal processes to facilitate various transactions with the citizens, such as employing accounting software to track marriage, hunting, dog, and other license fee records instead of previously-used and laborious manual tracking from voluminous ledger books. In addition, internal meeting reports were distributed electronically to the staff instead of hard copies.

E-democracy constitutes using electronic communication channels, such as e-mail and the Internet, to facilitate the participation of citizens in the decision making processes on matters that affect them and their society as a whole. Apart from e-voting, which is one of the most advanced implementations of e-democracy and e-government, the commonly employed public participation activities are online feedback or opinion polls, online political campaigns and fundraising activities, and accessing various meeting agendas and minutes from government websites. Panagopoulos (2004) considers e-democracy as a special category of e-government initiatives based on five principles for information management:

access, convenience, awareness, communication and involvement in political processes. Such programs aim to transition citizens from passive information access to active participation. (p.119)

E-commerce, as the term implies, involves financial transactions over the Internet for goods and services involving any government agency and other parties. Cook et al. (2002) summarize examples of e-commerce activities and matters related to their study area in the United States:

Citizens paying taxes and utility bills, renewing vehicle registrations, and paying for recreation programs are all examples of e-commerce. Government buying office supplies and auctioning surplus equipment online are also examples of local government e-commerce. (p. 9)

Heeks (2001) categorizes e-government services into three domains that are more or less analogous to Cook et al.'s (2002) dimensions. These include e-administration which involves improvement of the government's administrative services; e-citizens and e-services, which basically involve online participation of citizens in various activities including their interaction with government agencies. The third domain posited by Heeks is e-society which emphasizes the involvement of civil society at large in the interaction with their governments to bring about policy decisions for mutual benefit. Martin and Byrne (2003), citing Caldow (1999) contend that e-government can be viewed as a phenomenon that entails "multiple and overlapping dimensions including the regulatory and policy dimensions, the citizen services and community dimensions and the digital dimension" (p. 14). It appears that this view embraces entire implementation strategies of e-government but it clearly covers dimensions described above from Cook et al. (2002).

These dimensions can also be viewed from the perspective of three major categories relating government-user interactions; these are described in the following section.

GOVERNMENT-USER INTERACTIONS IN E-GOVERNMENT IMPLEMENTATION

Since the present study also seeks to draw corresponding possible models of relationships or interactions between government agencies and users in e-government implementation in the study area, it is necessary to review previous studies on the subject. Thus, e-government implementation plans can be considered to constitute three major categories based on the nature of interactions between government agencies and other participants (Awan, 2007; Evans & Yen, 2005; Evaristo & Kim, 2005; UN, 2003; Lau, et al., 2008; Reddick, 2004; Yuan et al., 2004). These are Government-to-Government (G2G), Government-to-Business (G2B) and vice versa, and Government-to-Citizen/Consumer (G2C) and vice versa. These models are defined in the United Nations' 2003 Global E-Government Survey report as follows (UN 2003, p.10):

- Government-to-Government (G2G) involves sharing data and conducting electronic exchanges between governmental actors. This involves both intra- and inter-agency exchanges at the national level, as well as exchanges among the national, provincial and local levels.
- Government-to-Business (G2B) involves business-specific transactions (e.g. payments with regard to sale and purchase of goods and services) as well as provision on line of business-focussed services.
- Government-to-Consumer/Citizen (G2C) involves initiatives designed to facilitate people's interaction with government as consumers of public services and as citizens. This includes interactions related to the delivery of public services as well as to participation in the consultation and decision-making process.

Evans and Yen (2005) consider these models as categories of government-organized opportunity for web-enabled services to its citizens, and they include a fourth category called 'intra-government internal efficiency and effectiveness' (IEE), while Yong and Leong (2003) add government-to-employee (G2E) category, but these could be argued to fit within the G2G model. Reddick (2004) relates such categories to the stages of e-government development of which he conceptualizes two: cataloguing of information online and the completion of transactions online. He conducted an empirical study to examine how the three key categories apply to US cities in relation to his two-stage e-government growth model. His findings revealed that the G2C relationship corresponds with the first stage and as the e-government develops it relates to G2G category.

Additionally, Reddick noted that the most advanced e-government stages correspond to G2B category, especially online procurement transactions. DeBenedictis et al. (2002) make similar observations in emphasizing the importance of G2G in e-government implementation. The present study attempts to draw such relationships based on the observed extent of services and users in the study area. Regarding the status of e-government implementation in the study region, a content analysis of government websites of Tanzania, along with the related East African nations of Kenya and Uganda, indicated that they correspond to the first and second stages of e-government development taking into account the UN's model. The UN Model includes the following five stages: I: Emerging Presence, II: Enhanced Presence, III: Interactive Presence, IV: Transactional Presence, and V: Networked Presence (Kaaya, 2004; UN 2003, p. 13), The assumption is that the available e-government services with associated information, their delivery and their access tend to correspond with the above-mentioned implementation model. However, this needs to be ascertained through present empirical study since no such study has been reported from the study area.

RESEARCH METHODS

In this study, therefore, a cross sectional survey was conducted in Tanzania in 2004-2005 through a self-administered mail questionnaire to explore the extent of e-government services and corresponding users. One of strengths of this survey is that the answers that respondents gave measured variables effectively and in a standardized manner. For more characteristics of this technique see Babbie (2007).

Study Site

Geographically, Tanzania—formally known as the United Republic of Tanzania—is located in East Africa, bordered by the Indian Ocean on the east; Kenya and Uganda on the north; Burundi, Democratic Republic of Congo, and Rwanda on the west; and by Malawi, Mozambique, and Zambia on the south. Table 1 provides some statistics related to basic characteristics of Tanzania.

Study Sample and Survey Instrument

The survey sample consisted of all known central government agencies that handle most government functions (as opposed to local government agencies) at the level of ministries, semi-autonomous bodies, and embassies abroad. Government directories and ministerial websites and researcher's own personal knowledge of government agencies in the study area were used as sources of information on existing government agencies. The researcher identified a total of 300 government agencies as potential respondents to the questionnaires (259 in Tanzania, and 41 diplomatic missions [DMs] of which 11 were honorary consulates).

In the process two agencies were merged and one ceased to exist; hence, the questionnaires were effectively sent to 298 agencies. As it was specified in the introductory note to the questionnaire, top government officials (who make deci-

sions on behalf of their agencies) completed the questionnaires whether or not their agencies had websites; these included permanent secretaries, heads of consulates abroad, directors of government departments, and heads of semi-autonomous government bodies. To avoid respondent bias, the researcher advised top-government officials to consult with their IT or related staff (e.g. the webmasters) in answering some of the questions.

The questionnaire (copy available from the author), which covered several studies, consisted of four parts but only two parts (I & III) are relevant to this study and thus outlined hereafter. Part I (government agency information) captured general information about responding government agencies, including covering agency's name; location; broad function; staff size; and the existence of a website (important for determining whether or not a government agency was implementing e-government), the website address and its establishment year; or, whether the agency intended to establish a website within six months, one year, two years, or whether they had no intention to establish one. Part III (categorization of e-government services and users as well as their communication channels) is the theme of this paper. The research questions were operationalized by asking the respondents to choose from lists of likely available and planned (future) services, user groups and communication channels (seven-point Likert scale from a list of likely user groups and communication channels as shown in boxes 1, 2 and 3.

As noted in the variable lists (boxes 1-2), each question also had an open-ended space for additional information on services or users ('others, please specify'), as well as specifying types of licenses and registrations.

Pre-testing of the questionnaire was made early 2004 using 15 respondents (5 based in Tanzania and 10 in the United States) to assess and ensure reliability and validity of the questionnaire. This small sample of respondents for pre-testing was sufficient because the questionnaire was adapted from previous related studies (e.g. Ho, 2002), thus no problems were anticipated. After some revisions, the questionnaire underwent necessary Institutional Review Bureau's (IRB) process through the UCLA Office of Protection for Research Subjects, and a certificate of exemption from IBR review was issued end of April 2004. The researcher mailed the questionnaires in May 2004. On receipt of the completed questionnaires the data were cleaned and analyzed in May 2005 using descriptive statistics—which were enough for extrapolating findings and addressing objectives of this exploratory study (via SPSS package Version 13).

FINDINGS

Overview of Government Agencies based on the Survey Data

The returned 96 questionnaires corresponded to 32.2% response rate (32.6% for home agencies and 30% for DMs) and from these, 44 (45.8%) had websites while 52 (54.2%) did not have

websites. Thus a slight majority of government agencies (45.2% home agencies and 50% DMs) did not have websites and hence had not adopted e-government strategies based on the definition of e-government adopted in this study. The sample is somewhat marginal but is acceptable since it involves institutions as units of analysis. To offset the limitation, supplementary information was obtained from content analysis of 83 websites (both respondents and non-respondents) which exhibited relatively similar patterns; as such non-respondent bias is believed to be minimized. In addition, the observed patterns showed that the analyzed survey sample could represent government agencies in the study area. Among government agencies without websites, 25% intended to create websites within 6 months, 23% within one year, 37% within 2 years while15% had no intention, implying that they would not provide e-government services in the foreseeable future. The same figures imply that 85% of government agencies that had not adopted e-government strategies intended to do so in the near future. This means only 8% of the responding agencies

Box 1. Categorization of e-government services

Which of the following services your government agency provides or plans to provide through its website? (Please check appropriate box; Av=available now, Pl=not available now but planned)	Av.	Pl.
To disseminate official reports and speeches		
To describe mission, structure and functions of your government agency		
To announce job opportunities		
To disseminate scientific and technical information		
To announce tender information		
Passport and visa application information		
To provide tax forms and information		
To receive completed tax forms		
Licence application info/forms [Please list here type(s) of licence specific to your agency]:		
Registration info/forms [Please list here type(s) of registration specific to your agency]:		
To educate the general public (e.g. farmers)		
To provide a forum for interaction with the business community		
Online voting		
Others (please specify):		

Box 2. Categorization of e-government targeted users

How important are the following user groups in relation to your agency's provision of government information? Please mark the appropriate box for each user category (7 = top priority user groups, 1 = least priority user groups).	7	6	5	4	3	2	1
Government staff from your agency							
Government staff from other agencies							
Local business community/private sector							
Foreign business community and investors							
Ordinary citizens including farmers							
Various researchers							
Academic community (teachers, lecturers, students)							
Tourists							
Others (please specify):							

did not intend to establish websites within two years. The services and users discussed in these findings constitute both the existing and planned ones (i.e. 92% of the respondents or 29.5% effective response rate).

Available and Anticipated E-Government Services

Table 2 shows a list of e-government services that were available and planned in the study area during the study period and their importance in terms of percentages of the respondents who indicated that they provided those services. Nearly 90% of the responding government agencies indicated that their available e-government services involved the description of their missions, structures and

functions on the agencies' websites, followed by posting official reports (64%), disseminating scientific and technological information (STI), and educating the public (60%).

Function-oriented service trends were also observed for home agencies and diplomatic missions (DMs). For instance, from Table 2, while 8% of home agencies provided visa services, the function corresponds to 100% of DMs, while the figure for disseminating STI by home agencies (63%) is nearly twice that of DMs (33%). The reverse trend was observed for business-related services (32% home agencies and 68% DMs) mainly due DMs' duty of promoting foreign investors. Sophisticated services such as online voting and online tax filing were not available. Likewise, services requiring two-way and interac-

Box 3. Identifying user-government communication channels

To what extent have the current users of your agency's website used the following channels of communication to send inquires about your services? Please mark the appropriate box for each channel (7 = mostly used, 1 = least or not used).	7	6	5	4	3	2	1
Ordinary mail via postal services							
E-mail							
Telephone							
Face-to-face (in person) visits							
Not applicable (no enquiries from users)							
OTHERS (please specify)							

tive communication with users were well below 50% of the respondent agencies, although there was substantial interest in establishing a forum for interaction with the business community (52% home agencies and 75% DMs).

The observed trends for planned services were similar to those of the available services, although one can notice a shift in emphasis toward using the web to educate the public (from available 58% to planned 75%), to announce job opportunities and tender information (from 7% to 53% and 7% to 46%, respectively), and toward interactions with businesses (37% available, 54% planned). The reverse trend for planned services applies to the e-government services not in the agencies' plans (see last column of Table 2).

Targeted Users of E-Government Services and Related Communication Channels

The distribution of user groups of e-government services during the study period is shown in Table 3. On average, on a scale of 7, the top-rated (6.33) user group constitutes government staff within individual government agencies. Further, the corresponding standard deviation is relatively on the low side, indicating a narrow variation among government agencies as far as the importance of this user group is concerned. This is followed by the academic community user group and, again, government staff from other government agencies at average scales of 5.49 and 5.45 respectively. It

Table 2. Available and planned e-government services in Tanzania (% of government agencies)

Types of E-government Services	Available (% of Govt agencies)			Not available but planned (% of govt agencies)			Not available and not planned at all (% of govt agencies)		
	Min	DM	Overall	Min	DM	Overall	Min	DM	Overall
Describe mission, structure and functions of govt agencies	86.8	100.0	88.6	88.0	83.3	87.5	12.0	16.7	12.5
Disseminate official reports and speeches	65.8	50.0	63.6	68.4	66.7	68.2	31.6	33.3	31.8
Disseminate scientific and technical info	63.2	33.3	59.1	81.4	0.0	69.6	18.6	100.0	30.4
To educate the general public	59.5	50.0	58.1	78	55.6	75.0	22.0	44.4	25.0
Interaction with the business community	32.4	66.7	37.2	51.5	75.0	53.9	48.5	25.0	46.1
Registration info/forms	18.4	33.3	20.5	41.9	9.1	37.6	58.1	90.9	62.4
Passport and visa application information	8.1	100.0	20.9	19.7	71.4	24.1	80.3	28.6	75.9
Provide tax forms and information	8.1	16.7	9.3	24.7	27.3	25.0	75.3	72.7	75.0
Licence application info/forms	7.9	16.7	9.1	21.8	9.1	20.2	78.2	90.9	79.8
Announce tender information	5.3	16.7	6.8	50.6	9.1	45.6	49.4	90.9	54.4
Announce job opportunities	7.9	0.0	6.8	59.0	16.7	53.3	41.0	83.3	46.7
Receive / submit tax forms	0.0	0.0	0.0	23.8	0.0	20.9	76.3	100.0	79.1
Online voting	0.0	0.0	0.0	10.0	0.0	8.7	90.0	100.0	91.3

Note: The first primary column represents those agencies that have adopted and are already implementing e-government strategies. As such, this column stands alone. The second (middle) column represents most of the agencies that have not started implementing e-government strategies but intend to do so. It also includes some few agencies that are implementing but have not started to provide certain services even though they intend to do so. The last column reflects the agencies depicted in the second column but the figures show the services not planned at all. The percentage values in the second and the third columns add up to 100%. Min=home agencies; DMs=Diplomatic missions is worth mentioning that most of the academic-based users in the study area are affiliated to public academic institutions and may be categorized as government agencies. On the lower side, tourists and foreign business community scored, on average, 3.30 and 4.45 on a 7-point scale; while in between, there are local business communities and ordinary citizens.

After adjusting the data to generate top-two rated users groups and their aggregated values ranked, the government staff within the responding and other agencies were the top at 85% and 65% respectively. The second top-rated user groups were academic community and researchers (about 60% each), while foreign businesses and tourists were last (see last column, Table 3). However, from Table 3, the standard deviations for the last two user groups are relatively on the high side, implying relatively broad variations in the importance of these user groups among government agencies. To that end, when the data was split according to two major government agency types (i.e. home agencies and diplomatic missions, DMs) the trend for DMs changed in favor of tourists as their key user group (average score of 6.5 on a 7-point scale) as compared to the overall mean of 3.3 (for both agency types) and 2.9 for home agencies (see shaded cells of Table 3). The internal staff as a user group were second, and then the local and foreign businesses and investors.

Concerning communication channels that user groups employed for feedback to government agencies in relation to their use of e-government services through government websites, on average, the users' feedback was received via e-mail (5.82 score on a 7-point scale), followed by telephone (5.34), face-to-face (4.75) and ordinary mail (4.73). These trends are shown in Table 4.

DISCUSSION

This study explores types of available e-government services and associated targeted user groups with their frequently-used communication channels in Tanzania. These are discussed in this section along with how they relate to the various e-government dimensions and interaction models described earlier in this paper.

Table 3. Distribution and importance of key user groups of e-government services in Tanzania (7=most important, 1=least important)

E-govt Service User Groups	N	Mean (overall)	Mean (home agencies)	Mean (diplomatic missions)	Std. Dev.	Top & and 2nd priority scores combined (%)
Government staff from responding agency	96	6.33	6.38	5.67	1.083	85.4
Academic community (teachers, lectures, students)	96	5.49	5.52	4.75	1.436	59.4
Government staff from other agencies	96	5.46	5.40	5.50	1.486	64.6
Various researchers	96	5.36	5.38	4.67	1.616	59.4
Local business/private sector	96	5.25	5.13	5.58	1.536	50
Ordinary citizens incl. farmers	96	5.24	5.24	4.67	1.709	50
Foreign business community & investors	96	4.45	4.24	5.58	2.082	35.4
Tourists	96	3.30	2.87	6.50	2.340	25.1
Valid N (listwise)	96					

Table 4. Importance of communication channels for user feedback relating e-government services in Tanzania (7=most important, 1=least important and least used)

Communication Channel	N	Range	Min	Max	Mean	Std. Dev.	Top & and 2nd priority scores combined (%)
Ordinary mail via postal services	44	6	1	7	4.73	1.796	72.7
E-mail	44	6	1	7	5.82	1.795	54.5
Telephone	44	6	1	7	5.34	1.697	34.1
Face-to-face (in person) visits	44	6	1	7	4.75	1.740	31.1
Others - 1. Fax	1	0	7	7	7.00		4.2
Others - 2. In workshops (gatherings)	4	0	7	7	7.00		1.0

Note: N=44=government agencies with websites.

Types of Services and Targeted Users

The top-rated available services observed from the study's findings—describing missions, structures and functions of government agencies; posting downloadable official reports; disseminating scientific and technological information (STI); and educating the public—are closely related, as they all imply a one-way-communication delivery of e-government services through these agencies' websites. These types of services have also been associated with emerging e-government strategies (Layne & Lee, 2001; Silcock, 2001; Stowers, 2004; 2001; UN 2002, 2003, 2005), characterized by relatively simple websites as forums for publishing government information for public view or, according to Lau et al (2008), in their analysis of e-government adoption in Latin America, 'information dissemination stage' of e-government. The information may be contained in downloadable form that can be filled out offline. The survey findings also reveal function-oriented e-government services and targeted users that were especially evident in diplomatic missions, reflecting their key roles of providing services related to tourism and business investments from host countries.

On the other hand, the study's findings show 'top-rated' e-government services that were not in the plans of the responding government agencies, at least in the foreseeable future, for implementation. Such services, including online voting and tax filing, require sophisticated portal designs of government websites for effective implementation. Moreover, respondents indicated a great deal of interest in portal sites, online voting and tax filing though these were mixed with other functions and services. The extent of future implementation strategies relating to online voting and tax filing are difficult to gauge from these survey findings because taxation and election agencies did not return questionnaires. Nevertheless, as for other agencies, it was possible to judge their levels of available e-government services by assessing their respective websites. The tax administration agency's website (http://www.tra.go.tz) does provide some 112 downloadable tax forms although there is no evidence of direct online filing. However, one set of these forms concerns electronic transfer of revenue (i.e. tax returns) from taxpayer's financial institution to the central bank. This might be a precursor to online tax filing. On the other hand, the election coordinating agency created its website during the course of this study as the country was preparing for general elections in December 2005. The website (see http://www.nec.go.tz/homepage.asp) provides general information about that country's electoral process and results, including answers to FAQs on a new permanent national voter's register, as well as relevant contact information, all meant to increase voter participation.

Overall, the available e-government services entail one-way communication from government agencies to users. Also, the level of e-government implementation in the study area was oriented toward serving internal government structures. But this situation may change as technical skills diffuse within these government agencies, and as agencies that have not yet provided services online begin to recognize the potential benefits and imitate more effective agencies (Abramson & Means, 2001). There are many advantages, though, of using information and communication technologies (ICTs) within and between government agencies. Blackstone, Bognanno and Hakim (2005) contend that "[g]overnments can use the Internet to improve internal operations both by improving communication and collaboration among government agencies and by raising the accountability of government employees through the use of new technologies…" (p.4). Thus, improvement of internal government operations as part of e-government implementation strategies creates a good groundwork for improved online interactions with citizens, including the business community.

It is worth mentioning that this study did not investigate individual users per se (the information on user groups was provided by government agencies that returned survey questionnaires); but increasing Internet penetration may lead to increased demands from the citizenry for more interactive e-government implementation strategies. Even with low level of Internet penetration in Tanzania (see Table 1), demands from citizens via various web-blogs and other media necessitate that the government create a fairly interactive website (for instance, see http://www.wananchi.go.tz/),) as a forum for all citizens to present their varieties of grievances to be addressed by top government officials through the same website. From its homepage, the aims of creating it are

to "communicate with the government via the Web; send feedback, opinions, [and] complaints to the government; [and] to track and follow-up on queries sent to the government via this portal". It thus contains the following links: learn, submit, track, appeal, news, contacts and survey; and it is presented in both English and Swahili. In connection with this development and coming back to the Internet penetration data in Table 1, it is important to view the statistics with caution since the majority of Internet users in developing countries depend on public access points such as Internet cafes, government offices, restaurants, public libraries, schools, and so on, rather than on personal ownerships of computers and other ICTs (as are common trends in developed countries).

There is also a great potential for accessing the Internet via cellphones because of their high rate of penetration in Africa and all over the world. For example, recent figures for cellphone penetration in Tanzania (Table 1) indicate that the cellphone density has increased from 4% in 2004 (1,640,000 subscribers) to 24% (9,358,000 subscribers) in 2008 (CIA, 2005, 2008). This potential has been observed from OECD e-government studies. According to OECD (2005),

[I]n other countries the rapid and widespread adoption of mobile phones, WiFi and similar wireless technologies is causing governments to start looking at the role that these platforms might play in delivery of mobile e-government services (so called "m-government") or ubiquitous government ("u-government"). How such levels of access are viewed by the public is culturally circumscribed; for many, universal access may seem liberating, while others may worry about opportunities for government control. (p.35)

Exploring potentials of m-government in the study area and related regions warrants consideration by e-government and other researchers.

Despite the relative inexperience of many Tanzanian government agencies with providing online services, e-mail is emerging as an important communication channel among the government employees, researchers and academics that constitute most of the user base. The cultural implications of the shift to e-mail from traditional channels like face-o-face and surface mail is one of the possible directions for future study; for example, in a recent survey of individual user behaviour of e-government services in Australia—one of the top ten countries in the world in terms of e-government presence—it was revealed that "the most common channel is in person (52%) followed by the telephone (26%) and Internet (15%). The mail is the least popular channel used to contact government" (Australian Government Information Management Office, 2005, p.8). The diffusion of e-mail in Tanzania and other developing countries is more pronounced among business communities. For instance, Kenny (2006) notes that "58 percent of businesses in Tanzania used e-mail for interacting with clients and suppliers" (p. 63). Obviously, e-commerce implementation strategies preceded e-government; however, this study's findings reveal some potential for government agencies to reach out to business communities and other citizens in enhancing provision of and access to government information and services.

Government-User Interactions vs. E-Government Services

Having noted the types of e-government services and targeted users with communication channels used, it is possible to relate them to various e-government dimensions and interaction models discussed before. If we reflect on the dimensions posed by Cook et al. (2002) or Heeks (2001), the findings presented here indicate that most of the e-government activities in the country during the study period can be characterized as e-services and e-management (or e-administration). E-commerce and even farther e-democracy do not yet play more than a minimal role.

To date, most of the available and planned sites have focused on the delivery of government information (i.e. describing missions, structures and functions; disseminating official reports and speeches; and disseminating scientific and technical information; tourism; and other function-oriented services). The official government website (see http://www.tanzania.go.tz/), for instance, contains the texts of speeches, tax information, a government directory, training and scholarship announcements, a population census and its database, and several downloadable policy documents. These types of information clearly constitute e-services in the sense discussed by Cook et al. (2002) or Heeks (2001). In terms of government-user interaction models, the observed e-services correspond to G2C relationship.

E-management displayed an importance more or less in line with e-services. Although examining specific e-management strategies was beyond the scope of this study, the findings on targeted users depict the importance of this dimension. Available services were geared toward the staff of government agencies as a vital user group since, as noted above, improving flow of information within the government structures is crucial for successful provision of other e-government services to the citizens (DeBenedictis et al., 2002; Relay, 2000; Stahl, 2005). As also noted from the results, e-mail has emerged as an important communication channel for government-user interactions. This is especially related to e-management as a means of speeding up intra-, inter-, and extra-departmental communication in order to better serve the other users at large. Another important aspect of e-management that emerged in the present study is government budgeting and accounting systems. Such networked systems help to integrate the fiscal control infrastructure among new and existing agencies and departments (Michael, 2005). We can thus associate this dimension with the G2G model.

Even though e-commerce and e-democracy dimensions did not feature importantly in the results, they seem to gain significance gradually and may be developed in the future. As noted before, 67% of the diplomatic missions indicated that their then current e-government services targeted businesses and the overall importance of this user group for both government agency types is expected to increase from 37% to 54% within two years. There are also signs of interactive features and downloadable materials (submitted offline) for the general public; thus a G2B relationship is emerging.

Apart from online voting, which was included in the questionnaire and for which the results show very low e-government involvement, other potential e-democracy features can be discerned on the websites of government agencies. These include online feedback forms with 'clickable' provisions for submission of opinions or suggestions relating to services used (in other words, a less stringent definition of e-democracy would encompass an interactive feedback between citizens and government agencies such as one that was created recently in Tanzania). However, since the study did not include the users as respondents we are not sure if the available feedback forms have been effective. Nevertheless, the implication here is that C2G and B2G interactions are minimal since the study's results have shown one-way communication features between government agencies and users.

Noticeably, G2C and G2G models are dominant in the study area. These findings support previous observations made by Reddick (2004), who noted that the G2C level of interaction relates to initial stages of e-government implementation and that increased sophistication leads to G2G, while even more complex e-government relates to G2B interaction. DeBenedictis et al. (2002) draw similar deductions although they play down the importance of the G2C relationship owing to its attachment to the lowest level of e-government development and emphasize the importance of G2G. This is echoed by Riley (2000: xiii), who notes:

New technologies are being used not only to deliver services to the public but to enhance government administration and facilitate businesses.

As seen before, the level of e-government services in the study area (predominantly one-way communication between government agencies and users) as well as user groups (predominantly government employees) reflect the extent of e-government development at the time of the survey and this can also be assessed and reflected in various ways. For example, the United Nations ranks its member countries based on e-government readiness and e-participation indices (UN, 2005; 2008), among other measures. E-government readiness evaluates how ready the governments worldwide are "employing the opportunities offered by ICT to improve the access to, and the use of, ICTs in providing basic social services" (UN, 2005, p. xi), and uses government websites, telecommunication infrastructure and human resource capacity as its key indicators.

E-participation Index, on the other hand, "assesses the quality and usefulness of information and services provided by a country for the purpose of engaging its citizens in public policy making through the use of e-government programs" (UN, 2005, p. 35). The values for both indices range from 0 to 1 (for detailed indicators and the methodologies for deriving the indices see UN, 2005 & UN, 2008). While the global average for e-government readiness is 0.4514 (it was 0.4267 in 2005 when this study was carried out), with the top-ranked Sweden at 0.9157 while the United States was top-ranked at 0.962 in 2005 and now it scores at 0.8644. These trends in the index values over time are subject to future studies (and further 2005 figures below relative to current values are presented in brackets). The value for Tanzania is 0.2929 [0.3020] and the other two East African countries, Kenya (0.3474 [0.3298]) and Uganda (0.3133 [0.3081]), compared to such developing countries in other regions as Chile (0.5819 [0.6963]), Colombia

(0.5317 [0.5221]), India (0.3814 [0.4001]) and Pakistan (0.3160 [0.2836]). Top-ranked African countries in terms of e-government readiness are South Africa (0.5115 [0.5075]), Mauritius (0.5086 [0.5317]) and Seychelles (0.4942 [0.4884]). Africa generally lags behind other regions in terms of e-government presence since the latter correlates economic development.

In the case of e-participation, the United States is top-ranked in 2008 at 1.000 [0.9048] (in 2005 the United Kingdom was at the top at 1.000 but now it ranks 25th at 0.4318 due its shift from national to local governments that are not included in the UN surveys) followed by the Republic of Korea at 0.9773 [0.8730] and Denmark at 0.9318 [0.7619], compared to Tanzania at 0.0227 [0.0317], and Kenya (0.0455 [0.0317]) and Uganda (0.0909 [0.0476]). Mozambique remains top-ranked in Africa since 2005 (0.4318 [0.3333]). In other regions, Colombia stands at 0.4318 [0.5873], Chile (0.1818 [0.5873]), India (0.2500 [0.1587]) and Pakistan (0.0909 [0.1270]). At the time of this study, the United Nation's E-Government Report (2005) provided the extent of e-government service delivery by the UN's member countries in terms of percentage utilization of the stages of e-government growth. The report shows that Tanzania had utilized 100% of Stage I and 26% of Stage II compared with Kenya (75% Stage I, 29% Stage II), Uganda (50% Stage I, 31 Stage II), Mauritius (100% Stage I, 80% Stage II and the top in Africa), Chile (100% Stage I, 93% Stage II), India (100% Stage I, 77% Stage II), Pakistan (100% Stage I, 62% Stage II), U.S. (100% Stage I, 99% Stage II). The UN's report for 2008 shows % utilization for Tanzania at 8% (Stage I), 18% (II), 41.5 (III), 0 (IV & V), totalling 67.5%. It seems different assessment approaches were used but these percentages imply some shift to advanced stages of service delivery even though it also conflicts with the above trends of e-government readiness and e-participation indices for Tanzania in 2005 and 2008.

All in all, these figures reflect this study's findings that most of e-government services and corresponding interactions with users in Tanzania involve mostly one-way communication between government agencies through dissemination of government information on the websites (Stage I: G2G, G2C, G2B relationships) with the minimum user interaction through e-mail and a few downloadable materials to be submitted offline (Stage II: G2G, G2C, G2B, C2G, B2G relationships). Likewise, Karim and Khalid (2003), in their delineation of e-government implementation blue print in Malaysia, categorize e-government services into transactional and informational; these signify all stages of e-government implementation. However, most African countries are still at the emerging stages of e-government implementation [see, for example, studies by Ifinedo (2007) in Nigeria and Weerakkody et al. (2007) in Zambia].

CONCLUSION AND FUTURE RESEARCH

The present study is one of very few empirical investigations that have endeavored to explore the extent of e-government services and corresponding users in the East African nation of Tanzania. The survey findings have revealed that the services that have been effectively implemented are those that involve one-way communication, disseminating government information agency websites. The findings have also shown that services that require sophisticated portal designs for two-way interaction with users, such as online submission of completed forms have not been established.

The government staff constitutes the main user group of the available and planned e-government services in the study area, followed by the general public and the business community. Considering the dimensions of e-government services, we can conclude that e-services and e-management predominate in the study area. These correspond with G2C and G2G categories of e-government interactions. From the findings, e-commerce and e-democracy dimensions have not been established but there are indicators of their gradual increase in importance as the government agencies gain experiences in implementing e-government strategies. These correspond to G2B interactions as well as C2G and B2G models. These findings also mirror the United Nation's assessment of the e-government presence in terms of e-government readiness and e-participation, as well as the effective utilization of Stage I of e-government and partial utilization of Stage II.

One of the limitations of the present study is that it only surveyed top officials to get the insights of the current and future e-government services and associated users. The results are thus based on the perspectives of government officials. This is a good start for exploring the extent of the adoption of e-government strategies in the study area. However, there is a need to go further by examining the services from users' perspectives and exploring the extent of their awareness of and access to e-government services—hence some degree of empowerment. The survey variables could take the model of the European Union's e-government surveys (eEurope 2004) but there is a need to supplement survey data with qualitative data.

In addition, there is a need to go beyond identifying the dimensions of e-government services and go deeper into the specific activities and mechanisms involved in effecting such services, as well as the role of communication channels. For instance, what is the level of back-office integration (back-end government systems) in the provision of government-to-government services? E-mail is an important means of communication for Tanzanian government employees: to what extent do they perceive their messages as government records that must be preserved? A qualitative approach will also assist researchers when they explore the reasons why about 15% of government agencies indicated that they do not intend to adopt e-government in the foreseeable

future; a qualitative or ethnographic approach might be most useful for addressing this question. One other policy implication relating to present findings is that those services that the government agencies can best deliver be strengthened, namely, e-services and e-management (once these are improved, the provision of government services through other media will also be more efficient even without ubiquitous Internet access) and solutions be recommended for establishing those that are not effectively delivered but have potential beneficial impact, such as e-commerce and e-democracy.

The present study bears a wider implication to other African countries and related regions since they share many common characteristics and are in relatively early stages of implementing e-government strategies. The research could also be replicated in these regions and repeated in the same study area for comparative purposes and for relating the findings to the United Nations' and other surveys.

ACKNOWLEDGMENT

The author wishes to thank her former members of dissertation committee under Prof. Leah Lievrouw for their guidance and support during her graduate studies at UCLA that led to successful implementation of this study; to Dr. Tunu Ramadhani for her advice on statistical analysis techniques; and to Dr. J.M. Haki for his encouragement during the data collection process. Also, many thanks and appreciation to Sara McGah and Ken Roehrs for their invaluable editorial help, and to anonymous reviewers and IJEGR editors for their insightful comments and constructive criticisms that greatly helped to improve this paper. Initial draft of this paper was presented at the International Conference on E-Government and the participants' comments are gratefully acknowledged.

REFERENCES

Abramson, M. A., & Means, G. E. (2001). *E-Government 2001*. New York: Rowman & Littlefield Publishers, Inc.

Andersen, K. V. (2004). *E-government and public sector process rebuilding: Dilettantes, wheel barrows, and diamonds*. Boston: Kluwer Academic Publishers.

Australian Government Information Management Office (2005). *Australian's use of and satisfaction with e-government services*. Department of Finance and Administration, Australian Government Information Management Office. Barton: Commonwealth of Australia

Awan, M. A. (2007). Dubai e-Government: An Evaluation of G2B Websites. *Journal of Internet Commerce, 6*(3), 115–129. doi:10.1300/J179v06n03_06

Babbie, E. R. (2007). *The practice of social research* (11th ed.). Belmont, CA: Thomson Wadsworth.

Bertot, J. G., Jaeger, P. T., & McClure, C. R. (2008). Citizen-centered e-government services: benefits, costs, and research needs. *ACM International Conference Proceeding Series, 289,* 137-142 (Proceedings of the 2008 International Conference on Digital Government Research).

Blackstone, E., Bognanno, M., & Hakim, S. (2005). Electronic government: Review, evaluation, and anticipated impact. In E. Blackstone, M. Bognanno & S. Hakim (Eds.), *Innovations in E-Government: The Thoughts of Governors and Mayors*. New York, Boulder, Oxford, Lanham: Rowman & Littlefield Publishers, Inc.

CIA. (2005). *The world factbook 2005*. Retrieved November 3, 2005, from Central Intelligence Agency's Website https://www.cia.gov/cia/publication s/factbook/index.html

CIA. (2008). *The 2008 world factbook*. Retrieved September 29, 2008, from Central Intelligence Agency's Website https://www.cia.gov/library/publications/ the-world-factbook/index.html

Cook, M. E., La Vigne, M. F., Pagano, C. M., Dawes, S. S., & Pardo, T. A. (2002). *Making a case for local E-government*. Retrieved September 29, 2008, from Center for Technology in Government, University at Albany, SUNY Website http://www.ctg.albany.edu/publications/ guides/ making_a_case/making_a_case.pdf

Dadidrajuh, R. (2007). Towards measuring true e-readiness of a Third-World country: A case study on Sri Lanka. In Al-Hakim, L. (Ed.), *Global E-Government: Theory, Applications and Benchmarking. (* (pp. 185–199). London: Idea Group.

DeBenedictis, A., Howell, W., Figueroa, R., & Boggs, R. A. (2002). *E-government defined: An overview of the next big information technology challenge*. Retrieved September 29, 2008, from http://www.zeang.com/RobertFig/egov.pdf

eEurope (2004) *Top of the web: User satisfaction and usage survey of eGovernment services*. Retrieved October 25, 2005, from http://www. europa.eu.int/egovernment_research (also available at http://www.cisco.at/pdfs/publicse ctor/ egov_service-survey_02-05.pdf accessed on July 30, 2008)

Evans, D., & Yen, D. C. (2005). E-government: An analysis for implementation: Framework for understanding cultural and social impact. *Government Information Quarterly, 22*(3), 354–373. doi:10.1016/j.giq.2005.05.007

Evaristo, R., & Kim, B. (2005). A strategic framework for a G2G e-government excellence center. In Huang, W., Siau, K., & Wei, K. K. (Eds.), *Electronic government strategies and implementation* (pp. 68–83). London: Idea Group.

Garson, G. D. (2004). The promise of digital government. In Pavlichev, A., & Garson, G. D. (Eds.), *Digital government: Principles and best practices* (pp. 2–15). London: Idea Group Publishing.

Heeks, R. (Ed.). (2001). *Reinventing government in the information age: International practice in IT-enabled public sector reform.* London: Routledge.

Heeks, R. (2002). E-Government in Africa: promise and practice. *Information Polity, 7*(2, 3), 97-114.

Ho, A. T. (2002). Reinventing local governments and the e-government initiative. *Public Administration Review, 62,* 434–444. doi:10.1111/0033-3352.00197

Holliday, I. (2002). Building e-government in East and Southeast Asia: Regional rhetoric and national (in)action. *Public Administration and Development, 22,* 323–335. doi:10.1002/pad.239

Ifinedo, P. (2007). Moving towards e-government in a developing society: Glimpses of the problems, progress, and prospects in Nigeria. In Al-Hakim, L. (Ed.), *Global E-Government.*

International Telecommunication Union. (2008). ITU/ICT Statistics. Retrieved September 29, 2008 from ITU website http://www.itu.int/ITU-D/ict/statistics/

Internet World Stats. (2008): *Usage and population statistics.* Retrieved September 29, 2008, from http://www.internetworldstats.com/

Kaaya, J. (2004). Implementing e-government services in East Africa: Assessing status through content analysis of government websites. *Electronic Journal of E-Government, 2(1),* 39-54. Retrieved September 29, 2008, from EJEG Website http://www.ejeg.com/volume-2/volume2-issue-1/v2-i1-art5-kaaya.pdf

Kalu, K. N. (2007). Capacity building and IT diffusion: A comparative assessment of e-government environment in Africa. *Social Science Computer Review, 25*(3), 358–371. doi:10.1177/0894439307296917

Karim, M. R. A., & Khalid, N. M. (2003). *E-government in Malaysia: Improving responsiveness and capacity to serve. Selangor D.E.* Malaysia: Pelanduk Publications.

Kenny (2006). *Overselling the web? Development and the Internet.* London: Lynne Rienner Publishers, Inc.

La Porte, T. M., Demchak, C. C., & de Jong, M. (2002). Democracy and bureaucracy in the age of the web. *Administration & Society, 34,* 411–446. doi:10.1177/0095399702034004004

Lau, T. Y., Aboulhoson, M., Lin, C., & Atkin, D. J. (2008). Adoption of e-government in three Latin American countries: Argentina, Brazil and Mexico. *Telecommunications Policy, 32*(2), 88–100. doi:10.1016/j.telpol.2007.07.007

Layne, K., & Lee, K. (2001). Developing fully functional E-government: a four stage model. *Government Information Quarterly, 18,* 122–136. doi:10.1016/S0740-624X(01)00066-1

Martin, B., & Byrne, J. (2003). Implementing e-government: Widening the lens. *Electronic Journal of E-Government, 1*(1), 11–22.

Michael, B., & Bates, M. (2005). Implementing and assessing transparency in digital government: Some issues in project management. In Huang, W., Siau, K., & Wei, K. K. (Eds.), *Electronic Government Strategies and Implementation* (pp. 20–43). London: Idea Group.

Michael, G. (2005, August 6). Gov[ernmen]t uses US software to control use of revenue. *The Guardian* (Tanzania). Retrieved August 6, 2005, from http://www.ippmedia.com

Mkonya, J. (2007, March 21). Deputy PS roots for e-government agenda. *The Guardian* (Tanzania). Retrieved March 21, 2007, from http://www.ippmedia.com

Mutula, S. M. (2002). Africa's web content: Current status. *Malaysian Journal of Library & Information Science, 7*(2), 35–55.

Norris, D. F., & Moon, M. J. (2005). Advancing e-government at the grassroots: Tortoise or hare? *Public Administration Review, 65*(1), 64–75. doi:10.1111/j.1540-6210.2005.00431.x

OECD. (2007). *e-Government for better government. OECD e-Government Studies.* Paris: OECD Publishing.

Okello, D. (1999). *Towards sustainable regional integration in East Africa: Voices and visions.* Konrad Adenauer Stiftung Occasional Papers: East Africa, 1/1999. Nairobi: Konrad Adenauer Foundation.

Panagopoulos, C. (2004). Consequences of the cyberstate: The political implications of digital government in international context. In Pavlichev, A., & Garson, G. D. (Eds.), *Digital government: Principles and best practices* (pp. 116–132). London: Idea Group Publishing.

Reddick, C. G. (2004). A two-stage model of e-government growth: Theories and empirical evidence for U.S. cities. *Government Information Quarterly, 21*(1), 51–64. doi:10.1016/j.giq.2003.11.004

Relyea, H. C., & Nunno, R. M. (2000). *Electronic government and electronic signatures.* Huntington, NY: Novinka Books.

Riley, T. B. (2000). *Electronic Governance and Electronic Democracy: Living and Working in the Wired World.* London: Commonwealth Secretariat.

Rycroft, R. W., & Kash, D. E. (2004). Self-organizing innovation networks: implications for globalization. *Technovation, 24,* 187–197. doi:10.1016/S0166-4972(03)00092-0

Salem, J. A. Jr. (2003). Public and private sector interests in e-government: a look at the DOE's Pub-SCIENCE. *Government Information Quarterly, 20,* 13–27. doi:10.1016/S0740-624X(02)00133-8

Silcock, R. (2001). What is e-government? *Parliamentary Affairs, 54,* 88–101. doi:10.1093/pa/54.1.88

Singh, S., & Naidoo, G. (2005). Towards an e-government solution: A South African perspective. In Huang, W., Siau, K., & Wei, K. K. (Eds.), *Electronic Government Strategies and Implementation* (pp. 325–353). London: Idea Group.

Stahl, B. C. (2005). The paradigm of e-commerce in e-government and e-democracy. In Huang, W., Siau, K., & Wei, K. K. (Eds.), *Electronic Government Strategies and Implementation* (pp. 1–19). London: Idea Group.

Stowers, G. N. L. (2004). Issues in e-commerce and e-government service delivery. In Pavlichev, A., & Garson, G. D. (Eds.), *Digital government: Principles and best practices* (pp. 169–185). London: Idea Group Publishing.

Theory, Applications and Benchmarking. (pp 143-166. London: Idea Group.

UNDP. (2008). *Human Development Report 2007/8—Fighting climate change: Human solidarity in a divided World.* Retrieved September 29, 2008, from the United Nations Development Program's Website http://www2.unpan.org/egovkb/ global_reports/08report.htm

United Nations. 2001. *E-Commerce and Development Report 2001.* Prepared by United Nations Conference on Trade and Development. New York & Geneva: United Nations. [online edition available at http://www.unctad.org/en/docs/ecdr01ove.en.pdf]

United Nations. (2002). *Benchmarking e-government: A global perspective—Assessing the progress of the UN member states.* 81pp. Retrieved September 29, 2008, from the United Nations, Division for Public Economics and Public Administration & American Society for Public Administration Website http://unpan1.un.org/intradoc/groups/ public/documents/un/unpan021547.pdf

United Nations. (2003). *UN global e-government survey 2003.* Retrieved September 29, 2008, from the United Nations, Department of Economic and Social Affairs, Division for Public Administration and Development Management Website:http://www2.unpan.org/egovkb/ global_reports/03survey.htm

United Nations. (2004). *Global e-government readiness report 2004: Towards access for opportunity.* Retrieved September 29, 2008, from the United Nations, Department of Economic and Social Affairs, Division for Public Administration and Development Management Website: http://www2.unpan.org/egov kb/global_reports/04report.htm

United Nations. (2005). *Global e-government readiness report 2005: From e-government to e-inclusion.* September 29, 2008, from the United Nations, Department of Economic and Social Affairs, Division for Public Administration and Development Management Website: http://www2.unpan.org/eg ovkb/global_reports/05report.htm

United Nations. (2008). *UN e-government survey 2008: From e-government to Connected governance.* Retrieved May 30, 2008, from the United Nations, Department of Economic and Social Affairs, Division for Public Administration and Development Management Website: http://unpan1.un.org/intradoc/group s/public/documents/un/unpan028607.pdf

URT-MoCT. (2003). *National Information and Communications Technologies Policy.* Dar es Salaam: Ministry of Communications and Transport, The United Republic of Tanzania.

Weerakkody, V., Dwivedi, Y.K., Brooks, L., Williams, M. & Mwange, A. (2007). E-government implementation in Zambia: contributing factors. *Electronic Government, an International Journal, 4(4),* 484-508.

Yong, J. S. L., & Koon, L. H. (2003). E-government: Enabling public sector reform. In Yong, J. S. L. (Ed.), *Enabling Public Service Innovation in the 21st Century E-Government in Asia* (pp. 3–21). Singapore: Times Editions.

Yong, J. S. L., & Leong, J. L. K. (2003). Digital 21 and Hong Kong's advancement in E-government. In Yong, J. S. L. (Ed.), *Enabling Public Service Innovation in the 21st Century E-Government in Asia* (pp. 97–116). Singapore: Times Editions.

Yuan, Y., Zhang, J., & Zheng, W. (2004). Can e-government help China meet the challenges of joining the World Trade Organization? *Electronic Government, 1*(1), 77–91. doi:10.1504/EG.2004.004138

Chapter 8

Acceptability of ATM and Transit Applications Embedded in Multipurpose Smart Identity Card:
An Exploratory Study in Malaysia

Paul H.P. Yeow
Multimedia University, Malaysia

W.H. Loo
Multimedia University, Malaysia

ABSTRACT

The study investigates the user acceptance of automated teller machine (ATM) and transit applications (Touch 'n Go) which are embedded in Malaysian multipurpose smart identity card named as MyKad. A research framework was developed based on a well known user acceptance model i.e. Unified Theory of Acceptance and Use of Technology (UTAUT) model. Five hundred questionnaires were randomly distributed in the Multimedia Super Corridor, Malaysia. The data were analyzed using descriptive and inferential statistics. The results show that Malaysians do not have strong intentions of using the two applications. This can be explained by factors shown in the research framework: performance expectancy, effort expectancy, social influence, facilitating conditions, perceived credibility, and anxiety. Malaysians have little understanding of their benefits and the efforts needed to use them. In addition, they have the misconception that there are insufficient facilities to support the usage of the applications. Consequently, there is no social support to use the applications. Moreover, they perceive that the applications do not have credibility. Besides, they are unsure if they use of the applications would cause anxieties. As a result, few Malaysians have intentions of using the applications. Recommendations were given to increase the acceptance and to resolve the discovered issues. The present research can be replicated to study user acceptance of other applications in MyKad/smart identity cards in other countries (e.g. Hong Kong, India and the sultanate of Oman).

DOI: 10.4018/978-1-60960-162-1.ch008

INTRODUCTION

The inventions of various technologies have dramatically changed many aspects of contemporary society. The influence of such technologies on human life, as well as the way we perceive and use them, has attracted the interest of many researchers from various streams. One of the streams is studying technology acceptance in electronic government (E-Gov) initiative. It is crucial to understand the factors influencing user acceptance in E-Gov initiative because investment in E-Gov initiative is usually huge and can be considered worthwhile only if it is accepted and used by the citizens (Dillion and Morris, 1996). Many studies have been conducted in this area, e.g., Reddick (2008) conducted a study of technology acceptance of E-Gov in the US and found that the E-Gov usage was positively related to managerial effectiveness, having a champion of E-Gov and perceived effectiveness of citizen access to online information. In the industrially developing countries, Kannabiran et al. (2008) conducted a technology acceptance study on an E-Gov initiative in Tamilnadu, India, which is used to provide rural access to government services. They found issues such as lack of government support, non-scaleable technology, and ownership problems. Similarly, Al-Fakhri et al. (2008) did a study of E-Gov initiatives in Saudi Arabia and found poor adoption issues such as lack of Internet facilities, poor awareness from the public and government employees, lack of legal framework for secure e-transactions, etc. The present study investigates the technology acceptance of two Malaysian E-Gov initiatives, i.e. Touch 'n Go (a transportation card) and automated teller machine (ATM) applications embedded in Malaysian national identity smartcard (called MyKad).

Nowadays, one might find that his/her wallet probably has several smartcards such as credit card, identity card, ATM card, etc. A smartcard can be defined as a plastic card, usually similar in size and shape to a credit card, containing a microprocessor and memory (which allows it to store and process data) and complying with ISO 7816 standard (Dhar, 2003). Smartcards began to be used in the last 10 years due to several reasons: (1) most smartcard-related patents expired in 1995, (2) the high fraud rate associated with magnetic-striped cards compelled many companies to search for a more secure and cost-effective alternative, (3) technological advancements have made possible the existence of smartcard technology through the availability of cost-effective equipment and interoperability of different smartcards (Gail et al., 1995), and the enhanced security and flexible features of smartcards make them an ideal solution for today's technology-savvy and demanding consumers (Yates, 2005).

Following the global trend towards smartcard implementation, many countries in Western Europe, Asia, and the Middle East have begun introducing smart identity card. Some countries such as Hong Kong, India, the sultanate of Oman, and Malaysia had implemented smart identity card with multipurpose applications (Yeow et al., 2007). As such, smart identity card with multiple applications will become more prevalent in the near future. Many countries are expected to spend huge amount of money to implement such card, e.g. Malaysia had spent RM500 million (USD139 million) on its MyKad project (Singh, 2006). Therefore, it is of paramount importance to investigate the acceptability of the applications in multipurpose smart identity cards. In fact, the success of a multipurpose smart identity card project (e.g. MyKad) largely depends on how frequently and actively people use the card (i.e. each application embedded in the card), in comparison with other available options.

MyKad is the world's first multipurpose smart identity card. It was officially launched in September, 2001. The initial roll-out of MyKad project is in the major urban area of Malaysia, i.e. Multimedia Super Corridor (MSC). MyKad is a piece of plastic card with an embedded microchip and dimensions of a standard credit card. Its first version contained a 32 kilobytes EEPROM (Electrically Erasable Programmable Read-Only Memory) chip running on the proprietary M-COS (MyKad Chip Operating System) operating system. In November 2002, the memory capacity was increased to 64 kilobytes which allowed more applications to be added into the card. The M-COS operating system enables different applications to be stored separately in the chip and prevents unauthorized access. MyKad utilizes biometric technology with thumbprints being encrypted in its chip. In addition, it is a hybrid card with dual interface for both contact and contactless applications. To date, there are nine applications embedded in MyKad: national identity card (NIC), driving license (DL), passport information (PI), health information (HI), electronic purse, ATM card, transit application (i.e. 'Touch 'n' Go'), Public Key Infrastructure (PKI), and Frequent Traveler Card (FTC). In the near future, the government intends to merge MyKad with debit and credit card functions and other financial applications (Thomas, 2004).

Currently, all MyKads are loaded with NIC application by default. The other applications depend on cardholders' initiative to go to various places to activate the applications, such as the National Registration Department (for DL, PI, and FTC applications), hospitals (for HI application), Touch 'n Go counters (for transit application), local banks (for ATM card and electronic purse applications), and certificate authority (for PKI application).

Among all, NIC and DL are the two most popular applications of MyKad (Yeow and Miller, 2005). The major advantages of MyKad as NIC include the ability to detect terrorist, identify false NIC, and enable more productive and effective government services, whereas MyKad with DL application can be used to reduce fake license, improve the efficiency of driving license verification, increase the accuracy of road summons, and enable convenient summon payment through MyKad electronic purse. However, other applications in

MyKad remain untapped. Several constraints have been identified, chiefly two of which include lack of awareness and lack of infrastructure. Many are just using MyKad for NIC and DL only and the other applications are not so popular among Malaysians (Ishahak, 2006). In fact, Yeow and Miller's (2005) study discovered that there is low usage of MyKad transaction-based applications (ATM, Touch 'n Go, MEPS, and PKI).

Yeow and Miller (2005) are probably the first researchers who investigated Malaysians' attitudes toward the multipurpose smart identity card as the study was conducted in 2003 during the early introduction of the card. The study discovered that Malaysians were aware and had a positive attitude toward MyKad. A follow-up study was conducted by Yeow et al. (2007) which focused on issues related to the acceptability of MyKad NIC and MyKad DL applications. The study discovered that Malaysians strongly accept MyKad as NIC and DL. Nevertheless, they are facing issues related to privacy, civil liberties, fine, and cost of upgrading when adopting the NIC application; and issues related to ease of update, ease of reading, recognition, and availability of card readers when adopting the DL application. So far, there is no other study conducted in Malaysia or other countries that examine the acceptability of the other applications embedded in smart identity card. To fill this gap, the present study explores the acceptability of MyKad Touch 'n Go and MyKad ATM applications as these were identified as the top two transaction-based applications in the previous studies (Yeow and Miller, 2005; Yeow et al., 2007).

LITERATURE REVIEW

It is important to address specifically the question of why people use a system in a particular context, because it will lead to insights as to what factors in the system that cause people to choose it over other available options (Hu et al., 1999; Sun, 2005). To explain why people use information system (IS), researchers in user acceptance research stream have applied various established theories from social psychology to IS user behavior, such as the Theory of Reasoned Action (TRA) (Fishbein and Azjen, 1975), the Social Cognitive Theory (SCT) (Bandura, 1986), the Theory of Planned Behavior (TPB), and the Innovation Diffusion Theory (IDT). Based on these theories, various models were developed, such as the Technology Acceptance Model (TAM/TAM2), the Model of Personal Computer Utilization (MPCU), and a model combining the Technology Acceptance Model and the Theory of Planned Behavior (CTAM&TPB). Each model has its own independent and dependent factors for user acceptance although there are some overlaps (Dillion and Morris, 1996).

Although TAM has received extensive empirical support through validations, applications, and replications for its power to predict use of information systems (Davis et al., 1982; Davis et al., 1989; Davis 1993; Davis and Venkatesh, 1996; Taylor and Todd, 1995; Venkatesh and Morris, 2000; Horton et al., 2001; Lu et al., 2003), and is believed to be the most robust, parsimonious, and influential model explaining information system adoption behavior (Davis, 1993; Davis, 1989; Igbaria et al., 1995) some researchers recognize its limitations. TAM has failed to supply meaningful information about the user acceptance of a particular technology due to its generality. Furthermore, TAM does not consider barriers such as lack of expertise, and time or money constraint as factors that would prevent an individual from using an information system. Consequently, TAM omits these important sources of variance (Mathieson et al., 2001). Given this background, a number of modified TAM models were proposed which are applicable to contemporary technologies (Agarwal and Prasad, 1998; Chau and Hu, 2001; Chau, 1996; Horton et al., 2001; Hu et al., 1999; Jiang et al., 2000). However, researchers are confronted with a choice among a multitude of models and they have to "pick and choose" constructs across the models, or choose a "favored model" and largely ignore the contributions from alternative models. Hence, a new model was developed to address these limitations, which is named as the Unified Theory of Acceptance and Use of Technology (UTAUT) model (Venkatesh et al., 2003).

UTAUT Model

The UTAUT model captures the essential elements of eight previously established models (i.e. the TRA the TAM/TAM2, the Motivational Model (MM), the TPB, the CTAM&TPB, the MPCU, the IDT, and the SCT) to explain behavioral intentions and actual use of technology. It has been found to be outperforming the abovementioned theoretical frameworks as it is able to account for a high percentage of the variance (adjusted R^2) in usage intention (Venkatesh et al., 2003). The UTAUT model's constructs are performance expectancy, effort expectancy, social influence, and facilitating conditions. It is confirmed as a definitive model that synthesizes what is known and provides a foundation to guide future research in user acceptance area (Venkatesh et al., 2003). By encompassing the combined explanatory power of the many models above and key moderating influences, the UTAUT model advances cumulative theory while retaining a parsimonious structure.

The present study has developed a research framework by adapting the UTAUT model framework, adding two variables i.e. perceived credibility and anxiety, to explore factors affecting user acceptance in the context of MyKad Touch 'n Go and MyKad ATM applications. The determinants of the models and the added variables (with their justifications) are explained in the sections below.

User Acceptance

User acceptance is defined as a person's psychological state with regard to his or her voluntary use and intention to use a technology (Dillion and

Morris, 1996). Many researchers (e.g. Compeau and Higgins, 1995; Davis et al., 1989; Taylor and Todd, 1995; and Venkatesh and Davis, 2000) in user acceptance stream have been using intention to use or actual usage as a dependent variable. In the present model, behavioral intention to use is used as the dependent variable. Behavioral intention to use MyKad ATM and MyKad Touch 'n Go applications is measured by three items adapted from Venkatesh et al. (2003) (refer to Table 3: Nos. 7.1, 7.2 and 7.3, and Table 5: Nos. 7.1, 7.2 and 7.3).

Performance Expectancy

Performance expectancy is defined as the degree to which an individual believes that using a system will help him or her to attain gains in job performance (Venkatesh et al., 2003). In the present study, performance expectancy refers to MyKad holders' perception that using MyKad ATM and MyKad Touch 'n Go MyKad will help them attain gains in their daily life. The attributes of performance expectancy of the respective applications are presented in Table 3: Nos. 1.1–1.4 and Table 5: Nos. 1.1–1.4. Venkatesh et al. (2003) stated that "performance expectancy is the strongest predictor of intention" (p.447). This is consistent with prior findings that performance expectancy has more significant impact on intention to use than effort expectancy (Agarwal and Prasad, 1998; Compeau and Higgins 1995; Davis et al., 1992; Taylor and Todd 1995; Thompson et al., 1991; Venkatesh and Davis, 2000). Therefore, it is expected that performance expectancy will positively influence intention to use both MyKad's applications. To test such relations, the following hypotheses are suggested:

H1a: Performance expectancy will have a positive influence on behavioral Intention To Use MyKad Touch 'n Go application.

H1b: Performance expectancy will have a positive influence on behavioral Intention To Use MyKad ATM application.

Effort Expectancy

Effort expectancy is defined as the degree of ease associated with the use of a system (Venkatesh et al., 2003). In fact, three constructs from the models mentioned earlier capture the concept of effort expectancy, i.e. perceived ease of use (the TAM/TAM2), complexity (the MPCU), and ease of use (the IDT). It was found that there are similarities among the construct definitions and measurement scales which have been noted in prior research (Davis et al., 1989; Moore and Benbasat, 1991; Plouffe, 2001) e.g. all constructs defined effort expectancy as the perceived ease of use which refers to the usability of a computerized interface in transaction-based application (Dillion and Morris, 1996). In the present context, effort expectancy refers to the perceived ease of use of MyKad ATM and MyKad Touch 'n Go applications. The attributes of effort expectancy of the respective applications are shown in Table 3: Nos. 2.1–2.6 and Table 5: Nos. 2.1–2.7. Prior studies (e.g. Venkatesh et al., 2003; Carlsson et al., 2006; He and Lu, 2007) revealed that users' intention of using technology will increase if users perceive that the particular technology is easy to use. Similarly, the present study postulated that MyKad holders will use MyKad Touch 'n Go and ATM applications if both applications are easy to use, we therefore posit that:

H2a: Effort expectancy will have a positive influence on behavioral Intention To Use MyKad Touch 'n Go application.

H2b: Effort expectancy will have a positive influence on behavioral Intention To Use MyKad ATM application.

Social Influence

Social influence is defined as the degree to which an individual perceives others' belief (particularly their close ones) that they should use a new system (Venkatesh et al., 2003). Three constructs from earlier theories have attempted to measure social influence. These include subjective norm from the TRA, the TAM2, the TPB, and the CTAM&TPB, social factors from the MPCU, and image from the IDT. In the present study, social influence refers to an individual altering his or her intention to use MyKad ATM and MyKad Touch 'n Go applications in response to social pressure to comply with social norm. The attributes of social influence of the respective applications are presented in Table 3: Nos. 3.1–3.4 and Table 5: Nos. 3.1–3.4. Empirical findings support that social influence exerts positive influence on intention to use (Venkatesh and Davis, 2000; Venkatesh et al., 2003; He and Lu, 2007). In view of this, the following hypotheses are tested:

H3a: Social Influence will have a positive influence on behavioral Intention To Use MyKad Touch 'n Go application.

H3b: Social Influence will have a positive influence on behavioral Intention To Use MyKad ATM application.

Facilitating Conditions

Facilitating conditions are defined as the degree to which an individual believes that an organizational and technical infrastructure exists to support the use of a system (Venkatesh et al., 2003). Three constructs from earlier theories have attempted to measure facilitating conditions. They include perceived behavioral control from the TPB and the CTAM&TPB, facilitating conditions from the MPCU, and compatibility from the IDT. In the present context, facilitating conditions refer to the objective factors (e.g. campaigns, infrastructure, and recognition) in the environment that facilitates the usage of MyKad ATM and MyKad Touch 'n Go. The attributes of facilitating conditions of the respective applications are shown in Table 3: Nos. 4.1– 4.5 and Table 5: Nos. 4.1– 4.5. Prior literature revealed that facilitating conditions do have influence on intention to use (Wu et al., 2007). To test such relation, the following hypotheses are suggested:

H4a: Facilitating Conditions will have a positive influence on behavioral Intention To Use MyKad Touch 'n Go application.

H4b: Facilitating Conditions will have a positive influence on behavioral Intention To Use MyKad ATM application.

Perceived Credibility

Many studies identified perceived credibility as a direct determinant which significantly affects the adoption of a technology (e.g. Internet banking and e-commerce) (Warrington et al., 2000; Wang et al., 2003; Metzger et al., 2003; Poston et al., 2007). In these studies, perceived credibility factor was related to security and privacy concerns that affect user acceptance. Perceived credibility is relevant in the present study because MyKad ATM and MyKad Touch 'n Go holders may have perception of security and privacy of using the applications. Perceived credibility covers issues such as security, privacy, misuse, etc. as shown in Table 3: Nos. 5.1–5.3 and Table 5: Nos. 5.1–5.3. From a theoretical point of view, it is reasonable to expect MyKad holder's intention to use MyKad Touch 'n Go and ATM to increase if he/ she perceives the card to be free from privacy and security threats. Thus, the following hypotheses are suggested:

H5a: Perceived credibility will have a positive influence on behavioral Intention To Use MyKad Touch 'n Go application.

H5b: Perceived credibility will have a positive influence on behavioral Intention To Use MyKad ATM application.

Anxiety

Anxiety is defined as the evoking of anxiety or emotional reactions when it comes to performing a behavior (such as using MyKad applications) (Compeau and Higgins, 1995). Yeow et al.'s (2007) study found that fear of losing and damaging MyKad, uncertainty of the card's durability, and lack of expertise in using the card might cause anxiety in using two of its applications, i.e. National Identity Card and Driving License. By identifying with these issues, the present study added anxiety as a direct determinant of intention to use MyKad ATM and MyKad Touch 'n Go applications. The attributes of anxiety in using the two applications include fear of losing the card, fear of card damage, and intimidating experience as shown in Table 3: Nos. 5.1–5.3 and Table 5: Nos. 5.1–5.3, respectively. Anxiety has been identified as a negative factor where it is unlikely for a person to adopt a new technology

if they have any aversion towards anxiety (Gilbert et al., 2003; Sam et al., 2005; Doyle et al., 2005). The present study intends to investigate if MyKad holders' intentions of using both applications are inversely related to their anxiety level. Thus, the following hypotheses are suggested:

H6a: Anxiety will have a negative influence on behavioral Intention To Use MyKad Touch 'n Go application.

H6b: Anxiety will have a negative influence on behavioral Intention To Use MyKad ATM application.

Figure 1 shows the research framework for user acceptance of MyKad Touch 'n Go and MyKad ATM applications.

METHOD

The present study employed a cross-sectional approach to collect survey data at one point in time since this is a new field focusing on user acceptance

Figure 1. User acceptance of MyKad Touch 'n Go and MyKad ATM applications

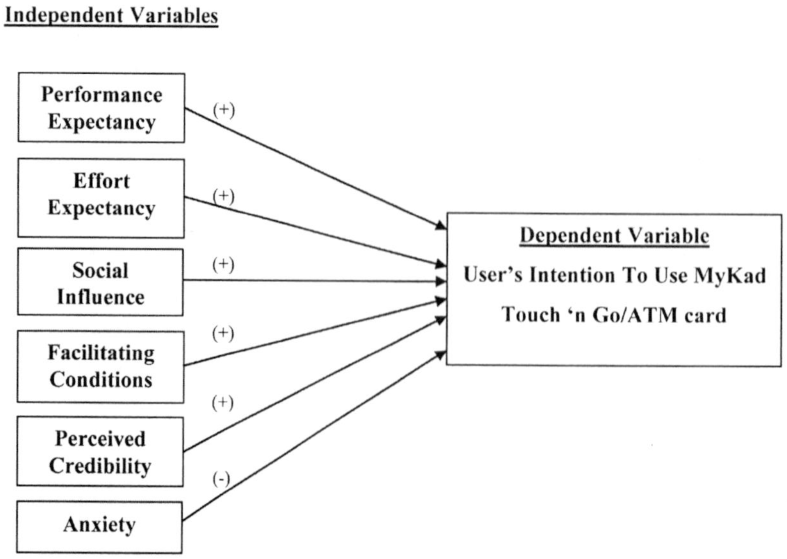

of MyKad ATM and Touch 'n Go applications. Questionnaires were distributed randomly in year 2006 (in public places, e.g. bus stations, shopping complexes, streets, etc.) to 500 MyKad holders residing in the Multimedia Super Corridor (MSC), Malaysia. According to Israel (1992), 200 - 500 is considered a good sample size for data analyses like multiple analyses. Five years (from 2001 to 2006) were given for users to familiarize with MyKad applications. The MSC is the "Silicon Valley" of Malaysia where the information and communication technology industry is focused. An area of approximately 15x50 km², it stretches from Petronas Twin Towers to the Kuala Lumpur International Airport. It includes major cities such as Kuala Lumpur, Putrajaya, and Cyberjaya. The study is confined to the MSC because the government has chosen this area for MyKad's pilot project. In addition, all up-to-date infrastructure and facilities needed for implementing MyKad Touch 'n Go and MyKad ATM exist here.

A questionnaire comprising four sections was developed based on the research framework: (1) respondent's demographic information, (2) 27 questions (Table 5: Nos. 1.1–6.4) pertinent to factors affecting respondent's intention to use MyKad Touch 'n Go as shown in the research framework in Figure 1, (3) 26 questions (Table 3: Nos. 1.1–6.4) related to factors affecting respondent's intention to use MyKad ATM as shown in the same framework, and (4) 6 questions measuring the extent of respondent's intention of using MyKad Touch 'n Go and ATM (Table 5: Nos. 7.1–7.3 and Table 3: Nos. 7.1–7.3). The items (in the questionnaire) were taken from the questionnaires in prior studies (see below). Since the items were validated in previous studies, performing another validation was unnecessary. Items measuring performance expectancy, effort expectancy, social influence, facilitating conditions, and intention to use were taken from Venkatesh's et al. (2003) questionnaire. Items measuring perceived credibility and anxiety were taken from Wang's et al. (2003), and Compeau and Higgins's et al. (1999) questionnaires,

respectively. To suit the context of the study, some words have been modified (e.g. "system" is changed to "MyKad ATM" or "MyKad Touch 'n Go") accordingly. In sections (2), (3) and (4), Likert's five-point scale is used, i.e. 1=strongly disagree, 2=disagree, 3=neither agree nor disagree, 4=agree, 5=strongly agree. The data collected were analyzed using descriptive and inferential statistics.

RESULTS

Table 1 presents the percentages of respondents who have activated MyKad ATM and MyKad Touch 'n Go applications, respectively.

Table 2 presents the results of the regression analysis between the independent and dependent variables for MyKad ATM application. The adjusted R^2 of 0.817 indicates that approximately 82% variance of intention to use MyKad ATM application is explained by the five independent variables i.e. anxiety, effort expectancy, social influence, perceived credibility, and performance expectancy (except facilitating conditions). The five independent variables are found to significantly influence Malaysian citizens' intention to use MyKad ATM application (dependent variable). Thus, H1b, H2b, H3b, H5b, and H6b are accepted. H4b is rejected.

Table 3 shows the respondents' ratings of their intention to use MyKad ATM (dependent variable) and the factors affecting their intention to use (independent variables).

The results of regression analysis between independent and dependent variables of MyKad

Table 1. Activated MyKad applications

MyKad applications	Frequency	Percent
Touch 'n Go	106	21.2
ATM	74	14.8

Sample size= 500

Table 2. Multiple Linear Regression between the independent and dependent variables for MyKad ATM application

Independent variable	B	Standard Error	Standardized Beta Coefficient	t Statistic	p-value
	.036	.063			
Anxiety	-.195	.013	-.307	-15.484	<.05
Effort Expectancy	.212	.012	.264	9.996	<.05
Social Influence	.187	.016	.264	11.955	<.05
Performance Expectancy	.151	.015	.255	10.093	<.05
Perceived Credibility	.195	.019	.250	10.289	<.05
Facilitating Conditions	.019	.017	.027	1.135	.257

F = 523.890 (p < 0.05); Adjusted R^2 = 0.817

Touch 'n Go application are shown in Table 4. The adjusted R^2 of 0.559 implies that approximately 56% of the effect on the dependent variable (i.e. Intention To Use MyKad Touch 'n Go application) is explained by six independent variables, namely anxiety, performance expectancy, facilitating conditions, perceived credibility, social influence and effort expectancy. Hence, H1a, H2a, H3a, H4a, H5a and H6a are accepted.

Table 5 shows the respondents' ratings of their intention to use MyKad Touch 'n Go (dependent variable) and the factors affecting their intention to use (independent variables).

DISCUSSIONS

The results reveal that the minority of the respondents have activated their MyKad Touch 'n Go and MyKad ATM applications (see Table 1). The results are consistent with Yeow and Miller (2005) and Yeow's et al. (2007) findings which revealed that the transaction-based applications in MyKad (e.g. Touch 'n Go, MEPS, ATM and PKI) were underutilized (Yeow and Miller, 2005; Yeow et al., 2007).

MyKad Touch 'n Go Application

Overall, the respondents gave low ratings for their intention of using MyKad Touch 'n Go (Table 5 Nos. 7.0–7.3). This is not surprising as there are many other alternate modes of payment for expressway toll, public transportation, and parking, e.g. Touch 'n Go card, credit card with built-in Touch 'n Go (such as Touch 'n Go Zing), and cash. Furthermore, the results of multiple linear regression (Table 4) show that respondents' anxiety is the strongest (negative) predictor of intention to use the application.

Performance Expectancy

Performance expectancy is the most important predictor of intention to use MyKad Touch 'n Go (Table 4). However, descriptive statistics results show that the respondents were uncertain about the performance (or functions) of MyKad Touch 'n Go (Table 5: No.1.0). This explains their low intention to use the application. MyKad Touch 'n Go are prepaid transportation cards that can be used for payments of all highway tolls in Malaysia, major public transportations in Klang Valley and selected parking sites. Nevertheless, Malaysians prefer to use cash for transportation payment as there are only 3 million users of Touch 'n Go

Table 3. Factors affecting users' intention to use MyKad Automated Teller Machine (ATM) application

No.	Statement presented in the questionnaire	Mean	SD
	Independent variables		
1.0	**Performance expectancy**	**3.05**	**.84**
1.1	With MyKad ATM, I do not have to bring a lot of ATM cards with me.	3.19	.93
1.2	The ATM card loaded in MyKad performs just as well as the original ATM card.	3.07	.89
1.3	MyKad with additional ATM application is convenient.	3.01	1.04
1.4	MyKad ATM is useful to me.	2.95	1.11
2.0	**Effort expectancy**	**3.07**	**.62**
2.1	It is easy to remember how to use MyKad ATM.	3.12	.76
2.2	It is very easy to use MyKad ATM.	3.11	.81
2.3	It is easy to learn how to use MyKad ATM.	3.09	.77
2.4	I can complete my banking transactions (e.g. withdrawing money, checking balances of savings account) within seconds by using MyKad ATM.	3.08	.77
2.5	ATM machines will not be confused if I load more than one bank account's information into MyKad.	3.02	.72
2.6	MyKad ATM seldom incurs any errors when I use it.	3.00	.77
3.0	**Social influence**	**2.77**	**.70**
3.1	I am aware that MyKad can be used as ATM card.	2.85	.99
3.2	Many Malaysians are using MyKad ATM for their banking transactions.	2.78	.90
3.3	I have been influenced to apply for MyKad ATM because it is becoming a popular application.	2.77	.87
3.4	My peer group affects me to apply for MyKad ATM.	2.70	.87
4.0	**Facilitating conditions**	**3.17**	**.69**
4.1	Extensive campaign has been organized to promote the usage of MyKad ATM.	3.51	.98
4.2	Many instructions or guidelines are given to guide users to use MyKad ATM.	3.31	.89
4.3	Process of activating MyKad ATM is simple.	3.07	.75
4.4	Most ATM terminals accept MyKad ATM.	3.05	.84
4.5	Most banks provide service for activating MyKad ATM.	2.94	.84
5.0	**Perceived credibility**	**2.90**	**.64**
5.1	MyKad ATM card is difficult to forge.	2.98	.81
5.2	Others cannot view my bank account information embedded in MyKad.	2.93	.85
5.3	MyKad ATM is more secure than ATM card.	2.83	.85
6.0	**Anxiety**	**3.31**	**.78**
6.1	I feel apprehensive about using MyKad ATM, fearing that my MyKad will be swallowed by the ATM terminal.	3.43	1.04
6.2	I hesitate to use MyKad ATM, fearing that MyKad may be damaged due to extensive use.	3.40	.95
6.3	MyKad ATM is somewhat intimidating to me.	3.23	.91
6.4	I am afraid to use MyKad ATM because I don't know how to use it.	3.19	.96
	Dependent variable		
7.0	**Intention to use MyKad ATM**	**2.89**	**.50**
7.1	I plan to use MyKad ATM in the near future.	2.93	.67
7.2	I predict I would use MyKad ATM in the near future.	2.87	.55
7.3	I intend to use MyKad ATM in the near future.	2.86	.62

Sample size= 500; Mean= mean based on Likert's five-point scale: 1= strongly disagree, 2=disagree, 3=neither agree nor disagree, 4=agree, 5=strongly agree; SD=standard deviation; ATM= Automated Teller Machine

Table 4. Multiple Linear Regression between the independent and dependent variables for MyKad Touch 'n Go application

Independent Variable	B	Standard Error	Standardized Beta Coefficient	t Statistic	p-value
	1.466	.080		18.363	
Anxiety	-.248	.018	-.458	-13.813	<.05
Performance Expectancy	.153	.019	.334	8.261	<.05
Facilitating Conditions	.217	.026	.314	8.270	<.05
Perceived Credibility	.165	.019	.306	8.587	<.05
Social Influence	.100	.015	.216	6.472	<.05
Effort Expectancy	.096	.028	.148	3.442	<.05

$F = 213.290$ ($p < 0.05$); Adjusted $R^2 = 0.559$

card out of the 23 million population, of which 1.5 million are active users. The respondents were unaware that MyKad Touch 'n Go holders (travelers and commuters) can save queuing time (Table 5: No. 1.1) at ticketing counters or toll booths. In addition, some of them disagree with the usefulness of MyKad Touch 'n Go (Table 5: No. 1.4, SD above 1.00).

Effort Expectancy

Effort expectancy is one of the predictors of intention to use MyKad Touch and Go (Table 4). However, the respondents were neutral about effort expectancy (Table 5: No. 2.0), which indicates that respondents were unclear about the effort to use the application. In fact, it is easy to use MyKad Touch 'n Go. A MyKad holder only needs to touch the reader with the card and a beep will indicate a successful transaction. The card holder can then check the card balance through a display screen.

Social Influence

Social influence is one of the predictors of intention to use MyKad Touch 'n Go (Table 4). However, the respondents did not get strong support from their social environment in using MyKad Touch 'n Go as shown by the low MR (Table 5: No.

3.0). They were unsure that MyKad can be used as Touch 'n Go and MyKad Touch 'n Go is a popular application (Table 5: No2. 3.2-3.4). In fact, MyKad Touch 'n Go is not a popular application compared to Touch 'n Go card as there are about 1 million MyKad Touch 'n Go users and about 3 million Touch 'n Go users (Rangkaian Segar Sdn Bhd [RSSB], 2006).

Facilitating Conditions

Facilitating conditions is one of the predictors of intention to use MyKad Touch 'n Go (Table 4). However, the result indicates that there were no strong facilitating conditions for the use of MyKad Touch 'n Go (Table 5: No. 4.0). Even though the Malaysian government and some private companies have organized a few campaigns (such as Ekspo and CardEx) to promote MyKad, these are small-scale and conducted once a year. This explains the respondents' poor awareness about the campaigns (Table 5: No. 4.1). The use of MyKad Touch 'n Go is the same as Touch 'n Go card; however, the respondents were unsure of the recognition of MyKad Touch 'n Go (Table 5: No. 4.2). In addition, they were uncertain of the procedure of activating MyKad Touch 'n Go (Table 5: No. 4.3). In fact, the procedure is simple, i.e. load cash value into MyKad at any Touch 'n

Table 5. Factors affecting users' intention to use MyKad Touch 'n Go (TNG) application

No.	Statement presented in the questionnaire	Mean	SD
	Independent variables		
1.0	**Performance expectancy**	**3.24**	**.90**
1.1	MyKad TNG saves queuing time.	3.33	.98
1.2	MyKad TNG can fit well into my lifestyle (e.g. I can use it to pay for tolls, parking, or public transport in my daily life).	3.22	.97
1.3	MyKad with additional TNG is convenient.	3.22	.96
1.4	MyKad TNG is useful to me (e.g. pay for tolls, parking, or public transport).	3.19	1.01
2.0	**Effort expectancy**	**3.18**	**.64**
2.1	It is easy to learn how to use MyKad TNG.	3.29	.86
2.2	It is easy to use MyKad TNG.	3.28	.83
2.3	It is easy to remember how to use MyKad TNG.	3.23	.84
2.4	My car can immediately pass through tolls after MyKad TNG is detected by the reader.	3.19	.86
2.5	MyKad TNG seldom incurs errors when I use it.	3.16	.77
2.6	I do not think that I will lose control of my transportation expenses if I use MyKad TNG.	3.05	.90
2.7	I do not have any problems tracking the amount stored in MyKad TNG.	3.03	.78
3.0	**Social influence**	**2.61**	**.89**
3.1	My peer group affects me to use MyKad TNG.	2.69	.93
3.2	I am aware that MyKad can be used as TNG.	2.68	1.07
3.3	I perceive that most Malaysians like to load their money into MyKad TNG.	2.62	.95
3.4	I perceive that most Malaysians use MyKad TNG rather than TNG card.	2.60	1.02
4.0	**Facilitating conditions**	**3.21**	**.60**
4.1	Extensive campaign has been organized to promote the usage of MyKad TNG.	3.50	.93
4.2	Toll booths in all states already recognize MyKad TNG.	3.17	.80
4.3	It is easy to activate MyKad TNG.	3.17	.77
4.4	It is easy to find toll booths that provide services for topping-up credits into MyKad TNG.	3.13	.82
4.5	I use MyKad TNG because the activation fee is excluded.	3.08	.81
5.0	**Perceived credibility**	**2.59**	**.77**
5.1	It is secure to load money into MyKad TNG.	2.76	.92
5.2	Others cannot use my MyKad TNG to pass through tolls if it is lost.	2.53	.94
5.3	It is secure to use MyKad TNG rather than using a separate TNG card.	2.49	.91
6.0	**Anxiety**	**2.77**	**.76**
6.1	I hesitate to use MyKad TNG, fearing that MyKad may be damaged due to extensive use.	2.94	.96
6.2	I fear the tollgate, and parking and ticketing machine will automatically deduct more than they should if I use MyKad TNG.	2.87	.90
6.3	I am afraid to use MyKad TNG because I do not know how to use it.	2.68	.97
6.4	I feel apprehensive about using MyKad TNG because I fear that I will lose my MyKad.	2.60	.96
	Dependent variable		
7.0	**Intention to use MyKad TNG**	**2.86**	**.41**
7.1	I plan to use MyKad TNG in the near future.	2.89	.66
7.2	I predict I would use MyKad TNG in the near future.	2.87	.51
7.3	I intend to use MyKad TNG in the near future.	2.82	.52

Sample size= 500; Mean= mean based on Likert's five-point scale: 1= strongly disagree, 2=disagree, 3=neither agree nor disagree, 4=agree, 5=strongly agree; SD=standard deviation; TNG= Touch 'n Go

Go sales and customer service counter located at major toll plazas.

Perceived Credibility

Perceived credibility is an important predictor of intention to use MyKad Touch 'n Go. The standardized beta coefficient of perceived credibility is high and very close to that of performance expectancy (see Table 4). However, the descriptive statistics results show that the respondents did not perceive MyKad Touch 'n Go as a credible application (Table 5: No. 5.0). The respondents did not think that loading money into MyKad Touch 'n Go is secure and using MyKad Touch 'n Go is more secure than Touch 'n Go card (Table 5: Nos. 5.1 & 5.3). In fact, loading money into MyKad Touch 'n Go is very safe because the data in MyKad are encrypted with a 128-bit triple-DES key which makes it very difficult for anybody to steal money through accessing the card from other MyKad's applications, e.g. ATM, DL, etc.

Anxiety

Anxiety is the strongest (negative) predictor of intention to use MyKad Touch 'n Go (see Table 4). The results show that the respondents faced anxiety when using MyKad Touch 'n Go (Table 5: No. 6.0). The respondents were afraid of using MyKad Touch 'n Go as it would increase the risk of losing MyKad (Table 5: No. 6.4), e.g. MyKad Touch 'n Go holders might accidentally leave the card at the reloading machine or ticketing counter. As a matter of fact, losing the card is very serious as it is also a national identity card. To get a replacement is troublesome, requiring a police report, an oath in the court, payment of a hefty fine (RM100 or USD27), and a long wait for the new card. The respondents perceived that extensive use of MyKad Touch 'n Go would result in card damage (Table 5: No. 6.1). Moreover, the respondents had anxiety over the error of over

deduction (Table 5: No. 6.2) even though to date, no such incidence was reported in the press.

MyKad ATM Application

Overall, the respondents' intention to use MyKad ATM is low based on the MRs scored (Table 3: Nos. 7.1–7.3). This can be explained by the five factors (i.e. anxiety, effort expectancy, social influence, performance expectancy and perceived credibility– see Table 2) that influence intention to use MyKad ATM. It is interesting to note that all the positive predictors of intention to use (i.e. effort expectancy, social influence, performance expectancy and perceived credibility) have very close standardized beta coefficient values, indicating that they are equally important.

Performance Expectancy

Although performance expectancy is an important predictor of intention to use MyKad ATM, the results from descriptive statistics show no strong performance expectancy. The MR of 3.05 (Table 3: No. 1.0) obtained from the attributes of performance expectancy indicates that the respondents were uncertain of the performance of MyKad ATM. The respondents were not convinced by the added advantage of MyKad ATM application in reducing the number of bankcards in their wallet (Table 3: No. 1.1). Actually, the application can reduce the number of bankcard because three bank accounts can be loaded into the application. Besides, some of them did not think that the application is useful to them or convenient to use (Table 3: No. 1.3 & 1.4). Moreover, they were unsure whether MyKad ATM could perform just as well as a bankcard (Table 3: No. 1.2, MR of 3.07).

Effort Expectancy

Effort expectancy is also an important predictor of intention to use MyKad ATM. However, the results show that the respondents were unsure of the degree of effort needed to use MyKad ATM application (Table 3: Nos. 2.1 – 2.6). Actually, little effort is needed to use MyKad ATM if the respondents have some experience using it because its use is exactly the same as that of ATM card. Nevertheless, many respondents (85.2%) were non-users.

Social Influence

Social influence is an important predictor of intention to use MyKad ATM with the highest standardized beta coefficient value (Table 2). However, results from descriptive statistics reveal a lack of social support for the respondents to use MyKad ATM (Table 3: No. 3.0). Prior research conducted by Venkatesh and Davis (2000), and Warshaw (1980) revealed that compliance mechanism would cause an individual to simply alter his or her intention to use a technology in response to social pressure. Currently, the status quo is to use the ordinary ATM card issued by banks for any ATM transaction. To comply with social norms, the respondents would probably use ATM card instead of MyKad ATM as they think that few Malaysians and few of their peers (Table 3: No. 3.2 & 3.4) use MyKad ATM.

Facilitating Conditions

Facilitating conditions is not a predictor of intention to use MyKad ATM (Table 2). This is consistent with prior literature which showed that the effect of facilitating conditions on intention to use is insignificant (Venkatesh et al., 2003) if users are inexperienced. As previously mentioned, most of the respondents were non-users; thus, they did not perceive facilitating conditions as a determinant of intention to use. However, descriptive statistics results show that the respondents perceived that there is insufficient infrastructure (e.g. activating services, ATM machines that accept MyKad, etc.) to support MyKad ATM usage (Table 3: Nos. 4.1–4.5). Besides, only few banks provide

activation service, i.e. CIMB Bank, Maybank, Public Bank, and Bank Islam. In addition, the respondents were uncertain about whether most ATM machines accept MyKad ATM as bankcard (Table 3: No. 4.4). Actually, about 4,590 (99.2%) ATM machines have been upgraded to accept MyKad ATM (MSC Malaysia, 2006).

Perceived Credibility

Perceived credibility is an important predictor of intention to use MyKad ATM. However, the respondents gave low MRs for perceived credibility of MyKad ATM (Table 3: Nos. 5.1–5.3). They did not perceive that using MyKad ATM is secure and they thought that others can easily view their bank account information if the card is used (Table 3: No. 5.3 & 5.2). Moreover, they perceived that MyKad ATM can be easily forged (Table 3: No. 5.1). In reality, MyKad ATM is very secure, and difficult to read and forge as the data in MyKad are encrypted (as previously mentioned).

Anxiety

Anxiety is the strongest negative predictor of intention to use MyKad ATM (with the largest standardized beta coefficient absolute value – see Table 2). Descriptive statistics showed that some respondents feared that ATM machine failure might "swallow" their MyKad (Table 3: No. 6.1, SD of above 1.00), which will cause much inconvenience. Actually, this should not be a cause for fear since this rarely happens and there has been no such report thus far. In addition, the respondents were unsure whether they had sufficient knowledge of using MyKad ATM (Table 3: No. 6.4). Furthermore, the respondents were uncertain whether extensive use of the application would damage the card, particularly when the card goes through the ATM card slot numerous times (Table 3: No. 6.2).

CONCLUSION

The study has enriched the existing user acceptance of technology literatures as the UTAUT model is applied on a new technology, i.e. MyKad. The study has adapted the model by adding two additional independent variables, i.e. perceived credibility and anxiety, to measure the user acceptance of MyKad ATM and MyKad Touch 'n Go. The study discovers that Malaysians do not have strong intention to use MyKad ATM and MyKad Touch 'n Go leading to the low usage of these applications. Five determinants of intention to use MyKad ATM are identified. Anxiety is found to have the greatest impact on intention to use MyKad ATM, followed by effort expectancy, social influence, performance expectancy and perceived credibility. However, facilitating conditions are found to be non-significant in influencing intention to use MyKad ATM. As for MyKad Touch 'n Go application, six determinants influence MyKad holders' intention to use the application. Anxiety is found to have the highest impact, followed by performance expectancy, facilitating conditions, perceived credibility, social influence and effort expectancy. Overall, 1) Malaysians lack understanding about the benefits and the efforts needed to use both applications, 2) they perceive that there are no strong facilitating conditions to use them, 3) there is no strong social support, 4) there are misconceptions about their credibility and 5) they face anxiety in using the applications. These findings have very important implications on the measures that can be taken to increase the acceptability of MyKad as Touch 'n Go and ATM card.

To date, the present study is the first and only empirical study that investigates the user acceptance of multipurpose smart identity card. Therefore, its results can be used as a case study to aid countries in their struggle to design and implement multipurpose smartcard applications that is acceptable to their citizens. They could increase the acceptance of the applications embedded in the smartcard by learning from Malaysia's experience, avoiding the mistakes made and taking on the recommendations of the study.

RECOMMENDATIONS

The major issue is performance expectancy, i.e. Malaysians lack awareness of the benefits of using MyKad ATM and MyKad Touch 'n Go. This is not surprising since the extensive government campaigns have only focused on changing the existing paper-based identity card to MyKad (Yeow et al., 2007). Thus, the Malaysian government, in collaboration with banks and Rangkaian Segar Sdn Bhd (RSSB – the owner of Touch 'n Go system), should promote the benefits of using MyKad Touch 'n Go and MyKad ATM, e.g. having fewer cards in wallet, shorter queuing time, faster payment for parking, transit ticket, and toll, lesser need of carrying cash, and savings of the original cost of Touch 'n Go or ATM card. However, it is possible to be informed yet be skeptical. As an example, the "Forum on Risks to the public in computers and related systems" (RISKS Forum) (an ACM-affiliated list) has repeatedly included skeptical comments about the benefits of smart cards such as those introduced in Malaysia and Bahrain. Some of the risks highlighted include the risks of having all personal, medical, financial, and educational data in one convenient, centrally informatic location (Mannes, 2003). Possibly, this problem cannot be overcome as skeptics are likely to exist in any computerized information system.

The study has found that Malaysians are unsure of how much effort is needed to adopt MyKad Touch 'n Go and MyKad ATM. Hence, it is recommended that MyKad holders should be informed that the use of MyKad Touch 'n Go and MyKad ATM is just like that of Touch 'n Go and ATM cards. With this knowledge, it will eliminate the need for new adopters to learn about their usage and thus encourage more adopters.

As for social influence, the status quo is using Touch 'n Go and ATM cards because Malaysians perceive MyKad Touch 'n Go and MyKad ATM to be unpopular. RSSB and banks should encourage and attract more users to adopt MyKad Touch 'n Go and MyKad ATM by selling the benefits such as savings of RM10, the convenience of having the applications always (as citizens carry their MyKad wherever they go), etc.

As for facilitating conditions, Malaysians are unsure of the procedure for activating MyKad Touch 'n Go and MyKad ATM. They are also uncertain whether the applications are recognized in most toll booths and ATM machines. Moreover, few banks allow activation of MyKad ATM. Thus, it is recommended that these issues be resolved by increasing awareness about the activation procedure and recognition of the applications and having more activating location for MyKad ATM.

As for perceived credibility, Malaysians believe that people can steal money from both applications by just possessing their MyKad. This misconception can be corrected by educating MyKad holders on the triple-DES encryption technology used in MyKad applications, i.e. it has 2^{128} combinations and breaking the combinations is almost impossible. They should also be informed that they can get a refund of the balance of their MyKad Touch 'n Go and a new MyKad ATM if they lose/damage their MyKad. Besides, banks could increase perceived credibility by introducing biometric verification in MyKad ATM because MyKad stores thumbprint templates which can be used to conduct the verification.

Some Malaysians experience anxiety of losing MyKad while using its applications because the card is also a National Identity Card. Malaysians are unsure if extensive use of the card would cause damage. These fears can be allayed if they are better informed about the durability of the card (approximately 10 years / 100,000 read-write cycles) and the ease of changing to a new card in the event of card damage. They just submit their damaged card (without requiring a police report or an oath in the court) to the National Registration Department in exchange for a new card. The processing is quick and the charge is nominal, i.e. RM10 (USD2.7).

If all these recommendations are taken, a conducive environment will be created for the use of MyKad ATM and MyKad Touch 'n Go, thus they will be more acceptable for Malaysian citizens.

LIMITATIONS OF THE STUDY AND FUTURE STUDIES

Due to the constraint on the length of paper, moderating variables such as education level, income, age, car ownership and other relevant moderating factors were not examined. The results are only applicable to Malaysian as all the subjects were from the country.

Future studies can be conducted to discover the effects of these moderating factors on the acceptability of the applications. For example, investigating the differences between non-users and users' rating on MyKad ATM and Touch 'n Go acceptance factors as well as their behavioral intention to use both applications. Moreover, user acceptance models for the other applications of MyKad such as passport information, health information, electronic purse, public key infrastructure, and frequent traveler card could be developed based on the model presented in this paper. The study can be replicated in other countries that are implementing multipurpose smart identity card which is similar to that of MyKad (such as Hong Kong, India, and the sultanate of Oman) by using the models developed in the study. They may measure user acceptance, identify possible technology adoption problems, and find solutions using the proposed model. The present study found that Malaysians lack understanding of MyKad. Thus, future studies can include training and advertisement levels as independent variables. Last but not least, the study could be improved by generating a dialogue with members of the public, through focus groups or other open-ended discussion. This would provide a rich source of data about public attitudes.

ACKNOWLEDGMENT

The authors thank the Ministry Of Science, Technology, and Innovation (MOSTI), Malaysia for providing the Industrial Research Prioritize Area (IRPA) fund, the respondents, and Multimedia University for their cooperation and support. The authors also thank Christina Tong for editing the paper.

REFERENCES

Agarwal, R., & Prasad, J. (1998). A conceptual and operational definition of personal innovativeness in the domain of information technology. *Information Systems Research*, *9*(2), 204–215. doi:10.1287/isre.9.2.204

Al-Fakhri, M. O., Cropf, R. A., Kelly, P., & Higgs, G. (2008). E-Government in Saudi Arabia: Between Promise and Reality. *International Journal of Electronic Government Research*, *4*(2), 59–85.

Bandura, A. (1986). *Social foundations of thought and action: a social cognitive theory*. Englewood Cliffs, NJ: Prentice Hall.

Carlsson, C., Carlsson, J., Hyvonen, K., Puhakainen, J., & Walden, P. (2006). Adoption of mobile devices/services-searching for answers with the UTAUT. Proceedings of the 36th Hawaii International Conference on System Sciences (2006) IEEE.

Chau, P. Y. K. (1996). An empirical assessment of a modified technology acceptance model. *Journal of Management Information Systems*, *13*(2), 185–204.

Chau, P. Y. K., & Hu, P. J. H. (2001). Information technology acceptance by individual professionals: a model comparison approach. *Decision Sciences*, *32*(4), 699–719. doi:10.1111/j.1540-5915.2001.tb00978.x

Compeau, D., Higgins, C., & Huff, S. (1999). Social Cognitive Theory and Reactions to Computing Technology: A Longitudinal Study. *Management Information Systems Quarterly*, *23*(2), 145–158. doi:10.2307/249749

Compeau, D. R., & Higgins, C. A. (1995). Computer self efficacy: development of a measure and initial Test. *Management Information Systems Quarterly*, *19*(2), 189–211. doi:10.2307/249688

Davis, F., Bagozzi, R., & Warshaw, P. (1992). Extrinsic and Extrinsic Motivation to Use Computers in the Workplace. *Journal of Applied Social Psychology*, *22*(14), 1111–1132. doi:10.1111/j.1559-1816.1992.tb00945.x

Davis, F. D. (1989). Perceived usefulness, perceived ease of use, and user acceptance of information technology. *Management Information Systems Quarterly*, *13*(3), 319–339. doi:10.2307/249008

Davis, F. D. (1993). User acceptance of information technology: system characteristics, user perceptions and behavioral impacts. *International Journal of Man-Machine Studies*, *38*, 475–487. doi:10.1006/imms.1993.1022

Davis, F. D., Bagozzi, R. P., & Warshaw, P. R. (1982). Extrinsic and Intrinsic Motivation to Use Computers in the Workplace. *Journal of Applied Social Psychology*, *22*(14), 1111–1132. doi:10.1111/j.1559-1816.1992.tb00945.x

Davis, F. D., Bagozzi, R. P., & Warshaw, P. R. (1989). User acceptance of computer technology: a comparison of two theoretical models. *Management Science*, *35*(8), 982–1002. doi:10.1287/mnsc.35.8.982

Davis, F. D., & Venkatesh, V. (1996). A critical assessment of potential measurement biases in the technology acceptance model: three experiments. *Internet Journal of Human-computer Studies, 45*(1), 19–45. doi:10.1006/ijhc.1996.0040

Dhar, S. (2003). Introduction to smart card. Data Security Management <http://sumitdhar.blogspot.com/2004/11/introduction-to-smart-cards.html> (Accessed March 1, 2007).

Dillon, A., & Morris, M. (1996). User acceptance of new information technology: theories and models. In Williams, M. (Ed.), *Annual Review of Information Science and Technology*. Medford, NJ: Information Today.

Doyle, E., Stamouli, I., & Huggard, M. (2005). Computer anxiety, self-efficacy, computer experience: an investigation throughout a computer science degree. October 19-22, 2005, Indianapolis, IN 35th ASEE/IEEE Frontiers in Education Conference S2H-3.

Fishbein, M., & Ajzen, I. (1975). *Belief, attitude, intention and behavior: an Introduction to theory and research*. Reading, MA: Addison-Wesley.

Gail, E. T., Marshall, I. M., & Jone, S. (1995). One in the eye to plastic card fraud. *International Journal of Retail and Distribution Management, 23*, 3–11. doi:10.1108/09590559510089195

Gilbert, D., Kelly, L. L., & Barton, M. (2003). Technophobia, gender influences and consumer decision-making for technology-related products. *European Journal of Innovation Management, 6*(4), 253–263. doi:10.1108/14601060310500968

He, D., & Lu, Y. (2007). Consumers' perceptions and acceptances towards mobile advertising: an empirical study in China. IEEE. 3770-3773.

Horton, R. P., Buck, T., Waterson, P. E., & Clegg, C. W. (2001). Explaining intranet use with the technology acceptance model. *Journal of Information Technology, 16*(2), 237–249. doi:10.1080/02683960110102407

Hu, P. J., Chau, P. Y. K., Sheng, O. R. L., & Tam, K. Y. (1999). Examining the technology acceptance model using physician acceptance of telemedicine technology. *Journal of Management Information Systems, 16*(2), 91–112.

Igbaria, M., Guimaraes, T., & Davis, G. B. (1995). Testing the determinants of microcomputer usage via a structural equation model. *Journal of Management Information Systems, 11*(4), 87–114.

Ishahak, J. (2006). Smart card 'success' - gauged by number of cards issued or card usage? Available at <http://www.frost.com/prod/servlet/market-insight-top.pag?docid=4813276> (Accessed November 30, 2006).

Israel, G. D. (1992). Sampling the evidence of extension program impact. Program Evaluation and Organizational Development, IFAS, University of Florida. PEOD-5. October.

Jiang, J. J., Hsu, M. K., Klein, G., & Lin, B. (2000). E-commerce user behavior model: an empirical study. *Human Systems Management, 19*(4), 265–276.

Kannabiran, G., Xavier, M. J., & Banumathi, T. (2008). E-Governance and ICT Enabled Rural Development in Developing Countries: Critical Lessons from RASI Project in India. *International Journal of Electronic Government Research, 4*(3), 1–19.

Lu, J., Yu, C. S., Liu, C. and Yao, J. E. (2003). Technology acceptance model for wireless internet. Internet research: electronic networking application and policy 13(3), 206-222.

Malaysia, M. S. C. MSC flagship Applications Updates. Available at: <http://www.msc.com.my/updates/flagships.asp> (Accessed December 2, 2006).

Mannes, G. (2003). Bahrain's proposed smart ID cards. The Risk Digest: Forum on risks to the public and computers and related systems, 22 (89). Available at: <http://catless.ncl.ac.uk/Risks/22.89.html#subj8.1> (Accessed June 10, 2008).

Mathieson, K., Peacock, E., & Chin, W. W. (2001). Extending the Technology Acceptance Model: the influence of perceive user resources. *The Data Base for Advances in Information Systems*, *32*(3), 86–112.

Metzger, M. J., Flanagin, A. J., & Zwarun, L. (2003). College student web use, perceptions of information credibility, and verification behavior. *Computers & Education*, *41*, 271–290. doi:10.1016/S0360-1315(03)00049-6

Moore, G. C., & Benbasat, I. (1991). Development of instrument to measure the perception of adopting an information technology innovation. *Information Systems Research*, *2*(3), 192–222. doi:10.1287/isre.2.3.192

Plouffe, D. R., Hulland, J. S., & Vandenbosch, M. (2001). Research Report: Richness versus parsimony in modeling technology adoption decisions-understanding merchant adoption of a smart card-Based Payment System. *Information Systems Research*, *12*(2), 208–222. doi:10.1287/isre.12.2.208.9697

Poston, R. S., Akbulut, A. Y., & Looney, C. A. (2007). Online advice taking: examining the effects of self-efficacy, computerized sources, and perceived credibility, Available at: <http://sigs.aisnet.org/SIGHCI/Research/ ICIS_workshop_2005.html> (Accessed March 4, 2007).

Rangkaian Segar Sdn Bhd (RSSB). Facts and figures. Available at: <http://www.touchngo.com.my/MediaCentre_FF.html> (Accessed November 30, 2006)

Reddick, C. G. (2008). Perceived Effectiveness of E-Government and its Usage in City Governments: Survey Evidence from Information Technology Directors. *International Journal of Electronic Government Research*, *4*(4), 89–104.

Sam, H. K., Othman, A. E. A., & Nordin, Z. S. (2005). Computer Self-Efficacy, Computer Anxiety, and Attitudes toward the Internet: A Study among Undergraduates in Unimas. *Journal of Educational Technology & Society*, *8*(4), 205–219.

Singh, S. (2006). "RM500m letdown," New Strait Times (December 25, 2006).

Sun, J. (August 2005). User readiness to interact with information systems- a human activity perspective. Doctoral Dissertation, Texas A&M University

Taylor, S., & Todd, P. A. (1995). Understanding information technology usage: a test of competing models. *Information Systems Research*, *6*(2), 144–176. doi:10.1287/isre.6.2.144

Thomas, M. (2004). Is Malaysia's MyKad the 'One Card to Rule Them All'? The Urgent Need To Develop a Proper of Legal Framework for the Protection of Personal Information in Malaysia, Melbourne University Law Review.

Thompson, R. L., Higgins, C. A., & Howell, J. M. (1991). Personal Computing: Toward a Conceptual Model of Utilization. *Management Information Systems Quarterly*, *15*(1), 124–143. doi:10.2307/249443

Venkatesh, V., & Davis, F. D. (2000). A Theoretical extension of the technology acceptance model: four longitudinal field studies. *Management Science*, *45*(2), 186–204. doi:10.1287/mnsc.46.2.186.11926

Venkatesh, V., & Morris, M. G. (2000). Why don't men ever stop to ask for directions? Gender, social influence, and their role in technology acceptance and user behavior. *Management Information Systems Quarterly*, *24*, 115–139. doi:10.2307/3250981

Venkatesh, V., Morris, M. G., Davis, G. B., & Davis, F. D. (2003). User acceptance of information technology: toward a unified view. *Management Information Systems Quarterly*, *27*(3), 425–478.

Wang, Y. S., Wang, Y. M., Lin, H. H., & Tang, T. I. (2003). Determinants of user acceptance of internet banking: an empirical study. *International Journal of Service Industry Management*, *14*(5), 501–519. doi:10.1108/09564230310500192

Warrington, T. B., Abgrab, N. J., & Caldwell, H. M. (2000). Building trust to develop competitive advantage in e-business relationships. *Competitive Review*, *10*(2), 160–168.

Warshaw. (1980). A New Model for Predicting Behavioral Intentions: An Alternative to Fishbein. Journal of Marketing Research, 17 (2), 153-172.

Wu, Y. L., Tao, Y. H., & Yang, P. C. (2007). Using UTAUT to explore the behavior of 3G mobile communication users. IEEE. 199-203.

Yates (2005). Emerging technology scan: smart cards. Available at <http://www.emory.edu/smartcard.htm> (Accessed December 12, 2006).

Yeow, P. H. P., Loo, W. H., & Chong, S. C. (2007). User acceptance of multipurpose smart identity card in a developing country. *Journal of Urban Technology*, *14*(1), 23–50. doi:10.1080/10630730701259862

Yeow, P. H. P., & Miller, F. (2005). The Attitude of Malaysians towards MyKad. In N. Kulathuramaiyer, A. W. Teo, Y. C. Wang & C.E. Tan (Eds.), Proceedings of the 4th International Conference on Information Technology in Asia 2005 (CITA'05), Hilton, Kuching, 12-15 Dec 2005 (pp.39-44). Kuching, Malaysia: Faculty of Computer Science and Information Technology, Universiti Malaysia Sarawak.

This work was previously published in International Journal of Electronic Government Research (IJEGR), edited by Vishanth Weerakkody, pp. 37-56, copyright 2009 by Information Science Reference (an imprint of IGI Global)

Chapter 9
Digital Disempowerment in a Network Society

Kenneth L. Hacker
New Mexico State University, USA

Shana M. Mason
New Mexico State University, USA

Eric L. Morgan
New Mexico State University, USA

ABSTRACT

The objective of this article is to examine how the inequalities of participation in network society governmental systems affect the extent that individuals are empowered or disempowered within those systems. By using published data in conjunction with theories of communication, a critical secondary data analysis was conducted. This critical analysis argues that the Digital Divide involves issues concerning how democracy and democratization are related to computer-mediated communication (CMC) and its role in political communication. As the roles of CMC/ICT systems expand in political communication, existing Digital Divide gaps are likely to contribute to structural inequalities in political participation. These inequalities work against democracy and political empowerment for some people, while at the same time producing expanded opportunities of political participation for others. This raises concerns about who benefits the most from electronic government in emerging network societies.

INTRODUCTION

As the roles of computer-mediated communication (CMC)/information and communication technology (ICT) systems expand in political communication, existing Digital Divide gaps are likely to contribute to structural inequalities in political participation. [1] This is true for both within-nation and across-nation gaps. These inequalities work against democracy and political empowerment and produce social injustices at the same time as they produce expanded opportunities to political participation. Rather than assuming that increasing networking of societies leads to democratization, the broader relationship between the two needs to be examined.

Our examination responds to the larger question of how the structures of advanced societies

DOI: 10.4018/978-1-60960-162-1.ch009

are becoming increasingly networked and the role that CMC plays in both creating new social networks and restructuring existing ones, particularly in the political arena. We first present these structures followed by a discussion of the existing global Digital Divide, in which we point out the ethical concerns raised by allowing groups who could most benefit from connectivity to remain disconnected. Finally, we raise the important point that universal access may not be enough to solve the structural inequalities created by allowing segments of the population to remain disconnected. Rather, it is important to go beyond access and ensure that technology is used to reduce structural inequalities in the best ways possible by marginalized groups. By using published data in conjunction with theories of communication, a critical secondary data analysis was conducted. In this critical analysis, we conclude by offering recommendations for electronic government analysis and research from existing data and theories.

Network Society

Jan van Dijk (2006) defines network society in terms of communication networks that shape the most important forms of organization in a society. In what we have known for decades as mass society, citizens have been informed and entertained by mass media and somewhat disconnected from people outside of their primary (e.g. family, friends) and secondary social groups (e.g. workplace). In those nations that appear to have an emergence of network society characteristics, increasing numbers of social structures involve interconnected individuals using computer networks to seek out information, relationships, and networks of influence. In these societies, political power and politics are more about relationships among people than characteristics of individuals (van Dijk, 2006). Dimensions of geographical space are accompanied by a technological space. This space is sometimes referred to as social

geography, wherein social networks rather than physical space become the basis for closeness or distance. Political systems, which traditionally have been modeled as top-down organizational charts, may be changing into polycentric systems of power in which political power is based more on network position than traditional roles (van Dijk, 2006).

The consequences of people being connected to the new communication networks of network society are becoming more significant as participation in these networks is increasingly linked to tangible benefits. Network society perspectives of social organization and communication technologies include economics as well as politics. Indeed, economic reorganization is seen historically as the main impetus for the emergence of network societies (Stalder, 2006). Globalizing trade and finance make up an informational economy with the center of the global economy as finance (Stalder, 2006). Organizations become more flexible to meet changing markets and governments where changes and discontinuities constitute a new focus. As Castells (2001) notes, the organizational changes were enabled but not caused by communication technology innovations.

Power in network society social, economic, and political contexts can be viewed more as matters of position and network relations than of material or content advantages. Power in previous paradigms like Fordism or Weberian organizational assumptions was about getting others to do one's will. In contrast, power in networks is more about flows of influence, investment, and planning (Stalder, 2006). Barney (2004) argues that "access to networks and power to determine what flows over them is a significant marker of systemic advantage and disadvantage domestically and globally," (p. 178).

Communication technologies have always been central to both the exercise of power by the state and to the formation of public spheres of deliberation made available to citizens (Barney, 2004). While there is little evidence that CMC yet

139

has strong empowerment effects for those without extant power, there is a sense of democratic potential that does have some empirical support.

The Annenberg Digital Futures project notes several interesting trends in the use of the Internet by Americans. Sixty-five percent say that they are more involved with social activism (Digital Futures Project, 2007). Approximately 75% view government websites as reliable sources of information. A majority (59%) of Americans online believe that Internet usage can help them learn about their political system (Digital Futures Project, 2007). However, only 19% believe that their Internet usage gives them more voice in government (Digital Futures Project, 2007). The Pew Research Center (2008) reports that about 24% of Americans are now using the Internet as a major source of information about the 2008 United States presidential election—a number that is nearly double that of the election of 2004.

The Embedded Infrastructure

As Wellman and Haythornwaite (2002) indicate, the Internet is increasingly becoming embedded in the everyday lives of its users. This means that the Internet is incorporated into daily routines and provides a platform for numerous personal, social, economic, and political forms of communication and action. Its convenience facilitates many of the activities that were previously done offline. Thus, those who use the Internet are afforded an additional avenue of communication to facilitate important activities such as working at home, doing research for school, contacting friends, conducting commercial transactions, and communicating with government representatives. Howard, Rainie and Jones (2002) show that levels of usage experience accounts for the most significant differences between access and use of the Internet across groups. Those who have been using the Internet the longest are most likely to have access to it and to use it more heavily (Wellman & Haythornwaite, 2002). Longer-term users

tend to find ways to incorporate the Internet into all aspects of their lives, including personal and work environments.

The critical realization regarding CMC embeddedness is that a means of communication that was once necessary for a minority of citizens in a given population is now important for many, if not most, people in both developed and developing societies. This assumes that these societies are taking on the characteristics of network society.

As the Internet and CMC become embedded with economic, social, and political activities, citizens are likely to develop stronger needs to use the networks in order to maximize their abilities to participate in online opportunities or social formations. Those who become most skilled and active with CMC networking are more likely to gain power than those without these skills and activities. This means there may be accelerating gaps in network sophistication. As van Dijk (2002) notes, digital skills are cumulative. Accordingly, the inequalities resulting from their increasingly embedded nature are cumulative as well. Holderness (1998) argues that the Digital Divide gaps that we have been discussing may become self-reinforcing. Those individuals and nations who accelerate their use of CMC systems build their communication capital at rates that perpetuate how far they stay ahead of others in networking.

Exclusionary Forces

It is generally accepted that the increasing organization of societies with the use of CMC technologies facilitates the importance of information and knowledge for economic growth and a shift of importance from densely-knit bounded groups to computer-supported social networks (United Nations, 2004). The emergence of network societies entails social and organizational formations that are constructed in relation to flows of symbolic interaction more than in relation to traditional institutional, governmental, and organizational boundaries (Contractor & Monge, 2003).

Networks are comprised of nodes; these nodes are connected by communication and join together to become influence networks (Castells, 2000). When a node does not connect to other nodes, it may be dropped from the network. Such nodes are then excluded from exercising influence on social organization. Those who are part of the networks that exert influence on society can work to increase the impact of their influence by stimulating changes in, or reinforcing, existing patterns in the social structures that are beneficial to them.

Those with the most power and resources tend to be the early adopters of new technologies, and their influence shapes the evolution of technological changes in society (van Dijk, 2006). Thus, social inequalities may be perpetuated as those who use the technologies are increasingly organizing social networks around them. The inability to access or make effective use of the Internet and computers becomes increasingly significant as those with power make their use increasingly prominent in all areas of society. Those who do not have access to new forms of communication technology are increasingly excluded from the organization of society on many levels. This suggests that increasing networking that is accompanied by increasing gaps in usage for government and political communication may disempower many citizens in any nation moving toward network society status.

The notion of disempowerment with such a participatory means of communication may seem paradoxical at first blush. However, once one realizes that new power in network societies is strongly linked to influence over system configurations, position within networks and control over information flows, it is no longer surprising that those with greater connectivity, centrality, and interactivity are those in a society that will benefit the most from network technologies of communication. Moreover, simply being connected to new communication networks does not assure any degree of influence or power. In fact, connection without power is likely to assure that

the connected person is subjected to new forms of domination by those with more control over the information flows and configurations of the networks (Barney, 2004).

POLITICAL IMPLICATIONS OF THE DIGITAL DIVIDE FOR ELECTRONIC GOVERNMENT

There are two major categories of electronic political communication that are gaining increasing attention today: electronic government (e-government) and digital democracy. According to van Dijk (2006), e-government concerns service provision and communication between government agencies and citizens, while digital democracy concerns participation in political deliberation.

There is little indication that CMC is drawing new people into democratic political processes, but there is substantial evidence that people who already participate are becoming more enabled in their participation (Bimber & Davis, 2005; van Dijk & Hacker, 2000). It is easier for CMC users to contact governmental officials, obtain government documents, and join political discussions with people they do not know (Pratchett, Wingfield, & Polat, 2006; van Dijk, 2004). Indeed Weerakkody & Dhillon (2008) note that in the U.K., electronic government now exists in a transformational phase that could allow for increased civic participation. Bimber & Davis (2005) argue that CMC is providing effective tools for political activities and mobilization, but that "the divide between those who are political activists interested in electoral campaigns and those who are not will expand" (p. 168).

Without the knowledge and ability to evaluate policies and potential leaders, citizens cannot engage in the democratic process in its true sense (Barber, 1984; Yankelovich, 1991). However, as Yankelovich (1991) maintains, information given to citizens in a downward flow means that they possess only that information passed onto them by elites. Receiving information in this type of downward flow pattern does not necessarily empower citizens; rather, it can serve to reinforce existing power structures as citizens maintain the passive role of consumers of information generated by the elite, who maintain control over all information (Bordewijk & van Kaam, 1986; van Dijk, 1996). If high CMC users have more multilateral political communication than low CMC users, the latter are less likely to develop empowering roles for themselves in the polycentric power structures that appear to be part of network societies.

Political movements have been employing the Internet to organize their struggles, and some of these users are developing a practice known as "self-directed networking" (Castells, 2000, p.55). Self-directed networking involves people inventing personal ways of organizing and disseminating information. As more formal political structures such as civic organizations have less public membership today, political movements can employ CMC to effectively mobilize political action (Castells, 2001). Those who are involved with online politics have an advantage over those with less involvement since online politics are becoming more common and influential.

E-government can be used in any type of political system. Issues of digital disempowerment may not be weighted highly in totalitarian societies that use e-government solely for efficient delivery of information and collection of information from citizens. In democratic systems, those who administer e-government must confront the expectations of citizens concerning active influence of the people on how their government reaches decisions, sets policies, and interacts with citizens. Structural inequalities work against democratic governance because structural inequality is related to positions in networks that privilege various groups of citizens over others (van Dijk, 2006).

THE NONTRIVIAL NATURE OF LOW PARTICIPATION IN NETWORK SOCIETY

Communication researchers now know that old media concepts, theories and research paradigms are no longer adequate to explain networks of electronic communication, CMC, and the interconnected and interoperable systems of communication technologies that make up the Internet and World Wide Web. The old media, which assumed a linear source-receiver process, led to deterministic accounts of media effects. These accounts fail to explain newer media technologies, which are characterized by interactivity, interconnectedness, and complexity that is simply not present in the isolated old media paradigm. With increasing use of the Internet and network communication systems, individuals, groups, organizations, and nations are able to develop many-to-many forms of interaction with networks that have shifting configurations. The essential point in understanding network society as a sociological and communicative concept is that individuals and groups are now creating new social formations. This is sometimes referred to as social morphology in the ways people make contacts, establish connections, and regulate interactivity and feedback. This allows people to build affiliations that facilitate the flow of social, cultural, and political capital. As van Dijk (2006) notes, the statistical pattern thus far for network societies is that a small percentage of a population becomes the center of the most important economic and political networks. These people have the most power and influence in the network society and unless the center expands to include increasing numbers of citizens, electronic government systems may empower those already empowered. At the same time, however, the cumulative nature of CMC skills acquisition and development leaves out many people. This leads to a stabilizing of structural (network) inequalities over time, ultimately resulting in the disempowerment of those with little or no connectivity.

People with influential network positions and flows are more likely to be more interest-bound than place-bound in how they associate and work with others. Those who are left as more place-bound become less and less important to those who are sharing capital flows in cosmopolitan networks of association and influence. Barney (2004) argues that the less connected are likely to be passive consumers of communication content rather than active creators of messages and content.

E-governments can exacerbate the problems just described when they do not add political value to service provision. In other words, e-governments have opportunities to encourage more political participation in governance as well as more transparency in decision-making, but they rarely seize these opportunities. For example, Norris (2001) observes that government websites rarely publish information like citizens' reactions to policy proposals. E-government is more likely to be used to enhance the efficiency of information access than the democratization of governance (Barney, 2004).

THE GLOBAL DIGITAL DIVIDE

United Nations research indicates that the Digital Divide has narrowed dramatically between member nations of the Organisation of Economic Co-operation and Development (OECD) and developing nations, from 80.6:1 to 5.8:1 in the past decade. The gaps persist, however. "In 2005, half of all OECD citizens were Internet users, compared with just one in every twelve citizens in developing economies and one in every one hundred in Least Developed Countries (LDC's)" (International Telecommunications Union, 2007, p.22).

Additionally, as gaps close in one area, they shift to others. The "quantity" gap is being replaced by the "quality" gap, a phenomenon addressed by early Digital Divide researchers who argued that closing one gap would simply open another.

Although the gap in the ratio of broadband users in OECD and developing countries has shrunk, "the absolute gap measured in percentage points has grown almost tenfold between 2000 and 2005," (International Telecommunications Union, 2007, p.23). Thus, with the increased dominance of broadband in the marketplace, this gap becomes increasingly significant (OECD, 2007).

There are dangers in the acceleration of a broadband divide that follows the existing Digital Divide gaps among people with the same forms of Internet access. Broadband is projected to become more important as Web sites will increasingly be designed for broadband, and services like Internet telephony become more commonly used (Vanston, Hodges & Savage, 2004). Along with increasing bandwidth capability and speed, CMC users need to have personal computers with increased amounts of processing speed and memory (Vanston, Hodges & Savage, 2004). As computing and CMC become more ubiquitous, devices will continue to become more sophisticated, interconnected, and operable as nodes in personal communication networks.

Users in high-income countries accounted for 74% of broadband users globally in 2005 (International Telecommunications Union, 2007). China alone accounted for 87% of broadband subscribers in lower-middle income economies, while India and Vietnam accounted for 94% of subscribers in low-income countries. Though there is availability in many of these regions, the price makes access difficult. For example, the 2007 World Information Society report maintains that broadband access in Cape Verde is available for over 2,000 USD per 100 kbits/month, while the same access in Japan is available for less than 10 USD/month. On average, users in low-income economies can expect to spend 900 times their average income on broadband access, while those in high-income countries spend about 2% of their average monthly income for the same access. While Internet access will not fix non-technological problems, it can increase information sharing, knowledge accumulation, and work collaboration through networking. Indeed, the United Nations report states that "developing countries risk being left further behind in terms of income, equality, development, voice and presence on an increasingly digitized world stage" (International Telecommunications Union, 2003, p. 4).

It is important to recognize the fact that there are many areas of Digital Divide gaps that involve much more than the commonly referenced ones of physical access (computer and net access). Kotamraju (2004), for example, notes that women tend to be employed in Web sites design more than in Web sites programming even if they have both sets of skills. While schools are more connected to the Net, studies show that few teachers know how to use the technology to augment their classroom instruction. The students attending Internet-wired schools may not be developing the skills they need to function well in an Internet-based economy. The gaps in ethnic and social class levels of learning may be worsened by this pattern of poor teaching proficiency. While there is expanding diversity, there are also gaps in usage and skills as well as in abilities to pay for what is becoming less free in new media and moving toward conditional access (pay-per-usage) models of network access (van Dijk, 2004).

Norris (2001) argues that access to the information and communication opportunities offered by the Internet may be most consequential in the poorest nations. The lack of distance barriers and relatively cheap implementation of the Internet (once access is possible) allow business owners in countries such as Mexico the opportunity to participate in the global marketplace. Health information and education are available via the Internet in areas like Kolkata, India as they are to doctors in New York. Physicians in developing nations can network and share information and resources with those in more developed nations through the Internet. Distance education allows increased access to sophisticated educational tools, enabling universities in disenfranchised nations to offer educational tools and training comparable to those in industrialized nations.

According to the OECD (2004), the results of natural disasters, such as the earthquake and accompanying tsunami that struck nations around the Indian Ocean in 2004, are also lessened by new communication technologies. These are thought to provide important tools to warn of the impending catastrophe, mitigate its impact by speeding information and relief efforts, and provide a place for victims and family members to post messages and pictures regarding the missing.

Additionally, Norris (2001) maintains that the Internet may increase the mobilization of grassroots campaigns and their visibility, enabling groups to network and share resources in order to impact policy makers at a higher level. "Foreign policymakers…can no longer assume that the usual diplomatic and political elites can govern political affairs with a passive 'permissive consensus' without taking account of the new ability for public information, mobilization, and engagement engendered by the new technology" (Norris, 2001, p. 2). In the former Soviet Union, for example, the Internet network Relcom is credited with playing a significant role in the dissemination of information during the coup attempt of 1991 (Press, 1993).

Marginalized societies can become more marginalized as societies become more globalized and information is increasingly the most valuable commodity (Norris, 2001; Norris & Curtice, 2006). The differences in economic growth between those nations that have reliable, high-speed access to the Internet and those who do not may be exacerbated as the affluent nations are able to profit from increased visibility and productivity. Low literacy levels, language barriers, and income are key obstacles to Internet adoption for those in developing countries (OECD, 2004). Floridi (2001) argues that members of these societies are marginalized by the Digital Divide because they "live in the shadow of a new digital reality, which allows them no interaction or access, but which profoundly affects their lives" (p. 3).

Education and attitudes about the importance of connectivity may be as significant as lack of access globally. In China, about 10% of the population uses the Internet, and the majority of these are young (70% under age 30) and male (60%) (Fallows, 2007). Farmers and peasants comprise about 0.4% of Internet users. One in three non-users in China reports insufficient skills to use the Internet, and one-third of non-users lack access. Fallows (2007) quotes a farmer from rural China who compares computers to aircraft carriers because neither has any significant relevance to his life.

Implications of the Gaps

van Dijk (2004, 2006) argues a "Matthew Effect" (2004, p. 20) for CMC adoption. This effect (based on the Bible passage "unto every one who hath shall be given") indicates that those who already have high-quality Internet and CMC access and usage patterns are gaining more and more network power while those who do not are losing their ability to catch up (van Dijk, 2004). As information becomes more important in jobs and everyday routines, the Matthew Effect becomes more deleterious for those with less CMC usage experience. Digital skills and usage are becoming more important for increasing numbers of professions and jobs. Thus, those with access and enhanced usage tend to become more valuable to their employers in the workplace. As distance education and online learning become more common and accepted, those with online usage and skills have easier access to educational courses and degrees. Research shows that those who combine online communication with offline social interaction expand their social networks and increase their social capital (van Dijk, 2004; Wellman and Haythornwaite, 2002).

Because of the distributed nature of network organizations like international corporations, division of labor becomes more selective, which means that employers can hire people in remote locations. This is why an American insurance company can hire typists in Ireland and save wage costs in contrast to hiring typists in the U.S. (van Dijk, 2006). The best quality jobs will be those that involve activities related to what people most connected to the centers of emerging networks are doing. The jobs of the least quality will continue to exist on the periphery of networks and will involve individuals with low connectivity, usage, and positioning (van Dijk, 2006). Digital disempowerment is likely to increase for those employees who fall behind in learning the technical skills for which there is increasing demand.

The data regarding Digital Divide issues show three important generalizations which when added together indicate the likely digital disempowerment trend for many people in the world. First, CMC usage continues to accelerate the expansion of networks that link people to economic and political influence. Second, CMC usage is related to tangible benefits such as increasing one's social capital. Third, CMC usage gaps and poor positioning in network society networks are related to diminished opportunities in advanced societies when compared to high usage and effective positioning.

THE EMERGENCE OF STRUCTURAL INEQUALITIES

Globalization increases as economic, political and cultural activities of nations become more interdependent. Within one globalization structure, a nation's position can be determined by its pattern of interactions with other nations (Barnett, 2001). This formulates a three-tiered structure of nations and societies such that those with increased interconnectivity and potential for interconnectivity represent a core group, with other nations representing "semi-peripheral" and "peripheral" groupings accordingly (Chase-Dunn & Grimes, 1995). Those nations that are most central in the global network are also those with the highest

GDP. Barnett's (2001) network analysis of international telecommunications from 1978 to 1996 indicates that the global network has become more centralized and more integrated. Moreover, the study showed that more information is flowing through the core nations (USA, Canada, Japan, and Western European nations) rather than being exchanged with nations at more peripheral network positions (Barnett, 2001).

The inability of subpopulations to have access to the global network infrastructures is diminishing their abilities to be as competitive and influential as those populations which do have input and position in the expanding networks of capital, influence, and power. Each developing economy becomes more dependent on CMC networks for commerce, government, education, and various social services (Montagnier, Muller & Vickery, 2002). The most educated citizens may also leave these countries for the economic opportunities offered by core nations, causing a "brain drain" that further inhibits progress (Bridges.org, 2003/2004).

The reality of structural inequalities that produce what we are calling disempowerment is seen in the evidence that a) CMC and Internet skills are cumulative, and b) Digital Divide gaps persist and regenerate with each new communication technology innovation (van Dijk & Hacker, 2000; van Dijk, 2004). Those citizens who could reap the most benefits from the democratic potential of the Internet are those who are already politically marginalized. In other words, the people who need these benefits most are those who have the least amount of access and skills (Hacker & Mason, 2003).

In Europe, citizens with few or no skills, as well as the unemployed, comprise the majority of those who use government services, yet are a minority of Internet users (O'Donnell, 2002). Thus, the increased information, communication, and access to these programs afforded by electronic government enterprises in Europe go unused by the majority of those whose need is greatest. We see a similar pattern in the United States among disabled people. The common picture is that those with the greatest needs for CMC are those with the least usage.

There is some support for the mobilization hypothesis (Norris, 2001), which asserts that some traditionally less active groups may be mobilized to engage in political activity by the low communication costs of the Internet. For example, Muhlberger (2002) found that, if given the opportunity, online discussion is employed at a slightly higher rate by those with less education, women, those who do not own a home, and those who are young, all of whom are generally less involved in political activities. Thus, there is evidence that previously uninvolved citizens might take a more active political role if access and usage obstacles did not exist. If left without access, however, those members of uninvolved and marginalized groups will continue to lag behind those of other groups, creating new forms of inequality as the opinion of those who participate in online discussion influences policymakers.

When a new avenue of access becomes available that would facilitate citizens' ability to make informed decisions about policy, to communicate with representatives, and allow for more equal opportunities to influence decision-making, it would seem to follow that governments should take measures to enable access to this important platform of social and political communication, serving as a check to ensure equal access to the process. Muhlberger (2002) argues that if the Internet enables citizens to exert political influence and obtain political information, then its representativeness is at issue. "Those concerned with the development of a democratic public sphere need to be aware of the representativeness of Internet political activity," because "...an Internet that over-represents some political views advantages those views relative to others" (Muhlberger, 2002, p. 2). If we accept that the possibility of increased political influence exists via the Internet, then we must consider that the potential for power imbalances to be created (or exacerbated) also exists when some members of a society may exercise this influence, while others are excluded due to economic, educational, and other social factors.

It is important to note that closing Digital Divide gaps might do more for e-commerce than for democracy in situations where there is no strong political will for democratization. We should also recognize that political will to close gaps in power exists at various levels of a political system, including those who govern and those who are governed. When both of these agree that increasing political participation is necessary for democratization, CMC can be substantially useful for democracy. Democratic systems without strong political will of their citizens are not likely to benefit from political CMC. If CMC is not politically useful, the gaps in various divides do not raise the ethical issues that they might otherwise. In other words, the more important CMC is for the democratic nature of a system, the more unethical it is to have social exclusion for CMC access, usage, and content.

The research on the Digital Divide makes it clear that connectivity remains an unsolved problem for realizing digital democracy. Within the United States, there are pockets of Americans who are living more and more on the periphery of the network society. Hoping that digital democracy can repair the problems of offline democracy is a strong issue for intellectual debate. However, the longer significant groups of people lack meaningful participation in their political system, the more likely that the system will not change for the better and that structural inequalities will stabilize.

Hacker (2002a, 2002b, 2004) argues that the issues of Digital Divide gaps, whether national or global, will not be resolved without political will that is deliberately aimed at increasing citizen participation in digital democracy. Political will stems from political culture and the abilities and willingness of leaders and citizens to make practices match values. Naïve notions about digital democracy can emerge when one does not address political culture and the differences in democratic

systems. For example, the political system in the United States contains a form of elitism by which most Americans remain mildly involved in politics and trust their leaders to do most of the actual policy making. Thus, to understand why most American leaders are not encouraging digital democracy past the point of e-government and freedom for citizen discussions, one has to examine American political culture and its history. Today, we usually think of a political system as being democratic if political decisions ultimately must be accounted for to the people of the nation in question (Scruton, 1982).

A global economic infrastructure, as envisioned by Bill Gates and others, is not the same thing as the public spheres for democratic communication envisioned by scholars of political communication. Couldry (2003) argues that most developed national governments have focused more on global digital economies than on digital democracies. This focus holds more concern for expanding markets than concern for making sure that citizens are not socially excluded from important spaces of political deliberation. This focus also neglects the need for content that helps disadvantaged people find sources and spaces to improve their social and political positions by helping them with job training, job searches, and other information that is useful for them. As Menou (2002) maintains, the focus of many efforts by the private sector to close the Digital Divide is to make consumers out of the poor. "What should really be at stake is social change and not the marketing of ICT's" (Menou, 2002, para. 3).

Couldry (2003) observes a scale-extension/scale-reduction effect of CMC. While CMC expands communication, it also create a gap between the literate and nonliterate. Just as the nonliterate people would stay in the market squares while the literate deliberated in the coffee houses, experienced CMC users may develop exclusive spaces for deliberation that, by their nature, simply are not inviting to inexperienced CMC users. Couldry draws attention to thical concerns regarding presence light CMC users do different things than heavy users. Heavy users, for example, are less passive in their use of the Internet and are more likely to disseminate information and create content (Couldry, 2003).

One problem with research done on the Digital Divide as well as with governmental approaches to implementing e-government has been a lack of communication theory as well as political theory (van Dijk & Hacker, 2000). A deliberative design model of political CMC could build upon theories of deliberative democracy from which ethics concerns emerge which say that it is wrong to have people non-connected, absent, or socially excluded by hierarchies in political CMC. Deliberative democracy theory says that citizens should have the opportunities to actively participate in decisions made about policies that affect them (Couldry, 2003; Dryzek, 1990). Dryzek's (1990) deliberative design principle says that citizens should have spaces for recurrent social interaction about politics where they can communicate only as citizens and not as representatives of any governmental, corporate or hierarchical organization. This principle can now be realized more easily today than in the past with the increasing prevalence of CMC such as political blog usage. This concept differs from Habermas' concept of the ideal speech situation in that it recognizes that much of deliberation about politics will involve emotional interaction and not always appear rational (Couldry, 2003).

Hacker and Mason (2003) argue a strong nexus which links issues of political power and issues of ethics. Political policy is often formulated on the basis of factual information and observation, but values serve as the filters through which those facts are used to implement policy. Research is done and facts are generated about social problems, but values inform what is done about them. Ethics considerations are a necessary component of policy making because ethics establish whether or not something is a problem and, if it is, what the best course of action is to remedy it.

Those who argue that digital exclusion is not a problem because some groups do not actually need access take an ethical position that says it is morally acceptable to allow some groups to be excluded from the social networking that the Internet enables. Social inequities are legitimized by arguments that some groups do not need access or are not being adversely affected by digital exclusion in the face of documented and potential benefits of connectivity. Additionally, policies implemented to facilitate access are not free of ethical considerations. It may be unethical, for example, to argue that some groups are unable to become digitally connected on their own, without governmental assistance. This may also further extend stereotypes about some groups among the groups themselves and society in general.

DIRECTIONS AND RECOMMENDATIONS

Public spheres of deliberation are vital to democratic political systems. Electronic government technologies which add more citizen deliberation, political interactivity between leaders and citizens, and greater debate about various issues, are likely to help citizen motivation for getting involved with electronic government and digital democracy (Chen & Dimitrova, 2006). However, studies show that governments rarely use their network technologies to do these things (Barney, 2004).

Market-based arguments assume that digital inequities go away with continued adoption and diffusion of communication technologies. This ignores the fact that computer-based communication technologies are more interdependent and more cumulative in usages, networking, and required skills than old media which were functionally independent (van Dijk, 2004). Universal-access arguments assume that governments must provide access to everyone because they cannot function in modern society without such access, and the markets are insufficient to provide affordable access. These arguments ignore the fact that some people can prosper without CMC and that market independence does, in fact, help high-technology companies innovate new communication products and services.

Where particular groups of people appear to be marginalized in CMC networks and creation of content, there should be efforts to give them voice from a perspective developed here that brings together political theory and communication theory. The United Nations 2004 Human Development Report argues that "unless people who are poor and marginalized—who more often than not are members of religious or ethnic minorities or migrants—can influence political action at local and national levels, they are unlikely to get equitable access to jobs, schools, hospitals, justice, security, and other basic services" (United Nations Development Programme 2004, pg. v).

According to Bennett and Entman (2001), "access to communication is one of the key measures of power and equality in modern democracies" (p.2). As a form of communication that offers democratic potential unique from previous types of media (Bentivegna, 2002), such as the telephone, access to CMC and the Internet is arguably such a measure. CMC and the Internet offer citizens the opportunity to exercise control over content, offer opinions, exert pressure on the government, and actively participate in its politics. Additionally, they offer both citizen-to-citizen and citizen-to-official communication opportunities, reduce the role of the media as gatekeepers of information and allow citizens access to previously unavailable (or very difficult to obtain) information. Also unique from previous forms of media, the Internet and CMC allow small groups and movements to acquire visibility that would have been unavailable to them in media such as television due to high costs. Finally, the speed and absence of boundaries offered by the Internet allow for quick mobilization of citizens with similar concerns and unlimited contact and communication among them. However, if groups most in need of

these access opportunities continue to be excluded, their marginalization may be increased. In such a case, digital disempowerment is realized.

High CMC exclusion does not mean that people have no voice in governance, but rather that they have less than they would if they were able to employ CMC as a key resource in creating or changing social structures related to political issues and causes. The provision of universal access, similarly, does not guarantee radical social restructuring. Menou (2002) argues that the focus of the Digital Divide debate should not be how to bring the technology to the marginalized, but to discover the best ways for those who need the technology to put it to use and improve their network positions. It is important to keep in mind that online inequalities often mirror offline ones, and existing social problems will not be undone by technology. It is also necessary to understand the role of CMC in political structuration and how it may magnify or mitigate inequalities.

CONCLUSION

In this article, we have attempted to present an argument saying that CMC/ICT systems continue to have democratic potential and can be useful for extending political deliberation that is necessary for democracy. This has profound implications for the conceptualization and implementation of e-government systems. This position holds that it is morally wrong to have these systems develop and expand in ways that give more political power to those who are already ahead in how much political influence they have while not providing more political access to those who tend to lag behind in political power. The key, we argue, is to have political will among leaders, among citizens, and within various social groups, to provide CMC access, training, content creation, usage opportunities, and encouragement in order to make e-government and digital democracy more open to newly participating citizens. This kind of will can also facilitate citizens having more meaningful political deliberation that has actual and viewable effects on political governance. By understanding the way that technology can be used to change and/or exacerbate existing power structures, we shed light on the Digital Divide that goes beyond issues of haves and have-nots and considers the implications of connectivity from a network society perspective. The argument presented here is not just that gaps should be closed, but that allowing the gaps to persist exacerbates structural inequalities, and this possibility is an important consideration for citizens and leaders alike in democratic societies.

REFERENCES

Barber, B. (1984). *Strong democracy: Participatory politics for a new age.* Berkely, CA: University of California Press.

Barnett, G. A. (2001). A longitudinal analysis of the international telecommunication network, 1978-1996. *The American Behavioral Scientist, 44*, 1638–1655.

Barney, D. (2004). *The network society.* Malden, MA: Polity Press.

Bennett, W. L., & Entman, R. (2001). *Mediated politics: Communication in the future of democracy.* New York: Cambridge University Press.

Bentivegna, S. (2002). Politics and new media. In Lievrouw, L. A., & Livingstone, S. (Eds.), *Handbook of new media* (pp. 50–61). London: Sage Publications.

Bimber, B., & Davis, R. (2005). *Campaigning online.* New York: Oxford University Press.

Bordewijk, J. L., & van Kaam, B. (1986). Towards a new classification of teleinformation services. *Intermedia, 14*, 16–21.

Bridges.org. (2003/2004). *Our perspective on the digital divide.* Retrieved August 7, 2005 from http://www.bridges.org/perspectives/digitaldivide.html

Castells, M. (2000). Toward a sociology of the network society. *Contemporary Sociology, 29*, 693–699. doi:10.2307/2655234

Castells, M. (2001). *The Internet galaxy: Reflections on the Internet, business, and society.* New York: Oxford University Press.

Chase-Dunn, C., & Grimes, P. (1995). World-systems analysis. *Annual Review of Sociology, 21*, 387–417. doi:10.1146/annurev.so.21.080195.002131

Chen, Y., & Dimitrova, D. (2006). Electronic government and online engagement: Citizen interaction with government via web portals. *International Journal of Electronic Government Research, 2*(1), 54–76.

Contractor, N., & Monge, P. (2003). *Theories of communication networks.* New York: Oxford.

Couldry, N. (2003). Digital Divide or discursive design: On the emerging ethics of information space. *Ethics and Information Technology, 5*, 89–97. doi:10.1023/A:1024916618904

Digital Futures Project. (2008). Available: http://www.digitalcenter.org/pages/current_report.asp?intGlobalId=19

Dryzek, J. S. (1990). *Discursive democracy.* New York: Cambridge University Press.

Fallows, D. (2007). *Chinese online population explosion: What it may mean for the Internet globally…and for US users.* Pew Internet & American Life Project. Retrieved November 11, 2007 from http://www.pewinternet.org/pdfs/China_Internet_July_2007.pdf

Floridi, L. (2001). Information ethics: An environmental approach to the digital divide. *Philosophy in the Contemporary World*, *9*(1), 1–7.

Hacker, K. (2002a). Network democracy and the fourth world. *European Journal of Communication Research, 27*, 235–260.

Hacker, K. (2002b). *Network democracy, political will and the fourth world: Theoretical and empirical issues regarding computer-mediated communication (CMC) and democracy. Keynote address to EURICOM.* The Netherlands: Nigmegan.

Hacker, K. (2004). The potential of computer-mediated communication (CMC) for political structuration. *Javnost/The Public, 11*, 5-26.

Hacker, K., & Mason, S. (2003). Ethics gaps in studies of the digital divide. *Ethics and Information Technology, 5*, 99–115. doi:10.1023/A:1024968602974

Holderness, M. (1998). Who are the world's information poor? In Loader, B. (Ed.), *Cyberspace Divide* (pp. 35–56). London: Routledge.

Howard, P., Rainie, L., & Jones, S. (2002). Days and nights on the Internet. In Wellman, B., & Haythornwaite, C. (Eds.), *The Internet in everday life* (pp. 45–73). Oxford, UK: Blackwell Publishers. doi:10.1002/9780470774298.ch1

International Telecommunications Union. (2003). World Summit. Retrieved November 1, 2007 at http://www.itu.int/wsis/index.html

International Telecommunications Union. (2007). *World Information Society Report 2007.* Retrieved November 2, 2007 at http://www.itu.int/osg/spu/publications/worldinformationsociety/2007/report.html

Kotamraju, N. (2004). Art vs. code. In Howard, P., & Jones, S. (Eds.), *Society online* (pp. 189–200). London: Sage Publications.

Menou, M. J. (2002, March). *Digital and social equity? Opportunities and threats on the road to empowerment.* Paper prepared for The Digital Divide from an Ethical Viewpoint, International Center for Information Ethics Symposium, Ausberg, Germany.

Montagnier, P., Muller, E., & Vickery, G. (2002, August). *The digital divide: Diffusion and use of ICTs.* Paper presented at the IAOS Conference, London, U.K.

Muhlberger, P. (2002, October). *Political values and attitudes in Internet political discussion: Political transformation or politics as usual?* Paper presented at the Euricom Colloquium: Electronic Networks & Democracy, Nijmegen, Netherlands.

Norris, P. (2001). *Digital divide: Civic engagement, information poverty, and the Internet worldwide.* Cambridge, NY: Cambridge University Press.

Norris, P., & Curtice, J. (2006). If you build a political website, will they come?: The Internet and political activism in Britain. *International Journal of Electronic Government Research, 2*(2), 1–21.

O'Donnell, S. (2002, October). *Internet use and policy in European union and implications for e-democracy.* Paper prepared for the European Colloquium, Nijmegen, Netherlands.

Organisation for Economic Cooperation and Development. (2004). *Regulatory reform as a tool for bridging the digital divide.* Retrieved June 26, 2005 from http://www.oecd.org/topic/0,2686,en_2649_37441_1_1_1_1_37441,00.html

Organisation for Economic Cooperation and Development. (2007). *OECD Communications Outlook 2007.* Retrieved March 11, 2008 from http://213.253.134.43/oecd/pdfs/browseit/9307021E.PDF

Pew Research Center. (2008). *The Internet's broader role in campaign 2008: Social networking and online videos take off.* Retrieved February 19, 2008 from http://people-press.org/report/384/internets-broader-role-in-campaign-2008

Pratchett, L., Wingfield, M., & Polat, R. K. (2006). Local democracy online: An analysis of local government web sites in England and Wales. *International Journal of Electronic Government Research, 2*(3), 75–92.

Press, L. (1993). *Relcom: An appropriate technology network.* Retrieved August 21, 2002 from ibiblio database, http://www.ibiblio.org/pub/academic/russian-studies/Networks/Relcom/relcom.history

Scruton, R. (1982). *A dictionary of political thought.* New York: Hill & Hwang.

Stalder, F. (2006). *Manuel Castells: The theory of the network society.* Malden, MA: Polity Press.

United Nations. (2004). *United Nations economic and social commission for Western Asia interim report. Foundations of ICT Indicators Database.* New York: United Nations.

United Nations Development Programme. (2004). *Human development report 2004: Cultural liberty in today's diverse world.* New York: United Nations Development Programme.

van Dijk, J. (1996). Models of democracy-- behind the design and use of new media in politics. *Javnost/The Public, 3,* 43-56.

van Dijk, J. (2002). A framework for Digital Divide research. *The Electronic Journal of Communication, 12*(1 & 2). Retrieved July 31, 2005 from http://www.cios.org/getfile/vandijk_v12n102

van Dijk, J. (2004). *The deepening divide: Inequality in the information society.* London: Sage.

van Dijk, J. (2006). *The network society* (2nd ed.). London: Sage.

van Dijk, J., & Hacker, K. (2000). Summary. In Hacker, K., & van Dijk, J. (Eds.), *Digital democracy: Issues of theory and practice.* London: Sage Publications.

Vanston, L., Hodges, R., & Savage, J. (2004). *Forecasts for higher bandwidth broadband services.* Retrieved August 21, 2005 from http://www.tfi.com/pubs/r/r02004_broadband.html

Weerakkody, V., & Dhillon, G. (2008). Moving from e-government to t-government: A study of process reengineering challenges in a U.K. local authority context. *International Journal of Electronic Government Research, 4*(4), 1–16.

Wellman, B., & Haythornwaite, C. (2002). The Internet in Everyday Life: An Introduction. In Wellman, B., & Haythornwaite, C. (Eds.), *The Internet in everyday life* (pp. 3–41). Oxford, U.K.: Blackwell Publishing. doi:10.1002/9780470774298

Yankelovich, D. (1991). *Coming to judgment: Making democracy work in a complex world.* Syracuse, NY: Syracuse University Press.

ENDNOTE

[1] We will use the term CMC only for the rest of this article.

This work was previously published in International Journal of Electronic Government Research (IJEGR), edited by Vishanth Weerakkody, pp. 57-71, copyright 2009 by Information Science Reference (an imprint of IGI Global)

Chapter 10
A Multi–Level Relational Risk Assessment Model for Secure E–Government Projects

Dionysis Kefallinos
National Technical University of Athens, Greece

Maria A. Lambrou
University of the Aegean Business School, Greece

Efstathios D. Sykas
National Technical University of Athens, Greece

ABSTRACT

In this chapter, the authors propose a model for a risk assessment tool directed towards and tailored specifically for e-government projects. The authors' goal is to cover the particular threats pertinent to the e-government project context and provide an interface between the broader philosophy of IT governance frameworks and the technical risk assessment methodologies, thus aiding in the successful and secure implementation and operation of e-government infrastructures. The model incorporates a wide range of applicable risk areas, grouped into eleven levels, as well as seven accompanying dimensions, assembled into a checklist-like matrix, along with an application algorithm and associated indices, which an evaluator can use to calculate risk for one or for multiple interacting projects.

INTRODUCTION

In this chapter we propose a risk assessment (RA) method and tool directed towards and tailored specifically for e-government projects. Given the diversity of concepts in e-government, creating a workable definition is becoming increasingly difficult (Roy, 2003). Generally, e-government refers to strategies, organizational forms and processes, as well as information technology employed so as to enhance access to and delivery of government information and services to citizens, businesses, government employees and other agencies. From a technical standpoint, e-government initiatives usually involve several types of digital technology and information systems, including databases, networking, collaboration services, multimedia, tracking and tracing, and and privacy technolo-

DOI: 10.4018/978-1-60960-162-1.ch010

gies (Snellen, 2002). In particular, we consider e-government projects as technical ventures that further the cause of modeling and transfer of G2G, G2B and G2C processes into the electronic world; they typically include and deal with (both in their development as well as their operational phase) public servants, private enterprises, professionals and the general public.

The important issues and impediments for the successful design, deployment and use of secure e-government (and in general e-service) infrastructures have been documented extensively (Curthoys & Crabtree, 2003; Gil-Garcia & Pardo, 2005; Jaeger, 2003; Löfstedt, 2005; Martin, 2005; Relyea, 2002; Vassilakis, Lepouras, Fraser, & Georgiadis, 2005), depicting the range of highly complex and diverse challenges public managers and security professionals must face in the their design, implementation and operation. It is generally accepted that success is less about selecting the right technology and more about managing organizational capabilities, facing regulatory constraints and environmental pressures and anticipating social, political and psychological issues of people involved; in other words effectively assessing risks and governing technological structures, within context.

Efficient management and security of a complex information and communications technology (ICT) system essentially depends on concise specification of requirements and security goals, their correct and consistent transformation into policies and appropriate deployment, enforcement and monitoring of these policies. This has to be followed-on by an incessant process to adapt the policies to changing contexts, environments, technologies, usage patterns and attack methods. To help understand the complex interrelations between security policies and ICT infrastructure and vulnerabilities, to validate security goals and especially to raise the assurance level of the RA process and the confidence level to the reviewed system, formal tool-based methodologies are necessary, which, as an additional benefit, also

guide towards a systematic evaluation and assist in determining exactly what really needs protection and which security policies to apply.

The RA tool that we model in this chapter can be viewed as an extension of established technical ICT RA approaches, aiming to: (i) better target the security and privacy goals in e-government projects, since a contextualized tool promotes improved formulation and facilitation of accurate security-related decisions, (ii) form a connection between technical ICT RA methodologies and Information Technology Governance (ITG) frameworks, (iii) increase security and privacy awareness by promoting the active involvement of a larger variety of non-technical personnel, and finally (iv) facilitate the application of baseline security and privacy policies.

Our motivation for the development of the presented model stems from experiences and observations in the RA field, whereupon a lack of adequate interaction between the technical methodologies (as well technology oriented practitioners and researchers) and the managerial ones has been clearly evident, especially in a context where public servants, the private industry and the general public interact with each other. Therein, a number of problematic aspects are witnessed in the development, implementation and production phases, precisely because of this "disregard", the result being a downgrade of the importance, confidence and acceptance of the results and suggestions of RA, which are often too narrow in scope anyway.

In section 2 of the chapter, a brief overview of the concepts and the background of our model is given, the techniques, standards, tools and practices upon which it is based, including general-purpose and ICT RA methodologies and tools, as well as ITG frameworks. We follow that with a detailed description of our model in section 3, and a discussion on its usage and our surmises in section 4. Finally, in section 5, we outline its possible future development.

RISK ASSESSMENT AND INFORMATION TECHNOLOGY GOVERNANCE

RA can be defined as the systematic process for the analysis, identification, control and communication of risks, their probability of happening and their effects to a system or an organization. RA should culminate with the specification of countermeasures to mitigate the risks and it should be an integral part of the design and implementation cycle of any system that is important to an organization, in order to enable educated and documented decisions for the allocation of appropriate resources to the areas that are most at risk. RA is only one (albeit a central one) of a broader set of risk management activities, other elements being the establishment of central management, the implementation of appropriate policies and related controls, awareness promotion, controls effectiveness monitoring and evaluation policy. The relationship between risk, threats, vulnerabilities, assets, consequences, security specification and controls is illustrated in Figure 1 (Gritzalis & Katsikas, 2004).

RA is based on statistical decision theory (Anand, 1993; Hansson, 1994); accepting the possibility that any information may be inaccurate, it implements a fundamental theorem of probability theory, Bayes' theorem. In practice, RA (of which vulnerability assessment is just one phase, but a critical one) is widely used in both the public and private sectors to support decision-making processes. Employing RA methodologies to drive decision-making processes around security and associated technology allows for consistent and effective use of decision-support data, as well as removal of technical bias from what are essentially business decisions. Moreover, RA should be a repetitive and self-evaluating process, as depicted in Figure 2 (Adar & Wuchnet, 2005).

For the phases of the RA process, several approaches have been proposed. The Software Engineering Institute identified five phases (identification, analysis, response development and control) (Tseng, Kyellberg, & Lu, 2003), while the Project Management Institute indicated four (identification, quantification, response development and control) (Durofee et al., 1996). Klein and Cork (1998) described a four phase process (identification, analysis, control and reporting),

Figure 1. RA relationships

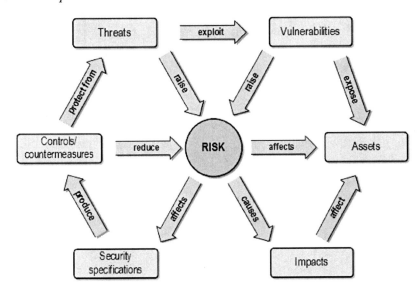

Figure 2. RA process cycle

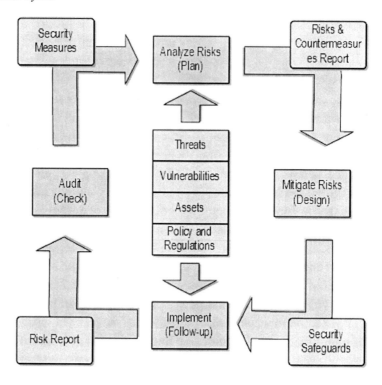

while Boehm (1991) suggested separate RA (identification, analysis and prioritization) and risk control phases (risk management planning, risk resolution and risk monitoring planning, tracking and corrective action). Moreover, Chapman (1998) proposed nine phases (definition, strategic approach, identification of risks, information structuring, ownership, uncertainty estimation, magnitude of risks, response, monitoring and controlling), whilst Fairley (1994) proposed seven (identification, assessment, mitigation, monitoring, contingency planning, managing the crisis and recovery from the crisis). RA as per Australia and New Zealand Risk Management Standard comprises of seven iterative processes; establishing the context, identification of risks, analysis of risks, evaluating risks, treating risks, communication and consultation and monitoring and review (AS/NZ Standard 4360, 1999). Similar to the above, the Riskman methodology pro-

vided a framework for risk analysis and control (Carter, Hancock, Morin, & Robins, 2001).

To summarize, RA methodologies are generally include the following elements:

a. Identification the system's constituents and the background upon which it operates.

b. Identification of the threats that could harm and, thus, adversely affect critical operations and assets. Threats may include such elements as intruders, criminals, disgruntled employees, terrorists, natural disasters etc.

c. Estimation of the likelihood that such threats will materialize based on historical information, research and the judgment of knowledgeable individuals.

d. Identification and categorization of the value, sensitivity, and criticality of the operations and assets that could be affected should a threat materialize, in order to determine which operations and assets are the most important.

e. Estimation, for the most critical and sensitive assets and operations, of the potential losses or damage that could occur if a threat materializes, including recovery costs.

f. Specification and justification of cost-effective controls to eliminate or mitigate the risks and facilitate business continuity. These actions can include constitution of new organizational policies and procedures as well as technical or physical controls.

g. Documentation of the results, development of an action plan and a RA iteration policy.

Risk Assessment Methodologies

There are many methodologies for conducting general-purpose RA and they can be grouped into *qualitative, tree-based and methodologies for dynamic systems.* Following we briefly review them as background reference, since their philosophy and particular aspects consist the basis of our model.

a. Qualitative methodologies
    i. *Preliminary risk analysis or hazard analysis* (PHA) is a qualitative technique involving a disciplined analysis of the event sequences which could transform a potential hazard into an accident (Andrews & Moss, 1993).
    ii. *Hazard and operability* (HAZOP) systematically examines processes and engineering intentions of new and existing facilities to assess the potential hazards that can arise from deviations from design specifications, as well as their consequential effects (Sutton, 1992).
    iii. *Failure mode and effects analysis* (FMEA) examines each potential failure mode in a system in order to determine its effect on the system

and classify it according to its severity (Stamatis, 1995). When FMEA is extended by criticality analysis, the technique is called *Failure Mode and Effects Criticality Analysis* (Bouti & Kadi, 1994). Price et al. (1992) proposed the automation of FMEA using a knowledge-base system, whereas Bell et al. (1992) offered the use of a causal reasoning model. Kara-Zaitri et al. (1991; 1992) presented an improved FMEA methodology which uses a single matrix to model the entire system and a set of indices derived from probabilistic combination to reflect the importance of an event relating to the indenture under consideration and to the entire system. A similar approach was made by Pelaez and Bowles (1995) to model the entire system using a fuzzy cognitive map.

b. Tree-based methodologies
    i. *Management oversight risk tree* (MORT) is a diagrammatic method which arranges safety elements logically and analyses them in a fault tree, where the root event is 'Damage, destruction, other costs, lost production or reduced credibility of the enterprise' (Frei, Kingston, Koornneef, & Schallier, 2002; Knox & Eicher, 1992).
    ii. *Safety management organization review technique* (SMORT) is a simplified version of MORT (Jouko & Rouhiainen, 1993). It is performed using structured analysis levels with associated checklists, instead of a comprehensive tree structure.
    iii. *Fault tree analysis* (FTA) builds a logical diagram which shows the relation between undesirable effects in the system and failures of the components of the system, by identifying the causal relationships of the failures as a tree

graph, using deductive logic (Aven, 1992).

iv. *Event tree analysis* (ETA) uses inductive logic to illustrate the sequence of outcomes which may arise after the occurrence of a selected initial event (Pate-Cornell, 1984).

v. *Cause-consequence analysis* (CCA) combines cause analysis (described by fault trees) and consequence analysis (described by event trees), with the purpose to identify chains of events that can result in undesirable consequences (Aven, 1992).

c. Dynamic systems methodologies

i. *Go method* (Siu, 1994) is a technique where a model of the system is constructed using independent and dependent GO operators, which are representations of the system's elements, along with logic operators to combine them into the success logic of the system.

ii. *Digraph matrix / fault graph analysis* (Fullwood & Hall, 1988; Siu, 1994) constructs graphs composed of system elements, AND/OR gates and loops.

iii. *Markov modeling* is a classic modeling method used for assessing the time-dependent behavior of dynamic systems (Pate-Cornell, 1993; Siu, 1994).

iv. *Dynamic event logic analytical methodology* (DYLAM) is an integrated framework which explicitly assesses time, process variables and system behavior (Cojazzi & Cacciabue, 1994).

v. *Dynamic event tree analysis method* (DETAM) uses a dynamic event tree where branching is allowed at different points in time and treats time-dependant evolution of hardware states, process variable values and operator states over the course of a scenario (Acosta & Siu, 1993).

These RA methodologies have been used to assess risk in a wide variety of sectors, such as nuclear energy production, oil production, semiconductor manufacturing, chemical industry and financial markets, among others.

Risk Assessment for Systems and E-Services

Apart from the RA methodologies that we have listed above, there are a number of methodologies developed specifically for ICT security, a product of ongoing research in this field. Most of them rely on model-based symbolic interpretation, simulation and analysis of systems using graphs or trees and attack or risk paths. These are efficient ways of dealing with the complexities and to precisely predict and analyze the behavior of interrelated systems that often change in order to meet functional requirements, advances in technology and diversified hostile environments. However, while scientifically superior, these methodologies require advanced study, are quite difficult to implement, do not carry the quality assurance of formal RA methodologies and do not necessarily conform to international security standards and best practices.

For that purpose, commercial methodologies have been drafted and implemented by government and private organizations, which, having evolved into *toolkits*, aid (essentially guide) security personnel and technical managers to carry out an effective RA process. Following, we list the most noteworthy of them:

a. *Consultative, Objective and Bi-functional Risk Analysis* (http://www.riskworld.net/); COBRA is a security RA, consultation and review tool. Developed by a private party to address the demands placed upon businesses and organization in these areas and to aid in ISO 27002/27001 compliance, COBRA tries to guide the security reviewer towards a formal RA methodology in order to ensure that the security controls for a system are fully commensurate with its risks.

b. *CCTA Risk Analysis and Method Management* (http://www.insight.co.uk/products/cramm.htm); originally a methodology developed in 1987 by the central Agency of Data Processing and Telecommunications of the UK government, CRAMM is now a commercial tool currently in its 5th edition. CRAMM is a three-staged process: asset identification and valuation, threat and vulnerability assessment and countermeasure selection and recommendation. Each stage is supported by objective questionnaires and guidelines.

c. *End-to-End Security Assessment* (Adar, 2002); EESA deals with Critical Information Infrastructure Protection (CIIP), analyzing "Security Quality of Service" (SQOS) along the path of critical processes within the business environment or system and evaluates whether the security mechanisms along it are adequate for protecting against likely threats. The analysis covers both strategic issues as well as very detailed technical security design issues.

d. *Méthode Harmonisée d'Analyse de Risque* (http://www.clusif.asso.fr/en/production/mehari/); MEHARI provides a formal RA model, formula and parameter based analysis, modular components and processes, asset classification, vulnerability discovery through auditing, optimal selection of corrective actions and compliance measures to ISO/IEC 27002. It replaced a previous tool named MARION in 2007.

e. *Operationally Critical Threat, Asset and Vulnerability Evaluation* (http://www.cert.org/octave/); developed by Software Engineering Institute at Carnegie Mellon University, OCTAVE is a systematic way for an organization to address its information security risks, sorting through the complex web of organizational and technological issues. The OCTAVE approach includes a set of criteria that defines the requirements for a comprehensive, self-directed information security risk evaluation, and a set of methods consistent with the criteria.

f. *RiskWatch* for Information Systems and ISO 27002 (http://www.riskwatch.com/ISRiskWatchProduct.html); a tool to conduct automated RA and vulnerability assessment, it provides user-customizable knowledge databases, including the ability to create new asset, threat and vulnerability categories, safeguards and question categories and sets. Also, it supports the creation of what-if scenarios and financial impact analysis.

These RA tools have been widely used to assess risk in a large variety of ICT projects, both in the private and the public sector, covering technical and organizational IT-related risk quite well. Nevertheless, as we assert in the next section, they hardly take into account pertinent issues that arise in the context of e-government projects. However, as they commonly allow for user-defined RA elements (risks, vulnerabilities, countermeasures), they provide us with the opportunity and motive to extend them with our own tool, using their knowledge base of risks, impact and probability estimates and countermeasures within our model, where applicable.

Information Technology Governance

As we have noted, the above presented RA tools fail to face, explicitly and in detail, application-specific and management-oriented issues, which arise in the context of an e-service project, such as an e-government one. Such an approach is considered the domain of ITG, which reflects broader corporate governance principles while focusing on the management and use of IT to achieve corporate performance goals. It represents the policies and procedures necessary to ensure that an organization's information systems support its objectives, that they are used responsibly and that, most importantly, IT-related risk is minimized. Moreover, effective ITG is an increasingly important element of standards and legal compliance, since many civil regulations apply to an organizations' information, most of which resides within IT systems.

There are several well-established ITG models and related standards (Elieson, 2006), whereupon organizations can approach governance by adapting and adopting, and within these ICT security issues consist a central element. Usually, they are evolving international efforts maintained by governing bodies and they reflect the experience and best practices of a large number of organizations.

Firstly, ISO/IEC 27002:2005 – 'Code of Practice for Information Security Management' (ISO/IEC, 2005b) is the only security-specific international ITG standard available, which, being quite broad-scoped, can be applied in a wide variety of applications and organizations, including e-government settings. Many organizations strive to become certified with this standard, in addition to more common management and organizational standards such as ISO 9001. Complementary to the code of practice are ISO/IEC 27001 – 'a specification for an Information Security Management System' (ISO/IEC, 2005a), ISO/IEC 27003:2010 – 'Information security management system implementation guidance' (ISO/IEC, 2010), ISO/IEC 27004:2009 – 'Information security management – measurement' (ISO/IEC, 2009) and ISO/IEC 27005:2008 – 'Information security risk management' (ISO/IEC, 2008). The later deserves special attention as, while not a functional RA toolkit like the ones listed in the previous section, it is a standard that provides valuable guidelines for risk management, supports the general concepts specified in ISO/IEC 27001 and is designed to assist the satisfactory implementation of information security based on a holistic risk management approach, while being applicable to all types of organizations (e.g. commercial enterprises, government agencies, non-profit organizations) which intend to manage

risks that could compromise the organization's information security.

Secondly, Control Objectives for Information and related Technology (COBIT) (http://www.isaca.com/cobit/) aims at providing clear ITG policy and good practice towards a better understanding and management of risks associated with ICT. It accomplishes this by providing a formal ITG framework and detailed control objective guides for management, process owners, users and auditors. Its guidelines are generic but well structured and it is positioned to be more comprehensive for management and to operate at a higher level than pure technology standards for information systems. The COBIT Management Guidelines are one of its main strengths and include the following main dimensions, which we have imported and adapted within our RA model:

a. Critical success factors (CSFs) define the most important issues or actions that management should consider or undertake in order to achieve control over and within the IT processes. They are management-oriented implementation guidelines that identify the most important things to consider, strategically, technically, organizationally and procedurally.

b. Key goal indicators (KGIs) define indices that inform management whether an IT process has achieved its business requirements.

c. Key performance indicators (KPIs) define metrics to aid in determining how well the IT process is performing in goal achievement. They are lead indicators of whether a goal will likely be reached or not and gauges of capabilities, practices and skills.

Thirdly, IT Infrastructure Library (ITIL) (http://www.itil-officialsite.com/) is a set of concepts and techniques for the management of IT infrastructures. In its current revision, ITIL v3 takes a lifecycle approach to ITG guidance, as opposed to organising according to ICT delivery sectors.

It comprises a set of core texts, supported by additional complementary and web-based material, which define a set of best practices in 24 disciplines. The "library" currently consists of the service strategy, service design, service transition, service operation and continual service improvement books. ITIL is largely aimed at identifying the best practices for managing IT service levels.

Whereas COBIT emphasizes on IT controls and metrics, ISO/IEC 27002 mainly covers IT security and ITIL accentuates processes and services. While the above standards constitute very good ITG and ICT-security models, covering organization, procedural and implementation issues quite well, we have found that they have little correlation with the technical RA methodologies and tools, except that they do (at least in the case of ISO/IEC 27002) dictate the use of one in the design stage of a project. On the other hand, although some of the RA tools have special modules to assist in ISO/IEC 27002 compliance, we feel that there should be a more systematic interface between them, and that technical RA should widen its scope to further cover managerial issues, taking into account the broader managerial philosophy of COBIT. Finally, we believe that technical RA tools fail to cover important areas of risk that pertain to the e-government projects context. Against this background our model was devised based on the aforementioned requirements and to articulate the respective concerns and aspects.

THE RIPC[4] MODEL: EXTENDED RA FOR SECURE E-GOVERNMENT PROJECTS

In this section, we describe an enhanced model of RA, which we propose as more appropriate to the e-government project's context. We aim to cover additional risks to the undisrupted, effective and secure operation of a system beyond those of the ICT RA procedures (such as technology, infrastructure or malignant parties), risks that stem from the broader environment wherein e-government projects evolve and cover other critical dimensions such as cultural, social, political and psychological aspects of the users and operators of a system.

Challenges related to the overall institutional framework and policy environment in which organizations pursuing e-government efforts operate have become crucial. In this context, institutions must account not only to laws and regulations, but also to norms, actions, behavior and capabilities of a broad range of stakeholders (Lim, Tan, & Pan, 2007; Tan, Pan, & Lim, 2007). Policy agendas, political and management commitment, public readiness and education, as well as financial and commercial factors may affect the results of e-government initiatives (Titah & Barki, 2006). Unclear or contradictory laws and regulations, balances among the executive, legislative and judicial branches, or negative norms and behavior can constrain efforts and lead to resistance to change and internal conflicts. Furthermore, due to the technical and procedural complexity that new ICT technologies impose, as well as the novelty that these technologies and associated work practices present, not only at the end-user level but also at the designer/contractor level, it is imperative to ensure strong technical and managerial skills for the key members of the project.

The iterative process of analysis, design and decision making is central in an effort to build a good understanding of regulatory, management, operational, financial, organizational, users' and procurement related risks. The need for early dialogue with all classes of users, in order to address their concerns and proactively assess risks, must be stressed, as well as the importance of planning, effort, coordination and appropriate culture establishment. End-user, but also ICT, security actors and public management officer participation is also a major prerequisite. Against this background, concise and comprehensive, but most of all context-specific RA tools are necessary for efficient and secure implementation of e-government projects and applications.

Against this background, our model includes seventy three risk areas grouped into eleven additional levels: political, managerial, service staff, contractors, end-users, social, pre-existing operational policies/procedures, legal/regulatory, financial, procurement and interoperability. These have been consolidated from field experiences, research, reports from international projects (Durham, 2002; Hampshire, 2006; OECD, 2001; Tasmania, 2005), as well as associated literature (Ebrahim & Irani, 2005; Evangelidis, 2007; Evangelidis, Akomode, Taleb-Bendiab, & Taylor, 2002; NECCC, 2000; Vassilakis et al., 2005). Our collection methodology was focused on the particularities introduced by citizen and public servant participation and involvement. These additional risk areas are evaluated in parallel to the ones in a typical technical RA process.

In Table 1, we elaborate our proposal of the risks areas that a RA process of an e-government project should incorporate, along with seven accompanying dimensions, namely (a) impact on the effective and secure deployment and operation of the system, (b) probability to happen, (c) associated CSFs, (d) countermeasures taken towards their effective mitigation, (e) countermeasure cost (qualitative), (f) coverage threshold and (g) countermeasure risk mitigation coverage estimate, assembled in a compact matrix. Each risk is listed with an implied "lack-of" preposition. A few of the risks appear in more than one risk level; this is due to differing points of view that cause a separate risk according to the people or area it affects. These risk areas can be further developed into numerous risk elements, in the scope of a specialized software application that we will propose in the directions for further research; however in this chapter we will evaluate each risk area as a single homogenous risk.

The proposed Risk–Impact–Probability–CSF–Countermeasures–Cost–Coverage (RIPC[4]) matrix serves as a checklist assessment tool aimed to be used by key ICT and security governance actors, during the planning, design and implementation phases, in a periodic and iterative manner. It is envisioned as a lightweight, qualitative and complementary module to the established RA methodologies geared towards the requirements of e-government projects and intended to be used on a focus group or personal interviews data collection technique. Alternatively, Delphi method rounds (Delbecq, Van de Ven, & Gustafson, 1975; Schmidt, Lyytinen, Keil, & P. Cule, 2001) can be applied, where the panel of experts evaluate - based on the RIPC[4] matrix dimensions - the risk level of the examined e-government application, towards a unified assessment.

The values of the seven normative dimensions are as follows:

a. Impact (risk-bound): ranges from high to low, on a ten point numerical scale; it represents the footprint that the realization of the risk will have on the success of the project, a value of one meaning almost no effect and a meaning of ten meaning failure of the project.

b. Probability (risk-bound): ranges from high to low on a zero to one numerical scale; it represents the possibility that the risk may happen.

c. CSFs: critical success factors are descriptive and differ among the various risk areas and e-government projects. In essence, they consist of a breakdown of the nature and causality of the risks considered in each level, the factors that should be considered to assess successful implementation and goal completion.

d. Countermeasures are of descriptive nature and consist of the proper actions to be taken based on the nature and severity of the risk (impact-probability), as well as known best practices in the particular context. Technical countermeasures are rather adequately covered by most formal RA tools, whereas political and social countermeasures include awareness campaigns, consensus building

Table 1. Risk–Impact–Probability–CSF–Countermeasures–Cost–Coverage (RIPC⁴) Matrix

N	Risk	Impact	Probability	CSFs	Countermeasures	Countermeasure cost	Coverage threshold	Coverage
A	*Political level*							
1	Supporting resolve and decision making							
2	Mid/long – term unified project planning							
3	Social and commercial placement and justification (how well is the project placed to cover social and commercial needs)							
4	Support for / support by other complementary or sequential policies and projects (political level)							
5	Technology culture of political leadership							
6	Project familiarization of political leadership (high-level)							
7	Funds allocation (high-level)							
8	Jurisdiction resolution between government agencies							
9	Placement and interaction with/within national policies and short/mid – term strategies (political level)							
10	Proper placement and interaction of the project with regional and global international directives and projects (political level)							
11	Presentation, justification and promotion of the project to the lower-level personnel and the public (political level)							
B	*Managerial level*							
12	Short/mid – term unified project support							
13	Business case planning and justification							

continued on following page

Table 1. Continued

14	Appropriate and complete project strategy planning and supervision							
15	Functional, goal-oriented, technology-aware and user-friendly project specifications compilation and supervision							
16	Technology culture of management personnel							
17	Project familiarization of management personnel							
18	Legal, regulatory, policy and procedural framework adaptation to the project and compliance by the project							
19	Appropriate authority delegation along the governance and management hierarchy							
20	Cost justification, business case development							
21	Productivity, progress monitoring and accountability methodology							
22	Partners (external organizations) technological, procedural and administrative readiness							
23	Support for / support by other complementary or sequential policies and projects (managerial level)							
24	Personnel allocation and qualification assurance and development							
25	Presentation, justification and promotion of the project to the lower-level personnel and the public (managerial level)							
C	*Service staff level*							

continued on following page

Table 1. Continued

26	Familiarization and training of the allocated personnel							
27	Friendliness of the system to the staff							
28	Suitable technical background (can greatly affect staff performance in unforeseen events)							
29	Incentives for secure and effective use of the system							
30	Incentives for prompt, effective and courteous support of the end-users							
31	Shift of power due to use of new technologies and knowledge (can create barriers, closed groups, competition)							
32	Change of duties which may be required by organizational or procedural modifications							
33	Fears for job loss by the adoption of new technologies and procedures							
34	Staff communication skills (can greatly affect staff effectiveness) – staff point of view							
35	Staff-to-end-user interface (procedures, ease of access) – staff point of view							
D	*Contactor(s) level*							
36	ICT skills							
37	Technology expertise							
38	Past experience specific to the project							
39	Communication skills							
40	Incentives/bindings for proper implementation and support							
41	Accurate implementation of the project business case, planning and specification							

continued on following page

Table 1. Continued

42	Accurate and complete documentation of implementation procedures and specifics							
E	***End-users level***							
43	Familiarization with the project							
44	Technology culture							
45	Friendliness of the system							
46	Trust towards the system, especially where personal/private data storage and use is concerned							
47	Incentives to use (financial, time)							
48	Incentives to use responsibly/securely							
49	Ease of access to requisite materials and services							
50	Costs of materials and/or use of the system							
51	Staff-to-end-user interface (procedures, ease of access) – user point of view							
52	Staff communication skills (can greatly affect user view of the system) – user point of view							
53	Meet expectations of users							
54	Multi-lingual and multi-cultural issues							
F	***Social level***							
55	Shift of power due to use of new technologies and knowledge (can create barriers, closed groups, competition)							
56	Penetration to closed or discreet social groups (e.g. ethnicities, genders, rural, nomadic)							
G	***Existing operational policies/procedures level***							

continued on following page

Table 1. Continued

57	Compatibility with current policies and procedures							
58	Planning for rapid response to non-anticipated issues							
H	***Legal/regulatory level***							
59	In-advance study of legal conformity							
60	Cross-county harmonization of legislation							
61	Formal and unambiguous jurisdiction assignment							
62	Legal framework adaptation to the needs of the project							
63	Regulatory framework for operation and use by staff and users							
64	Placement and interaction with/within national policies and short/mid – term strategies (legal/regulatory level)							
65	Planning for rapid response to non-anticipated legal and regulatory issues							
I	***Financial level***							
66	Business case planning and development							
67	Appropriate budget allocation							
68	Funds management and tracking							
J	***Procurement level***							
69	Timely procurement procedures							
70	Warehousing and asset management							
K	***Interoperability level***							
71	Technology integration planning							
72	Lack of standards							
73	Backwards compatibility with existing systems							

mechanisms, training and education programs and stakeholders' empowerment; financial countermeasures include ensuring stakeholders' commitment by incentives, negotiation strategies, and employment of modern financial engineering tools.

e. Countermeasure cost (countermeasure-bound): since cost is not always measured in monetary terms, and may involve other factors, such as more complex design, organizational or social procedures, this factor is of qualitative nature and ranges from low to high on a ten point numerical scale, for all the countermeasures of each risk. As such, it is not directly usable to determine the actual cost of the countermeasures; rather it is included so as to aid the selection of the most cost-effective countermeasures and it articulates the "trade off" approach of more rigid and secure procedures vs. user-friendliness, functional requirements and ease of use.

f. Coverage threshold (risk-bound): this term represents the minimum acceptable risk coverage that the designers estimate they should aim for; it ranges from high to low on a ten point numerical scale. 'Coverage threshold' is clearly dependent on 'Impact', 'Probability' and 'Cost'; the higher the impact of the risk and the more probable it is, the closer to 'Impact' 'Coverage threshold' should be. The higher the 'Cost' is, the more we may lower 'Threshold'; in this case it may be viewed as a fail-safe low boundary of 'Coverage' and a measure of the coverage margin that we have.

g. Coverage estimate (countermeasure-bound): ranges from high to low on a ten point numerical scale and represents the coverage of the risk by the countermeasures. 'Coverage' works similar to 'Threshold', but it signifies the actual effect of the countermeasures. It should never be lower than the 'Threshold'

and at least equal to 'Impact', unless the designers take a calculated risk due to 'Cost'.

The first two dimensions (Impact, Probability) function as typical RA pillars, whereas the next two (CSFs, Countermeasures) function merely as risk management pillars; nevertheless their incorporation into an e-government RA methodology phase is aimed at offering a quick, qualitative pre-estimation of important and available measures. Being used on an iterative basis, these two dimensions give an additional insight on the impact of the risks under consideration, while the last three (Cost, Threshold and Coverage) provide a self-check feedback mechanism to aid the evaluators in the proper formation of the countermeasures and the balancing of Cost and Coverage.

In the process of a RA evaluation, a multi-disciplinary team of qualified security experts, e-government specialists, ICT engineers and public administration professionals would perform a checklist-based evaluation. Firstly, the evaluators would select the items (risks) from the RIPC[4] matrix that apply to the particular e-government application and their field of expertise, along with the project CSFs relevant to this risk. Secondly, the evaluators would select a value for the impact and the probability of each risk, as well as their view of what the minimum mitigation coverage should be. In order to utilize previous knowledge to evaluate the real consequence and likelihood of risk items, risk evaluation typically consists of decision support systems utilizing the analytical hierarchy process (AHP) (Saaty, 2001), bayesian belief networks (BBN) (Press, 1989), or a commercial RA tool, with additional support by historical data or personal field expertise. Thirdly, they would select the countermeasures that can alleviate each risk, towards the realization of the CSFs. Fourthly, for each risk they would input their estimation of the total cost for the selected countermeasures. And finally, in the last column, they would input their estimation of risk mitigation coverage offered by the countermeasures for each risk, reevaluating them in the process; evaluators should strive to achieve a coverage value as close as possible to the impact value and certainly higher than the threshold.

Having populated the table, the evaluators would then calculate KGI risk index Ri, and KPIs coverage index Ci, coverage margin index Mi and total cost index Co as follows:

$$Ri = \frac{\sum_{j=1}^{n} P_j \left(I_j - C_j \right)}{n},$$

$$Ci = \frac{\sum_{j=1}^{n} \left(I_j - \bar{I} \right) \left(C_j - \bar{C} \right)}{\sqrt{\sum_{j=1}^{n} \left(I_j - \bar{I} \right)^2 \sum_{j=1}^{n} \left(C_j - \bar{C} \right)^2}},$$

$$Mi = 1 - \frac{\sum_{j=1}^{n} \left(C_j - \bar{C} \right) \left(Ct_j - \bar{Ct} \right)}{\sqrt{\sum_{j=1}^{n} \left(C_j - \bar{C} \right)^2 \sum_{j=1}^{n} \left(Ct_j - \bar{Ct} \right)^2}}$$

$$Co = \sum_{j=1}^{n} Co_j$$

where: $\bar{I} = \frac{1}{n} \sum_{j=1}^{n} I_j$, $\bar{C} = \frac{1}{n} \sum_{j=1}^{n} C_j$,

$$\bar{Ct} = \frac{1}{n} \sum_{j=1}^{n} Ct_j$$

where n is the total number of evaluated risks, P_j is the Probability value, I_j is the Impact value, C_j is the Coverage value, Ct_j is the Coverage Threshold and Co_j is the Countermeasure Cost for each risk j ($j = 1 \dots n$).

If Ri is close to 0 then the risk can be considered as adequately mitigated, provided that the risk factors are sufficiently covered, i.e. Ci is close to 1 (Ci is actually an indication of how closely Coverage follows Impact); the higher Ri is, the higher the non-alleviated risk is. Mi is an indication of how close to the margin (i.e. the coverage threshold) coverage is. The evaluators may tweak the countermeasures in order to lower total cost Co (and of' course Coverage), even if Ri rises, as long as Ci does not fall too far below 1 and Mi does not fall to 0.

Since at this point risk and its coverage are anticipatory and conjectural according to the experiences, perceptions, judgment, professional knowledge and skills of the evaluators, Ri and Ci are consequently subjective to their views; along with the RIPC[4] matrix, they consist an estimation tool, albeit one well suited to aid in covering the subject efficiently and straightforwardly. Our risk index system has similarities to the ones used by CRAMM, COBRA and some of the qualitative RA methodologies that we have reviewed; however we have made an effort to simplify and adapt it to the e-government project context.

Having completed the calculation, the RA team would proceed with documentation and implementation of the countermeasures within the normal procedure of project execution. However, as we have already pointed out, RA is a cyclic process; risk mitigation evaluation follows, during as well as after project completion, followed itself by either countermeasure re-selection, if optimum risk mitigation was not achieved, or risk re-evaluation if it was achieved. The complete RIPC[4] application algorithm is depicted in Figure 3; therein one can visualize the positioning of our model within the RA process, as well as interaction with and inputs from commercial RA tools and ITG frameworks.

A short example of a partial application of the RIPC[4] matrix, used on end-user issues of a Public Key Infrastructure project for the Greek Public Administration (GPA), is shown in Table 2.

In the aforementioned example, $Ri = 0.325$, $Ci = 0.905$, $Co = 74$ and $Mi = 0.083$. Interpreting this result, we can deduce that in this case risk is low, because the countermeasures have been selected so that their coverage closely follows impact, but at a fairly high cost. As further analysis, one could create a graph of Co vs. Ri or

Table 2. Partial example of the RIPC[4] Matrix

Risk	Impact	Probability	CSFs	Countermeasures	Countermeasure cost	Coverage threshold	Coverage
End-users level							
Familiarization with the project	8	0.9	End-users basic knowledge of what the project is about and how it will improve their life and interaction with government	Familiarization campaign, advertisements, distribution of leaflets, white papers	6	7	8
Technology culture	6	0.9	End-users have a basic understanding of the associated technology of the project	Easy to understand tech guides, technology penetration methods (e.g. cooperation with ISPs, schools, local authorities)	8	5	6
Friendliness of the system	7	0.5	Users are able to understand basic navigation and procedures of the system without reading too much material	Easy-to-use user-oriented GUI with context sensitive help system	5	5	6
Trust towards the system (especially where personal/private data storage and use is concerned)	10	0.7	Users entrust their personal data and transactions to the authority that manages and/ or operates the system	System certification by a data protection authority and/or control bodies, use of tamperproof security mechanisms, inform users about security within the project familiarization methods	5	8	10
Incentives to use (financial, time)	9	0.8	Users use the system because it costs less money and time than physical presence and gives them more abilities	Ease of access to e-services that lower the need for physical presence, lower administrative requirements, give access to advanced e-services if users use the system (i.e. smart cards and PKI infrastructure)	4	7	9
Incentives to use responsibly/ securely	8	0.8	Users protect the smart cards and pins, feel that forgery is nearly impossible	Easy-to-use, tamperproof security mechanisms, regulatory/legal penalties for improper use, inform users	5	6	8
Ease of access to requisite materials and services	10	0.6	Users can readily access materials and can make applications for smart-cards somewhere close-by	Widespread point-of-sales (citizen service centers, municipalities, other public authorities as commissioned offices), e-shops	10	6	8
Costs of materials and/or use of the system	10	0.6	User cost is low, or at least reasonably priced in relation to the services provided	Financing of smart cards and e-services from GPA and European funds	8	7	9
Staff-to-end-user interface (procedures, ease of access) – user point of view	8	0.6	User can has easy access to automated and human help desk, on-line support	On-line and call center-based automated and manual help system	6	5	8
Staff communication skills (can greatly affect user view of the system)	7	0.6	Support and service staff is knowledgeable and can efficiently guide users to any problems or difficulties	Focused training of personnel, financial incentives for performance to personnel	7	5	6
Meet expectations of users	6	0.5	Users feel that the functionality of the system is complete and it meets all their needs for the e-service	Survey and analysis of user expectations, pilot deployment of system to selected user groups, user satisfaction survey	5	4	5
Multi-lingual and multi-cultural issues	7	0.5	Foreign groups and minorities feel also at home using the system	Implementation of appropriate GUI for English, Turkish and Albanian	5	5	6

Figure 3. RIPC⁴ application algorithm

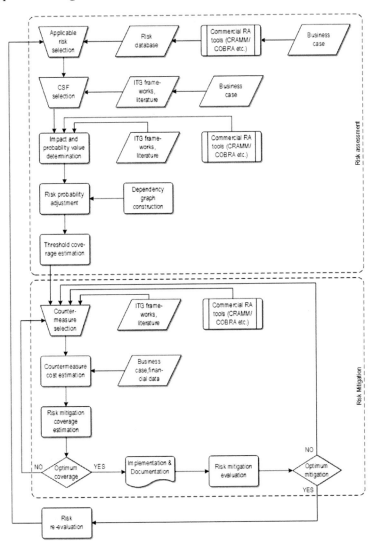

Ci, in order to determine the best combination of countermeasures and cost using an iterative process. However, this can be quite intricate and time consuming, since countermeasures and their risk coverage are not simple adjustable variables, but are complex quantities, often incorporating lengthy processes with uncertain results. Finally, since Ri has an unknown probability distribution, one could apply optimal control techniques on the Cj variables in order to control it, provided that Ij are considered constant for a given system.

RISK DEPENDENCY IN THE RIPC⁴ MODEL

In the above discussion, we assumed that the individual risks are independent from each other. However, this is seldom the case; it is quite common that some risks have dependencies with others, forming some form of hierarchy, and a realization of one produces the realization of others, or at least influences their probability. In reverse, the probability of one risk is affected by the probabilities of others, either in the same group or even across groups.

In order to calculate the new probabilities, we will use Bayes' formula for conditional probability, whereupon the (prior) probability of an event is modified by the likelihood of some hypothesis (or newly acquired data) to produce a new (posterior) probability of the event. In our case, having determined the prior probability of a dependent risk, we have to determine the likelihood of realizing the independent risks, given the prior realization of the dependent risk.

For example, in the case of two risks j and k, and their prior probabilities P_j and P_k, where the realization of risk j is affected by risk k, so that probability P_j is affected by probability P_k, we first have to determine the likelihood $L_{k,j}$ that risk k *will* be realized, having accepted the conjecture that risk j *has already been* realized. As the original risk probabilities, this can be determined using decision support systems utilizing AHP or BBN, with additional support by historical data and personal field expertise. Having determined this likelihood, the posterior probability $P_{j,k}$ of risk j is given by the expression:

$$P_{j.k} = \frac{L_{k.j}P_j}{P_k}$$

In the case of the dependency of risk j by two other risks k and m with prior probabilities P_j, P_k and P_m respectively, having determined the likelihoods $L_{k,j}$, $L_{m,k}$ and $L_{m,k,j}$ of realizing risks k, m and m respectively, given that the risks j, k, and k and j respectively *have been* realized, the posterior probability $P_{j,k,m}$ of risk j is given by the expression:

$$P_{j.k.m} = \frac{P_j L_{k.j} L_{m.k.j}}{P_k L_{m.k}}$$

This shows that beyond two dependencies, the calculation becomes complex, requiring specialized tools. Therefore, it is suggested that the assessment team construct a risk dependency graph, where they would try to limit the maximum path to two to three segments and the maximum branches to two to three dependencies, limiting them to major ones. This graph is a reverse tree diagram, where the more dependent risks are to the left and the less dependent to the right. The task of graph construction fits right after risk selection and initial probabilities determination, the second step of the RIPC[4] matrix compilation in the RIPC[4] application algorithm. The selection of the dependent risks and their dependencies can be accomplished using a commercial RA tool, associated literature, personnel surveys and field expertise. Having calculated the posterior probabilities of the dependent risks, the assessment team would proceed with the third step of the RIPC[4] application algorithm, countermeasure selection. Figure 4 shows a partial graph for the RA example that we portrayed previously.

Continuing the RA example, in Table 3 the new risk probabilities are shown in the additional P$_{post}$ column (whereas the original Probability column is renamed P$_{pri}$). Included are a few more risks from other categories from which stem dependencies for the end-user risks; for these we include only the necessary for the calculation prior probabilities P$_{pri}$. In this example we can see that the probability of two risks has risen significantly – namely risk 53, "meet expectations of users" and risk 46, "trust towards the system, especially where personal/private data storage and use is concerned" – because of the sequential dependency from two other risks with fairly high probabilities, depicting the importance of the countermeasures for this particular risk, as well as the usability of this extension of our model.

In this calculation, Ri takes the value of 0.355, showing that the overall project risk index has risen somewhat, because of the risk dependencies taken into account.

Figure 4. Risk dependency example

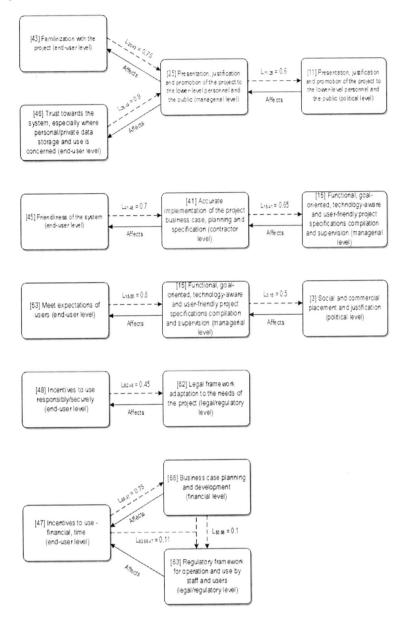

INTO THE THIRD DIMENSION: ACCOUNTING FOR OTHER PROJECTS

In the original RIPC[4] model we assumed that the scrutinized e-government project is being executed and assessed singly and autonomously; in effect the probabilities of the risks, upon which the total project risk index Ri is calculated, are independent factors, from each other and from other similar or identical risks of other projects. However, as we have already pointed out, this is seldom the case; most of the time projects are being executed in parallel, overlapping or in sequence, but in any case highly affecting each other. In the previous section we showed how risk dependencies can be

Table 3. Partial example of posterior probabilities calculation

N	Risk	Im-pact	P_{pri}	P_{post}	CSFs	Countermeasures	Coun-ter-mea-sure cost	Cover-age thresh-old	Cov-erage
3	Social and commercial placement and justification (how well is the project placed to cover social and commercial needs)		0.6						
11	Presentation, justification and promotion of the project to the lower-level personnel and the public (political level)		0.4						
15	Functional, goal-oriented, technology aware and user friendly project specifications compilation and supervision		0.6						
25	Presentation, justification and promotion of the project to the lower-level personnel and the public (managerial level)		0.5						
41	Accurate implementation of the project business case, planning and specification		0.7						
43	Familiarization with the project	8	0.9	0.9	End-users basic knowledge of what the project is about and how it will improve their life and interaction with government	Familiarization campaign, advertisements, distribution of leaflets, white papers	6	7	8
44	Technology culture	6	0.9		End-users have a basic understanding of the associated technology of the project	Easy to understand tech guides, technology penetration methods (e.g. cooperation with ISPs, schools, local authorities)	8	5	6
45	Friendliness of the system	7	0.5	0.46	Users are able to understand basic navigation and procedures of the system without reading too much material	Easy-to-use user-oriented GUI with context sensitive help system	5	5	6
46	Trust towards the system (especially where personal/private data storage and use is concerned)	10	0.7	0.84	Users entrust their personal data and transactions to the authority that manages and/or operates the system	System certification by a data protection authority and/or control bodies, use of tamperproof security mechanisms, inform users about security within the project familiarization methods	5	8	10

continued of following page

Table 3. Continued

47	Incentives to use (financial, time)	9	0.8	0.94	Users use the system because it costs less money and time than physical presence and gives them more abilities	Ease of access to e-services that lower the need for physical presence, lower administrative requirements, give access to advanced e-services if users use the system (i.e. smart cards and PKI infrastructure)	4	7	9
48	Incentives to use responsibly/securely	8	0.8	0.9	Users protect the smart cards and pins, feel that forgery is nearly impossible	Easy-to-use, tamperproof security mechanisms, regulatory/legal penalties for improper use, inform users	5	6	8
49	Ease of access to requisite materials and services	10	0.6		Users can readily access materials and can make applications for smart-cards somewhere close-by	Widespread point-of-sales (citizen service centers, municipalities, other public authorities as commissioned offices), e-shops	10	6	8
50	Costs of materials and/or use of the system	10	0.6		User cost is low, or at least reasonably priced in relation to the services provided	Financing of smart cards and e-services from GPA and European funds	8	7	9
51	Staff-to-end-user interface (procedures, ease of access) – user point of view	8	0.6		User can has easy access to automated and human help desk, on-line support	On-line and call center-based automated and manual help system	6	5	8
52	Staff communication skills (can greatly affect user view of the system)	7	0.6		Support and service staff is knowledgeable and can efficiently guide users to any problems or difficulties	Focused training of personnel, financial incentives for performance to personnel	7	5	6
53	Meet expectations of users	6	0.5	0.9	Users feel that the functionality of the system is complete and it meets all their needs for the e-service	Survey and analysis of user expectations, pilot deployment of system to selected user groups, user satisfaction survey	5	4	5
54	Multi-lingual and multi-cultural issues	7	0.5	0.45	Foreign groups and minorities feel also at home using the system	Implementation of appropriate GUI for English, Turkish and Albanian	5	5	6
62	Legal framework adaptation to the needs of the project		0.4						
63	Regulatory framework for operation and use by staff and users		0.6						
66	Business case planning and development		0.7						

incorporated into the calculations for the overall risk rating of one project. In this section we will extend our model to more than one project.

In our model, multi-project analysis can be viewed as multiple RIPC[4] matrices in a three-dimensional space, where each two-dimensional plane represents a single project and where the probabilities of the risks in one project-plane affect the probabilities of the risks in one or more other planes. This can form an extremely complex cause-and-effect scheme, making the calculation of the posterior probabilities of the risks difficult. While it is possible (with a specialized tool) to calculate the posterior probabilities taking into account cross-project risk dependencies (as per the previous section) by including risks from other projects in the risk dependency graph, in this section we will introduce a simplification; we will assume that a risk of one project can only affect the same risk of another project. This is based on the observation that, when running multiple projects within the same organization, particular types of tasks / services are assigned to / used by specific groups of people, usually based on specialty, job description and functionality. For example, the redaction of network architecture and hardware specification is assigned to hardware engineers, while the organization and structure of a help desk to the HR personnel. This simplification is optional; a RA team can instead chose to undertake the task of taking into account all possible dependencies by extending the risk dependency graph of the previous section across planes, incorporating risks from multiple projects.

As in the previous section, it is suggested that the assessment team construct a project dependency graph, a reverse tree diagram where the more dependent projects are to the left and the less dependent to the right, while trying to limit the maximum path to two to three segments and the maximum branches to two to three dependencies. In this case, the dependencies in this graph represent the dependencies of the risks for the projects. However, each dependency may not apply to all the risks of a project; many risks may not have cross-project dependencies. Thus, the likelihoods of Figure 4 would be replaced by vectors of likelihoods, one value for each risk pair "participating" in this dependency.

The team of RA experts would follow the same steps of the RIPC[4] model, determining the risks, their probabilities, their impact, the CSFs, countermeasures, countermeasure costs, coverage thresholds and estimates, as well as indexes Ri, Ci, Mi and Co. In an additional step, right after step 2, they would construct the dependency graph and modify the probabilities of the risks of a project, taking into account the probabilities of the risks of the project it depends upon. As in the previous section, this can be accomplished using Bayes' formula for conditional probability. In the case of a two project dependency:

$$P_{1.2.3.j} = \frac{P_{1.j} L_{2.1.j} L_{3.1.2.j}}{P_{2.j} L_{3.2.j}}$$

where $P_{1.2.j}$ is the posterior probability of risk j in project 1, given that it is dependent on the same risk in project 2, $P_{1.j}$ is the prior probability of risk j in project 1, $P_{2.j}$ is the prior probability of risk j in project 2 and $L_{2.1.j}$ is the likelihood that risk j *will* be realized in project 2, given that risk j *is* realized in project 1. For three interrelated projects, where the first has dependencies with two others, the expression is:

$$P_{1.2.j} = \frac{L_{2.1.j} P_{1.j}}{P_{2.j}}$$

where $P_{1.2.3.j}$ is the posterior probability of risk j in project 1, given that it is dependent on the same risk in projects 2 and 3, $P_{1.j}$ is the prior probability of risk j in project 1, $P_{2.j}$ is the prior probability of risk j in project 2, $L_{2.1.j}$ is the likelihood that risk j *will* be realized in project 2, given that risk j *is* realized in project 1, $L_{3.1.2.j}$ is the likelihood that risk j *will* be realized in project 3, given that risk j *is* realized in projects 1 and 2 and $L_{3.2.j}$ is the likelihood that risk j *will* be realized in project 3 given that risk j *is* realized in project 2.

As an example of project dependency, we submit the relation between a project for a Public Key Infrastructure for the Ministry of Economics of Greece and two projects for integrated web-based services for private accounting firms (Taxisnet for Accounting Firms) and private customs offices (Taxisnet for Customs Offices). Because of its nature, as well as the functional dependency (issuance, usage and management of end-user certificates), PKI project risks affect similar risks of the two other projects, especially in the areas of user trust and incentives to use. Focusing on the end-user level of the RIPC[4] matrix, Figure 5 depicts the project dependency, as well as same-risk dependency and likelihood values for the three projects. In this figure we follow the notation of the previous paragraph, where project 3 is the PKI project, project 1 is Taxisnet for Accounting Firms and project 2 is Taxisnet for Customs Offices. The two vectors of likelihoods represent both the risks participating in the project dependency, as well as the values of the probabilities.

DISCUSSION

In order to be knowledgeable, insightful and effective in their initiatives, e-government professionals and researchers need to be attentive to pertinent challenges and risks and are expected to use appropriate strategies and tools to overcome them. While modern security and risk management standards, methodologies, frameworks and toolsets cover the RA process and technical ICT security modeling quite well, it is the view of the authors that their employment in the domain of e-government projects is suboptimal, primarily due to the inability to formally take into account risks that stem from non-technical organizational, social and psychological issues of government employees, executives, contractors and the general public.

Against this background, our RIPC[4] model aims to aid in the design and management of e-government systems beyond technology or platforms, towards secure, effective and usable systems, services and processes. The novelty of

Figure 5. Project dependency example

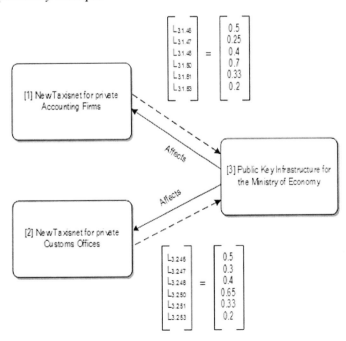

our approach lies in the integration of as many as possible non-technical e-government related factors into an easy-to-use checklist-based iterative RA process that promotes self-check and evaluation, beyond the bounds of technical RA tools and into the realm of ITG frameworks and effective e-government practices. Essentially, we propose our model as an extension of the tool-based commercial ICT RA methodologies; however, complementary to these, which are geared more towards technical aspects of systems, it attempts to specifically address non-technical areas which nonetheless we deem very important in the context of an e-government project. Therefore, RIPC[4] attempts to provide an interface between the broader managerial philosophy of COBIT, ISO/IEC 27002 and ITIL and the technical methodologies, by adding and integrating dimensions, upon which the attention of key e-government stakeholders can be drawn and respective actions or measures can be undertaken.

In posterior assessment runs of our model in e-government projects, we have found that it is well-suited to cover most of the common risks to their successful implementation and acceptance, and that it helps to promote evaluation self-check and important factors insight. The critical success factors of the model itself are: (a) selection of the appropriate risks for the project, (b) inclusion of the essential, for the purposes of the project, CSFs and (c) selection of effectual, attainable and cost-effective countermeasures. Its main weakness is that the model's performance and effectiveness rests upon the determination and experience of the professionals who will use it complementary to other more integrated and established toolkits, since in its current form it does not comprise a self-contained toolkit, one that can autonomously guide and assist in a systematic evaluation, determine a detailed security approach for assets needing protection and suggest the security policies to apply.

DIRECTIONS FOR FURTHER RESEARCH

As a subject of further research and development into this area, we propose full provisioning and incorporation of e-government factors in technical RA standards, methods and tools. Towards this goal, we aim to pursue the advancement of our model into a formal and functional e-government RA software toolkit, incorporating interfaces with the databases of particular RA toolkits, as well as the inclusion of its own knowledge base of risks (an elaboration of the RIPC[4] risk areas into many risk elements), CSFs, countermeasures, a risk dependency designer and probabilities calculator supporting multiple interrelated projects, and an impact estimator, all integrated into an easy to use web-based application with reporting capabilities.

ACKNOWLEDGMENT

The RIPC[4] model was inspired from our work for the SYZEFXIS project, a nation-wide ongoing initiative aimed to provide a modern ICT infrastructure for the GPA.

REFERENCES

Acosta, C., & Siu, N. (1993). Dynamic event trees in accident sequence analysis: application to steam generator tube rupture. *Reliability Engineering & System Safety*, *41*, 135–154. doi:10.1016/0951-8320(93)90027-V

Adar, E. (2002). *End-to-End Security Assessment*. Paper presented at the Analysis & Assessment for Critical Infrastructure Protection, Brussels.

Adar, E., & Wuchnet, A. (2005). *Risk Management for Critical Infrastructure Protection (CIP) Challenges, Best Practices & Tools*. Paper presented at the First IEEE International Workshop on Critical Infrastructure Protection (IWCIP'05), Darmstadt, Germany.

Anand, P. (1993). *Foundations of Rational Choice Under Risk*. Oxford University Press.

Andrews, J. D., & Moss, T. R. (1993). *Reliability and Risk Assessment* (1st Ed. ed.). Longman Group UK.

Aven, T. (1992). *Reliability and Risk Analysis* (1st Ed. ed.): Elsevier Applied Science.

Bell, D., Cox, L., Jackson, S., & Schaefer, P. (1992). *Using Causal Reasoning for Automated Failure & Effects Analysis (FMEA)*. Paper presented at the Annual Reliability and Maintainability Symposium.

Boehm, B. W. (1991). Software risk management: principles and practices. *IEEE Software, 8*, 32–41. doi:10.1109/52.62930

Bouti, A., & Kadi, D. A. (1994). A state-of-the-art review of FMEA/FMECA. *International Journal of Reliability Quality and Safety Engineering, 1*(4), 515–543. doi:10.1142/S0218539394000362

Carter, B., Hancock, T., Morin, J. M., & Robins, M. (2001). *Introducing Riskman Methodology – The European Project Risk Management Methodology*. Oxford, UK: NCC Blackwell Ltd.

Chapman, R. J. (1998). The effectiveness of working group risk identification and assessment techniques. *International Journal of Project Management, 16*(6), 333–343. doi:10.1016/S0263-7863(98)00015-5

Cojazzi, G., & Cacciabue, P. C. (1994). *The DYLAM Approach for the Reliability Analysis of Dynamic System*. Berlin, Heidelberg: Springer-Verlag.

Curthoys, N., & Crabtree, J. (2003). SmartGov: Renewing Electronic Government for Improved Service Delivery, ISociety Report, Available from http://www.pwc.com/uk/eng/about/ind/gov/smargovfinal.pdf

Delbecq, A. L., Van de Ven, A. H., & Gustafson, D. H. (1975). *Group Techniques for Program Planning: A Guide to Nominal Group and Delphi Processes*. Glenview, Illinois: Scott, Foresman and Company.

Durham, C. C. (2002). *Implementing Electronic Government Statement*. Retrieved. from http://www.durham.gov.uk/.

Durofee, A. J., Walker, J. A., Alberts, C. J., Higuera, R. P., Murphy, R. L., & Williams, R. J. (1996). *Continuous Risk Management Guidebook*. Pittsburg, PA: Carnegie Mellon University.

Ebrahim, Z., & Irani, Z. (2005). E-government adoption: architecture and barriers. *Business Process Management Journal, 11*(5), 589–611. doi:10.1108/14637150510619902

Elieson, B. D. (2006). Construction of an IT Risk Framework, Available from http://www.isaca.org/ContentManagement /ContentDisplay.cfm?ContentID=33595

Evangelidis, A. (2007). FRAMES – A Risk Assessment Framework for e-Services. *Electronic. Journal of E-Government, 2*(1), 21–30.

Evangelidis, A., Akomode, J., Taleb-Bendiab, A., & Taylor, M. (2002). *Risk Assessment & Success Factors for e-Government in a UK Establishment*. Paper presented at the Electronic Government, First International Conference, Aix-en-Provence France.

Fairley, R. (1994). Risk management for software projects. *IEEE Software*, 57–64. doi:10.1109/52.281716

Frei, R., Kingston, J., Koornneef, F., & Schallier, P. (2002). NRI MORT User's Manual, Available from http://www.nri.eu.com/NRI1.pdf

Fullwood, R. R., & Hall, R. E. (1988). *Probabilistic Risk Assessment in the Nuclear Power Industry* (1st ed.). Pergamon Press.

Gil-Garcia, J. R., & Pardo, T. A. (2005). E-government success factors: mapping practical tools to theoretical foundations. *Government Information Quarterly, 22*, 187–216. doi:10.1016/j.giq.2005.02.001

Gritzalis, D., & Katsikas, S. (2004). *Autonomy and political disobedience in cyberspace*. Athens: Papasotiriou.

Hampshire, C. C. (2006). *Section 5: Risk Assessment*. Retrieved. from http://www.hants.gov.uk/egovernment/IEG2-sec5.html.

Hansson, S. O. (1994). *Decision Theory, A Brief Introduction*, Available from http://www.infra.kth.se/~soh/decisiontheory.pdf

ISO/IEC. (2005a). *27001:2005 Information security management systems - Requirements,* Current Stage 90.92 Available from http://www.iso.org/iso/iso_catalogue/ catalogue_tc/catalogue_detail.htm?csnumber=42103

ISO/IEC. (2005b). *27002:2005 Code of practice for information security management,* Current Stage 90.92 Available from http://www.iso.org/iso/iso_catalogue/ catalogue_ics/catalogue_detail_ics.htm?csnumber=50297

ISO/IEC. (2008). *27005:2008 Information security risk management,* Current Stage 90.92 Available from http://www.iso.org/iso/iso_catalogue/ catalogue_tc/catalogue_detail.htm?csnumber=42107

ISO/IEC. (2009). *27004:2009 Information security management -- Measurement,* Current Stage 60.60 Available from http://www.iso.org/iso/iso_catalogue/catalogue_tc/catalogue_detail.htm?csnumber=42106

ISO/IEC. (2010). *27003:2010 Information security management system implementation guidance,* Current Stage 60.60 Available from http://www.iso.org/iso/iso_catalogue/ catalogue_tc/catalogue_detail.htm?csnumber=42105

Jaeger, P. T. (2003). The endless wire: E-government as global phenomenon. *Government Information Quarterly, 20*, 323–331. doi:10.1016/j.giq.2003.08.003

Jouko, S., & Rouhiainen, V. (1993). *Quality Management of Safety and Risk Analysis*. New York: Elsevier Science Publishers B.V.

Kara-Zaitri, C., Keller, A. Z., Barody, I., & Fleming, P. V. (1991). *An Improved FMEA methodology*. Paper presented at the Annual Reliability and Maintainability Symposium.

Kara-Zaitri, C., Keller, A. Z., & Fleming, P. V. (1992). *A Smart Failure Mode and Effect Analysis Package*. Paper presented at the Annual Reliability and Maintainability Symposium.

Klein, J. H., & Cork, R. B. (1998). An approach to technical risk assessment. *International Journal of Project Management, 16*(6), 345–351. doi:10.1016/S0263-7863(98)00006-4

Knox, N. W., & Eicher, R. W. (1992). *MORT User's Manual, rev. 3: US Department of Energy*. System Safety Development Center EG&G Idaho Inc.

Lim, E. T. K., Tan, C. H., & Pan, S. L. (2007). E-Government Implementation: Balancing Collaboration and Control in Stakeholder Management. *International Journal of Electronic Government Research, 3*(2), 1–28.

Löfstedt, U. (2005). E-Government – Assessment of Current Research and Proposals for Future Directions, Available from http://www.hia.no/iris28/Docs/ IRIS2028-1008.pdf

Martin, N. (2005). Why Australia needs a SAGE: a security architecture for the Australian government environment. *Government Information Quarterly, 22*, 96–107. doi:10.1016/j.giq.2004.10.007

NECCC. (2000). *Risk Assessment Guidebook for e-Commerce/e-Government*. Available from http://www.ec3.org/Downloads/2000 /Risk_Assessment_Guidebook.pdf

OECD. (2001). *The Hidden Threat to E-Government: Avoiding large government IT failures.* Retrieved. from http://www.oecd.org/dataoecd/19/12/1901677.pdf.

Pate-Cornell, M. E. (1984). Fault Tree vs. Event Trees in Reliability Analysis. *Risk Analysis, 4*(3), 177–186. doi:10.1111/j.1539-6924.1984.tb00137.x

Pate-Cornell, M. E. (1993). Risk Analysis and Risk Management for Offshore Platforms: Lessons from the Piper Alpha Accident. *Journal of Offshore Mechanics and Arctic Engineering, 115,* 179–190. doi:10.1115/1.2920110

Pelaez, C. E., & Bowles, J. B. (1995). *Applying Fuzzy Cognitive-Maps Knowledge- Representation to Failure Modes Effects Analysis.* Paper presented at the Annual Reliability and Maintainability Symposium.

Press, S. J. (1989). *Bayesian Statistics: Principles, Models and Applications.* New York: Wiley.

Price, C. J., Hunt, J. E., Lee, M. H., & Ormsby, R. T. (1992). A Model-based Approach to the Automation of Failure Mode Effects Analysis for Design. *IMechE, Part D: the Journal of Automobile Engineering, 206,* 285-291.

Relyea, H. C. (2002). E-gov: Introduction and overview. *Government Information Quarterly, 19*(1), 9–35. doi:10.1016/S0740-624X(01)00096-X

Roy, J. (2003). E-government. *Social Science Computer Review, 21*(1), 3–5. doi:10.1177/0894439302238966

Saaty, T. L. (2001). *Decision Making for Leaders – The Analytical Hierarchy Process for Decisions in a Complex World.* Pittsburgh, PA: RWS Publications.

Schmidt, R. C., Lyytinen, K., Keil, M., & Cule, P., P. (2001). Identifying software project risks: an international Delphi study. *Journal of Management Information Systems, 17*(4), 5–36.

Siu, N. (1994). Risk Assessment for dynamic systems: An overview. *Reliability Engineering & System Safety, 43,* 43–73. doi:10.1016/0951-8320(94)90095-7

Snellen, I. (2002). Electronic governance: Implications for citizens, politicians and public servants. *International Review of Administrative Sciences, 68*(2), 183–198. doi:10.1177/0020852302682002

Stamatis, D. H. (1995). *Failure Mode and Effect Analysis - FMEA from Theory to Execution.* ASQC Quality Press.

Sutton, I. S. (1992). *Process Reliability and Risk Management* (1st Ed. ed.): Van Nostrand Reinhold.

Tan, C. W., Pan, S. L., & Lim, E. T. K. (2007). Managing Stakeholder Interests in E-Government Implementation: Lessons Learned from a Singapore E-Government Project. *International Journal of Electronic Government Research, 3*(1), 61–84.

Tasmania. (2005). *Risk Management Resource Kit.* from http://www.egovernment.tas.gov.au/themes/ project_management/risk_management_resource_kit

Titah, R., & Barki, H. (2006). E-Government Adoption and Acceptance: A Literature Review. *International Journal of Electronic Government Research, 2*(3), 23–57.

Tseng, M. M., Kyellberg, T., & Lu, S. C. Y. (2003). Design in the new e-manufacturing era. *Annals of the CIRP, 52*(2). doi:10.1016/S0007-8506(07)60201-7

Vassilakis, C., Lepouras, G., Fraser, J., & Georgiadis, P. (2005). Barriers To Electronic Service Development. *e-Service Journal, 4*(1), 41-63.

Chapter 11
A Multiple Case Study on Integrating IT Infrastructures in the Public Domain

Muhammad Mustafa Kamal
Brunel University, UK

ABSTRACT

Local Government Authorities (LGAs) are complex organisations whose heterogeneous operational structures can be greatly enhanced by effectively using of Information Technology (IT) to support improvements in the quality of services offered to citizens. While the benefits of IT cannot be disputed, there are several concerns about its success as LGAs are confronted with the challenges of synchronising their cross-departmental business processes and integrating autonomous IS. This article examines a potentially important area of IT infrastructure integration in LGAs through Enterprise Application Integration (EAI). The adoption of EAI solutions is a burgeoning phenomenon across several private and public organisations. Nevertheless, where EAI has added efficacy to the IT infrastructures in the private domain, LGAs have also been slow in adopting cost-effective EAI solutions. The shortage of research studies on EAI adoption in LGAs presents a knowledge gap that needs to be plugged. The research methodology followed consisted of an in-depth analysis of two case studies by using the research tools of interviews, observation and referring to archival documents. This research is timely as the demand for integrated service delivery increases, the issues of harmonising business processes and integrating IS becomes pertinent. The conclusion and lessons that can be learnt from this research is that integrating IT infrastructures through EAI achieves significant efficiency in delivering end-to-end integrated electronic Government (e-Government) services.

INTRODUCTION

With the emergence of Information and Communication Technologies (ICTs), and e-Government, it is possible to improve the efficiency and effectiveness of operational activities within LGAs and to reposition LGA services at regions closer to the citizens (Reddick, 2009; Gichoya, 2005; Wimmer, 2004). Beynon-Davies (2005) and Beynon-Davies and Williams (2003) support that the rapid

DOI: 10.4018/978-1-60960-162-1.ch011

developments in technology is contributing to the growth of interest in the use of IT as an effective tool to enable and aid transformation in LGAs. On the other hand, the motivations for e-Government broadly include reduction of internal costs, increase of transparency, and the improvements in service delivery (Irani *et al.*, 2006). However, prior IS research exhibits several difficulties impeding the IT-enabled organisational transformation in LGAs such as including among others: (a) non-integrated nature of their IT infrastructure do not allow LGAs to deliver end-to-end integrated services (Lam, 2005; McIvor *el al.*, 2002), (b) lack of a single approach for implementing IS instead developing IS independently to provide specific business solutions (Janssen and Cresswell, 2005), (c) inflexible IS security requirements further constraining integration (Weerakkody *et al.*, 2007).

This has resulted in a wide range of different technologies and disparate IS with incapability to interoperate and eventually developing islands of information (McIvor *et al.*, 2002). The inaccessibility of substantial data archives and business processes in the isolated IS within LGAs, is at the heart of the foremost pressing challenges facing the architects of today's IT infrastructures in transforming LGAs (Weerakkody *et al.*, 2007; O'Toole, 2007; Janssen and Cresswell, 2005). Despite the growing interest in this area, in-depth enquiry into how LGAs overcome the several impediments in their way to manage IT-enabled transformations has remained relatively limited (Weerakkody and Dhillon, 2008; Tan *et al.*, 2005). A possible explanation for the scarcity of research interest is the pessimistic impression of LGAs as rigid, risk-averse and having insignificant desire for improvement (Ongaro, 2004; McIvor *et al.*, 2002; Bozman and Kingsley, 1998). However, despite this unfavourable belief, recent years have witnessed a rush of the implementation of e-Government to re-invent LGA services using IT (Weerakkody and Dhillon, 2008; Kawalek and Wastell, 2005). Themistocleous *et al.*, (2005) argues that e-Government platform should not

been merely seen as a stand-alone system but as a solution that communicates with back office applications through an integrated infrastructure. E-Government transformation is one of the biggest challenges within the IT-related sector from the perspective of scale and complexity, especially when it comes to adapting existing e-Government to new computing requirements based on the citizens' new service concept (Cheng-Yi Wu, 2007). Integrated e-Government IS can efficiently automate the business processes of the public domain and increase citizens' satisfaction. However, to achieve such a solution, LGAs need to integrate their IT infrastructures to provide a common and shared view of their information and services (Beynon-Davies, 2005; Lam, 2005). The benefits of integration have not been attained due to incompatible IS, platforms, and high maintenance costs coupled with a lack of understanding of the true purpose, value and power of integrated IS (McIvor *et al.*, 2002).

During the recent years, EAI has emerged to support organisations to integrate their IT infrastructures and deliver high quality of services (Lam, 2005). EAI can be used to piece together LGA information systems with packaged and legacy systems. In other words, EAI acts as a software data translator that takes information from, for example, organisational Enterprise Resource Planning (ERP) systems and convert it into formats that other applications can understand (Linthicum, 2000). Organisations that have integrated their IT infrastructures through EAI have reported significant benefits (Themistocleous *et al.*, 2005; Bass and Lee, 2002). For example, EAI assists in business process integration, support in collaborative decision-making, results in reduced integration cost, securing and providing privacy of citizens' data, and results in developing flexible, and maintainable integrated IT infrastructures (Themistocleous and Irani, 2001). Kamal *et al.*, (2009) proposed and validated an EAI adoption model in the area of LGAs. The model presents several factors (factors – as one of the four components of the model) influencing the decision making process for EAI adoption in LGAs and includes among others: project champion, citizen's satisfaction, critical mass, market knowledge, top management support. These factors have been well analysed in the literature (e.g. Kamal *et al.*, 2008a; 2008b; Kamal and Themistocleous, 2007). These factors can be used to understand EAI adoption for improving IT infrastructures in LGAs. The rest of the article is structured as follows: Section 2 develops the theoretical foundation and discusses on the factors influencing the decision making process in LGAs. The research methodology used to conduct the research is presented in Section 3 with Sections 4 and 5 presenting the case organisations and analysing the findings of the EAI adoption factors. Section 6 summarises the conclusions, lessons learned and usefulness of the research presented in this article.

THEORETICAL FOUNDATION: FACTORS INFLUENCING EAI ADOPTION IN LGAS

Technological adoption has been an important area for IS research and practice (Fichman, 1992). Several studies on integration technologies e.g. existing EAI adoption models like those proposed by Mantzana *et al.*, (2007); Khoumbati *et al.*, (2006); Chen (2005) and Themistocleous (2004) are domain specific (e.g. healthcare sector, private organisations). These models investigate the adoption of EAI through a set of influential factors. Some of these factors can be considered as common (e.g. benefits, barriers, costs) and therefore can be reused when adopting EAI in LGAs. Other factors like patients' satisfaction or physician and patient relationship are domain specific and thus are of no use by LGAs. In addition to this, there are differences indicating that the factors influencing the decision-making

process for EAI adoption differ from one type of organisation to the other depending among others on the nature and size. For instance, one set of factors is used to support EAI adoption in Small and Medium-Sized Enterprises (SMEs) and another in large organisations or healthcare organisations. However, the applicability and validity of these models is arguable and under research in LGAs, as these were proposed to support the decision-making process in other sectors and not in LGAs.

As a result of these distinctions, additional factors may be required particular to a context. Fichman (1992) also supports that such factors merely by themselves are unlikely to be strong predictors for technology adoption and thus, need additional factors according to a particular environment. Taking into consideration this argument, the authors incorporated the factors proposed by Kamal *et al.*, (2009) into this research as reported in Tables 1 to 5. These factors may support LGAs in developing an understanding while adopting EAI solutions and help them transforming their functions. These factors are categorised all according to Pressure Factors (PF), Technological Factors (TF), Support Factors (SF), Financial Factors (FF) and Organisational Factors (OF).

These factors make novel contribution at conceptual level. These factors are a combination of common factors identified from the previous studies on EAI adoption. The authors extend the previous works and adapt them to EAI in the area of LGAs, thus, resulting in the development of five categories of factors. The authors suggest that while adopting EAI, these factors might provide a deeper understanding on EAI adoption process in LGAs and support their decision making process. In doing so, these factors might: (a) extend the current research on EAI adoption, (b) enhance the level of EAI adoption analysis and (c) support LGA decision makers to adopt EAI. To test these factors, the authors propose the following research methodology.

Table 1. Pressure factors influencing EAI adoption in LGAs

Pressure Factors	
Project Champion	Championship refers to the existence of a person in the organisation who is committed to introduce IT initiative to the organisation. Project champions are personnel who actively and vigorously promote their personal vision for using IT, pushing the project over or around approval and implementation hurdles (Beath, 1991). Norris (1999) report that within government organisations, the existence of a project champion is one of the most important facilitators in the adoption of technologies.
Citizen's Satisfaction	Citizen's satisfaction has a significant impact on the performance of LGAs and in the growing push towards accountability between the LGAs (Welch *et al.,* 2005). IT adoption is viewed as a central part of the modernisation of LGAs in improving the quality of LGAs services and achieving citizen satisfaction (Beynon-Davies, 2005). Welch *et al.,* (2005) report that IT appears to offer a useful opportunity to LGAs to enhance citizen satisfaction by improving procedural transparency, cost-efficiency and effectiveness.
Critical Mass	Research on critical mass has shown that central government and LGAs are affected by the actions of other governments in IT adoption (Akbulut, 2002). For example, cities adopting innovations were located in close proximity to other innovation-adopting cities. This showed that organisations were affected by the actions of other organisations that were similar in terms of size and budgetary constraints. The benefit of having a critical mass of organisations adopting same technology is one aspect of inter-organisational relationships and IT adoption.
Market Knowledge	A majority of successful IT adoption cases are referred to the recognition of demands in the market (Kamal *et al.,* 2008). Lee and Treacy (1988) report that an unstable organisational environment generates increased potential for IT adoption. This requires an organisation's intent on being up-dated and well informed about the changes in market environment. Contact with environment through development of external information system e.g. can reduce the insecurity for the individual organisation.

Table 2. Support factors influencing EAI adoption in LGAs

Support Factors	
Top Management Support	Top management support has been recognised as one of the most important elements necessary for the successful implementation of integration technologies and integrated packages (Kamal *et al.,* 2008). Beath, (1991) reports that one of the most successful factors associated with large-scale IT implementation projects is securing the support of top management. In addition, sustained top management support within LGAs in IT projects, is needed throughout the implementation project (Chen and Gant, 2001). The reason is as the project progresses, active involvement of top management remains critical in constantly monitoring the progress of the project and providing direction to the implementation teams.
IT Support	EAI requires organisations to invest considerable investment on their IT infrastructure (Stal, 2002). Therefore, it is essential for organisations to have support from vendors and consultants. For example, consultants support the IT departments to introduce and evaluate EAI in organisations (Themistocleous, 2004). Vendors' support also has a correlation with IT infrastructure since vendors provide services (e.g. maintenance) to the organisations. As reported in several case studies the: (a) close relationships between one organisation and its hardware vendors and, (b) the dependence of the other organisation on the vendor's solution (hardware), influenced the decision for purchasing EAI package from the vendor (Themistocleous, 2004).
Higher Administrative Authority Support	Improving LGA technological facilities depends on whether support from higher administrative authorities elected or appointed top administrators, LGAs and also the central government is available for IT managers who are in charge of implementing technology adoption process and its utilisation. Kim and Bretschneider (2004) report that even in the case that IT managers initiate technology adoption; support from higher administrative authorities may play a significant role.

PROPOSED RESEARCH METHODOLOGY

The authors have followed an interpretivism, qualitative multiple case study approach to conduct this research and test the factors. Interpretivism assumes that the knowledge of reality is gained only through social constructions such as consciousness, shared meanings, language, documents, tools and other artefacts (Saunders *et al.,* 2000). Interpretivism stance was adopted, as the aim of this article is to understand how LGAs

Table 3. Financial factors influencing EAI adoption in LGAs

Financial Factors	
Return on Investment	In the context of LGAs, ROI is important, as technology budgets of LGAs at times are much lower as compared to other private and public organisations. Within LGAs budgets are often reduced and sometimes allocated with appropriations. Lam (2005) also reports that government organisations face difficulties in obtaining the level of funding requested, especially if funding is drawn from a funding pool that is meant to serve multiple initiatives. As a result, they do not want to invest more in technology, without significant ROI.
Cost	Literature indicates cost as a vital factor and many organisations perform a cost benefit analysis before taking any important decision regarding their investment for technology adoption (Themistocleous, 2004). In the context of EAI, Lee *et al.,* (2003) report that the basic concept of EAI is mainly in it externality of enterprise integration with lower cost and less programming using existing applications, whereas a significant benefit of EAI is the reduction of overall integration cost (Puschmann and Alt, 2001).

Table 4. Technological factors influencing EAI adoption in LGAs

Technological Factors	
Evaluation Frameworks	Themistocleous (2004) proposed two evaluation frameworks such as: (a) framework for evaluating integration technologies and (b) framework for evaluating EAI packages, which can be used by organisations to assess EAI packages and technologies. These frameworks highlight a possible combination of integration technologies and tools and EAI packages that can be used to integrate IT infrastructure. Therefore, these evaluation frameworks facilitate organisations to overcome the confusion regarding the selection of EAI products.
Technological Risks	Technological risks can make risk-averse managers require higher, not lower, rates of return before they invest. Studies on risk factor in IT projects have described issues e.g. organisational fit and technology planning. In LGAs, the degree of technological risks is escalating as the use of public networks increases together with data-bases that hold citizen's profiles and government information (Ebrahim and Irani, 2005).
IT Infrastructure	IT infrastructure consists of the computer systems and the supporting software needed to develop, manage and operate IT applications e.g. operating systems, development and management tools (Shaw, 2000). However, the non-integrated nature of IT infrastructure causes integration problems. Thus, raising the need for integration in the organisations. As a result, several researchers have reported IT infrastructure as a factor in their integration technologies adoption models.
Personnel IT Knowledge	Personnel IT knowledge refers to the IT capabilities of an organisation (Akbulut, 2002). The available skill set of the personnel is an important factor that constraints the introduction of new technologies. One of the most important factors in the adoption of computer applications by LGAs is staff competence. Norris (1999) reports that employees in LGAs were not very well trained in using information technologies and this inadequate training resulted in resistance to change, resistance to use, and under utilisation of computers.
IT Sophistication	IT sophistication is reported as an influential factor in integration technologies adoption models. The research findings on integration technology adoption represent that organisations with sophisticated IT resources will be likely to be adopters of integration technology. Themistocleous (2004) has reported IT sophistication as a factor for EAI adoption. This is due to the level of understanding in addressing technical problems at an enterprise and cross enterprise level.
Data Security and Privacy	Security and privacy of citizens' data has always been important (Signore *et al.,* 2005). Applications that have evolved autonomously rather than as part of an overall architecture inevitably end up having their own security architecture that is incompatible. Key security functions, such as authentication, authorisation, and confidentiality are managed according to the application's own specific set of rules, and thus present a significant challenge to the definition of a single security administration function across an integrated e-Government solution (Lam, 2005). In addition, citizens' concern on privacy and confidentiality of the personal data has been a critical obstacle in implementing e-Government projects.

integrate their IT infrastructures and make EAI adoption decisions. An interpretivism stance allows the authors to navigate and better explain this phenomenon. It is also anticipated that as the social world cannot be reduced to isolated variables, such as space and mass, it must be observed in its totality. Therefore, the authors assert that, there is a need for a research approach that may

Table 5. Organisational factors influencing EAI adoption in LGAs

Organisational Factors	
Centralisation	Centralisation refers to the degree of power or decision-making authority in organisations and encompasses participation in decision-making and authority hierarchy. In such organisations, decision-making is typically concentrated at the top level of hierarchy while in decentralised structures decision-making is distributed across different hierarchical levels. Since the decision-making for technology adoption is typically concentrated at top level of management in public sector organisations (Kamal, 2006; Ebrahim *et al.*, 2004), hence, the degree of centralisation may influence EAI adoption in LGAs.
Managerial Capability	Managerial capability refers to the ability of managers to identify problems of the current systems, and to develop and evaluate alternatives to improve the IT infrastructure of the organisation appears to be important. However, Senyucel (2005) argues that some managers are not realistic in their demands regarding IT infrastructure, which can be traced back to an inward-looking approach in the literature. Several managers were seen as highly suspicious of new initiatives and unsupportive. Such managerial shortsightedness has the potential to jeopardise the success of e-Government facilitation.
Barriers	Literature indicates that EAI adoption presents a similar case to ERP systems in terms of its barriers (Themistocleous, 2004). Like ERP systems, EAI: (a) promises to integrate IT infrastructures, (b) introduce changes to the organisational structure and the way of doing business, (c) influences the employees tasks as well as inter-organisational relationships, (d) it costs a lot of money and (e) is more likely adopted by big organisations. Since there are a lot of failures on ERP adoption, organisations tend to estimate the possible impact of EAI adoption before proceeding to its adoption.
Benefits	Benefits refer to the level of recognition of the advantages that the integration technologies could provide to the organisation. Iacovou *et al.*, (1995) classified perceived benefits into direct and indirect. Direct benefits were mostly operational saving-related (e.g. reduced transaction cost) and indirect benefits were mostly tactical and competitive advantages that had an impact on business process and relationships (e.g. increased operational efficiency). Bradford and Florin (2003) report benefits as organisational benefits that include the facilities for the integration problems, real-time planning, user satisfaction and support to quick customer response.
Formalisation	Formalisation refers to the existence of clear procedures, norms and formal processes for carrying out organisational tasks more effectively and efficiently (Lee *et al.*, 2003). Ebrahim *et al.*, (2004) reports formalisation as an organisational factor that is internal to the public sector, which influences the adoption and design of e-Government applications. IT adoption provides challenges to organisations as it not only addresses changes in technology and systems but also deals with the need for changing the way an organisation runs its business in terms of processes, workflows, policies, procedures, and structure (Ebrahim and Irani, 2005).
Size	Akbulut (2002) measures size in terms of the size of the community served and the number of the services provided by the organisation. Norris (1999) reported that larger organisations would adopt more sophisticated and advanced IT compared to smaller as larger organisations: (a) have greater financial resources, (b) are in more need of technologies and, (c) have superior institutional ability such as IT departments, to support the technologies. This is also because larger organisations input sufficient volume to justify the adoption of new technology to accommodate variations in input even when variations occur infrequently. Smaller organisations, however, experience many types of input variations so rarely that they could not reasonably expect to benefit from making similar accommodations.

allow LGAs to be viewed in their entirety and permits the authors to get close to participants (i.e. the interviewees), penetrate their realities, and interpret their perceptions. Hence, the authors consider interpretivism as more appropriate for the research reported herein. Having justified the use interpretive research approach, the authors describe the nature of qualitative research approach in order to justify its relevance to the research presented in this article. Qualitative research is multi-method in focus, involving an interpretive, naturalistic approach to its subject matter (Denzin and Lincoln, 1994). This implies that the qualitative researchers study things in their natural environment, and they comprehend events in terms of meanings that people bring to them. The qualitative paradigm recommends that researchers observe human behaviour and action as it occurs in mundane everyday life (Schutz, 1967). Thus, the authors suggest that in the context of this research a qualitative approach is more appropriate as such approach can be used to: (a) investigate less acknowledged phenomena like EAI adoption in the local government authorities, (b) examine the in-depth complexities and processes e.g. analysing the factors influencing the decision making process for EAI adoption, (c) examine the phenomenon in its natural setting, (d) provide considerable flexibility during interviews and observations and (c) learn from practice

A case study examines a phenomenon in its natural setting, employing multiple methods of data collection to gather information from one or a few entities e.g. people, groups, or organisations (Yin, 1994). Case studies can be single or multiple – a single case study may enable the researchers to investigate a phenomenon in depth, getting close to the phenomenon, providing rich primary data and revealing its deep structure within the organisational context (Cavaye, 1996). However, a single case may not provide sufficient insight into the phenomenon of EAI adoption in LGAs. The reason is that most research efforts require multiple cases, as single case studies are only useful in specific instances e.g. useful at the outset of theory generation and late in theory testing (Bonoma, 1985), which is not the case for this research. Instead, the authors suggest that multiple case studies might be more appropriate, as multiple case studies may enable the researchers to examine and 'cross-check' findings and may provide the research with a more 'robust' investigation of cause and effect relation of the units of analysis (Herriot and Firestone, 1983). Data was collected via interviews, observation, and archival documentation. Interviews are regarded as the main tool of qualitative research for data collection process (Denzin and Lincoln, 1994). In this research, interviews constituted the main data source in the case organisations. Six participants from each LGA_ABC and LGA_XYZ were interviewed using structured interviews. Structured interviews were based on the interview agenda. Using the interview agenda, the interviewees replied to specific questions on EAI adoption. Semi-structure interviews also took place but

without the use of an interview agenda. Using this type of interviews the authors attempted to clarify some issues that derived from structured interviews.

All the structure or semi-structured interviews took place at interviewees' office. Unstructured interviews dealt with discussions that the authors had with interviewees but without using a structured or semi-structured type of interview. The authors had unstructured interviews during lunches, coffee breaks, out of office hours. Using unstructured interviews some important data on the case studies were collected (e.g. information about resistance to change). All of the interviews were tape recorded and transcripts prepared as soon as possible after each interview. Tape recording supported the authors in collecting accurate data and interpreting them without time pressures.

CASE ORGANISATIONS: LGA_ABC AND LGA_XYZ

To test the influential factors (Tables 1 to 5), a series of case studies were undertaken with this section reporting the data of two of them. Both the case organisations are situated in the region of England and are responsible for providing services through various sectors such as: social and environmental services, housing, education and health. For confidentiality reasons the real names of these case organisations can not be reported. Instead the authors use the coded-names LGA_ABC and LGA_XYZ to refer to the case organisations. As reported earlier, structure, semi-structured and unstructured interviews were conducted in both case organisations to investigate the decision making process for EAI adoption. In achieving this, those factors considered to influence the decision-making process are identified, when seen from a multiple-stakeholder perspective. These perceptions were seen from those stakeholders that were involved in the EAI adoption, implementation and evaluation process. From LGA_ABC, the stakeholders that were interviewed included the: Head of Information Technology (HIT), Project Manager (PM), Development Service Manager (DSM), Principle Systems Developer (PSD), Senior Development Engineer (SDE), and Service Delivery Manager (SDM). At LGA_XYZ stakeholders interviewed included: Head of Information Communication and Technologies (HICT), Senior Systems Developer (SSD), Service Delivery Manager for Applications (SDMA), Web Manager (WM), Project Manager (PM) and Senior Software Engineer.

Integration Problems

LGA_ABC

LGA_ABC was faced with considerable pressures to cope with the extensive social regeneration of the borough, while meeting statutory requirements for integrated service delivery targets, performance indicators, e-Government targets, and legislation changes. In addition, LGA_ABC faced funding pressures and challenges in terms of improved resource and asset management. LGA_ABC was also confronted with pressures to reduce the cost of maintaining non-integrated IT infrastructure, providing better service delivery, IT infrastructure integration, and support improved ways of working through collaboration and remote/home working capabilities. The interviewees at LGA_ABC mutually agreed that:

... their IT infrastructure was very much fragmented with different IS all over the borough with no integration, there was no communication and lack of transparency and silo mentality prevailed ....

LGA_ABC's efforts to modernise have been hindered by an IT infrastructure that has grown in a piecemeal over the years. LGA_ABC implemented various IS to enhance their service delivery. These information systems did not solve all the problems as they used a variety of hardware of different ages, running different operating systems and software applications. Thus, LGA_ABC turned to integrated applications by developing manual point-to-point connections. However, such an approach has also led to applications spaghetti, which increases the complexity of the integration solution as the number of interconnected applications rise thus, preventing in overcoming the limitations of their IT infrastructure. These problems became an obstacle for LGA_ABC as they prevented it from implementing its business goals. For instance, LGA_ABC could not support its goal of closer collaboration and coordination of inter-organisational business processes due to the non-integrated nature of its applications. This held LGA_ABC back from achieving cost reductions.

LGA_XYZ

LGA_XYZ is a big borough with several service areas. Each service area has its own IT infrastructure with numerous heterogeneous information systems that were based on a diversity of platforms, operating systems, data structures and computer languages. Most of these systems were legacy applications that still run today on mainframe environments. Since there was a lack of common IT infrastructure, and a lack of central coordination of IT, the majority of LGA_XYZ departments adopted their own applications to support their business activities. These individual applications were not developed in a coordinated way but instead evolved as a result of the latest technological innovation. This led to incompatible systems with integration problems. LGA_XYZ has attempted to overcome this problem by integrating their systems. For example, LGA_XYZ implemented ERP systems to overcome their integration problems and automate their business processes. Although ERP systems partially addressed the problems of LGA_XYZ, nevertheless, they simply provide some degree of solution for the integration problems.

This is because ERP systems were not designed to integrate disparate systems but rather to replace them to achieve integration. The need for an integrated and flexible IT infrastructure has been necessitated with the existing infrastructure causing numerous problems. The interviewees at LGA_XYZ illustrated that their IT infrastructure has been underdeveloped and not integrated and thus, several limitations existed e.g. the interviewees mutually agreed that:

... IT infrastructure was constructed in a departmental way. Each of the major service areas within this borough had their own IT infrastructure ....

EAI Adoption Process

LGA_ABC

The limitations in the IT infrastructure led LGA_ABC to take a decision to significantly advance in service delivery by adopting EAI solution to develop an integrated IT infrastructure. Project manager reported that:

... the reasons for adopting EAI was reduction in duplication of data and cost of implementing an integrated IT infrastructure, improvements in business process reengineering, savings and efficiency, streamlining processes, accuracy of data output and up-to-date information ....

LGA_ABC was faced with the option of withdrawing their heterogeneous systems away and procuring new systems, or finding a method of migrating to a new generation of systems, which would support integrated service delivery. Due to the rich source of information contained in them and to make development more manageable, the second option was chosen to work on an integration project (SoftVendor [their software vendor] and CRM [Customer Relationship Management] integration project) for the environmental health department. The aim of the project was to pro-

vide citizens with better services and respond to their waste collection queries quickly. SDE and SDM reported that in the first three months LGA_ABC logged approximately 13,000 jobs using the integrated system. Also by comparing the pre SoftVendor business processes with the post integration processes, LGA_ABC estimates an improvement in business processes across all service areas over the next 12 months. For other areas, LGA_ABC is gradually moving towards an EAI hub and spoke methodology to develop a global integrated IT infrastructure.

Later during the interview sessions, the interviewees were asked to highlight the importance of factors that influenced the decision making process for EAI technological solution adoption (Table 6) in this project. The level of importance as presented in Table 6 follows a scale similar to the one used by Miles and Huberman (1994) i.e. scale of less important (O), medium important (⊙) and most important (●).

As highlighted in Table 6, two factors were not tested, i.e. evaluation frameworks and IT support. Regarding the evaluation frameworks, the project manager reported that:

"...we had cheaper in house solution with expertise, knowledge and skills. Basically we just conducted a market survey for cost evaluation, not to procure a solution that may have cost us lots of money. We needed a solution that could assist us in implementing the project quickly, thus we evaluated the solution from the options based how quickly we can implement this project ...".

The interview discussions also illustrate that the project team did not use any support from external consultant. The service delivery manager reported the reason for not selecting IT support from external consultant:

"... we wanted to be able to use our existing staff skills, support the system better and it is better value for money ...".

Table 6. Importance of factors influencing EAI adoption in LGA_ABC

	Factors	LGA_ABC					
		HIT	PM	DSM	PSD	SDM	SDE
PF	Project Champion	●	●	●	●	●	●
	Citizen's Satisfaction	●	◉	◉	●	●	○
	Critical Mass	◉	◉	●	●	●	●
	Market Knowledge	◉	●	◉	●	●	◉
SF	Top Management Support	●	●	●	●	●	●
	IT Support	◉	○	◉	○	○	◉
	Higher Administrative Authority	●	●	●	◉	◉	●
FF	Return on Investment	●	◉	◉	●	●	●
	Cost	●	●	●	●	●	●
TF	Evaluation Frameworks	○	○	◉	○	◉	◉
	Technological Risks	◉	◉	●	●	●	●
	IT Infrastructure	●	◉	●	●	●	●
	Personnel IT Knowledge	●	●	●	●	●	◉
	IT Sophistication	●	●	◉	●	◉	◉
	Data Security and Privacy	●	●	●	●	●	●
OF	Centralisation	◉	●	◉	◉	●	●
	Managerial Capability	●	●	●	●	●	●
	Barriers	●	●	●	●	●	●
	Benefits	●	●	●	●	●	●
	Formalisation	◉	◉	●	◉	●	●
	Size	●	●	◉	●	◉	●

The remaining factors as highlight had varied findings as such preference on the importance of factors by the interviewees is simply based on the interviewee's observation, understanding and involvement during their project at LGA_ABC.

LGA_XYZ

To overcome their integration problems, LGA_XYZ initiated a plan for working on the top level electronic Forms (e-Forms) and CRM system integration pilot project. The motivation behind this project was to address the limitations of its existing systems, and to meet the targets set by the central government. The managing board made the decision for this project after discussing this issue with their project manager and other senior managers involved. The objective of this project was to demonstrate to LGA_XYZ and to other LGAs that investing in a long-term programme of integration between packaged systems and legacy applications is necessary. On this basis the adoption of such integration architecture within LGA_XYZ and other London boroughs will deliver measurable business benefit. The objectives of the project were to: (a) demonstrate and deliver the benefits of integrating cash receipting (i.e. via online payment system), CRM and e-Forms, (b) re-establish and re-energise development and investment in CRM and (c) demonstrate the benefits of business process re-engineering.

Prior to start working on this project, the project team was working on an Electronic Service Delivery (ESD) project – an enterprise wide project. ESD is the strategic view that the whole department is undertaking, whereas, this is a tactical pilot project to achieve the purpose as aforesaid. This pilot project focused on re-engineering five specific business processes such as: (a) issue and administration of green waste bins, (b) bulky item collection, (c) vehicle crossover applications, (d) skip license applications and (e) trade waste sack applications. In doing so, it would help LGA_XYZ in justifying the decision to adopt EAI solution for ESD project. Later during the interview sessions, the interviewees were asked to highlight the importance of EAI adoption factors (Table 7) in this project. The level of importance as presented in Table 7 follows a similar scale as used for testing the factors used in Table 6.

As illustrated in Table 7, one factor was not tested, i.e. return on investment. The project improved data collection with intelligent, dynamic top level e-Forms. Cost fell because quality data reduced the form rejection rates and received e-Forms that did not need to be manually keyed-in, whereas, ROI was not validated. The head of ICT reported that:

"... due to silo mentality in LGA_ABC we were not able to prove the return on our investment in this project...".

Table 7. Importance of factors influencing EAI adoption in LGA_XYZ

	Factors	LGA_ABC					
		HIT	PM	DSM	PSD	SDM	SDE
PF	Project Champion	•	•	•	•	⊙	•
	Citizen's Satisfaction	•	•	⊙	⊙	•	•
	Critical Mass	⊙	⊙	•	•	•	⊙
	Market Knowledge	⊙	⊙	•	•	•	⊙
SF	Top Management Support	•	•	•	⊙	•	•
	IT Support	•	⊙	⊙	•	⊙	⊙
	Higher Administrative Authority	•	•	•	⊙	⊙	•
FF	Return on Investment	O	⊙	O	O	⊙	⊙
	Cost	•	⊙	•	•	⊙	•
TF	Evaluation Frameworks	⊙	⊙	•	•	⊙	⊙
	Technological Risks	•	⊙	•	•	•	•
	IT Infrastructure	•	•	O	⊙	•	•
	Personnel IT Knowledge	•	⊙	⊙	•	•	•
	IT Sophistication	⊙	•	⊙	•	⊙	•
	Data Security and Privacy	•	⊙	•	•	•	•
OF	Centralisation	•	•	•	⊙	•	⊙
	Managerial Capability	•	⊙	•	•	•	⊙
	Barriers	•	•	•	•	⊙	•
	Benefits	•	•	•	•	•	•
	Formalisation	•	•	•	⊙	•	⊙
	Size	•	•	•	⊙	⊙	⊙

The remaining factors illustrate varied answers from the interviewees, however, comparing the outcome of Tables 6 and 7, it is clear that most of the factors have influenced the decision making process for EAI adoption in LGA_ABC and LGA_XYZ. Thus, validating the factors influencing EAI adoption in LGAs as reported in Tables 1 to 5. These findings are in accordance with the literature findings (e.g. Kamal 2006; Khoumbati *et al.,* 2006; Chen 2005; Themistocleous, 2004) that present these factors for the adoption of different integration technologies in various types of organisations e.g. public sector to private sectors and SMEs to large organisations.

SUMMARISING THE FINDINGS ON THE EAI ADOPTION FACTORS

This section presents the findings regarding EAI adoption factors derived from the case studies conducted in two case organisations. In doing so, the authors develop an evaluation matrix that depicts the similarities and differences of the proposed EAI adoption factors across the two case studies. Tables 8 to 12 illustrate the synthesis of the EAI adoption factors using the findings derived from the case organisations during the interview discussions. These tables confirm the validation of the EAI adoption factors (Tables 1 to 5) with new factors that are derived from the empirical findings. The new factors (indicated with a '*' symbol in the following tables) played an important role in the EAI adoption process in the case organisations.

The data collected from the two case organisations was confirmed to be of relatively similar significance with marginal differences, therefore, it can be said that selecting another case study would have provided comparatively similar results. Therefore, the empirical findings illustrated in Sections 4 indicate that the factors presented

Table 8. Findings of the pressure factors across the case organisations

Factors	LGA_ABC	LGA_XYZ
Project Champion	Project champion has been an important and key player in leading this integration project.	An important player that leads both technological and business projects also championed this project.
Project Delivery Time*	There has been pressure from management to deliver the project on time.	An important factor as there was pressure from the top management to deliver the project on time.
Data Consistency*	Data duplication and inconsistencies has been an important internal pressure.	Data duplication and inconsistencies in matching data with other systems has been a pressure.
Citizen's Satisfaction	A prime factor influencing the decision makers for EAI adoption.	A prime factor influencing the decision makers for EAI adoption.
Critical Mass	The project team initially investigated and analysed the solutions of other boroughs and how it benefited them.	The project team investigated and assessed the solutions of other boroughs and analysed their outcomes.
Market Knowledge	An important factor that influenced EAI adoption process as without knowledge on different integration technologies it is not possible to proceed.	One of the prime important factors that influenced EAI adoption in LGA_XYZ.
Stakeholders Pressure*	Pressure from different stakeholder e.g. peers from other departments and citizens for information sharing and improving service provision.	Pressure from different stakeholders e.g. suppliers, partners, community, etc for information sharing, shared services and improving service provision.
Competition*	Competition between neighboring boroughs to be more productive in service delivery and responsive to citizen queries.	Pressure from central government to achieve the e-Government targets developed a competitive environment for improving services.

in Tables 1 to 5 can be used for improving the decision making process for EAI adoption in LGAs. The main findings drawn from investigating EAI adoption in two LGAs are summarised below:

- **Finding 1:** Empirical findings suggest that IT adoption in LGAs has been through a considerable continuous process. Most successful developments of IT in LGAs in the past have been centred on supporting and improving infrastructure and internal processes. However, in the 1990's the focus of IT shifted in improving LGA business processes and service delivery. Furthermore in 2000 and onwards, the focus of IT usage resulted in LGAs adopting

several IT applications e.g. CRM, to improve legacy business processes, service delivery to citizens with-to-date information and improving IT infrastructure. The evident support from the case studies and the documents provided for the essential focus was to see the computer not just as a tool to provide information, but rather as a communication and integration tool.

- **Finding 2:** The empirical findings suggest that the IT implementation decisions in the UK local government domain have gone through several phases. As a result, the IT infrastructure of two case organisations resulted as non-integrated. Consequently, the case organisations faced integration problems while working with other LGAs, part-

Table 9. Findings of the support factors across the case organisations

Factors	LGA_ABC	LGA_XYZ
Top Management Support	Has always provided support moreover, there was also pressure from the top management to deliver the project on time.	Top management support has been very important in backing this integration project.
IT Support	Because the project team wanted to use in-house expertise, thus this factor was *NOT VALIDATED*.	Their software vendor provided the necessary IT support along with the market survey to improve their IT capabilities.
Higher Administrative Authority Support	Has been substantially important in providing support for data sharing and improving services.	Vital in providing support for sharing of data with other departments and in improving services to citizens.
Stakeholders Support*	General administrative and availability of skilled staff proved to be supporting factors for the integration projects.	General support from other neighbouring boroughs to consult service delivery issues and information sharing.

Table 10. Findings of the financial factors across the case organisations

Factors	LGA_ABC	LGA_XYZ
Return on Investment	As for ROI, comparing the pre business processes with the post integration processes, the borough estimates a vital improvement in the next 12 months.	Sometimes it is difficult to prove because of the silo mentality that still prevails in the department. Thus, this factor was *NOT VALIDATED*.
Cost	Several costs were identified during EAI adoption e.g. cost of training staff for developing EAI skills, maintenance cost.	Important but an issue because if integration projects not well implemented then it increases implementation cost.
Central Government Grant*	Better the functioning of borough, then central government gives more grants to work on other integration projects.	To support community, central government provides sufficient grants and funding to improve to integrated services to citizens.

ners and other government bodies. Thus, it was difficult for the case organisations to reconfigure and integrate all the applica- tions that run on the mainframe and non-mainframe platforms. In addition, there was a redundancy of data and functionality

Table 11. Analysis of the technological factors across the case organisations

Factors	LGA_ABC	LGA_XYZ
Evaluation Frameworks	The project team followed the suggestions of their SoftVendor and no evaluation framework was used to assess the EAI solution adopted. Thus, it was *NOT VALIDATED.*	Used Best Value as evaluation tool to asses EAI. In addition, the project team followed a pilot evalua- tion method to test the information flow between CRM system and the e-Forms.
Technological Risks	Crucial factor and faced several risks e.g. EAI may not deliver the benefits, EAI may not work, lack of EAI skills, lack of commitment to EAI projects etc.	Important factor and faced EAI risks while imple- menting the CRM system and e-Forms integration project i.e. selection of supplier and identifying the business needs.
IT Infrastructure	IT infrastructure was fragmented with different IS and no integration, communication, lack of trans- parency and silo mentality prevailed.	IT infrastructure was constructed based on silo men- tality i.e. applications were developed with different operating systems and platforms.
Personnel IT Knowledge	Reported as vital factor but IT knowledge among the staff was extremely limited.	An important requirement to be able to work on integration projects
IT Sophistication	IT sophistication has influenced because it has as- sisted in securing data through latest technologies.	Faced IT sophistication problems with lack of staff with EAI skills and knowledge and lack of resources.
Data Security and Privacy	An important issue as citizens' data may contain vital information and to use citizen data for sharing with other departments their consent is required.	Has been one of the most important problems to meet because citizens' data may contain important information.

Table 12. Analysis of the organisational factors across the case organisations

Factors	LGA_ABC	LGA_XYZ
Centralisation	Important factor as without the authority of single person data standards cannot be embed- ded in the organisation.	Important factor and the decision-making was cen- tralised by the head of ICT to adopt an EAI solution.
Managerial Capability	An important driving factor behind this project for EAI adoption.	Lack of competency regards to IT among managers and there is a need for corporate level competencies within the managers.
Barriers	Several barriers identified during EAI adop- tion with among others lack of EAI skills a major barrier.	Several barriers were experienced during the EAI adoption with silo mentality as an important barrier.
Benefits	Several benefits influenced EAI adoption e.g. efficiency of business processes, reliable data transfer, increase in flexibility of systems, and support in reducing citizen's data errors.	Several benefits influenced EAI adoption e.g. ef- ficiency of business processes, reliable data transfer, increase in flexibility of systems, and support in reduc- ing citizen's data errors as important benefits.
Formalisation	As there was pressure from HIT for standar- disation in the work processes, this factor influenced EAI adoption.	Pressure from head of ICT to have standardisation in work processes across the department.
Size	It is not only the organisational and com- munity size but also the amount of resources, the capital the borough has and the amount of funding the borough gets from the central government. This factor has influenced the decision to adopt EAI.	An important factor as a large part of the resources is driven by the organisational and community size. Funding from the central government depends on com- munity size as more the community the more funding borough gets.

as many applications stored similar data or run systems overlapping in functionality. Additionally, the non-integrated infrastructure caused many problems, since it could not achieve integration. As a result, the case organisations could not take advantage of IT and support closer collaboration with their various stakeholders. Therefore, the IT infrastructure limitations motivated the case organisations for integration.

- **Finding 3:** The findings from the case organisations confirm that external pressures from stakeholders for the provision of integrated service delivery to the citizens, information sharing and shared services with other LGAs represented a highly influencing factor that resulted EAI adoption in the case organisations. The stakeholder pressure is from peers, residential, IT suppliers, private sector, and competition also represent external pressures. In addition, top management pressure for project delivery on time and pressure from the head of department to have standardisation of work processes and work without conflict in the organisation represented internal pressures. All these external and internal pressures influencing EAI adoption represented as decisive factors.

- **Finding 4:** The findings from the case organisations illustrate that benefits and barriers represent important factors during EAI evaluation. Both the case organisations achieved several benefits with the availability of right information at the right time and right place. For example, benefits include rationalising technical skills requirements, reduced data errors, citizen satisfaction, integration of business processes, support in the provision of better service delivery, improving data quality, flexibility of work place, allows organisations to do business more effectively. Similarly, these LGAs also experienced

several barriers during the adoption EAI process. For example, funding from central government to work on EAI projects, lack of employees on EAI skills, weak vendor support for EAI, resistance to change, high cost required for EAI implementation, security and confidentiality concerns about citizen data, reluctant to share data.

CONCLUSION, FURTHER WORK AND RESEARCH LIMITATIONS

While private organisations have continued to take advantage of EAI solutions to develop integrated IT infrastructures and improve their business processes, services offered by government organisations have remained deficient over the years. Implementing EAI is a challenging problem for the local government authorities, as it requires understanding and reengineering of the LGA structure, gaining stakeholder commitment, sharing of knowledge, and negotiating the division of costs and benefits. On the other hand, the concept of e-Government has emerged as a credible solution to improve such services as it allows people to access public services from within their own homes or offices. The importance of electronic service delivery has been widely recognised with 30 European ministers agreed upon a plan to speed up the development of e-Government applications in an attempt to modernise the European public sector. During the recent years LGAs have adopted CRM applications to improve their services and the relationships with their citizens. The application of CRM is beneficial for LGAs as it results in improvements in information sharing and cost reduction (as also illustrated from the case studies presented in this article).

Recently, many LGAs have attempted to link together their e-Government information systems and CRM applications to gain more advantages and deliver better services. Nonetheless, LGAs have realised that they can gain significant advantages when they integrate their CRM and e-Government information systems with their disparate back office solutions. Thus, they are seeking ways to integrate their applications and IT infrastructures. The authors suggest that LGAs can focus on integration technologies like EAI to incorporate their systems and processes, and thus, achieving their goals. Yet, the adoption of EAI is still in its infancy with LGAs, researchers need to understand the issues surrounding this technology. This article analysed and presented the EAI adoption practices by two LGAs, namely LGA_ABC and LGA_XYZ. Empirical data for the present study were extrapolated through various sources of data like interviews, documentation and observation from these case organisations. The purpose of this data collection was to test and validate the factors (Tables 1 to 5) influencing the decision making process for EAI adoption. Data was collected until there was enough data to test the proposed EAI adoption model. As highlighted in Tables 6 and 7, most of the factors were validated through the case studies, thus, supporting the authors' literature findings on the proposed EAI adoption factors. Although the empirical research tested the factors, the research work presented in this article is no exception; as a result this research can be further developed. In the light of the reflections and the limitations it is recommended that further work could usefully be pursed as follows:

• **Recommendation:** The theoretical and empirical data collected are confined to the limited context of the LGAs within the region of England. The structure of LGAs varies in different parts of the United Kingdom (UK). There are five different types of authorities in the UK and these are divided into single-tier and two-tier authorities with differences in the organisational structure, nature and size of each authority. In this context, it may be difficult to generalise the results of this research to other parts of the UK and other countries.

As reported in Tables 8 to 12, six new factors (indicated with '*' symbol) were identified from both the case organisations influencing EAI adoption. However, these factors have yet to be tested. A recommendation for the future study may be to test all the factors including the new factors as identified in this research in another LGA across other parts of the UK.

The empirical findings on the factors and the observations derived from the case organisation presented in this article cannot be generalised. Nevertheless, it is not the intention of this article to offer prescriptive guidelines for EAI adoption in LGAs, but rather to describe case organisation perspectives that allow others to relate their experiences to those reported. Hence, this article offers a broader understanding of the phenomenon of EAI adoption in LGAs.

REFERENCES

Akbulut, A. Y. (2002). An Investigation of the Factors that Influence e-Information Sharing Between State and Local Agencies, *Proceedings of the 8th Americas Conference on Information Systems*, USA, (pp. 2454-2460).

Bass, C. & Lee, J.M. (2002). Building a Business Case for EAI, *EAI Journal*, 18-20.

Beath, C. M. (1991). Supporting the Information Technology Champion . *Management Information Systems Quarterly, 15*(3), 355–372. doi:10.2307/249647

Beynon-Davies, P. (2005). Constructing Electronic Government: The Case of the UK Inland Revenue . *International Journal of Information Management, 25*(1), 3–20. doi:10.1016/j.ijinfomgt.2004.08.002

Beynon-Davies, P., & Williams, M. D. (2003). Evaluating Electronic Local Government in the UK . *Journal of Information Technology, 18*, 137–149. doi:10.1080/0268396032000101180

Bonoma, T. (1985). Case Research in Marketing: Opportunities, Problems and a Process . *JMR, Journal of Marketing Research, 22*, 199–208. doi:10.2307/3151365

Bozeman, B., & Kingsley, G. (1998). Risk culture in public and private organizations. *Public Administration Review, 58*(2), 109–118. doi:10.2307/976358

Bradford, M., & Florin, J. (2003). Examining the Role of Innovation Diffusion Factors on the Implementation Success of ERP Systems. *International Journal of Accounting Information Systems, 4*(3), 205–225. doi:10.1016/S1467-0895(03)00026-5

Cavaye, A. L. M. (1996). Case study research: a multifaceted research approach for IS. *Information Systems Journal, 6*(3), 227–242. doi:10.1111/j.1365-2575.1996.tb00015.x

Chen, H. (2005). *Adopting Emerging Integration Technologies in Organisations*, PhD Thesis, Department of Information Systems and Computing, West London, UK, Brunel University.

Chen, Y. C., & Gant, J. (2001). Transforming Local E-Government Services: The Use of Application Services Providers. *Government Information Quarterly, 18*(4), 343–355. doi:10.1016/S0740-624X(01)00090-9

Cheng-Yi Wu, R. (2007). Enterprise Integration in e-Government. *Transforming Government: People . Process and Policy, 1*(1), 89–99.

Denzin, N. Y. K., & Lincoln, Y. (1994). *Handbook of Qualitative Research*. London: SAGE Publications.

Ebrahim, Z., & Irani, Z. (2005). E-Government Adoption: Architecture and Barriers. *Business Process Management Journal, 11*(5), 589–611. doi:10.1108/14637150510619902

Ebrahim, Z., Irani, Z., & Sarmad, S. (2004). Factors Influencing the Adoption of E-Government in Public Sector. *European & Mediterranean Conference on Information Systems*, Tunis Tunisia.

Fichman, R. (1992). Information Technology Diffusion A Review of Empirical Research, *13th International Conference on Information Systems*, (pp. 195-206).

Gichoya, D. (2005). Factors Affecting the Successful Implementation of ICT Projects in Government. *The Electronic . Journal of E-Government, 3*(4), 175–184.

Herriot, R. E., & Firestone, W. A. (1983). Multisite Qualitative Policy Research: Optimising Description and Generalisability. *Educational Researcher, 12*, 14–19.

Iacovou, C., Benbasat, I., & Dexter, A. (1995). Electronic Data Interchange and Small Organisations – Adopting and Impact of Technology. *Management Information Systems Quarterly, 19*(4), 465–485. doi:10.2307/249629

Irani, Z., Al-Sebie, M., & Elliman, T. (2006). Transaction stage of E-Government systems: Identification of its location & importance. *Proceedings of the 39th Hawaii International Conference on System Sciences, IEEE*, (pp. 1-9).

Janssen, M., & Cresswell, A. (2005). Enterprise Architecture Integration in E-Government. *38th Hawaii International Conference on System Sciences*, Hawaii, (pp. 1-10).

Kamal, M. M. (2006). IT Innovation Adoption in the Government Sector: Identifying the Critical Success Factors. *Journal of Enterprise Information Management, 19*(2), 192–222. doi:10.1108/17410390610645085

Kamal, M. M., & Themistocleous, M. (2007). Investigating EAI Adoption in LGAs: A Case Study Based Analysis. *Proceedings of the 13th Americas Conference on Information Systems*, Keystone, Colorado, USA, (pp. 1-13).

Kamal, M. M., Themistocleous, M., & Elliman, T. (2008). Extending IT Infrastructures in LGAs through EAI, *Proceedings of the 14ᵗʰ Americas Conference on Information Systems*, Toronto, Canada, (pp. 1-12).

Kamal, M. M., Themistocleous, M., & Morabito, V. (2008). *Evaluating Information Systems: Public and Private Sector*. Published by Butterworth-Heinemann.

Kamal, M. M., Themistocleous, M., & Morabito, V. (2009). Justifying the Decisions for EAI Adoption in LGAs: A Validated Proposition of Factors, Adoption Lifecycle Phases, Mapping and Prioritisation of Factors. *Proceedings of the 42ⁿᵈ Hawaii International Conference on System Sciences*, Hilton Kaikoloa Village Resort, Hawaii, (pp. 1-10).

Kawalek, P., & Wastell, D. (2005). Pursuing radical transformation in information age government: Case studies using the SPRINT methodology. *Journal of Global Information Management*, *13*, 79–101.

Khoumbati, K., Themistocleous, M., & Irani, Z. (2006). Evaluating the Adoption of Enterprise Application Integration in Healthcare Organisations. *Journal of Management Information Systems*, *22*(4), 69–108. doi:10.2753/MIS0742-1222220404

Kim, H. J., & Bretschneider, S. (2004). Local Government Information Technology capacity: An Exploratory Theory. *Proceedings of the 37ᵗʰ Annual Hawaii International Conference on System Sciences*, (pp. 121-130).

Lam, W. (2005). Barriers to E-Government Integration. *Journal of Enterprise Information Management*, *18*(5), 511–530. doi:10.1108/17410390510623981

Lam, W. (2005). Investigating Success Factors in Enterprise Application Integration: A Case Driven Analysis. *European Journal of Information Systems*, *14*(2), 175–187. doi:10.1057/palgrave.ejis.3000530

Lee, C. S., Chandrasekaran, R., & Thomas, D. (2003). Examining IT Usage across Different Hierarchical Levels in Organisations: A Study of Organizational, Environmental, and IT Factors. *9ᵗʰ Americas Conference on Information Systems*, (pp. 1259-1269).

Lee, S., & Treacy, M. E. (1998). Information technology impacts on innovation. *R & D Management*, *18*(3), 257–271. doi:10.1111/j.1467-9310.1988.tb00592.x

Linthicum, D. (2000). *Enterprise Application Integration*. Massachusetts, USA: Addison-Wesley.

Mantzana, V., Themistocleous, M., Irani, Z., & Morabito, V. (2007). Identifying Healthcare Actors Involved in the Adoption of Information Systems. *European Journal of Information Systems*, *16*(1), 90–102. doi:10.1057/palgrave.ejis.3000660

McIvor, R., McHugh, M., & Cadden, C. (2002). Internet Technologies Supporting Transparency in the Public Sector. *International Journal of Public Sector Management*, *15*(3), 170–187. doi:10.1108/09513550210423352

Miles, M., & Huberman, A. (1994). *Qualitative Data Analysis: An Expanded Sourcebook*. Newbury Park, California: Sage.

Norris, D. F. (1999). *Leading Edge Information Technologies and their Adoption: Lessons from US Cities*. Hershey, PA: Idea Group Publishing.

O'Toole, K. (2007). E-governance in Australian local government: spinning a web around community? *International Journal of Electronic Government Research*, *3*(4), 58–75.

Ongaro, E. (2004). Process management in the public sector: the experience of one-stop shops in Italy. *International Journal of Public Sector Management*, *17*(1), 81–107. doi:10.1108/09513550410515592

Puschmann, T., & Alt, R. (2001). Enterprise Application Integration – The Case of the Robert Bosch Group. *Proceedings of the 34th Hawaii International Conference on System Sciences* (pp. 1-10) Maui, Hawaii, IEEE.

Reddick, C. G. (2009). Factors that Explain the Perceived Effectiveness of E-Government: A Survey of United States City Government Information Technology Directors. *International Journal of Electronic Government Research, 5*(2), 1–15.

Saunders. M., Lewis, P., & Thornhill, A. (2000). *Research Methods for Business Students'*. Essex, Pearson Education Ltd.

Schutz, A. (1967). *The phenomenology of the social world*. Evanston, IL: Northwestern University Press.

Senyucel, Z. (2005). Towards Successful e-Government Facilitation in UK Local Authorities. *e-Government Workshop* (pp. 1-16).

Shaw, N. G. (2000). Capturing the Technological Dimensions of IT Infrastructure Change. *Journal of the Association for Information Systems, 2*(8).

Signore, O., Chesi, F., & Pallotti, M. (2005). E-Government: Challenges and Opportunities. *CMG Italy – XIX Annual Conference*, Florence, Italy.

Stal, M. (2002). Web Services: Beyond Component – Based Computing. *Communications of the ACM, 45*(10), 71–76. doi:10.1145/570907.570934

Tan, C., Pan, S., & Lim, E. (2005). Managing stakeholder interests in e-government implementation: Lessons learned from a Singapore e-government project. *Journal of Global Information Management, 13*(1), 31–53.

Themistocleous, M. (2004). Justifying the decisions for EAI implementations: A validated proposition of influential factors. *Journal of Enterprise Information Management, 17*(2), 85–104. doi:10.1108/17410390410518745

Themistocleous, M., & Irani, Z. (2001). Benchmarking the Benefits and Barriers of Application Integration, *Benchmarking. International Journal (Toronto, Ont.), 8*(4), 317–331.

Themistocleous, M., Irani, Z., & Love, P. E. D. (2005). Developing E-Government Integrated Infrastructures: A Case Study. *Proceedings of the 38th Annual Hawaii International Conference on System Sciences* (pp. 1-10). Big Island, Hawaii: IEEE.

Weerakkody, V., & Dhillon, G. (2008). Moving from E-Government to T-Government: A Study of Process Reengineering Challenges in a UK Local Authority Context. *International Journal of Electronic Government Research, 4*(4), 1–16.

Weerakkody, V., Janssen, M., & Hjort-Madsen, K. (2007). Realising Integrated E-Government Services: A European Perspective. *Journal of Cases in Electronic Commerce, 3*(2), 14–38.

Welch, E. W., Hinnant, C. C., & Moon, M. J. (2005). Linking Citizen Satisfaction with E-Government and Trust in Government. *Journal of Public Administration: Research and Theory, 15*(3), 371–391. doi:10.1093/jopart/mui021

Wimmer, M. A. (2004). A European perspective towards online one-stop government: the e-Government project. *Electronic Commerce Research and Applications, 1*(1), 92–103. doi:10.1016/S1567-4223(02)00008-X

Yin, R. K. (1994). *Case Study Research Design and Methods*. London: Sage.

This work was previously published in International Journal of Electronic Government Research (IJEGR), edited by Vishanth Weerakkody, pp. 1-20, copyright 2009 by Information Science Reference (an imprint of IGI Global)

Chapter 12
Implementing Free Wi-Fi in Public Parks:
An Empirical Study in Qatar

Shafi Al-Shafi
Brunel University, UK

Vishanth Weerakkody
Brunel University, UK

ABSTRACT

This article examines the adoption of free wireless Internet parks (iPark) by Qatari citizens as a means of accessing electronic services from public parks. The Qatari government has launched the free wireless Internet parks concept under their national electronic government (e-government) initiative with a view to providing free Internet access for all citizens whilst enjoying the outdoors. By offering free wireless internet access, the Qatari government hopes to increase the accessibility of e-government services and encourage their citizens to actively participate in the global information society with a view to bridging the digital divide. The adoption and diffusion of iPark services will depend on user acceptance and the availability of wireless technology. This article examines an extended technology acceptance model (TAM) that proposes individual differences and technology complexity in order to determine perceived usefulness and perceived ease of the iPark initiative by using a survey-based study. The article provides a discussion on the key findings, research implications, limitations, and future directions for the iPark initiative in Qatar.

INTRODUCTION

The Qatari free-wireless- internet-park (iPark) concept was launched in 2007 under the banner of e-government in order to make public services more accessible to citizens. The iPark initiative is implemented in a number of public parks and

provides free connection to the citizens' Internet ready devices, such as laptops and Personal Digital Assistants (PDA) at any time of the day. The first such initiative was launched in the city of Doha in Qatar and it is the first of its kind in the Western Asia region. The primary goal of this initiative is to increase internet usage by establishing "hot spots" in public parks (IctQATAR, 2007). There are currently three designated wireless internet hot-

DOI: 10.4018/978-1-60960-162-1.ch012

spots throughout selected public parks in the city (ibid). This article explores citizens' acceptance of the iPark concept as part of the e-government initiative in Qatar.

The Qatari E-government initiative was launched in 2000 and in global terms the UN E-government readiness report (2008) ranked Qatar's e-government project as number 53 worldwide. As in many countries, the national e-government focus for Qatar is to achieve the highest performance in executing governmental transactions electronically, through streamlined business processes and integrated information technology solutions (IctQATAR, 2007). Moreover, the Qatari government hopes that free internet access offered through the iPark concept will encourage more citizens to use e-government services and help bridge the digital divide.

The Internet, whilst being the primary mode of access to e-government services, has not been adapted globally at the same time or rate; some countries are considered as leaders (such as the US, UK and Singapore) and others simply follow (i.e. the Arabian Gulf region) (Gupta and Gupta, 2005). More recently, wireless technologies have become a useful means of internet connectivity and access to electronic services (e-services). Wi-Fi for 'Wireless Fidelity' is a set of standards for wireless local area networks (WLAN) and provides wireless access to the Internet (Lehr and McKnight, 2003). Hotspots providing such access include Wi-Fi cafes where services may be free to customers. In fact, a hotspot need not be limited to a confined location; as illustrated in this article, public parks can be used to offer free wireless internet access to citizens.

In particular, Wi-Fi has opened up new opportunities for electronic commerce (e-commerce) and e-government by allowing citizens to build connectivity 24 hours a day, seven days a week (24/7). Moreover, it helps to increase accessibility of services and to expand social, government and business networks. The European Commission estimated the number of Wi-Fi users to be around 125 million worldwide in the year 2006, and that there will be more than 500 million Wi-Fi users worldwide by 2009 (JiWire, 2006). However, wireless security remains the most important factor that challenges wireless internet hot spots. As Wi-Fi grows the security threat also increases rapidly and therefore the need to protect information becomes imperative (Peikari and Fogie, 2003). The security risk remains largely from hackers, who are individuals that access the system without any authorization for personal gain.

Given the above context, the rationale for this research is to gain a better understanding of the free wireless internet park "iPark" initiative in Qatar. Using a pilot survey questionnaire, this study aims to explore the intention of citizens to use iPark services in Qatar. This is achieved by examining their perceptions of 'ease of use' and 'usefulness' in relation to internet access in the iPark. To pursue this line of inquiry, this research uses the Technology Acceptance Model (TAM). TAM theorizes that an individual's behavioural intention to use a technology is determined by two factors: perceived usefulness and perceived ease of use (Gardner and Amoroso, 2004).

The article is structured as follows. In the next section the national Qatari e-government background is presented. Then a literature perspective of e-government is offered, followed by an outline of the theoretical model used for the research. Next, the empirical background to the research is presented. This is followed by the methodology used for the research and a presentation of the empirical results. Finally, the article concludes by analysing the empirical results, discussing the research implications and identifying areas for future research.

E-GOVERNMENT IN THE STATE OF QATAR

The State of Qatar is a peninsula with a strategic position at the centre of the west coast of the Arabian Gulf. The total land area is approximately 11,437 sq km. The population is estimated to be around 1,500,000 (The Peninsula, 2008; Al-Shafi and Weerakkody, 2009); however, only a minority of the population are Qatari citizens by birth, whilst the rest are residents who live or work in Qatar (Al-Shafi and Weerakkody, 2009; Al-Shafi, 2008).

E-government was launched in Qatar in July 2000 and the initial period of strategy formulation and implementation was slow-moving compared to e-government efforts during the same period in developed countries. However, the establishment of ictQATAR in 2004 and their consequent takeover of the national e-government initiative a year later resulted in accelerated progress over the last three years. Parallel programmes were introduced in key areas such as health, interior affairs and education. The vision of ictQATAR states that its aim is to "…Serve as an independent and fair regulator, protecting consumers and businesses from unfair practices as Qatar transitions to a competitive telecoms market. ictQATAR aims to lead the government's ICT strategy, nurture innovative technologies to benefit those who live and work in Qatar, and help make people from all walks of life become comfortable with technology" (ictQATAR, 2008).

The Qatari e-government site offers many services, ranging from student registration and paying traffic violations to applying online for visas and permits (Al-Shafi and Weerakkody, 2009; 2007). In global terms the UN Global E-government readiness report (2008) ranked Qatar's e-government project as number 53 worldwide, where as in 2005 it was ranked as number 62 worldwide (Al-Shafi and Weerakkody, 2009; 2007). In addition to this, the UN (2005) report considered the Qatari e-government project to be regional (West Asia) best practice. This implies that major improvements and developments have been made during recent times. As part of the Qatar government's ongoing efforts to increase accessibility to e-government services and bridge the digital divide, the free wireless internet access

in public parks (iPark) initiative was launched in March 2007; this concept provides "Broadband for all" and aims to foster a knowledge-based society. The primary goal of the initiative is to increase internet usage by establishing "hot spots" in public parks (IctQATAR, 2007). There are currently three designated wireless internet hotspots throughout selected public parks in the city; these parks are targeting visitors who have internet access available on their laptops, PDAs, and other internet-ready devices (The Peninsula, 2008).

E-Government Adoption and Diffusion: A Literature Perspective

With the popularity of e-government growing, various researchers have offered different definitions to explain the concept (Seifert and Petersen, 2002). However, these definitions differ according to the varying e-government focus and are usually centred on technology, business, citizens, government, process or a functional perspective. (Wassenaar, 2000; Wimmer and Traunmuller, 2000; Bonham et al., 2001; Seifert and Petersen, 2002; Zhiyuan, 2002; and Irani et al., 2006). The definition considered to be most suitable for the purpose of this article is one that defines e-government as making full use of the potential of technology in order to help put its citizens at the centre of the e-services provided and which makes its citizens its intention (Waller et al., 2001).

Like e-business, e-government promises to deliver a number of benefits to citizens, businesses and governments. The most significant benefits of e-government, according to the literature, are delivering electronic and integrated public services through a single point of access to public services, 24 hours a day, seven days a week (Reffat, 2003); bridging the digital divide so that every citizen in society will be offered the same type of information and services from government (InfoDev, 2002); rebuilding customer relationships by providing value-added and personalised services to citizens (Davison et al., 2005); fostering economic development and helping local businesses to expand globally; and creating a more participative form of government by encouraging online debating, voting and exchange of information (Reynolds and Regio-Micro, 2001; Bonham et al., 2001; InfoDev, 2002; and Davison et al., 2005).

Research has shown that the introduction of e-government to a country will also result in a number of challenges for the citizens and the government alike (Margetts and Dunleavy, 2002; Seifert and Petersen, 2002; Zakareya and Irani, 2005). Lack of access to e-services (Chircu and Lee, 2005), security concerns (Harris and Schwartz, 2000), trust (Al-Sebie and Irani, 2005), individual differences (Reffat, 2003) and the digital divide (Carter and Bélanger, 2005; Chen et al., 2006) are challenges that can impact on participation and thereby obstruct the further take-up of e-government services.

As previously stated, this research will focus particularly on the influence that Wi-Fi technology complexity and attitudes have on the intention to use a new technology, in this instance iPark. Similarly it will also examine perceived usefulness and ease of use of, and attitude to iPark services and the influence that a citizen's individual differences have on the intention to use such services.

In terms of adoption and diffusion, several studies have explored e-government acceptance in the United States (US), (Carter and Bélanger, 2005) and the UK (Choudrie and Dwivedi, 2005). However, no studies exist which examine the factors that influence Qatari citizens' adoption of e-government services. In this respect Lee et *al.*, (2005) state that cross-national research on e-government is sparse in the literature and Dwivedi et *al.*, (2006) highlight the need for studies that investigate the adoption rate and behaviour of e-services. Given this context, this study attempts to address this gap from a Qatari perspective by integrating the aforementioned constructs from the Technology Acceptance Model.

TECHNOLOGY ADOPTION: THE THEORETICAL BACKGROUND

Researchers in the field of Information Systems and Technology have long been interested in investigating the theories and models that have power in predicting and explaining behaviour (Venkatesh et al, 2003). For example, these theories and models include, Theory Of Reasoned Action (TRA), Theory Of Planned Behaviour (TPB), Innovation Diffusion Theory (IDT), (Rogers, 1995), Unified Theory of Acceptance and Use of Technology (UTAUT), and the Technology Acceptance Model (TAM) (Venkatesh et al., 2003).

The Technology Acceptance Model is adapted from the Theory of Reasoned Action (TRA) to the field of information systems. Davis developed TAM in 1989 (Davis, 1989) and uses TRA as a theoretical basis for specifying the linkages between two key beliefs: perceived usefulness and perceived ease of use and users' attitudes, intentions and actual usage behaviour. According to Davis et al., (1989) the main goal of the model is to give an explanation of the determinants of computer acceptance which would result in an explanation of user behaviour across a broad range of end-user computing technologies and user populations (Davis et al., 1989). In addition, another key focus of TAM (Figure 1) is to provide a base for determining or exposing the impact of external variables on internal beliefs, attitudes and intentions.

In previous years, TAM has received extensive support through validation, applications and replications, for its power to predict the use of information systems (Cheng et al., 2006). Also, Davis et al., (1989), Venkatesh and Davis (2000) and Chau (1996) claimed that TAM is the most influential model in explaining IS/IT adoption behaviour. In this study, TAM will be used as the theoretical basis for examining user intention to use the iPark concept in Qatar. We have extended the TAM model to include two other relevant constructs; technology complexity in terms of wire-

Figure 1. Technology acceptance model (source Davis, 1989)

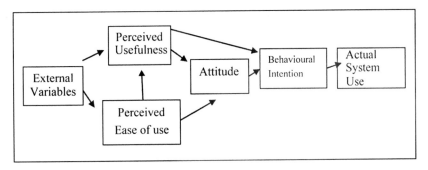

less internet response time, and individual differences that includes age, gender and qualifications.

Conceptual Model and Hypothesis

Attitude Towards iPark

Attitude has been defined as a cause of intention (June et *al.*, 2003). According to Fishbein and Ajzen (1975), attitude is classified into two constructs: attitude towards the object and attitude towards the behaviour. Investigation of the attitude construct towards using iPark services and examining its relationship with intention to use is appropriate for predicting usage behaviour.

Perceived Usefulness

According to Davis (1989) the perceived usefulness in TAM relates to productivity, performance, and effectiveness. This is an important construct as it provides a clear indication of how user attitude towards using and intention to use are influenced by perceived usefulness; perceived usefulness also has an indirect impact on intention to use via attitude (Davis et al,. 1989; June et al., 2003).

Perceived Ease of Use

Perceived ease of use is another important determinant of attitude in TAM (June et al., 2003). According to Davis (1989) perceived ease of use

is the individual's assessment of the level of effort involved in using the given system. A few authors have confirmed the impact of ease of use on attitude towards use (Al-Gahtani and King, 1999; Venkatesh and Davis, 1996).

Individual Differences

TAM primarily focuses on the development and identification of two key constructs 'perceived usefulness' and 'perceived ease of use'. However, the original TAM does not include any moderating influences, and research suggests incorporating these moderators to include gender and age into the original TAM in order to make better predictions and explanations associated with user behaviour for a particular technology (Venkatesh and Davis, 2000; Morris and Venkatesh, 2000). Furthermore, research has found that gender (Venkatesh et al., 2003; June et al., 2003) and educational background (Agarwal and Prasad, 1999) significantly moderates the influence of the determinants on behaviour intention.

Technology Complexities

The adoption and diffusion of wireless hotspots will depend very much on the speed of the Internet connection. Moreover, according to Pew Internet and American Life Project (2006), Wi-Fi has the potential to reshape customers' needs and conversely customers can reshape the technology.

Based on the aforementioned and the theoretical context offered, this article will test the strength of the hypothesized relationships mentioned in the theoretical model outlined in Figure 2 and the appropriateness of the model for predicting users' intention to use iPark in the State of Qatar.

RESEARCH METHODOLOGY

To explore the arguments set out above (see Table 1) and to understand the context of the iPark initiative in Qatar, brief informal open-ended interviews (Yin, 1994) were conducted from a convenience sample of three citizens and one Academic during the last quarter of 2007. The interviews lasted around 30 minutes and provided the context to formulate a detailed survey questionnaire that was to be used to investigate citizens' perceptions of iPark in Qatar. After the questionnaire was designed, limited testing was done using one researcher and four practitioners. This was important to improve the questions and to test respondents' comprehension and clarity before the actual survey was administered (Saunders et al., 2002). The pilot testing led to the removal of one question and the modification of another.

The protocol followed for the data collection was as follows. First, one of the researchers approached the iPark users, identified himself and provided a brief description of the research and the main purpose of the questionnaire. Then, the process of completing the questionnaire began by distributing the questionnaire to users and briefly explaining the contents of the questionnaire. Thereafter, the questionnaires were collected after a period of approximately between 20 and 30 minutes.

The survey questionnaire was distributed to a total of 55 iPark users between the period of November and December 2007; 54 usable responses were obtained. Overall, a survey questionnaire approach was selected as it is inexpensive, less time consuming and has the ability to provide both quantitative scale and qualitative data from a large research sample (Cornford and Smithson, 1997). The questionnaire used had 26 closed-ended questions and used Likert scale type (5-point scale) questions (Saunders et al., 2002).

Data Analysis

The proposed research model consisted of seven dependant and independent variables: i) individual differences in terms of age, gender and qualifications; ii) perceived usefulness; iii) perceived ease of use; iv) attitude; v) intention to use; and vi) technology complexity in terms of wireless internet speed.

Figure 2. Factors influencing iPark adoption

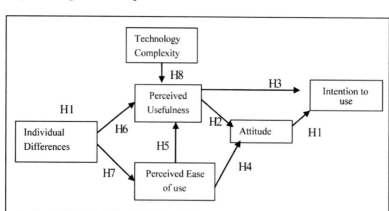

Table 1. Research hypotheses

NO.	Hypothesis
H1:	*Attitude towards iPark may have a significant positive impact on intentions to use iPark.*
H2:	*Perceived usefulness of iPark may have a significant positive impact on attitude towards using iPark*
H3:	*Perceived usefulness of iPark may have a significant positive impact on intention to use iPark.*
H4:	*Perceived ease of use of iPark may have a significant positive impact on attitude towards using iPark.*
H5:	*Perceived ease of using iPark may have a significant positive impact on perceived usefulness of iPark*
H6:	*Individual differences may have a significant positive impact on perceived usefulness of iPark in terms of gender, age and qualification.*
H7:	*Individual differences may have a significant positive impact on perceived ease of using iPark in terms of gender, age and qualification*
H8:	*Technology complexity may have a significant positive impact on perceived usefulness of iPark in terms of wireless internet speed.*

The authors generated the descriptive statistics (percentage and tables), used reliability of measurements and construct correlations by utilising SPSS (Version 15.0). Descriptive data analysis provides the reader with an appreciation of the actual numbers and values, and hence the scale that the researchers are dealing with (Dwivedi and Weerakkody, 2007).

RESEARCH FINDINGS

The demographic background of the 54 usable respondents was made up of 15% females and 85% males. Also, of the 54 respondents, 74% had an internet connection at their home/work or both. In terms of age, the results revealed that the largest occurrence of respondents (38%) were found to be in the age group 31-45, followed by the age group 19-30, constituting around (32%) of the total respondents. In contrast, the younger group (18 or less) and older age group (46-60) together consisted of 30% of the total respondents (see Figure 3). As far as education was concerned, the largest occurrence of respondents 38% hold postgraduate degrees (Masters and PhD), and 36% hold undergraduate level qualifications, and 26% hold secondary school certificates (see Figure 3).

In terms of the type of usage, Figure 4 shows that the majority of respondents used iPark for emails (76%); whilst other occurrences were chat (35%); e-government services (22%); fun (43%);

Figure 3. Age and qualification of the respondents

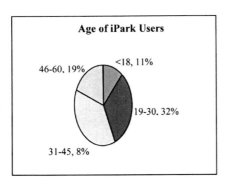
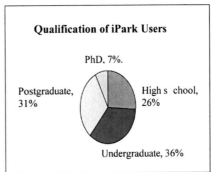

research (22%); news (56%); enquiry (26%); downloading information (54%); and other (6%).

As outlined in Table 2, the average scores of respondents for perceived ease of use ranged from 4.07 to 4.37. Descriptive statistics show that these scores are quite high. For perceived usefulness, the score ranged from 3.85 and 4.54, which is again quite high. Concerning technology complexity, the score was 3.70, indicating that the scale is average. In addition, for attitude towards iPark, the score ranged from 4.35 to 4.59, indicating that the scale is uniform. Also, the score ranged from 1.15 to 2.64 for individual differences towards iPark. The last score ranged from 3.30 to 3.72 for intention to use.

Reliability Test

Cronbach's coefficient alpha values were chosen to examine the internal consistency of the measure (Hinton et al., 2004) (Table 3). Cronbach's results varied between 8.26 for the intention to use and 8.60 for the individual constructs.

Hinton et al., (2004) have suggested four different points of reliability; excellent (0.90 and above), high (0.70- 0.90), high/moderate (0.50-0.70) and low (0.50 and below). The previously mentioned values in Table 3 show that all of the constructs achieved high reliability. The high Cronbach's values of the constructs mean that constructs were internally consistent and the reliability is measuring the same construct.

Table 4 also shows that the correlation is significant for one key factoring as per hypothesis (H3), *perceived usefulness* (0.417), which implies that it is correlated with intention to use, whereas attitude (0.143) was found to be insignificantly associated with intention to use as per hypothesis (H1).

Table 5 also shows that the correlation is significant for one key factor as per hypothesis (H2), *perceived usefulness* (0.286), which implied that it has an impact on attitude. Whereas, *perceived ease of use* (0.025) was found to have an insignificant correlation and a negative impact on attitude towards using iPark as per hypothesis (H4).

Table 6 also shows that the correlation is significant for the following four key factors: perceived ease of use (0.489) as per hypothesis (H5); technology complexity in terms of wireless internet speed (0.358) as per hypothesis (H8); individual differences in terms of age (0.478) and qualification (0.251) as per hypothesis (H6). These results imply that these constructs have a significant impact on perceived usefulness. In contrast, individual difference in terms of gender (0.113) was found to have an insignificant correlation. In addition, individual differences in terms of age and qualification had a negative

Figure 4. Primary uses of iPark user respondents, November & December, 2007

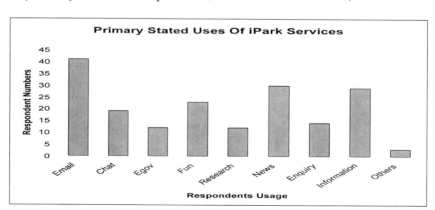

Table 2. Descriptive statistics of iPark usage

	Mean	Std. Deviation
Use iPark frequently	3.43	1.238
Encounter problem frequently in using the iPark	2.70	1.218
Individual Differences		
Gender	1.15	.359
Qualification	2.20	.919
Age	2.64	.922
Perceived Ease Of Use		
Using the iPark service is easy for me	4.07	.949
Interaction with the iPark services is clear and understandable	4.08	.917
Becoming skilful in the use of the iPark services is easy	4.37	.853
Overall, I find the use of the iPark services easy	4.17	1.060
Perceived Usefulness		
Using the iPark would enable me to accomplish my tasks more quickly	3.89	1.144
Using the iPark would make it easier for me to carry out my tasks	3.85	1.139
iPark is useful for me	4.54	.803
Overall, using the iPark is advantageous for me	4.50	.897
Technology Complexity Speed of Internet in the iPark is good	3.70	.964
Attitude		
Using the iPark is a good idea	4.59	.567
Using the iPark is pleasant	4.43	.742
In my opinion, it would be desirable to use the iPark	4.35	.828
In my view, using free wireless internet from public parks is a wise idea	4.42	.936
Intention To Use		
I would use iPark channels for my needs.	3.72	1.140
I would use iPark for my e-government services	3.30	1.207

Table 3. Reliability of measurements

Constructs	Number of items	Cronbach's Alpha (α)
Individual Differences	3	0.86
Perceived Ease Of Use	4	0.83
Perceived Usefulness	4	0.83
Technology Complexity	1	0.84
Attitude	4	0.84
Intention To Use	2	0.83

Table 4. Correlations of intention to use construct

Constructs		Intention to use
Perceived usefulness	Pearson Correlation	.417(**)
	Sig.(1-tailed)	.001
Attitude	Pearson Correlation	.143
	Sig.(1-tailed)	.155

Table 5. Correlations of attitude

Constructs		ATTITUDE
Perceived ease of use	Pearson Correlation	-.025
	Sig.(1-tailed)	.430
Perceived Usefulness	Pearson Correlation	.286(*)
	Sig.(1-tailed)	.021

Table 6. Perceived usefulness correlations

Constructs		Perceived Useful-ness
perceived ease of use	Pearson Correlation	.489(**)
	Sig.(1-tailed)	.000
Technology complexity	Pearson Correlation	.358(**)
	Sig.(1-tailed)	.005
Age	Pearson Correlation	-.478(**)
	Sig.(1-tailed)	.000
Gender	Pearson Correlation	.113
	Sig.(1-tailed)	.212
Level of Education	Pearson Correlation	-.251(*)
	Sig.(1-tailed)	.036

Table 7. Perceived ease of use correlations

Constructs		Perceived ease of use
Age	Pearson Correlation	-.140
	Sig.(1-tailed)	.159
Gender	Pearson Correlation	-.066
	Sig.(1-tailed)	.317
Level of Education	Pearson Correlation	.003
	Sig.(1-tailed)	.491

impact on perceived usefulness towards using iPark as per hypothesis (H6).

Table 7 shows that the correlation is insignificant for the key factor, *individual differences in terms of age (0.140)*, gender (0.066) and qualification (0.491) as per hypothesis (H7). This result implies that they have an insignificant impact on the perceived ease of use construct. In addition, individual differences in terms of age and gender constructs have a negative impact on perceived ease of using iPark as per hypothesis (H7), whereas the individual's educational background showed a positive impact on perceived ease of using iPark.

DISCUSSION AND CONCLUSION

E-government is widely accepted and seen as a growing trend worldwide. In the Middle East, most countries have implemented e-government services. However, the growth and adoption of e-government in a country will depend on basic prerequisites such as education, trust, marketing and awareness (Reffat, 2003; Navarra and Cornford, 2003; Bhattacherjee, 2002). The Qatari e-government project serves natural citizens, foreign residents and workers, and business and government agencies. As part of Qatar's wider national e-government program objectives to entice more users to access electronic government services, the free wireless internet access in public parks (iPark) initiative was launched in March 2007.

This article discussed the results of a survey targeted towards iPark users in Qatar and provides a representative account of citizens' perceptions of the iPark project. While the availability and use of Wi-Fi networks continue to increase, for users Wi-Fi facilitates greater mobility, information access, and flexibility in connectivity, improved efficiency, low cost, ease of use, and new applications, which could change the way they access electronic services.

The significant and non-significant factors found in the study and their influences on practice are outlined below.

Significant Factors

All of the adoption factors for H1, H2, H3, H5, H6 and H8 had a significant impact on intention to use iPark services in Qatar.

Non-Significant Factors

Regarding the constructs in H4 and H7, the analysis showed that they were not significant predictors of the behavioural intention to use iPark services.

Implications for Practice

Citizens and visitors using the iPark initiative may benefit from the services and consequently be encouraged to adopt e-government. If the government provides more benefits to its citizens in terms of convenient access and prompt services, compared to the old and traditional means, then possibly this practice (iPark) might spread the use of e-government services throughout the Qatari society and thus contribute to reducing any gaps that may exist in terms of the digital divide.

Furthermore, one conclusion that has emerged from the analysis in this study shows that although research exists that explores citizen adoption of wireless technologies and e-government services in different countries, the authors argue that currently there are no independent studies that examine citizens' perceptions about the adoption of free wireless Internet park facilities for accessing public services.

The results of this research showed that individual differences have a negative impact on perceived ease of use (H7) and perceived ease of use has a negative relationship with attitude (H4). The remaining hypotheses showed that they were consistent. This implies that contrary to expectations the citizens' intention to use new technological concepts such as iPark has no link with the individual differences in terms of age and educational background of the citizen in the Qatari context. Given these findings, it can be concluded that the iPark initiative in Qatar has been successful initially in promoting wider access to the Internet. This is encouraging from an e- government perspective. The authors suggest that public parks (iParks) can be used to advertise and market the Qatari e-government national website and raise e-government awareness among Qatari citizens. While the research findings are encouraging from a practical perspective for the Qatari government, from a theoretical perspective these results reconfirm that technology acceptance is influenced by key constructs such as perceived usefulness, attitude and technology complexity aspects of the services used. In terms of implications to practice, the findings show that the iPark concept, if promoted effectively, can be utilized to entice more citizens to access and use national e-government services in Qatar.

This research describes an early attempt to study the adoption of the iPark initiative by Qatari citizens and thus represents only the views of a limited sample of users (54). To ensure more thorough testing of the hypotheses proposed in this research, the authors have planned research in order to survey a larger number of iPark users.

REFERENCES

Agarwal, R., & Prasad, J. (1999). Are individual differences germane to the acceptance of new information technologies? *Decision Sciences*, *30*(2), 361–391. doi:10.1111/j.1540-5915.1999.tb01614.x

Al-Gahtani, S., & King, M. (1999). Attitudes, satisfaction and usage: factors contributing to each in the acceptance of information technology. *Behaviour & Information Technology*, *18*(4), 277–297. doi:10.1080/014492999119020

Al-Sebie, M., & Irani, Z. (2005). Technical and organisational challenges facing transactional e-government systems: an empirical study. *Electronic Government. International Journal (Toronto, Ont.), 2*(3), 247–276.

Al-Shafi, S. (2008). Free Wireless Internet park Services: An Investigation of Technology Adoption in Qatar from a Citizens' Perspective. *Journal of Cases on Information Technology, 10*, 21–34.

Al-Shafi, S., & Weerakkody, V. (2007). Implementing and managing e-government in the State of Qatar: a citizens' perspective. *Electronic Government, an International Journal, 4*(4), 436-450.

Al-Shafi, S., & Weerakkody, V. (2009). Examining the Unified Theory of Acceptance and Use Of Technology (UTAUT) Of E-Government Services Within The State Of Qatar. In *Handbook of Research on ICT-Enabled Transformational Government: A Global Perspective.* Hershey, PA: Information Science Reference.

Bhattacherjee, A. (2002). Individual Trust in On-line Firms: Scale Development and Initial Trust. *Journal of Management Information Systems, 19*(1), 211–241.

Bonham, G., Seifert, J., & Thorson, S. (2001). *The Transformational Potential of e-Government: The Role Of Political Leadership.* Paper presented at the 4th Pan European International Relations Conference of the European Consortium for Political Research, University of Kent, Canterbury, UK.

Carter, L., & Bélanger, F. (2005). The utilization of e-government services: citizen trust, innovation and acceptance factors. *Information Systems Journal, 15*, 5–26. doi:10.1111/j.1365-2575.2005.00183.x

Chau, P. Y. K. (1996). An Empirical Assessment of a Modified Technology Acceptance Model. *Journal of Management Information Systems, 13*(2), 185–204.

Chen, C., Tseng, S., & Huang, H. (2006). A comprehensive study of the digital divide phenomenon in Taiwanese government agencies. *International Journal Of Internet And Enterprise Management, 4*(3), 244–256. doi:10.1504/IJIEM.2006.010917

Cheng, T., Lam, D., & Yeung, A. (2006). Adoption of internet banking: An empirical study in Hong Kong. *Decision Support Systems, 42*, 1558–1572. doi:10.1016/j.dss.2006.01.002

Chircu, A.M., & Lee, D.H-D. (2005). E-government: Key Success Factors For Value Discovery And Realisation. *E- Government, an International Journal, 2*(1), 11-25.

Choudrie, J., & Dwivedi, Y. (2005). A Survey of Citizens Adoption and Awareness of E-Government Initiatives. In *The Government Gateway: A United Kingdom Perspective.* E-Government Workshop. Brunel University, West London.

Cornford, T., & Smithson, S. (1997). *Project Research in Information Systems: A Student's Guide.* London: Macmillan Press.

Davis, F. D. (1989). Perceived usefulness, perceived ease of use, and user acceptance of information technology. *Management Information Systems Quarterly, 13*, 319–340. doi:10.2307/249008

Davis, F. D., Bagozzi, R. P., & Warshaw, P. R. (1989). User Acceptance of Computer Technology: A Comparison of Two Theoretical Models. *Management Science, 35*(8), 982–1003. doi:10.1287/mnsc.35.8.982

Davison, R. M., Wagner, C., & Ma, L. C. (2005). From government to e-government: a transition model. *Information Technology & People, 18*(3), 280–299. doi:10.1108/09593840510615888

Dwivedi, Y., Papazafeiropoulou, A., & Gharavi, H. (2006). Socio-Economic Determinants of Adoption of the Government Gateway Initiative in the UK. *Electronic Government, 3*(4), 404–419. doi:10.1504/EG.2006.010801

Dwivedi, Y., & Weerakkody, V. (2007). Examining the factors affecting the adoption of broadband in the Kingdom of Saudi Arabia. *Electronic Government, an International Journal, 4*(1), 43-58.

Fishbein, M., & Ajzen, I. (1975). *Belief, Attitude, Intention and Behaviour: An Introduction to Theory and Research.* Reading, MA, USA: Addison-Wesley.

Gardner, C., & Amoroso, D. (2004). *Development of an Instrument to Measure the Acceptance of Internet Technology by Consumers.* Paper presented at the Proceedings of the 37th Hawaii International Conference on System Sciences, Hawaii, USA.

Gupta, V., & Gupta, S. (2005). *Experiments in Wireless Internet Security.* Statistical Methods in Computer Security.

Harris, J. F., & Schwartz, J. (2000. June 22). Anti drug website tracks visitors. *Washington Post,* (p. 23).

Hinton, P. R., Brownlow, C., McMurvay, I., & Cozens, B. (2004). *SPSS explained.* East Sussex, England: Routledge Inc.

IctQATAR. (2007). *Free wireless internet in Qatar's public parks.* Retrieved from http://www.ict.gov.qa/output/page422.asp

InfoDev. (2002). *The e-Government Handbook for Developing Countries.* Retrieved from http://www.cdt.org/egov/handbook.

Irani, Z., Al-Sebie, M., & Elliman, T. (2006). Transaction Stage of e-Government Systems: Identification of its Location & Importance. *Proceedings of the 39th Hawaii International Conference on System Science.*

Jiwire (2006). *JiWire Launches Worldwide Point-of-Connection Wi-Fi Hotspot Advertising Network.* Retrieved from http://www.jiwire.com/about/announ cements/press-advertising-network.htm

June, L., Chun-Sheng, Y., Chang, L., & James, E. (2003). Technology Acceptance Model for Wireless Internet. *Internet Research: Electronic Networking Application And Policy, 13*(3), 206–222. doi:10.1108/10662240310478222

Lee, S. M., Tan, X., & Trimi, S. (2005). Current practices of leading e-government countries. *Communications of the ACM, 48*(10), 99–104. doi:10.1145/1089107.1089112

Lehr, W., & McKnight, L. W. (2003). Wireless Internet access: 3G vs. WiFi? *Telecommunications Policy, 27,* 351–370. doi:10.1016/S0308-5961(03)00004-1

Margetts, H., & Dunleavy, P. (2002). *Cultural Barriers to E-Government* (Working Paper). University Collage of London and London School of Economics for National Audit Office.

Morris, M. G., & Venkatesh, V. (2000). Age differences in technology adoption decisions: implications for a changing work force. *Personnel Psychology, 53*(2), 375–403. doi:10.1111/j.1744-6570.2000.tb00206.x

Navarra, D. D., & Cornford, T. (2003). A Policy Making View of E-Government Innovations in Public Governance. *Proceedings of the Ninth Americas Conference on Information Systems.*

Peikari, C., & Fogie, S. (2003). *Maximum Wireless Security.* Retrieved from http://www.berr.gov.uk/files/file9972.pdf.

Pew Internet & American Life Project. (2006). *Home Broadband Adoption.* Retrieved from http://www.pewinternet.org/pdfs /PIP_Broadband_trends2006.pdf

Reffat, R. (2003). *Developing A Successful E-Government* (Working Paper). University Of Sydney, Australia.

Reynolds, M. M., & Regio-Micro, M. (2001). The Purpose Of Transforming Government-E-Government as a Catalyst In The Information Age. *Microsoft E-Government Initiatives*. Retrieved from http://www.netcaucus.org/books/egov2001/pdf/EGovIntr.pdf

Rogers, E. M. (1995). *Diffusion of innovations.* New York.

Saunders, M., Lewis, P., & Thornhill, A. (2002). *Research methods for business students* (3rd ed.). Harlow: Prentice Hall.

Seifert, J., & Petersen, E. (2002). The Promise Of All Things E? Expectations and Challenges of Emergent E-Government. *Perspectives on Global Development and Technology, 1*(2), 193–213. doi:10.1163/156915002100419808

The Peninsula Newspaper. (2008). *Bursting at the seams.* Retrieved from http://www.thepeninsulaqatar.com/Display_news.asp?section=Local_News&month=January 2008&file=Local_News200801296298.xml.

UN. (2008). *World public sector report: UN E-Government survey.* New York: From E-Government To Connected Governance.

Venkatesh, V., & Davis, F. D. (1996). A model of the antecedents of perceived ease of use: Development and test. *Decision Sciences, 27,* 451–481. doi:10.1111/j.1540-5915.1996.tb01822.x

Venkatesh, V., & Davis, F. D. (2000). A theoretical extension of the technology acceptance model: Four longitudinal field studies. *Management Science, 46*(2), 186–205. doi:10.1287/mnsc.46.2.186.11926

Venkatesh, V., Morris, M., Davis, G., & Davis, F. (2003). User Acceptance of Information Technology: Toward a Unified View. *Management Information Systems Quarterly, 27*(3), 425–478.

Waller, P., Livesey, P., & Edin, K. (2001). e-Government in the Service of Democracy. *ICA Information, 74.*

Wassenaar, A. (2000). *E-Governmental Value Chain Models. DEXA* (pp. 289–293). IEEE Press.

Wimmer, M., & Traunmuller, R. (2000). Trends in e-government: managing distributed knowledge. *Proceedings of the 11th International Workshop on Database and Expert Systems Applications.*

Yin, R. K. (1994). *Case Study Research - Design And Methods* (2nd ed.). London: Sage Publications.

Zakareya, E., & Irani, Z. (2005). E-government adoption: Architecture and barriers. *Business Process Management Journal, 11*(5), 589–611. doi:10.1108/14637150510619902

Zhiyuan, F. (2002). E-Government in Digital Era: Concepts, Practice and Development. *International Journal Of The Computer. The Internet and Management, 10*(2), 1–22.

This work was previously published in International Journal of Electronic Government Research (IJEGR), edited by Vishanth Weerakkody, pp. 21-35, copyright 2009 by Information Science Reference (an imprint of IGI Global)

Chapter 13

Integrating Public and Private Services:
Intermediaries as a Channel for Public Service Delivery

Bram Klievink
Delft University of Technology, The Netherlands

Marijn Janssen
Delft University of Technology, The Netherlands

ABSTRACT

The advent and widespread use of the Internet enables governments to connect directly to citizens and businesses. This results in a decrease of the transaction costs and a reduction of the administrative burden for governments, citizens and businesses. One consequence of this is the bypassing of existing parties that are interacting as intermediaries between the clients and provider. Based on intermediation theory, two case studies are analyzed which counter the argument of the bypassing of intermediaries. It is possible to adopt a re-intermediation strategy, in which intermediaries are used as a value-adding service delivery channel. The empirical evidence shows that private sector intermediaries can be employed for service innovation. The intermediaries provide service delivery channels that are closer to the natural interaction patterns of the users of public services than direct service delivery channels. Furthermore, the intermediaries can bundle public services with their own. For governments, this implies that only adopting a direct interaction strategy, which is often motivated by a desire to reduce transaction costs, is too narrow an approach from the point of view of demand-driven government. As such, direct interaction strategies need to be complimented by a re-intermediation strategy employing private sectors intermediaries in order to advance towards a truly demand-driven and client centered government.

INTRODUCTION

Governments are looking for ways to improve public service delivery and move towards mature service provisioning (Layne & Lee, 2001). They should offer integrated and executable services in a "one stop shop" (West, 2004; Wimmer, 2002). In the latter half of the 1990's, there has been a rapid transformation in the government functions

DOI: 10.4018/978-1-60960-162-1.ch013

(Devadoss, Pan, & Huang, 2002). Governments are trying to improve their service provisioning and are looking to redesign their service delivery channels. Organizational website(s), telephone, mail and front-desk are the obvious service delivery channels employed by government organizations. The use of innovative service delivery structures has the potential to improve access to groups of citizens and business segments that cannot be reached via existing service delivery channels (Klievink & Janssen, 2008). The Internet has made it possible to connect to citizens and businesses directly without the need for front-desks or other expensive channels. Creating an online presence, including transaction and interaction features, is facilitated by many software packages. Yet, ICT enables other innovative service delivery structures that have gained little attention.

As a consequence of technology advances, it is easy to connect to clients directly. This may ease the process of interacting with the citizens and businesses, as no complicated agreements with and activities by middlemen are necessary. *Disintermediation* is the removal of intermediaries in the service delivery channels. The argument in favor of disintermediation is often based on transaction cost reduction (Malone, Yates, & Benjamin, 1987). Transaction costs are the result of friction in the interactions among parties (Coase, 1937). The reasoning is that ICT enables a shift towards more direct interaction by lowering transaction costs. The Internet involves lower transaction costs than third parties that need to make money to ensure their existence. The basic idea is that by disintermediation, direct interaction between the providing party and the service requester becomes possible at lower costs. Unnecessary activities are eliminated and removed.

Government organizations can, however, also use other – non-direct – service delivery channels to interact with their clients. Often these channels are operated by other parties and go beyond organizational boundaries (Janssen & Klievink, 2009). In this way, third parties act as intermediaries between government organizations and their customers. As a result, the disintermediation view is challenged because merely focusing on costs underestimates the range of facilitating services offered by intermediaries (Janssen & Sol, 2000; Sarkar, Butler, & Steinfield, 1995). New intermediaries may arrive that provide value-adding activities. Integrating these in innovative service delivery structures can be called re-intermediation. The discussion regarding disintermediation and re-intermediation strategies becomes more important in the light of demand-driven government, with both strategies having their merits (Klievink & Janssen, 2008). In this study, we contribute to the debate concerning the removal or use of intermediaries. The *aim* is to investigate the potential value of intermediaries in e-government service delivery. The results of this study should support government agencies in developing a better intermediation, disintermediation and re-intermediation strategy. We begin by reviewing theories regarding intermediation, disintermediation and re-intermediation. Next, we discuss our research approach and present two case studies. We will then compare the case studies, discuss our main findings and provide recommendations for future research. Finally, we present our conclusions in the last section.

LITERATURE BACKGROUND

Advances in ICT influence the interactions among parties and can lead to changes in whole network structures (Clemons & Row, 1992; Malone, et al., 1987). The costs involved in this kind of interactions are called transaction costs. The transaction cost theory has been used to predict that advances in the use of ICT would reduce transaction costs by enabling organizations to connect directly to each other (Gellman, 1996; Malone, et al., 1987). Transaction costs result from the transfer of property rights between parties and exist because of friction in economic systems (Coase, 1937; Williamson, 1975). Malone, Yates and Benjamin (1987) use transaction cost theory as the theoretical background for the Electronic Markets Hypothesis (EMH) and the bypassing of intermediaries resulting in disintermediation. Their Electronic Markets Hypothesis holds that, by reducing the costs of coordination, information and communication technology (ICT) will support an overall shift towards a proportionally increased use of markets over hierarchies to coordinate economic activity. Malone, Yates and Benjamin (1987) argue that one of the effects of using electronic networks will be the bypassing of intermediaries in electronic markets due to lower transaction costs.

Because the disintermediation argument focuses solely on the cost of intermediation, it takes too narrow a view (Chircu & Kauffman, 1999; Giaglis, Klein, & O'Keefe, 2002; Janssen & Sol, 2000; Janssen & Verbraeck, 2005; Sarkar, et al., 1995). For example, Sarkar, Butler and Steinfield (1995) have found that the predictions about the disappearance of intermediaries made assumptions about the relative transaction costs that may not be warranted and that the range of value adding services offered by intermediaries was underestimated. Furthermore, the disintermediation argument does not take the benefits into account that can be gained by using intermediaries (Giaglis, et al., 2002; Janssen & Sol, 2000). This implies a need for a better understanding of how intermediary service delivery channels are used to interact with customers. Intermediation, disintermediation and re-intermediation strategies depend on the life cycle of an industry and on the strategies, resources and assets of players (Chircu & Kauffman, 1999).

Governments started decades ago with the widespread implementation of ICT's. For public service delivery, e-government service delivery started as an online presence – often in the form of a set of webpages provided by different departments, on different – but sometimes overlapping – topics. Citizens are expected to find their way around in this maze of government services. To

begin with, they need to determine which service they require, and then they have to look for the organization or department that provides the service in question and contact that organization or department. From a service perspective, the next step that organizations have undertaken (or are undertaking, depending on their level of e-government maturity) is to add transaction features in order to enable service consumers to actually do something with the websites, instead of only being able to read them (Layne & Lee, 2001). West (2004) says that this is where citizens can get partial service delivery.

Integrating the processes and applications of various departments within organizations is required for further improvement of (electronic) services. To begin with, with only a web presence it is difficult for citizens to manage all the various (sub)processes and agencies involved in handling a service request. Furthermore, integration leads to efficiency from an organizational point of view. Therefore, services should be integrated, at least within organizations. This calls for the orchestration of processes, technologies and information (Janssen, Gortmaker, & Wagenaar, 2006). Once organizations have orchestrated their internal processes and systems, they can take the next step. From a service delivery perspective, the integration of services is often envisaged in web portals. These web portals go beyond individual services of fragmented organizations, and should offer integrated and executable services in a "one stop shop" (West, 2004; Wimmer, 2002).

In the Netherlands, integrated service delivery is primarily realized at the organizational level, and it is slowly moving towards the national level. Many individual government organizations provide (online) one-stop shops for their own products and services. However, citizens and businesses still have to manage and coordinate their interactions with the various organizations that are involved in the services they require. To ensure more integrated government service delivery, governments have to deal with the problem of fragmentation of government by orchestrating the various steps in a service delivery process, in order to relinquish the citizen of this responsibility.

When government organizations focus on disintermediation to contact citizens more directly, organizations that can take over the complex task of selecting and enacting the right services are cut out of the chain. While this may result in more efficiency, the service quality might be reduced. Furthermore, if such intermediary organizations provide added value – for example by combining or integrating public services with their own – this added value is also cut out of the chain.

Although, in principle, government agencies are able to connect directly, the use of intermediaries may provide value. However, this is difficult to determine, which is why a better understanding of the phenomenon of intermediaries is needed. This quest has been done in the field of e-commerce and has resulted in the identification of recurring pattern of intermediation, disintermediation, and re-intermediation through an IDR framework (Chircu & Kauffman, 1999). This process is presented in Figure 1. Intermediation is the situation shown at the top of the figure. To the right, the disintermediation cycle is shown, with traditional channels being abandoned in favor of direct service provisioning. To the left, a re-intermediation cycle is shown, with new intermediaries appearing

that can be used as channels to interact with the customers.

Intermediaries focus on bridging the gap between the services being provided and the wishes and requirements of those who use the services. As such, the concept of electronic intermediaries refers to phenomena such as portals, which are necessary to provide integrated services in the front office. In essence, these types of electronic intermediaries focus on providing facilities for enabling coordination between users and service providers. As yet, no accurate definition of the concept of intermediary has been developed and literature suggests that virtually any entity on the Internet can be seen as intermediary from some point of view (Bailey & Bakos, 1997; Janssen & Sol, 2000; Resnick, Zeckhauser, & Avery, 1995; Sarkar, et al., 1995; Spulber, 1996). In this study, we define intermediaries as *"any public or private organization facilitating the coordination between public service providers and their users"*. Users are typical citizens and businesses, but might be other public organizations.

Intermediaries play a number of roles that cannot be easily replaced or internalized through direct interactions. Four roles of intermediaries are identified by Bailey and Bakos (1997) and used in several other studies.

Figure 1. Service delivery channels with and without an intermediary

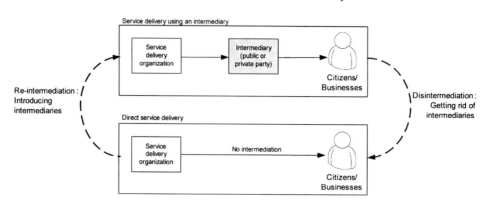

- *Matching demand and supply.* This involves matching a consumer's need to the services provided by the service providers;
- *Information aggregation.* Information aggregation involves the acquisition, processing and distribution of information. This includes information to monitor and control the execution of the service delivery process;
- *Providing trust.* This is aimed at ensuring the quality, the timely performance of activities and ensuring the accountability of decisions and the decision-making processes;
- *Facilitating, by providing the institutional infrastructure.* The institutional infrastructure concerns the providing of functionality that is used by multiple agencies in the PSN.

We use these roles to analyze the case studies in the remainder of this paper.

RESEARCH APPROACH

Due to the complex nature of innovative service delivery strategies and the need to gain a deeper understanding of the phenomenon, a qualitative approach based on case study research was adopted for this research (Yin, 1989). Case study research is one of the most common qualitative method used in Information Systems (IS) (Orlikowski & Baroudi, 1991). We examined two different types of case studies which were selected on the basis of their reputation in the field, including the Vehicle Licensing and Administration Agency (in Dutch abbreviated as RDW) and the Dutch Inland Revenue Service (IRS). The main reason to select these cases was that they are viewed as good examples of re-configuring a service channel by re-intermediation (e.g. Undheim & Blakemore, 2007).

We based our research primarily on document collection, content analysis and interviews with key representatives. At the RDW and IRS, two interviews were conducted. Furthermore, publicly available and internal documents relating to the history, current use of intermediaries and development were gathered and examined to gain an understanding of the operational, technical, management-related and organizational aspects. The publicly available documents provide a good overview of the problems and motives for changes, while internal documents contained information concerning the transformation, design choices and accomplished benefits.

CASE STUDIES BACKGROUND

The *Dutch vehicle administration and licensing agency* takes care of the administration of all cars and trucks in the Netherlands. They have a registry of cars and trucks that is used to supervise the vehicles on the roads, to monitor technical safety and collect the required taxes. In the past, the post office was used as the service delivery channel to transfer the car ownership between seller and buyer. Employees at the post office checked the identity of the persons, ensure that the forms were filled in correctly and ensured that the form arrives at the RDW in a secure and safe way. With the arrival of the Internet, it would make sense to pursue a disintermediation strategy and start interacting directly. The notification of change in car ownership via the post office often took 6 or 7 days. In addition, most of the registration fee had to be paid to the post office for checking the information and delivering it to the RDW (see Figure 2).

Rather than merely opting in favor of the disintermediation of channels like the post office, RDW decided to re-intermediate, adopting a multi-channel service delivery strategy; telephone inquiries are still supported and the post office is used to electronically transfer car ownership

Figure 2. Intermediation in the vehicle administration

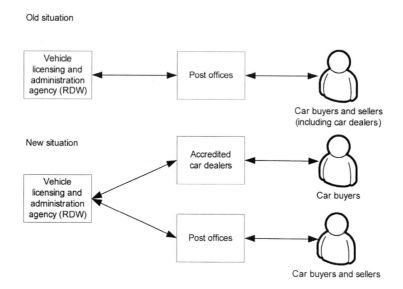

between individuals (without the intermediation of a car dealer) as shown in Figure 2. In addition, the car dealers were supported by allowing them to be accredited that allows them to register a change in car ownership directly with the RDW, the reason being that there are a limited number of car dealers, which are checked occasionally and fraudulent car dealers will lose their license. This enables car dealers to bundle their own services with the public services offered of the RDW.

Both post offices and accredited car dealers can transfer ownership documents using an electronic data link with the RDW. In the new situation, post offices remain an intermediary for the transfer of ownership between private individuals, as no car dealer is involved in this process, and for the transfer of ownership between a car buyer and a non-accredited dealer. Furthermore, post offices function as a back-up channel if a technical failure occurs in the data link between the dealer and the RDW. Car dealers already were an important element in the process of buying and selling cars, and they now became intermediaries between their customers and the RDW. In this way, ownership can be transferred instantly, and often the sellers

and buyers of cars are not aware of the process steps of the vehicle licensing and administration agency that are involved in their process of buying a car.

The *Dutch Inland Revenue Service* (IRS) has been actively involved in electronic service delivery for decades and one of its main objectives is making the filing of taxes for citizens and businesses easier (IRS, 2004). As of January 2005, all businesses have the legal obligation to file their tax forms electronically. The IRS provides various ways of interacting and has developed a software program that can be used by businesses and citizens to provide and submit the information, which means the information does not have to be re-entered by government employees. Over the last years, the IRS has begun to interact closely with the vendors of financial software. The relevant rules as developed by the IRS can be used by software vendors to extract the information from applications and submit it to the IRS in the eXtensible Business Reporting Language (XBRL) format. The software vendors play a major role in the discussions about the implementation of new policies and regulations in the software.

Figure 3. Intermediation by the IRS

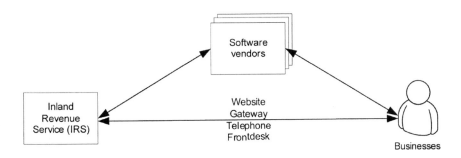

Software vendors employ the strategy to include tax filing as a part of their financial software. In this way they can offer a broader package of services and integrate taxes with other administrative functions. Whereas their core business is often to provide software in a certain area, like accountancy and financial management, this enables them to provide added value for their customers. Furthermore by directly collaborating with the IRS, the software vendors are more up-to-date about recent developments and new legislation they need to comply with. In this way they are able to better plan their software development and they enable their customers to comply with the newest regulations.

FINDINGS AND DISCUSSION

The aim of this study is to contribute to the discussion on re-intermediation and disintermediation. In the case studies, we found innovative strategies for the disintermediation of channels but also for the subsequent re-intermediation. Table 1 provides an overview of the two case studies, describing the past, the main drivers, the re-intermediation strategy and a brief description of the re-intermediation roles.

In both case studies the private parties see the advantage of bundling public services with their own services. They feel that they can provide an added value for their clients. In the two case studies, the roles of the intermediaries seem to be completely different. The RDW uses the car dealers for transferring ownership, with a focus on the facilitating and trusted roles. The IRS uses the software vendors as a vehicle to obtain tax information directly from the financial information stored in the systems already in use by the companies. In this way, it facilitates the collecting of information and aggregates the information in the correct format.

Not surprisingly, the matching role is absent in both case studies, as there is only one provider of the public service. The information aggregation played a minor role in the IRS case study, as the intermediaries were used to collect and/or distribute information. The intermediaries are especially useful to facilitate the information exchange with citizens and businesses and to a lesser extent to aggregate information.

The intermediaries are used especially to be closer to the processes of citizens or businesses and should facilitate the interactions between these processes and government services. In this way, the transaction costs between citizens/business and the government are reduced for both of the parties involved. The re-intermediation strategy is a way to become more demand-driven and at the same time reduce the transaction cost. Also, the trusted role is of importance in both case studies. The government agencies depend on the reli-

Table 1. Cross-case comparison

Case study name		Vehicle administration	Inland Revenue Service
Past		Post office intermediates between car buyers, sellers and vehicle administration.	Declare new regulations that businesses are to comply with.
Main driver		Improve supervision of the vehicles on the roads in the Netherlands to monitor technical safety and collect the required taxes. Reduce the interaction costs.	Standardize and unify formats for tax filing, Improve customer orientation and reduce tax administration costs.
Re-intermediation strategy		The post office is still being used to transfer car ownership between individuals. ICT makes this a direct service. The car dealers are used to transfer car ownership when buying or selling a car.	Support vendors of financial software to include (changes in) tax rules in their business software. In this way, businesses can use software packages to submit their tax information automatically.
Intermediation roles			
	Facilitating role	The car dealers facilitate the process of the transfer of the car ownership and in this way hide the interaction with this agency.	Facilitate collection of tax data. Combine taxing functionality with other financial and/or administrative functions in software packages.
	Trusted role	The identification and transfer of ownership.	Ensure that data is correct and new regulations are complied with.
	Matching role	-	-
	Information aggregation	Car dealers prevent the RDW having to deal with many individual car buyers. The information quality was improved. One key registry replaced distributed data.	Aggregates financial information in the correct format. Improving the information quality by avoiding transferring or re-entering of information.
Generated customer value		Not having to go to the post office. Direct transfer of ownership enables the buyer to drive away with the car directly. This brings service delivery closer to the natural process of citizens buying or selling a car. Reducing the complexity of finding how to transfer the ownership and how to fill in the form correctly, as this is done electronically.	Avoiding cumbersome collection and filing of tax data. In this way reducing the administrative burden for businesses. The use of financial software as intermediary bring service delivery closer to the businesses, as they already used these software to support their own processes.

ability and accuracy of the information provided by the intermediaries. Furthermore, the image of and trust in government depends on the quality of the intermediaries. In both case studies, the development of clear agreements and contracts, certification and randomly checking the performance of intermediaries is viewed as crucially important. A suggestion for further research is to analyze agreements and contracts in relation to the trusted role.

In literature, is was argued that the services (Sarkar, et al., 1995) and added value (Giaglis, et al., 2002; Janssen & Sol, 2000) of intermediaries should be analyzed. Our case studies show that the services and added values are inter-

related concepts, as services are used to create added value by improving service provisioning, increasing information quality and/or reducing costs. Because intermediaries may be closer to the customers, they are able to provide better services. By capturing information at the source, the quality of the information can be improved. Improving the quality of the information can be related to the facilitating role, as the intermediary provides the facility to capture information. In addition, by providing more services than individual government organizations, the costs involved may be reduced through economies of scale. Since the intermediaries have a close relationship with the customers, they support a demand-driven government and provide a better match between the service delivery chain and the customer chain.

The use of intermediaries within the interaction channels cannot be easily substituted or internalized through direct service delivery channels. Government organizations may pursue a variety of strategies to employ their channels. From a government perspective, channels have different characteristics and may be employed for different purposes, which would imply that a close relationship with customers requires the use of intermediaries in the service delivery channels. Governments may pursue a strategy in which all kinds of intermediaries are used to interact with citizens and businesses. This adds to the idea that government becomes less visible and uses other parties to provide services. To determine whether or not that is indeed the case, further and broader research is needed.

Transaction costs are often used to predict the disappearance of intermediaries (Malone, et al., 1987). In both case studies, the Internet was initially used to reduce transaction costs. This provides support for the disappearance of intermediaries. Although transaction costs are used to predict direct interaction between parties, the RDW and IRS case studies clearly illustrate that intermediaries can be used to reduce transaction costs in two ways. The first way is by improving the information quality and the second way is by reducing the transaction costs of the citizens or businesses. Whereas the original argument starts from the perspective of the service providers that use intermediaries as service delivery channels, it neglects the arguments that customers also incur transaction costs which can be reduced by making use of intermediaries. Car buyers and sellers only have to fill in one form containing all the relevant information, rather than having to fill in one form for the car dealer and another for the RDW. In addition, car buyers are hardly aware that the RDW is involved in the change of car ownership. In the case of the IRS, the companies

using financial software only have to press one button to ensure that the authentication mechanism is correct to submit their tax information to the IRS. The software has already extracted the correct information. This information does not necessarily need to be submitted to the IRS, it can be submitted to the accountant, who checks the information and then submits it to the IRS. This brings us to the proposition that intermediaries can reduce transaction costs and that transaction costs should be viewed from the government as well as from the citizens and businesses point of view.

It is possible that, in many other situations, intermediaries (like the post office) increase transaction costs. Consequently, the creation of added value by using intermediaries needs to be analyzed in greater detail. The circumstances in which intermediaries can help reduce transaction costs for government and for businesses or citizens require research attention as well. In addition, this confirms that there is a recurring pattern of intermediation, disintermediation, and re-intermediation (Chircu & Kauffman, 1999). Initially, intermediaries were used, after which, due to the rise of the Internet, the logical way to forward was disintermediation and then new ways to reduce costs and improve service delivery were sought through re-intermediation.

The use of intermediaries may have a negative impact on the immediate visibility of government. There should be mechanisms to ensure the overall quality, for example a fraudulent car dealer may impact the trust in the RDW or even in the government. Furthermore, citizens or businesses may not know if the government uses the intermediary and may be reluctant to interact with the government in this way. The objectives of the Dutch government include providing equal access for all and that all citizens are handled fairly and equally. Therefore, the quality and access should be guaranteed in order to avoid that certain groups, for example the disabled, have no access to these channels. This result in the proposition that the use of (private) intermediaries requires trusted mechanisms.

Further research is needed to determine the type of mechanisms that can be used.

CONCLUSION

Governments can use a variety of service delivery channels to interact with citizens and businesses. Front desks, call centers and websites are used to interact directly with the customer segments, bypassing intermediaries. The idea of disintermediation is often regarded as a way to cut costs and improve efficiency. As the budgets of government agencies are reduced, this appears to be a logical strategy for governments to pursue. In this study, we investigated two case studies employing a complimentary strategy which appears to be more client-centric. In a re-intermediation strategy, government agencies employ intermediaries that bundle public services with their own private services to improve the service delivery to their clients. This may help to reduce costs, improve the quality of information and create a demand-driven government. Four intermediation roles found in e-commerce literature were used to analyze the intermediation roles. We found that especially the facilitating and trusted role are of importance in e-government and the matching and information aggregation role are not or less relevant. Matching has to do with bringing together supply and demand, which was not an issue, as in both cases there was only one service provider. Information aggregation is less of an issue, as the service channels are used to interact with individual customers, whereas in e-commerce these channels may be employed to avoid having to deal with individual customers.

The intermediaries are viewed as an innovative service delivery channel and add value in four ways:

- Enable a more demand-driven government. The intermediaries are used to bring service delivery closer to the citizens or businesses. Often, customers already interact with the intermediary organization and including a government service is merely viewed as a natural part of the process. In one of the case studies, the service delivery agency becomes invisible, which can be viewed as a next step in demand-driven government in which customers are no longer aware that they are dealing with the government.

- Intermediaries can reduce the complexity for citizens or business for dealing with the government and the accompanying transaction costs. In this way intermediaries can reduce the administrative burden from the citizen of business point of view. This suggests that transaction cost should not only be viewed form the service provider point of view, but also from the service user point of view and that they not necessarily coincide.

- Intermediaries help improve the quality of information. They ensure that the information is correct and reduce the need to re-enter information or the need for customers to transfer information from one system to another.

- Intermediaries can help reduce transaction costs through economies of scale. Car buyers and sellers can use one form to buy/sell a car and provide the information to the vehicle administration and licensing agency, while in the IRS case the customers only have to press a single button to submit the information.

The value of the intermediaries illustrate that – from a customer point of view – intermediaries as service delivery channels cannot easily be substituted or internalized through direct service delivery channels. Disintermediation and re-intermediation strategies are not necessarily conflicting and can be complementary. To advance e-government, both strategies may need to be used. Disintermediation may be necessary to remove traditional channels in order to reduce costs and re-intermediation should result in the creation of new low-costs channels contributing to a demand-driven service provisioning. Especially intermediaries can augment interaction and transaction channels and should not be overlooked by government organizations. On the other hand, the use of other (especially private) organizations as interaction channels may be risky. Conflicts of interests, concerns regarding privacy, equal and fair access need to be addressed and contracts and trusted mechanisms need to be in place. Future research should focus on the services and added value provided by intermediaries to determine the circumstances in which intermediaries can help reduce transaction costs and improve customer orientation, and to generalize the findings.

REFERENCES

Bailey, J. P., & Bakos, J. Y. (1997). An Exploratory Study of the Emerging Role of Electronic Intermediaries. *International Journal of Electronic Commerce*, *1*(3), 7–20.

Chircu, A. M., & Kauffman, R. J. (1999). Strategies of Internet Middlemen in the Intermediation/Disintermedation/Reintermediation cycle. *Electronic Markets*, *9*(2), 109–117. doi:10.1080/101967899359337

Clemons, E. K., & Row, M. C. (1992). Information Technology and Industrial Cooperation: The Changing Economics of Coordination and Ownership. *Journal of Management Information Systems*, *9*(2), 9–28.

Coase, R. (1937). The Nature of the Firm. *De Economia*, *4*, 386–405. doi:10.1111/j.1468-0335.1937.tb00002.x

Devadoss, P. R., Pan, S. L., & Huang, J. C. (2002). Structurational analysis of e-government initiatives: a case study of SCO. *Decision Support Systems*, *34*(3), 253–269. doi:10.1016/S0167-9236(02)00120-3

Gellman, R. (1996). Disintermediation and the Internet. *Government Information Quarterly*, *13*(1), 1–8. doi:10.1016/S0740-624X(96)90002-7

Giaglis, G. M., Klein, S., & O'Keefe, R. M. (2002). The role of intermediaries in electronic marketplaces: developing a contigency model. *Information Systems Journal*, *12*(3), 231–246. doi:10.1046/j.1365-2575.2002.00123.x

IRS. (2004). *Visie op dienstverlening 2010*. Utrecht: Dutch Inland Revenue Service.

Janssen, M., Gortmaker, J., & Wagenaar, R. W. (2006, Spring). Web Service Orchestration in Public Administration: Challenges, Roles, and Growth Stages. *Information Systems Management*, 44–55. doi:10.1201/1078.10580530/45925.23.2.20060301/92673.6

Janssen, M., & Klievink, B. (2009). The Role of Intermediaries in Multi-channel Service Delivery Strategies. [IJEGR]. *International Journal of E-Government Research*, *5*(3), 36–46.

Janssen, M., & Sol, H. G. (2000). Evaluating the role of intermediaries in the electronic value chain. *Internet Research. Electronic Networking Applications and Policy*, *19*(5), 406–417. doi:10.1108/10662240010349417

Janssen, M., & Verbraeck, A. (2005). Evaluating the Information Architecture of an Electronic Intermediary. *Journal of Organizational Computing and Electronic Commerce*, *15*(1), 35–60. doi:10.1207/s15327744joce1501_3

Klievink, B., & Janssen, M. (2008). Improving Government Service Delivery with Private Sector Intermediaries. *European Journal of ePractice*, *1*(5), 17-25.

Layne, K. J. L., & Lee, J. (2001). Developing fully functional E-government: A four stage model. *Government Information Quarterly*, *18*(2), 122–136. doi:10.1016/S0740-624X(01)00066-1

Malone, T. W., Yates, J., & Benjamin, R. I. (1987). Electronic Markets and Electronic Hierarchies. *Communications of the ACM*, *30*(6), 484–497. doi:10.1145/214762.214766

Orlikowski, W. J., & Baroudi, J. J. (1991). Studying Information Technology in Organizations: Research Approaches and Assumptions. *Information Systems Research*, *2*, 1–28. doi:10.1287/isre.2.1.1

Resnick, P., Zeckhauser, R., & Avery, C. (1995). Roles for Electronic Brokers. In G. W. Brock (Ed.), *Towards a Competitive Telecommunication Industry* (pp. 289-306). New Jersey: Mahwah.

Sarkar, M. B., Butler, B., & Steinfield, C. (1995). Intermediaries and Cybermediaries: A Continuing Role for Mediating Players in the Electronic Marketplace. *Journal of Computer-Mediated Communication*, *1*(3).

Spulber, D. F. (1996). Market Microstructure and Intermediation. *The Journal of Economic Perspectives*, *10*(3), 135–152.

Undheim, T. A., & Blakemore, M. (Eds.). (2007). *A Handbook for Citizen-centric eGovernment*: cc:eGov.

West, D. M. (2004). E-Government and the Transformation of Service Delivery and Citizen Attitudes. *Public Administration Review*, *64*(1), 15–27. doi:10.1111/j.1540-6210.2004.00343.x

Williamson, O. E. (1975). *Market and Hierarchies, Analysis and Antitrust Implications. A study in the economics of internal organization*. New York: Macmillan.

Wimmer, M. A. (2002). Integrated service modeling for online one-stop Government. *EM - Electronic Markets, special issue on e-Government*, *12*(3), 1-8.

Yin, R. K. (1989). *Case Study Research: Design and methods*. Newbury Park, CA: Sage publications.

Chapter 14
Towards eGovernment in the Large:
A Requirements-Based Evaluation Framework

Thomas Matheis
German Research Center for Artificial Intelligence, Germany

Jörg Ziemann
German Research Center for Artificial Intelligence, Germany

Peter Loos
German Research Center for Artificial Intelligence, Germany

Daniel M. Schmidt
University of Koblenz-Landau, Germany

Maria A. Wimmer
University of Koblenz-Landau, Germany

ABSTRACT

An increasing level of cooperation between public administrations nowadays on national, regional and local level requires methods to develop interoperable eGovernment solutions and leads to the necessity of an efficient evaluation and requirements engineering process that guides the establishment of systems and services used by public administrations in the European Union. In this chapter, the authors propose a framework to systematically gather and evaluate requirements for eGovernment in the large. The evaluation framework is designed to support requirements engineers to develop a suitable evaluation and requirements engineering process with respect to interoperable eGovernment solutions. The methodology is motivated and explained on the basis of a European research project.

DOI: 10.4018/978-1-60960-162-1.ch014

INTRODUCTION

The European Union keeps growing and member states become more cross-linked every day. Some reasons are that governments are requested to work together more frequently, more intensely and in a vast and ever evolving environment. The drivers of change are manifold: modernization is requested by politics and citizen; public administrations face a huge gap between the burden of work and the available resources; new legal settings and strategic commitments are postulated; new ICT has to be introduced, keeping up with the change taking place in private business settings; customers of the public administrations have higher expectations for improved quality of service; enhanced public value generation is claimed, etc. One could list a large number of aspects implying the need for a smooth cooperation among public administrations and cooperation with their stakeholders on the basis and by means of advanced ICT (Ziemann, Kahl & Matheis, 2007). In this respect, eGovernment in the small means to implement concepts, technologies and tools to pave the way for eGovernment in the large which aims to make such visionary cross-organizational collaboration possible.

What could be considered a fact anyway, is the underdevelopment of the public sector compared to the business sector in terms of ICT adoption. Not to mention the lack of interoperability (IOP) at all levels. This leads on one side to a different business perception of the IOP problem (and of the different types of lacks of IOP) and to different requirements for the IOP solutions available. Regarding the development and application of ICT solutions the main challenges are the requirements specification and the management of customer requirements (Sommerville & Sawyer, 2003). Main objective of this paper is answering the question of

- How to gain a clear understanding of the interoperability problems or needs of public administrations and

- How to capture the stakeholder requirements in order to support the application of any IOP solution to public administrations.

The presented framework of this paper answers these questions by providing a methodology for analyzing the needs of public administrations to allow for eGovernment in the large. This is tackled by improving the process of discovering, documenting and evaluating requirements. In this context, three main action domains were defined: the problem space, the requirements space and the solution space.

The paper is structured as follows: First, the scope of IOP in eGovernment is presented including a snapshot of IOP in eGovernment. Based on this, we will discuss the differences between eGovernment in the large and eGovernment in the small and present an IOP lifecycle to support eGovernment in the large. The following section introduces the evaluation framework to discover, document and evaluate requirements for an eGovernment in the large taking into account the problem, requirement and solution space. Afterwards, we present the results of the application of the framework within the European research project R4eGov[1]. Finally, we provide a summary and describe future work.

SCOPE OF INTEROPERABILITY IN EGOVERNMENT

Within a growing Information Society as mentioned before, networked governments have become a crucial factor. A major challenge for Governments across Europe is to link up heterogeneous systems in a way that these can work together smoothly. The obstacles to overcome in the public sector are a vast amount of stand-alone solutions under local control, which need to work together to enable seamless government. Often, these legacy systems may not be changed and adapted (Werth, 2005).

As a consequence, other options have to be found to pave the way for a smooth cooperation and collaboration. To enable cooperation (either in terms of collaboration or coordination), two approaches can be identified: integration or interoperation. Integration can be defined as the forming of a larger unit of government entities in order to merge processes, systems, and/or shared information (Klischewski & Scholl, 2006). Integration is seen as not achievable across organizations for several reasons (Werth, 2005):

- The majority of eGovernment systems will always be heterogeneous; and
- The configuration of systems and definition of processes will always remain under local responsibility, management and control.

Since new emerging technologies allow loose coupling of systems by exploring web services, service-oriented architectures (SOA), etc., hugh monolithic systems integrating heterogeneous legacy systems are required no more. As a consequence, interoperation has become the primary focus of investigation. In a working document, the European Commission defined IOP as "*the means by which the inter-linking of systems, information and ways of working, whether within or between administrations, nationally or across Europe, or with the enterprise sector, occurs*" (European Commission, 2003, p. 6). This definition covers a wide understanding, addressing all levels of IOP (organizational, semantic, and technical, as well as across public/private/civic sectors). The European Interoperability Framework (EIF) of IDABC aligns IOP with "*the ability of information and communication technology (ICT) systems and of the business processes they support to exchange data and to enable the sharing of information and knowledge*" (European Communities, 2004, p. 3).

Klischewski and Scholl characterize interoperating systems and applications via independency, heterogeneity, and control by different jurisdictions/ administrations or by external actors; yet also cooperation in a predefined and agreed-upon fashion. Likewise, interoperation can only be reached by means of open standards (Wimmer, Liehmann & Martin, 2006).

To exploit the potentials of modern ICT to reach the vision of systems „*working in a seamless and coherent way across the public sector*" (Cabinet Office – Office of the E-Envoy, 2004, p. 4), proper mechanisms of cross-organizational IOP are required, which enable different governments, and software components and applications to smoothly communicate with each other and to work together in the given settings. The EIF and other literature stress that IOP needs to be addressed on different levels to enable communication and cooperation among systems and services (Bellman & Rausch, 2004; Benamou, 2006; European Commission, 2003; Guijarro, 2004; European Communities, 2004; Klischewski & Scholl, 2006; Sturm, 2007; Tambouris & Tarabanis, 2004; Wimmer et al., 2006):

- **Technical interoperability:** Linking computer services and systems together so that the systems and applications are able to communicate with each other based on standardised interfaces and commonly used open standards for metadata, document and data formats (e.g. XML, UTF), communication protocols (e.g. SOAP, HTTP, IP), and technologies (Web Services, etc.).
- **Semantic interoperability:** Establishing a unique meaning of exchanged data, information and procedures by adding semantics to the information objects, or by establishing glossaries, thesauri or even ontologies. Standards in the field of semantic interoperability are required to ensure the exchange of information without depending on interpretations of humans. Only if the involved parties interpret data and meta-information consistently in the same commonly agreed-upon unique un-

derstanding, the information can be processed automatically in a meaningful manner. Thereby, standardised data definitions (e.g. XML, RDF, OWL, etc.), process models and object description frameworks are being used.

- **Organisational interoperability:** This level of IOP – the most complex one – is concerned with aligning business processes and information architectures with organisational goals. Furthermore, overall agreements are settled on organisational and legal level to enable processes to co-operate beyond organisational and state borders.

All three levels of IOP deserve equal attention in order to make systems communicate with each other and to link up governmental systems and services beyond organisational and national borders. With the linkage of administrative processes and data a significant increase in efficiency and lower operational costs can be achieved. Sturm describes numerous potentials for IOP in eGovernment as e.g. faster processing in administration, improving quality and service, organizational improvements or reduction of costs (Sturm, 2007).

It also became clear, that IOP can only be reached step by step. As a consequence, the next phases of future IOP activities should investigate two perspectives of IOP: eGovernment in the large (longer-term strategy) and eGovernment in the small (implementations achievable in the next few years).

eGovernment in the Large vs. eGovernment in the Small

The overall aim of IOP is to provide tools and methodologies for enabling organizations to smoothly collaborate in different use-contexts thereby being supported with advanced ICT. Aiming at *IOP in the large* means to enable smooth collaboration horizontally (across organizations of the same level of government: e.g. municipality with municipality) and vertically (across organizations of different levels of government: e.g. local–national –European). Thereby, organizations are probably not any more fully mastering the coordination of the cross-organizational processes lined up across organizations. E.g. a European directive enables an authority to check the registry entry of a bidder from another member state. In IOP in the large, the authority contacts the portal of the home business register (or a European business register) to gather the registry script from the bidder. Full IOP is reached when the home business register's system can retrieve the company registry certificate from any other Member State's business registry without having to agree on a standard data format of the script and with clear understanding of the peculiarities of the legal forms of each Member State without needing to bilaterally negotiate the meaning of the form's characteristics. In this IOP in the large, the Member State's organization is not mastering any more all point-to-point interfaces with other Member State's business registers. Instead, one unique IOP format is agreed upon, which is used by all Member State's business registers and other public and private organizations. Such IOP in the large is not feasible in the next few years. However, it is a driving vision for long-term networked governments.

In this context, *IOP in the small* will investigate concepts, technologies and tools to pave the way for such visionary cross-organizational collaboration while preserving the ability – and testing the concept – for IOP in the large. IOP in the small is understood as the organizations aiming at collaborating across their organizations to agree upon common IOP means to enable cross-organizational process execution supported with ICT. In this way, the organizations are fully in control of when and how the organizations collaborate to execute a public service.

Towards an IOP Lifecycle for eGovernment in the Large

The following IOP lifecycle serves as an organizing mechanism for managing the development of IOP solutions. Further on, it provides a structure for analysing requirements in a more detail. The single phases of the IOP lifecycle have been derived from existing lifecycles (e.g. WfMC (Hollingsworth, 1995), FEAF (CIO Council, 2001), ArKoS (Hofer et al., 2005) or Matheis, Ziemann & Loos, 2006) according to their suitability to serve as a basis for an IOP lifecycle taking into account the needs of public administrations (e.g. by giving more importance to a data and document phase that is characteristic for eGovernment scenarios).

The IOP lifecycle will answer the question of how to achieve IOP between legacy applications as well as how to design novel applications to be interoperable. The lifecycle follows the goals to include a broad preparation phase (strategy) and a feedback providing phase (monitoring) as well as the development of elements found in each enterprise information system: data and processes. Taking into account the need to prepare IOP solution for use in SOA, it also included a phase for the development of services (interaction components). Thus, the proposed IOP lifecycle covers the following five phases:

- The cross-organizational strategy phase refers to the development of an overall strategy of how to achieve IOP. In comparison to detailed concepts for single areas of IOP, this part defines a coarse grained strategy for which concepts to apply to secure and safeguard IOP.
- The phase of developing interaction components identifies and adjusts the components being part of a cross-organizational process. In order to identify such components, organizational (e.g. interaction policies), functional (e.g. process chains, data exchange standards) as well as existing

technical components (e.g. interaction protocols, web services, or modules of a legacy system) have to be taken into account.

- The phase of cross-organizational business processes provides methods to develop and adjust interacting processes. This phase refers in a first step to the modelling of existing or intended cross-organizational processes by using modelling languages like the Event-driven Process Chain (EPC). Based on this, the phase aims in a second step at the execution of the modelled processes.
- The phase of cross-organizational data storage concentrates on collaborative management of data and documents. This comprises methods to automate document flows, to implement document standards, to annotate data etc.
- The phase of cross-organizational monitoring provides methods to supervise, monitor and analyze the cross-organizational processes, components and data repositories in order to improve the effectiveness and efficiency of the collaboration.

Note, that steps 2, 3 and 4 describe the design and implementation of distinct interaction dimensions but could be executed in parallel. Thus, seeing from a time perspective, the sequence of the IOP lifecycle would be step 1, step [2,3,4] in parallel followed by step 5.

FRAMEWORK FOR GATHERING AND EVALUATING REQUIREMENTS FOR EGOVERNMENT IN THE LARGE

In this section, the methodology for the elaboration of the requirements is described. The conceptual framework has the challenge to bring together the wide spread perspective of the different sources which requirements arrive from. The methodology takes into account existing requirements engineering approaches both from literature (e.g. Hull, Jackson & Dick, 2005; Sommerville & Sawyer, 2003) as well as research projects (e.g. ATHENA[2]). In this context, three main action domains have been identified: the problem space, the requirements space and the solution space. Figure 1 depicts the overall methodology for the requirements combination process and shows the different sources the requirements come from.

The *Problem Space* addresses the particular needs of public administration to solve their current IOP problems. Different perspectives (organizational, semantic, technical), implementations achievable in the next few years to longer-term strategy as well as different phases to realize IOP solutions are comprised by the analysis of the problem space. The general question to be answered here is: how is the maturity of IOP solutions perceived in the context of eGovernment and what are the required perspectives to enable public administrations to adopt a specific IOP solution to solve their problems?

Against the background of eGovernment, the problem space points out the main IOP lacks or needs at all levels within public administrations (see "Scope of IOP in eGovernment"). Thus, the requirements spanning perspectives of the problem space are represented by the IOP levels of the EIF and the phases of the IOP lifecycle considering both the eGovernment in the small and the eGovernment in the large approach. Further on, the adoption of the EIF framework and the IOP lifecycle permits an initial classification for the requirements of the requirement space.

The needs of public administrations that an IOP solution should take into account when trying to solve IOP problems are described in detail within the *Requirement Space*. The general question to be answered here is: what requirements (independent of specific IOP solutions) can be specified by public administrations to solve their IOP problems?

The requirement space represents the beginning of the methodical requirements elicitation process. Within a *first step* requirements are

Figure 1. Evaluation framework

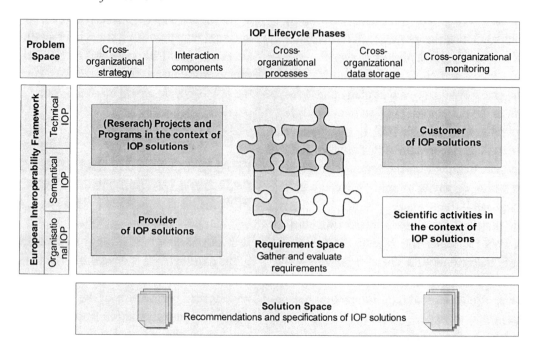

gathered and derived regarding different sources of requirements:

- Analysis from case studies and needs of the IOP solution customer
- Analysis of (research) projects and programs
- Analysis of the literature and experience form scientific partners
- Analysis of the products and experience from IOP solution provider

As a *second step* the requirements are presented in a synthesized format. The aim is to condense the great amount of gathered information to a form that allows taking measures in single phases of the IOP lifecycle and/or the EIF. In addition, weights put on the requirements will be the basis to drive and evaluate the (research) activities to develop IOP solutions within the solution space.

The *Solution Space* comprises a critical analysis and evaluation of the developed IOP solutions regarding the requirements identified in the requirements space. The basic question to be answered is: are the proposed IOP solutions (e.g. IOP architectures) suitable for public administrations? If not what are the necessary modifications to guarantee that the IOP solution fulfils the required needs?

Based on the analyzed and weighted requirements of the requirement space, the evaluation criteria and the baseline for the IOP solution and the needed tools, guidelines and concepts is defined within the solution space. The results should flow into the methodical and architectural specifications of the IOP solution. The tools, guidelines and concepts must embody a long-term view of eGovernment in the large, while their application and implementation regarding need to be scalable and customizable for a step-by-step advancement of reaching higher levels of IOP over time.

APPLICATION OF FRAMEWORK IN AN EUROPEAN RESEARCH PROJECT

Based on the described evaluation framework the requirements analysis process has been carried out within the R4eGov project in order to discover, document and evaluate IOP requirements for an eGovernment in the large. Taking into account the objective of R4eGov, the focus here is on the requirements that should be considered in order to develop methods for implementing IOP systems as well as for designing and evaluating an eGovernment in the large (Matheis et al., 2007). Note, that the results of the requirement analysis process derived from the application of the framework offer a first overview of the IOP situation of public administration. This data can be useful for obtaining general requirements that are applicable to most of IOP solutions regarding IOP in the large but do not provide requirements needed for all specific IOP and/or eGovernment topics.

Synthesis of Problem Space

To further analyze the problem space the IOP lifecycle phases and IOP levels were further detailed. Table 1 illustrates, that the elements covered by the EIF can be matched to the IOP lifecycle phases.

The reason why we do not only use the EIF within the problem space is that the lifecycle phases are closer aligned to need typically required in development of software systems. It explicitly focuses on the elements that constitute a software system and on the preparation and analysis of such systems. The identification of topics in Table 1 served as basis to structure the interviews and questionnaire in order to gather the requirements in an effective way. Based on this, the R4eGov user, industry and scientific partners were asked how far the different IOP lifecycle phases and IOP levels were important for their current and planned IOP activities in order to evaluate the

Table 1. Matching EIF and IOP lifecycle phases

EIF / IOP lifecycle phase	Organizational IOP	Semantic IOP	Technical IOP
Cross-organizational strategy	Business goals, legal constraints, Enterprise Architecture	Coarse grained specification of data exchange	IT Architecture
Interaction components	Identify services, making services available, describe service functionalities	Describe data being used by services	Open interfaces, interconnection services, WSDL, machine interpretable service level agreements
Cross-organizational business processes	Defining business processes; specifying organizational responsibilities	Describe data contained in processes	Business protocols (e.g. ebXML), executable processes (e.g. WS-BPEL), transactions (e.g. WS-atomic transaction)
Cross-organizational data storage	Describe data protection, specific data classification systems	Describe precisely meaning of exchanged data, ensure interpretability by different systems, document/form specification	Data integration, data presentation, XML schemata, technical data protection
Cross-organizational monitoring	Controlling of processes, performance measurement	Annotate data for monitoring/controlling purposes	XML based annotation for monitoring purposes

relevance of the single components. In the following the results of the survey (especially the prioritization of the methodical requirements) that was based on a questionnaire on methodical interoperability requirements are presented. In this context, we asked the R4eGov user partners, industry partners as well as scientific partners to prioritize the initial set of methodical requirements in view of their organization (e.g. specific use-case, or ICT provider/consultant of cross-organizational e-government solutions). A total of 19 R4eGov partner participated in the questionnaire among them 6 user partners, 8 industry partners and 5 scientific partners.

Figure 2 shows the results of the prioritization regarding the different phases of the initial R4eGov IOP lifecycle. The results illustrate that the cross-organizational business process activities are classified as most important on average (4,28). Even though this phase is most important for the industry partner (3,83) and the scientific partner (5,0), the user partner rated the strategy phase as most important (4,5). This confirms the impression captured during the interviews with the user partner that an overall IOP strategy is very important for administrations in order to convince

the stakeholders of the benefits of interoperability. Thus, the other phases should be supported by a corresponding IOP strategy. Due to the fact that most administrational processes are document-oriented (this is a main difference between the industry and the public sector) the user partner rated the phase of data and documents of high importance (4,33). In comparison with the other phases, interaction components received a low position by the user (3,83) and industry partner (2,33). An explanation for this result could be that interaction components mainly support the other phases and are thus not seen as main IOP activity (nonetheless interaction components play an important role in cooperations because they connect the components of the other phases). The user partner rated the monitoring phase with 4,0 due to the fact, that the cross-organizational analysis and evaluation of the processes and the exchanged data is important to improve the cooperation (e.g. reduce failures and exceptions, increase the transparency of the cooperation, enable a re-engineering). Up to now cross-organizational monitoring has been tackled only by a few initial approaches. This explains on the one hand the high importance for the scientific field

(4,5) and on the other hand low prioritization by the industry partner (2,33).

The results of the prioritization regarding the different IOP levels as defined in the EIF are shown Figure 3. As for the average the technical level was classified as most important (4,29). The organizational and technical level received comparable values on average. The user partner rated the organizational level as most important whereas the other levels received as well high values. This could be explained that in the first instance interoperability is seen by the user partner from a business (administration) perspective (4,67). On the other hand the user partner do not only need an organizational solution, but rather a holistic IOP solution regarding as well the other levels (4,33 and 4,5). Because the industry partner focus in general more on the development on products, the semantic (3,88) and technical level (3,63) were classified as more important as the organizational level (3,25). The fact that the semantic level is seen as more important than the technical level by the industry partner cannot directly explained. One reason could be that topics like for example semantic web or ontologies are on the top agenda of the industry partner. Due to the fact that in the scientific field topics like

Web Services, Service-oriented Architectures or Process-to-Application approaches (e.g. Model-driven Architectures) are getting more and more important, the technical level was ranked to a high position (4,75). The semantic level received a comparable low value (3,25), because one scientific partner is positioned more in the technical security field and ranked the semantic IOP level with a very low value.

Synthesis of Requirement Space

In a first step, requirements were gathered from different sources as mentioned above: requirements identified in current as well as finished IOP activities and developments in the public sector (e.g. EU-funded projects like ATHENA, national and regional initiatives and approaches like MODINIS (MODINIS program, 2006) or IOP frameworks like EIF (European Communities, 2004)), requirements from case studies and interviews provided by R4eGov user partners and requirements provided by R4eGov user partners, industry partners and scientific partners on the basis of a questionnaire. The identified requirements were classified according to the IOP levels

Figure 2. Prioritization of IOP lifecycle phases

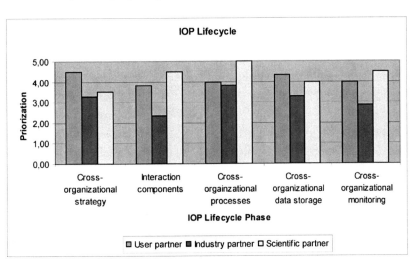

Figure 3. Prioritization of IOP levels

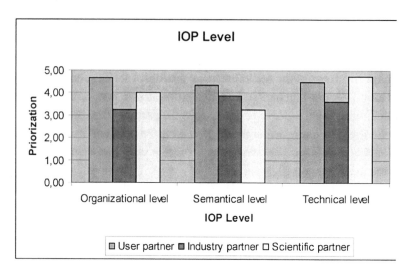

as defined in the EIF as well as to the different phases of the IOP lifecycle.

Secondly, the R4eGov user partners, industry partners as well as scientific partners were asked to prioritize the set of requirements in view of their organization (e.g. specific use-case, or ICT provider/consultant of IOP solutions). According to Table 1 requirements that were judged as "very important" (average value higher than 3,5 of 5) by the R4eGov partners were listed as exemplary shown in following sections. At the requirements level, the average values are used to indicate their relative significance within the same IOP level. Requirements that are judged to be of high relevance flow directly into the solution space in order to guide the further activities. Regarding the different phases of the IOP lifecycle and the listed requirements, the following (exemplary) synthesis can be made with respect to the IOP levels:

- **Cross-organizational strategy:** The listed requirements within the strategy phase illustrate that most of the requirements refer to the organizational and technical IOP level. This indicates that an IOP solution should mainly focus on organizational IOP (e.g. legal framework, compliance) and the

technical realization (e.g. enterprise architecture that is based on SOA) regarding the strategy phase. Semantic requirements are not in the main focus of this phase.

- **Interaction components:** Technical requirements like the use of technical protocols, web service technologies and the corresponding methods are the main objective within the phase of interaction components. Requirements regarding the organizational IOP level and the semantic IOP level should support the technical realization of the interaction components.

- **Cross-organizational business processes:** Regarding the listed requirements the organizational and the technical requirements dominate this group. Most important are methodologies to model cross-organizational processes and the consideration of different process types. On technical IOP level, the execution and synchronization of the modelled cross-organizational business processes are of high interest. The semantic IOP level should support the other levels by a common understanding of the cross-organizational processes.

- **Cross-organizational data storage:** The requirements of the semantic IOP level dominate the data and documents phase. The listed requirements illustrate the need for a semantic description of different views (internal and external view, view of the collaboration) of the involved data objects, the support of common data object definitions and the development of semantic techniques to enable a seamless exchange of the data objects. The requirements of the technical IOP level refer to the use of open standards (e.g. XML). Organizational aspects play a subordinated role.

- **Cross - organizational monitoring:** Organizational requirements (e.g. methods to analyze the critical cross-organizational processes) and semantic requirements (e.g. methods to analyze the status of document in a collaborative scenario) are in the focus of the monitoring phase. The fact that the technical IOP level of the monitoring phase contains nearly requirements can be explained due to the fact, that cross-organizational monitoring has been tackled only by a few initial conceptual approaches.

Synthesis of Solution Space

The finding for the solution space is the recommendation to support all phases of the lifecycle. But the effort spend on the different phases does not equally spread across the phases of the lifecycle. It becomes clearly visible, that the focus on different IOP levels differs among the phases of the IOP lifecycle. According to this result, a need for precise support for the application of the lifecycle can be anticipated. Thus, a high potential can be seen in the development of appropriate methods and tools that satisfy the mentioned IOP needs and requirements. The feedback and the requirements, especially the high ranked requirements, collected within the requirement space serve as the baseline for evaluation activities as well as for ongoing research and development activities. Within R4eGov this is conducted in order to ensure that the business needs will be fulfilled by the proposed IOP solution approach and that the R4eGov customer (user partners) see the potential for the approach to be used in daily business. Additionally, the first results of research and development activities are evaluated against the requirements in order to ensure the compliance with the requirement space.

The solution space is the action domain for evaluating and developing appropriate solutions which match and answer the requirements. In the course of the R4eGov project, the application of the framework gave way to elaborate a three-layered view on IOP which will further be detailed in this chapter. This architecture is designed based on the requirements of R4eGov user partners. On the top layer, we find conceptual descriptions of interoperability. On this level the alignment of requirements, directives and processes is secured. The descriptions on this level are transformed to a technical level of interoperability. On this level, technical complementarily is focus. On the execution level, the technical descriptions are deployed and executed.

In a scenario, where public services are in shared responsibility of a number of public sector organizations IOP has to be secured and visibility among partners has to be defined. To answer this requirement, a public view, which provides the necessary information and interfaces to the collaboration partners, was introduced. Only inside one organization, the private view is available, which gives detailed insight into the intra-organizational workflows.

'Joining-up' all the transactions and service information involved in delivering a service, and making them available to the public from one central access point can be done via service oriented architecture with a one-for-all service, as e.g. implemented in BundOnline, Germany.

Requirements defined with respect to services are for example:

- a proper mapping of corresponding service requesting and service providing
- ensuring the scalability of the service broker with respect to number of services and types of services
- managing different levels of trust and confidentiality for linking service providers with service requesters
- authentication mechanisms
- reliable data/archiving
- protection of citizens against misuse of data

In the same scenario partners know each other and so, they do create a common strategy for interoperability. This makes the tasks to be solved for setting up collaboration easier in a sense that a common interoperability environment can be set up. This includes a common virtual private network to be installed and an agreement on common business process execution and communication protocols as well as a common message exchange and data (document) exchange format. Requirements, which can be answered in respect to direct collaboration, are for example:

- setting up a common choreography for preparing interoperability
- extend the functionality of each partner's system in order to provide information to the other partner (e.g. interoperability gateway); preserve legacy systems
- identify the parameters of interaction and make them mutually accessible, e.g. what types of data will be exchanged, how can they mutually identified
- avoid errors that can be created due to collaboration, e.g. wrong access control due to inter-organizational rights delegation and revocation, mutual waiting for responses, signature checks, legal requirements

- time (stamping) concepts/time-dependent synchronization and agreement

Direct collaboration is essential even for service oriented interoperability since every linkage of service provider and service requester ends up in direct collaboration between them. This is why we focus on direct collaboration first and later extend our approach to more general interoperability architecture later, correspondingly to the statements on eGovernment in the small and on eGovernment in the large.

To ease the generation of executable code, to ensure compliance with conceptual model and for monitoring of collaborative processes the models will be used to transform them (automatically) into executable processes. The execution of the processes then send back required information to the model in order to visualize the execution during run-time and thus provide transparency and traceability and also to make auditing and analysis possible.

Figure 4 shows the model-driven approach followed in R4eGov to provide information for IOP. In particular, models (of processes, data, documents and organizations) are designed at the conceptual layer and transformed either directly to the executable layer or via the technical layer. The workflow architecture lying below executes these models and provides means to electronically exchange data.

To support the development of the model types needed in the architecture, a tool suite of complementary, mostly prototypical tools was created (Ziemann, 2010). Figure 5 provides an overview of the tool suite. The tool chain starts with the creation of private business processes on the conceptual level in the form of EPC.

Modeling of internal processes with ARIS or Kindler Tools. Various tools exist to create, validate and transform processes stored in the EPC Markup Language (EPML; Mendling & Nüttgens, 2005), including an open source tool from Cuntz & Kindler (2006), which can model the private

Figure 4. Architecture for IOP in the large (solution space)

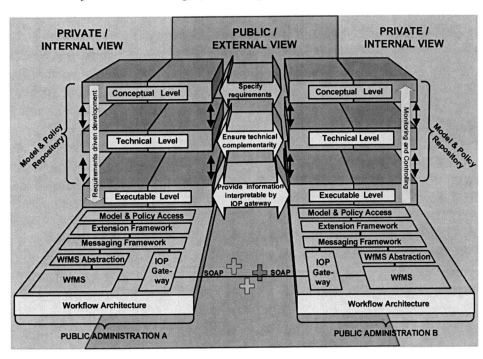

Figure 5. Tool chain supporting the modelling and enactment of IOP in the large (solution space)

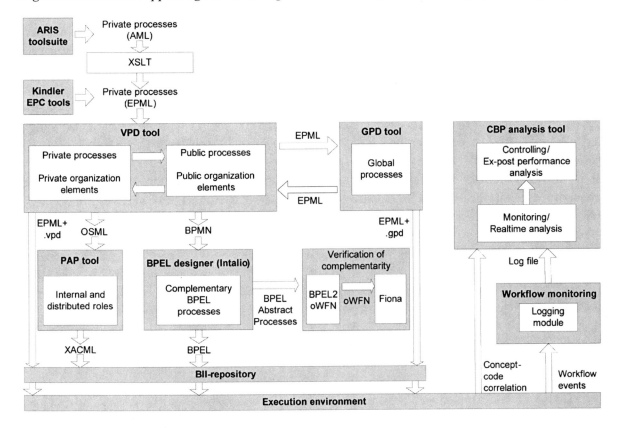

processes to be used inside the VPD tool. Another option is the ARIS toolset, a widespread industrial tool for enterprise modelling. The ARIS toolset exports EPC in the form of ARIS Markup Language (AML), which can automatically be transformed into EPML via an XSLT (Mendling & Nüttgens, 2004).

Deriving and Connecting Collaborative Views with the VPD/GPD. On the upper left, the modelling tools are shown that represent the core of the design solution: The View Process Demonstrator (VPD) is a tool to model processes and organization elements, both from a private and a public perspective. Taking the public models from the VPD as input, the Global Process Demonstrator (GPD) can connect complementary public process models into a global process model. Both VPD and GPD use an extended version of EPML as input and output format, and private EPML processes exported by the tools described above can be imported in the VPD.

Generation of BPEL. To generate the BPEL files corresponding to the public and private processes from the VPD, various commercial BPEL designers are available. Such tools also allow the import of conceptual models, for example in BPMN, and generate BPEL processes from them.[3] Since in the VPD, both EPC and BPMN models can be created, the BPMN models developed in the VPD in a BPEL designer can be used as a basis to generate BPEL.[4]

Verification of CBP. To ensure the complementarity of public processes that are supposed to form a global process, the abstract BPEL processes created in the BPEL designer can be validated using a Petri Net-based verification mechanism (Freiheit & Mondorf, 2007). Therefore, in a first step the BPEL processes are transformed into Open Workflow Nets (oWFN; Massuthe, Reisig & Schmidt, 2005) by using the BPEL2oWFN tool (Lohmann, Gierds & Znamirowski, 2007). The resulting oWFN can then be evaluated in the in the Fiona tool (Massuthe & Weinberg, 2007).

As a result, the complementarity of the public processes is confirmed or declined.[5]

Describing Organizational Roles and Rights with VPD and PAP. In the organization view of the VPD, organization structures are exported in form of the Organization Structure Markup Language (OSML), describing private, public and global views on organization elements as well as their correlation. These models constitute the basis for finer grained models, which describe the rights and roles inside collaborations, and as such can be used as input for the PAP tool. The Policy Administration Point (PAP) prototype was developed to support the concept of d-Roles and distributed XACML. More specifically, it can be used to model XACML policy instances that link local roles to distributed role sets or to link distributed role sets to access control lists (Lee & Luedemann, 2007). While the VPD tool tackles public, private and global roles on a conceptual level, the PAP tool refines private and global models to make them usable on the execution level.

Repository. The Business Interoperability Interface (BII) of each organization is stored and published via the BII-repository, which can be accessed both at design time and at run time of collaborative business processes (CBP).

Execution Environment. The execution environment complementary to the design time 3 elements is responsible for the execution of collaborative business processes, and comprises engines that execute internal processes as well as technical interoperability gateways that ensure the secure exchange of messages between the internal engines.

Monitoring and Controlling of CBP. To enable run time monitoring of processes, a workflow-monitoring component receives events from the execution environment. For further analysis of collaborative business processes, the CBP analysis tool imports the log files generated by the monitoring component; it also accesses the BII-repositories of the collaborating organizations in order to correlate the technical events with conceptual business processes (Matheis & Loos, 2008).

For a more detailed description of this solution and its application to an eGovernment scenario refer to Ziemann (2010).

CONCLUSION

Rapid advancements in technologies and regular emergence of new legal settings raise new challenges for public administrations. Thus, in recent years the development of interoperable eGovernment systems has gained importance due to the fact that more and more public administrations within Europe are challenged to work together and to adapt continuously to rapid technological changes. To keep pace with rapid evolving economic alterations and to gain a clear understanding of the problems and needs of public administrations for the application of ICT solutions, a suitable evaluation and requirements engineering process is of crucial interest. In this context, we presented a framework that enables a cohesive evaluation and requirements engineering process for eGovernment in the large.

The framework consists of three action domains: 1. the problem space for identifying the relevant objectives, 2. the requirement space for gathering requirements that serve as evaluation basis and 3. the solution space for evaluating and developing appropriate solutions. This framework was motivated and illustrated on the basis of the R4eGov research project. Based on this we described how the framework can be applied within the R4eGov research project to derive and evaluate requirements in different IOP levels and IOP lifecycle phases and demonstrated the transitions between the IOP levels and IOP lifecycle phases.

Future work remains to be done to further refine the proposed evaluation framework and validate its usefulness. Thus, future research should try to apply the framework for other projects and case studies as well as to apply it to different eGovernment scenarios. This could be realized by adapting the problem space, especially the IOP lifecycle dimension that is characteristic for the eGovernment in the large perspective, to other scenarios (e.g. security and/or public key infrastructure solutions for eGovernment, portal solutions for eGovernment). Another point for future research should be the comparison and connection of the presented framework with other evaluation and requirements engineering approaches. Further on, future work should focus on development of an open requirements library in order to make the requirements available for interested requirements engineers and to extend and consolidate the requirements library towards a holistic requirements and evaluation base for different perspectives of eGovernment in the large.

ACKNOWLEDGMENT

The work published in this paper is (partly) funded by the E.C. through the R4eGov project. It does not represent the view of E.C. or the R4eGov consortium, and authors are solely responsible for the paper's content.

REFERENCES

Bellman, B., & Rausch, F. (2004). Enterprise Architecture for eGovernment. In Traunmüller, R. (Ed) *Electronic Government*, conference proceedings,(LNCS # 3183, Springer Verlag Heidelberg et al,pp. 48 – 56).

Benamou, N. (2006). Bringing eGovernment Interoperability to Local Governments in Europe. *egovInterop'06 Conference*, Bordeaux, France.

Cabinet Office - Office of the E-Envoy. (2004). *E-Government Interoperability Framework (e-GIF)*, London, UK. Version 6.0, URL: http://www.alis-etsi.org/IMG/pdf/UK_e-Gov_V6_April_04.pdf, p. 4.

CIO Council (2001). *Practical Guide to Federal Enterprise Architecture.* Chief Information Officer Council, Version 1.0.

Cuntz, N., & Kindler, E. (2006): *EPC Tools.* http://wwwcs.upb.de/cs/kindler/Forschung/EPCTools/, last accessed: March 2009.

European Commission. (2003). *Linking up Europe: The importance of interoperability for e-government services.* Staff Working Document 2003, URL: http://europa.eu.int/information_s ociety/activities/eGovernment_research/ ar chives/ events/egovconf/doc/interoperability. pdf, p. 6.

European Communities IDABC – EIF. (2004). *European Interoperability Framework for Pan-European E-Government Services.* URL: http://europa.eu.int/id abc/en/document/3761, p. 3.

Freiheit, J., Mondorf, A. (2007*): Formal Analysis of web service standards.* Deliverable WP4-D4, R4eGov - Towards e-Administration in the large, project number IST-2004-026650.

Giorgini, P., et al. (2004). Requirements Engineering meets Trust Management: Model, Methodology, and Reasoning, In *Proc. of iTrust 2004,* (LNCS, vol. 2995, pp. 176–190). Springer-Verlag Heidelberg.

Guijarro, L. (2004). Analysis of the Interoperability Frameworks in eGovernment Initiatives, in Traunmüller, R. (Ed.) *Electronic Government,* conference proceedings (LNCS # 3183, pp. 36-39) Springer Verlag Heidelberg et al.

Hofer, A. (2005). Architektur zur Prozessinnovation in Wertschöpfungsnetzwerken. In Scheer, A.-W. (Ed.), *Veröffentlichungen des Instituts für Wirtschaftsinformatik (Vol. 181).*

Hollingsworth, D. (1995). *Workflow Management Coalition (WfMC) - The Workflow Reference Model.* URL: http://www.wfmc.org/stand ards/docs/tc003v11.pdf.

Hull, E., Jackson, K., & Dick, J. (2005). *Requirements Engineering.* London: Springer-Verlag.

Klischewski, R., & Scholl, H. J. (2006). *Information quality as a common ground for key players in e-government integration and interoperability.* In Proceedings of HICSS'06.

Lee, H., & Luedemann, H. (2007). A Lightweight Decentralized Authorization Model for Inter-domain Collaborations. *ACM Workshop on Secure Web-Services,* Fairfax, Virginia, USA, SWS '07. ACM, New York, pp. 83-89.

Lohmann, N., Gierds, C., & Znamirowski, M. (2007). *BPEL2oWFN - Translating BPEL Process to Open Workflow Nets.* Version 2.0.3. Http://www.gnu.org/software/ bpel2owfn/index.html, last accessed: November 2008.

Massuthe, P., Reisig, W., & Schmidt, K. (2005). An Operating Guideline Approach to the SOA. Annals of Mathematics. *Computing & Teleinformatics, 1*(3), 35–43.

Massuthe, P., & Weinberg, D. (2007). *Functional Interaction Analysis for open Workflow Nets.* Version 2.0. Http://www2.informatik.hu-be rlin.de/top/tools4bpel/fiona/, last accessed: November 2008.

Matheis, T. et.al. (2004). Methodical interoperability requirements for eGovernment, *Deliverable D8.1, R4eGov – Towards e-Administration in the large,* IST-2004-026650.

Matheis, T., & Loos, P. (2008): Monitoring cross-organizational business processes. *International Conference on E-Learning, E-Business, Enterprise Information Systems, and E-Government,* (EEE 2008), Las Vegas, USA.

Matheis, T., Ziemann, J., & Loos, P. (2006). A Methodical Interoperability Framework for Collaborative Business Process Management in the Public Sector. In *Proceedings of the Mediterranean Conference on Information Systems,* Venice, Italy.

Matheis, T., Ziemann, J., Schmidt, D., Freiheit, J. (2008). *R4eGov IOP tool suite for modelling cross-organizational processes and data.* Deliverable WP4-D10, R4eGov - Towards e-Administration in the large, project number IST-2004-026650.

Mendling, J., & Nüttgens, M. (2004): Transformation of ARIS Markup Language to EPML. In Nüttgens, M., Rump, F. (eds.). *Proceedings of the 3rd GI Workshop on Event-Driven Process Chains (EPK 2004).* Luxembourg, Luxembourg.

Mendling, J., & Nüttgens, M. (2005). *EPC Markup Language (EPML) - An XML-Based Interchange Format for Event-Driven Process Chains (EPC).* Technical Report JM-2005-03-10. Vienna University of Economics and Business Administration. MODINIS program (2006). *Study on interoperability at local and regional level.* URL: http://www.egov-iop.ifib.de/index.html.

Sommerville, J., & Sawyer, P. (2003). *Requirments Engineering – A good practice guide.* New York: Wiley.

Sturm, J. (2007). Interoperability and standards. In Zechner, A. (Ed.), *E-Government Guide Germany: Strategies, solutions and efficiency* (pp. 31–37). Stuttgart, Germany: Fraunhofer IRB Verlag.

Tambouris, E., & Tarabanis, K. (2004). Overview of DC-Based eGovernment Metadata Standards and Initiatives. In Traunmüller, R. (Ed.), *Electronic Government,* (LNCS # 3183, pp.40-47), Springer Verlag Heidelberg et al.

Werth, D. (2005). *E-Government Interoperability,* In Khosrow-Pour, M. (Ed.) *Encyclopedia of Information Science and Technology,* I-V., 985-989, Hershey, PA: Idea Group Inc.

Wimmer, M., Liehmann, M., & Martin, B. (2006). Offene Standards und abgestimmte Spezifikationen - das osterreichische Interoperabilittskonzept. In *Proceedings of MKWI.*

Ziemann, J. (2010). *Architecture of Interoperable Information Systems – An Enterprise Model-based Approach for Describing and Enacting Collaborative Business Processes.* Berlin: Logos Verlag.

Ziemann, J., Kahl, T., & Matheis, T. (2007). *Cross-organizational Processes in Public Administrations: Conceptual modeling and implementation with Web Service Protocols. 8.* Karlsruhe: Internationale Tagung Wirtschaftsinformatik.

ENDNOTES

[1] http://www.r4egov.eu

[2] http://www.athena-ip.org

[3] In the R4eGov project, for example, the Intalio BPEL designer was chosen that fulfils these functionalities, including a BPMN-based BPEL generation. Compare also Matheis et al. (2008).

[4] Since the scientific benefit of an automated BPMN (from the VPD) to BPMN (of the Intalio tool) transformation is questionable, we abstained from implementing an automated transformation from the VPD process export format to the Intalio import format. Therefore, the BPMN process from the VPD currently has to be re-modeled with the Intalio designer.

[5] A detailed example for such verification in the context of R4eGov's Europol-Eurojust scenario is provided in Freiheit & Mondorf (2007), pp. 60.

Chapter 15
Relating Acceptance and Optimism to E-File Adoption

Lemuria Carter
North Carolina Agricultural and Technical State University, USA

Ludwig Christian Schaupp
West Virginia University, USA

ABSTRACT

Electronic tax filing is an emerging area of e-government. This research proposes a model of e-filing adoption that identifies adoption factors and personal factors that impact citizen acceptance of electronic filing systems. A survey administered to 260 participants assesses their perceptions of adoption factors, trust and self-efficacy as they relate to e-file utilization. Multiple linear regression analysis is used to evaluate the relationships between adoption concepts and intention to use e-filing systems. Implications for practice and research are discussed.

INTRODUCTION

The United States (U.S.) government is one of the largest users of information technology (IT) systems in the world (Evans, 2006). A considerable percentage of the government's IT investment is allocated to e-government initiatives. E-government in the U.S. provides its citizens with convenient access to government information and services. The electronic filing of income tax returns (the e-file program) is an invaluable application that aids citizens with the process of collecting their personal tax information and provides them the ability to electronically transmit their return. Electronic filing (e-file) has the potential to improve the tax filing process for the individual and the government by reducing monetary and temporal costs for both taxpayers and tax collection agencies (Fu, Farn, and Chao, 2006).

The use of IRS endorsed e-file systems has continued to grow over the last couple of years with 52.9 million individual returns being filed in 2003 and approximately 68 million in 2005 (IRS, 2004). However, despite the numerous Internal Revenue Service (IRS) endorsed e-file systems that are available, this still only accounts for about 50% of the total number of returns. Congress

DOI: 10.4018/978-1-60960-162-1.ch015

wanted 80% of all tax and informational returns be filed electronically by 2007 (IRS, 2004). Thus, the U.S. has fallen short of the 80% goal and the lower than desired adoption rate continues to plague the IRS. Throughout the IS literature, the prediction of usage has always been a focus. With the growing interest in e-government and increased pressures to get to 80% utilization it raises the question of how to increase citizens' adoption of e-file.

One of the challenges to reaching the goal, is increasing citizen concerns related to the trustworthiness of electronic services. In 2007, the Federal Trade Commission received 20,782 complaints involving tax returns and identity theft. The number of complaints has risen 158% since 2003. Carter and Belanger (2005) found that citizens were still willing to adopt e-government systems, even when they perceived these systems to be precarious. One potential explanation for this counterintuitive behavior is optimisim bias. Optimism bias – which has been explored in social psychology (Weinstein 1980a), risk analysis (Sjoberg and J. Fromm 2001) and accident analysis (Dejoy 1989) - offers insight into how individuals handle risky situations. Building on previous technology adoption studies, we develop a model that depicts U.S. taxpayers' intention to use an e-file system to file their taxes. Specifically, a survey is conducted to examine taxpayers' intentions to use an IRS endorsed e-file system. The proposed model integrates concepts from technology adoption, optimism bias and self-efficacy to present a more comprehensive view of e-file adoption. This article is organized as follows: first, the background literature is presented; then, the research model and hypotheses are illustrated; next the methodology is discussed; finally the results, implications and suggestions for future research are presented.

BACKGROUND LITERATURE

Technology Adoption

Numerous studies have explored the factors that impact e-government adoption (Bélanger and Hiller, 2006; Burn, 2003; Choudrie and Dwivedi, 2005; Cross, 2007; Dwivedi, Papazafeiropoulou and Gharavi, 2006; Gefen, Rose, Warkentin, and Pavlou, 2005; Gilbert, Balestrini and Littleboy, 2004; Hackney and Jones, 2002; Huang, 2007; Thomas and Streib, 2003). Within this broad area of research there is a core of literature that focuses on intention to use an innovation. The Unified Theory of Acceptance and Use of Technology (UTAUT) is the most predominant theory existing in the literature to date. The UTAUT model is comprised of eight theoretical models: the theory of reasoned action (TRA), the technology acceptance model (TAM), the motivational model, the theory of planned behavior (TPB), a model combining the technology acceptance model and the theory of planned behavior, the model of PC utilization, the innovation diffusion theory, and the social cognitive theory. The goal of UTAUT is to understand intention/usage as the dependent variable (Venkatesh, Morris and Davis, 2003).

The UTAUT Model

Venkatesh et al.'s (2003) UTAUT model is composed of four core determinants: performance expectancy, effort expectancy, social influence, and facilitating conditions. Performance expectancy is defined as the degree to which individuals believe that using the system will help them improve their job performance (Venkatesh et al., 2003). Five variables comprise the performance expectancy construct: perceived usefulness, extrinsic motivation, job-fit, relative advantage, and outcome expectations (Venkatesh et al., 2003). Venkatesh et al. (2003) found the performance expectancy to be the strongest predictor of intention, which is consistent with previous model tests (Agarwal and Prasad, 1998; Compeau and Higgins, 1995; Taylor and Todd, 1995; Thompson et al., 1991; Venkatesh and Davis, 2000).

Effort expectancy is the level of difficulty associated with the use of the system (Venkatesh et al., 2003). Venkatesh et al. (2003) identify three constructs from the eight models which make up the concept of effort expectancy: perceived ease of use, complexity, and ease of use. Venkatesh et al. (2003) note that the similarity among these three variables has been documented in prior literature (Moore and Benbasat, 1991; Plouffe et al,. 2001; Thompson et al., 1991; Venkatesh et al., 2003). Venkatesh et al. (2003) found that their effort expectancy construct was significant in both voluntary and mandatory usage contexts, but only in the initial usage of the technology. It became insignificant after periods of extended use which is consistent with previous research (Agarwal and Prasad, 1999; Thompson et al., 1991; Thompson, Higgins and Howell, 1994; Venkatesh et al., 2003). Effort oriented constructs are usually prominent in the preliminary stages of a behavior (Szajna 1996; Venkatesh et al., 2003).

Social influence relates to an individual's perception that others who are important to him believe that he should use the system (Venkatesh et al., 2003). Social influence is comprised of subjective norms, social factors, and image. Thompson et al. (1991) use the term "social norms" to define their construct, and acknowledge its similarity to "subjective norm" within the Theory of Reasoned Action. According to Venkatesh et al. (2003) social influence contains the explicit or implicit notion that people's behavior is influenced by the way in which they believe others will view as a result of having used the technology. Venkatesh et al. (2003) found that none of the social influence constructs were significant in voluntary contexts; however, all of them were significant when usage was mandatory. Venkatesh and Davis (2000) suggest that these effects in a mandatory context

could be attributed to compliance that causes social influence to have a direct effect on intention. However, social influence in voluntary contexts, influences perceptions of the technology. E-file adoption for individuals is currently voluntary for individuals.

Facilitating conditions refer to an individuals belief that an organizational or technical infrastructure exists to support the system (Venkatesh et al., 2003). The authors found that when both performance expectancy constructs and effort expectancy constructs are present in the model, facilitating conditions are not significant predictors of system usage. Also, since facilitating conditions is predicted to have a direct effect on actual usage, not intention to use, it is not included in the proposed model (Venkatesh et al., 2003).

Trust of the E-Filer

The literature contains numerous definitions of trust. One popular definition was proposed by Rotter (1967). The author draws from social learning theory and defines trust as an expectancy that the promise of another can be relied upon. Rotter's research is referenced in numerous studies of trust (Castelfranchi and Pedone, 2000; Mayer, Davis, and Schoorman, 1995; Zucker, 1986). Trust of electronic services has been explored extensively in both e-commerce (Gefen and Straub, 2002; Gefen, Karahanna and Straub, 2003; Jarvenpaa, Knoll, and Leidner, 2000; McKnight, Choudhury, and Kacmar, 2002; Pavlou, 2003; Tan and Theon, 2001; Van Slyke, Belangerand Comunale, 2004) and e-government (Carter and Bélanger, 2005; Gefen et al. 2005; Welch, Hinnant and Moon, 2004; Warkentin and Gefen, 2002).

Citizens must possess trust in the entity providing the electronic service. E-file acceptance depends on the belief that e-file service providers are capable of providing electronic services effectively and confidentially. In e-commerce research, this concept is frequently referred to as the firm's reputation. Reputation effects the extent to which buyers believe an organization is honest and concerned about its customers (Jarvenpaa et al., 2000). Regarding e-government, citizens will be more likely to use Internet services provided by organizations with a good reputation. Hence, each citizens' individual level of trust in the e-file provider is an imperative element of e-file diffusion.

Optimism Bias

Prior research has found that despite high perceptions of risk citizens were still willing to adopt e-government services (Belanger and Carter, 2005). Citizens acknowledged the risk of completing electronic transactions; however, they were still willing to use e-government services. Optimism bias may explain this phenomenon. Optimism bias – which has been explored in risk analysis (Sjoberg and Fromm, 2001), accident analysis (Dejoy, 1989), social psychology (Weinstein, 1980a), and behavioral medicine (Weinstein, 1980b) - offers insight into how individuals handle risky situations. Prior literature has defined optimism bias as "a systematic error in perception of an individual's own standing relative to group averages, in which negative events are seen as less likely to occur to the individual than average compared with the group, and positive events as more likely to occur than average compared with the group (Weinstein, 1980b)." This concept suggests that although people identify situations as risky; they do not think they are as susceptible to the risk as the average person. Many people believe that their knowledge and abilities minimize their susceptibility to risk. For instance, optimism bias has been used to explain why some individuals drive under fatigue (Dejoy, 1989). Research shows that most people agree it is dangerous to drive when tired; however, the same people believe their driving abilities make them less likely to have an accident while fatigued than the average driver.

Regarding information technology (IT), previous studies asked respondents to rate the risk of certain events (e.g. identity theft) happening to the general public and to them individually (Sjoberg and Fromm, 2001). Participants scored the risk to the general public higher (Sjoberg and Fromm, 2001). The authors found that participants were aware of technology related risks. However, risks of technology were mostly seen as issues of concern for other people. In terms of e-file adoption, citizens who believe they are more competent than the average Internet user may not be deterred by the perceived risk of e-government transactions. Schaupp and Carter (2008) introduce optimism bias to the e-government literature by exploring it in conjunction with trust and risk. In this study, we examine the impact of optimism bias on use intentions when adoption factors are present.

Optimism bias is tested by asking participants to compare their ability to perform a given task to the ability of the average citizen. We adapted survey items to assess e-filing optimism bias. Participants in this study were asked to rate their ability to perform several tasks (download forms from the www.irs.gov website, complete a transaction with the IRS using an e-file system, etc.) compared to the average Internet user's ability. Citizens who believe they are more Internet savvy than the average user will be more likely to use an e-file service.

RESEARCH MODEL & HYPOTHESES

Based on the aforementioned literature, we propose the following research model (see Figure 1). Intention to use an e-file system is influenced by three technology acceptance factors – effort expectancy, performance expectancy and social influence – and three personal factors – trust of the e-filer, Web specific self-efficacy and optimism bias.

The research hypotheses are presented in Table 1.

Figure 1. Proposed e-file adoption model

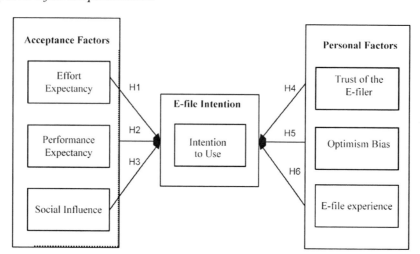

METHODOLOGY

Data was collected via an online survey. The literature states that surveys are a viable means for collecting data (Pedhazur and Schmelkin, 1991). To obtain study participants, an e-mail announcement was sent to MBA students, masters' level and upper level accounting students. The results were analyzed using multiple linear regression in SPSS 15.0.

Sample

The survey was completed by 260 MBA, upper level and graduate accounting students. 53% of the participants were female. The sample's age ranged from 18 – 54; 83% of the sample was in the 18-24 age group. 89% were Caucasians. 93% have completed an e-commerce transaction and 71% have completed an e-government transaction. 34% of the respondents used an e-file system last year.

Instrument Development

Questions were adapted from validated instruments (Carter and Belanger 2005, Fu et al. 2006, Pavlou 2003). Wording was modified to fit the e-filing context. The resulting items for each construct were then included in random order on the survey instrument. Questions were measured on a 7-point Likert-type scale, ranging form 1 (strongly disagree) to 7 (strongly agree). The instrument is available from the authors upon request.

Table 1. Research hypotheses

No.	Hypothesis
H1.	Effort Expectancy (EE) will have a positive effect on intention to use.
H2.	Performance Expectancy (EE) will have a positive effect on intention to use.
H3.	Social Influence (SI) will have a positive effect on intention to use.
H4.	Trust of the E-filer (TOE) will have a positive effect on intention to use.
H5.	Optimism Bias (OB) will have a positive effect on intention to use.
H6.	E-file use in the previous year (LSYR) will have a positive effect on intention to use.

Items were tested for reliability using Chronbach's alpha. The reliability analysis is presented in Table 2.

Factor analysis was conducted using principal component analysis with promax rotation. As illustrated in Table 3, items loaded on their respective factors.

DATA ANALYSIS

The research model was tested using multiple linear regression analysis. Regression analysis is used to relate a dependent variable to a set of independent variables. The goal of this study is to determine the relationship between use intentions (dependent variable) and citizens' perceptions of electronic filing systems (independent variables). The model includes six independent variables (effort expectancy, performance expectancy, social influence, trust of the e-filer, web-specific self-efficacy and previous use of an e-file system) and one dependent variable (intention to use).

RESULTS

The model explains a large percent of the variance in citizen adoption of e-filing systems; adjusted R Square equals.759. Since the overall model was significant (F=385.549 p=.000), we tested the significance of each variable. Four of the six hypotheses were supported. Performance expec-

tancy, social influence, trust of the e-filer and web-specific self-efficacy all have a significant impact on intention to e-file (see Table 4). Interestingly, effort expectancy and previous use of an e-file system did not increase one's intention to use an e-file system. On the contrary, those who e-filed last year were less likely to e-file in the future. Implications for practice and research are provided in the discussion section.

Table 3. Factor analysis

Item	EE	PE	SI	TOE	OB	USE
EE1	.846					
EE2	.806					
EE3	.673					
EE4	.707					
EE5	.758					
EE6	.464					
PE1		.745				
PE2		.650				
PE3		.724				
PE4		.600				
SI1			.844			
SI2			.816			
SI3			.623			
SI4			.713			
TOE1				.864		
TOE2				.855		
TOE3				.768		
TOE4				.853		
TOE5				.831		
OB1					.910	
OB2					.770	
OB3					.750	
USE1						.703
USE2						.768
USE3						.702
USE4						.758
USE5						.654

Table 2. Reliability analysis

Construct	# Items	Reliability
Effort Expectancy (EE)	6	.844
Performance Expectancy (PE)	4	.764
Social Influence (SI)	4	.783
Trust of the E-filer (TOE)	5	.897
Optimism Bias (OB)	3	.850
Intention to Use (USE)	5	.892

The resulting model is presented in (Figure 2). Two acceptance factors and two personal factors are significant.

DISCUSSION

Significant Results

Two of the three acceptance factors are significant: performance expectancy and social influence. Citizens who believe an electronic option will help them file their taxes more quickly and efficiently than traditional alternatives are more likely to adopt e-file systems. Hence, government agencies need to emphasize the benefits and advantages of e-file relative to paper-based and telephone alternatives. Regarding social influence, citizens with peers, mentors, bosses, etc. that use e-file are more likely to use e-file as well. Government agencies need to start grass root initiatives that get adopters to encourage their friends to give e-filing a try.

In addition to the acceptance factors, two personal factors were also significant: trust of the e-filer and web-specific self-efficacy. Trust is imperative when risk is present (Mayor et al., 1991). Citizen trust in the ability and integrity of the e-file provider is an important element of adoption. Hence, the IRS needs to make sure that the companies that it endorses (such as TurboTax) are trustworthy. If the e-filer is unreliable, citizens will be unwilling to use IRS endorsed e-file systems.

Prior studies have shown that web-specific self-efficacy is an important determinant of behavioral intention (Agarwal, 2000; Hsu and Chiu, 2004; Maraka et al., 1998). Specifically, WSE has been shown to have a significant direct effect on e-service usage (Hsu and Chiu 2004). It is for these reasons that WSE has been chosen for inclusion in the model predicting e-file intentions to use.

Table 4. Hypotheses testing

Hypothesis	Coefficient	t-value	Significance	Supported
H1(EE)	-.200	-0.531	.596	NO
H2 (PE)	.533	9.808	.000	YES
H3 (SI)	.246	6.211	.000	YES
H4 (TOE)	.273	5.118	.000	YES
H5 (OB)	.172	3.514	.001	YES
H6 (LSYR)	-.433	-4.651	.000	NO*
*Although the p-value is significant, the hypothesis is not supported since the coefficient is negative.				

Figure 2. Significant results

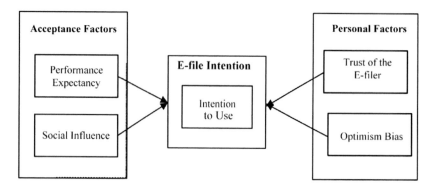

Non-Significant Results

Surprisingly, neither effort expectancy nor use of e-file last year had a positive impact on intention to use and e-file system. The significance of effort expectancy is inconsistent in the literature. Several studies suggest that performance expectancy is the most salient adoption factor (Benbasat and Barki, 2007). Venkatesh et al. (2003) posit that effort expectancy is only important at certain stages of adoption (the initial stages).

Even more surprising were the effects of e-filing in a previous year on intention to use an e-file system in the future. The results indicate that citizens who e-filed last year were less likely to e-file in the future. Perhaps, citizens are not satisfied with the current electronic options. Perhaps citizens have found that the benefits do not outweigh the costs (In the U.S. if your annual income is greater than $50,000 you have to pay a fee to e-file). Or perhaps, these results are a function of the sample. As stated earlier, the sample is composed of graduate and undergraduate business students who have a lot of confidence and experience with Internet systems. The sample's demographics are not representative of the population at large. Future studies should continue to evaluate the relationship between previous use and future intentions.

Implications for Research and Practice

The findings of the present study have various implications for practice as well as research. The present study confirmed that WSE had a significant effect on intention. This study provides an initial step toward the application of WSE to the study of citizens' intention to e-file. Our research confirms that WSE is a meaningful construct within the context of e-filing. The results indicate that citizens' with higher WSE are more likely to e-file. This implies that increasing citizens' WSE is critical to the success of e-file adoption. This study also serves as a bridge extending the e-services research into the specific domain of e-filing.

The relevance of optimism bias to e-service adoption raises many interesting research questions. For instance, does optimism bias impact trust or risk perceptions? Is there a relationship between optimism bias and self-efficacy? Is optimism bias more prevalent for certain users (beginner vs. skilled), genders, cultures or age groups? This study presents optimism bias as an important element of e-government diffusion. The findings highlight the need for further exploration of its influence on the adoption decision.

The implications of this study are also very relevant to the citizen e-filers themselves. This study has shown that in fact the effects of optimism bias do in deed exist presently with current and potential citizen e-filers. Citizens need to be aware of this bias and take the appropriate measures to ensure that indeed the e-file provider is a trustworthy vendor and that the appropriate controls are in place to ensure the privacy of their sensitive personal income tax records.

LIMITATIONS & SUGGESTIONS FOR FUTURE RESEARCH

There are a few limitations to this study that should be noted. The most notable is the diversity of the subjects that were evaluated. The sample was composed of graduate and undergraduate students. Previous research suggests that students have a higher affinity towards and access to technology than the average citizen. While valid results were produced from testing, there was limited diversity in the sample. Future research should attempt to validate the findings of this study by testing a more diverse array of participants to increase the variance on some variable dimensions. Another limitation is that the data for this study was collected through surveys, therefore allowing a potential of self-report bias from respondents. The survey was administered online which may also bias the results by capturing the views of those who may be more knowledgeable and comfortable with technology than the average citizen. Future research should consider using multiple-methods to collect and analyze data to test the proposed model.

CONCLUSION

In conclusion, this study presents a comprehensive yet parsimonious view of e-file adoption. Information Communication Technology (ICT) adoption is an important element of IS literature. This study highlights the acceptance factors and personal factors that impact the adoption of an emerging ICT: electronic tax filing. It integrates constructs from adoption, trust and self-efficacy literature to explain over seventy-five percent of the variance in intention to use an e-file system. This study uses very specific constructs such as Trust of the e-filer instead of a generic trust concept and web-specific self-efficacy instead of a general self-efficacy construct. The tailored e-file model can serve as a building block for future studies of e-file adoption. The constructs in the model are also applicable to other e-government systems, such as online license renewal. The proposed model adds to the current discourse on the evolution of e-government by presenting a very focused yet explanatory model of e-file utilization.

REFERENCES

Agarwal, R., & Prasad, J. (1999). Are individual differences germane to the acceptance of new information technologies. *Decision Sciences*, *30*(2), 361–391. doi:10.1111/j.1540-5915.1999. tb01614.x

Agarwal, R., Sambamurthy, V., & Stair, R. (2000). The evolving relationship between general and specific computer self-efficacy: An empirical investigation. *Information Systems Research*, *11*(4), 418–430. doi:10.1287/isre.11.4.418.11876

Bandura, A. (1982). Self-efficacy mechanism in human agency. *The American Psychologist*, *37*(2), 122–147. doi:10.1037/0003-066X.37.2.122

Bandura, A. (1997). *Self-Efficacy: The Exercise of Control*. New York: Freeman.

Becerra, M., & Gupta, A. (1999). Trust Within the Organization: Integrating the Trust Literature with Agency Theory and Transaction Costs Economics. *Public Administration Quarterly*, *23*(2), 177–203.

Bélanger, F., & Carter, L. (2008). Trust and Risk in E-government Adoption. *The Journal of Strategic Information Systems*, *17*(2), 165–178. doi:10.1016/j.jsis.2007.12.002

Bélanger, F., & Hiller, J. (2006). A Framework for E-Government: Privacy Implications. *Business Process Management Journal*, *12*(1), 48–60. doi:10.1108/14637150610643751

Benbasat, I., & Barki, H. (2007). Quo Vadis TAM. *Journal of the Association for Information Systems*, *8*(4), 211–218.

Burn, J., & Robins, G. (2003). Moving Towards E-government: A Case Study of Organizational Change Processes. *Logistics Information Management*, *16*(1), 25–35. doi:10.1108/09576050310453714

Carter, L., & Belanger, F. (2005). The Utilization of E-government Services: Citizen Trust, Innovation and Acceptance Factors. *Information Systems Journal*, *15*(1), 2–25. doi:10.1111/j.1365-2575.2005.00183.x

Chau, Y. K., & Hu, J. H. (2001). Information technology acceptance by individual professionals: a model comparison approach. *Decision Sciences*, *32*(4), 699–718. doi:10.1111/j.1540-5915.2001. tb00978.x

Choudrie, J., & Dwivedi, Y. (2005). *A Survey of Citizens Adoption and Awareness of E-Government Initiatives, The Government Gateway: A United Kingdom Perspective. E-Government Workshop*. Brunel University, West London.

Compeau, D., Higgins, C. A., & Huff, S. (1999). Social Cognitive Theory and Individual Reactions to Computing Technology: A Longitudinal Study. *Management Information Systems Quarterly, 23*(2), 145–158. doi:10.2307/249749

Compeau, D. R., & Higgins, C. A. (1995). Computer Self-Efficacy: Development of a Measure and Initial Test. *Management Information Systems Quarterly, 19*(2), 189–211. doi:10.2307/249688

Cross, M. (2007). £5m e-government awareness campaign flops. *The Guardian.* http://www.guardian.co.uk/technology/2006/oct/12/marketingandpr.newmedia

Davis, F. D. (1989). Perceived usefulness, perceived ease of use, and user acceptance of information technology. *Management Information Systems Quarterly, 13*(3), 319–340. doi:10.2307/249008

Dejoy, D. (1989). The Optimism Bias and Traffic Accident Risk Perception. *Accident; Analysis and Prevention, 21*(4), 333–340. doi:10.1016/0001-4575(89)90024-9

Dwivedi, Y., Papazafeiropoulou, A., & Gharavi, H. (2006). Socio-Economic Determinants of Adoption of the Government Gateway Initiative in the UK. *Electronic Government, 3*(4), 404–419. doi:10.1504/EG.2006.010801

Eastin, M. A., & LaRose, R. L. (2000). Internet self-efficacy and the psychology of the digital divide. *Journal of Computer-Mediated Communication, 6*(1).

Evans, K. (2006, December). *Expanding E-government: Making a Difference for the American People Using Information Technology.* Executive Office of the President, Office of Management and Budget.

Fu, J.-R., Farn, C.-K., & Chao, W.-P. (2006). Acceptance of electronic tax filing: A study of taxpayer intentions. *Information & Management, 43*(1), 109–126. doi:10.1016/j.im.2005.04.001

Ganesan, S., & Hess, R. (1997). Dimensions and Levels of Trust: Implications for Commitment to a Relationship. *Marketing Letters, 8*(4), 439–448. doi:10.1023/A:1007955514781

Gefen, D. (2000). E-commerce: The Role Of Familiarity and Trust. *Omega: The International Journal of Management Science, 28*(6), 725–737. doi:10.1016/S0305-0483(00)00021-9

Gefen, D., Karahanna, E., & Straub, D. W. (2003). Inexperience and experience with online stores: The importance of TAM and trust. *IEEE Transactions on Engineering Management, 50*(3), 307–321. doi:10.1109/TEM.2003.817277

Gefen, D., Rose, G., Warkentin, M., & Pavlou, P. (2005). Cultural Diversity and Trust in IT Adoption: A Comparison of USA and South African e-Voters. *Journal of Global Information Management, 13*(1), 54–78.

Gilbert, D., Balestrini, P., & Littleboy, D. (2004). Barriers and Benefits in the Adopiton of E-government. *International Journal of Public Sector Management, 17*(4/5), 286–301. doi:10.1108/09513550410539794

Hackney, R., & Jones, S. (2002, April). Towards E-government in the Welsh (UK) Assembly: an Information Systems Evaluation. *Proceedings of the ISOneWorld Conference and Convention,* Las Vegas, USA.

Hart, P., & Saunders, C. (1997). Power and Trust: Critical factors in the adoption and use of electronic data interchange. *Organization Science, 8*(1), 23–42. doi:10.1287/orsc.8.1.23

Holsapple, C. W., & Sasidharan, S. (2005). The dynamics of trust in B2C e-commerce: a research model and agenda. *Information System E-Business Management, 3*(4), 377–403. doi:10.1007/s10257-005-0022-5

Hsu, M., & Chiu, C. (2004). Internet self-efficacy and electronic service acceptance. *Decision Support Systems, 38*(3), 369–381. doi:10.1016/j.dss.2003.08.001

Huang, Z. (2007). A comprehensive analysis of U.S. counties' e-Government portals: development status and functionalities. *European Journal of Information Systems, 16*(2), 149–164. doi:10.1057/palgrave.ejis.3000675

IRS. (2004). *IRS e-Strategy for Growth.* http://www.irs.gov/pub/irs-pdf/p3187.pdf

Jarvenpaa, S. L., & Tractinsky, N. (2000). Consumer Trust in an Internet Store. *Information Technology Management, 1*(1-2), 45–70. doi:10.1023/A:1019104520776

Jeyaraj, A., Rottman, J., & Lacity, M. (2006). A review of the predictors, linkages, and biases in IT innovation adoption research. *Journal of Information Technology, 21*(1), 1–23. doi:10.1057/palgrave.jit.2000056

Lee, M. K. O., & Turban, E. (2001). A Trust Model for Consumer Internet Shopping. *International Journal of Electronic Commerce, 6*(1), 75–91.

Marakas, G. M., Yi, M. Y., & Johnson, R. D. (1998). The multilevel and multifaceted character of computer self-efficacy: Toward clarification of the construct and an integrative framework for research. *Information Systems Research, 9*(2), 126–163. doi:10.1287/isre.9.2.126

Mayer, R. C., Davis, J. H., & Schoorman, F. D. (1995). An integrative model of organizational trust. *Academy of Management Review, 20*(3), 709–734. doi:10.2307/258792

McKnight, D. H., Choudhury, V., & Kacmar, C. (2002). Developing and Validating Trust Measures for E-Commerce: An Integrative Approach. *Information Systems Research, 13*(3), 334–359. doi:10.1287/isre.13.3.334.81

McKnight, D. H., Cummings, L. L., & Chervany, N. L. (1998). Initial Trust Formation in New Organizational Relationships. *Academy of Management Review, 23*(3), 473–490. doi:10.2307/259290

Moore, G., & Benbasat, I. (1991). Development of an instrument to measure the perceptions of adopting an information technology innovation. *Information Systems Research, 2*(3), 192–222. doi:10.1287/isre.2.3.192

Pavlou, P. (2003). Consumer Acceptance of Electronic Commerce: Integrating Trust and Risk with the Technology Acceptance Model. *International Journal of Electronic Commerce, 7*(3), 69–103.

Pedhazur, E. J., & Schmelkin, L. P. (1991). *Measurement, Design, and Analysis: An Integrated Approach.* Hillsdale, NJ: Lawrence Erlbaum Associates, Inc.

Plouffe, C. R., Hulland, S. J., & Vandenbosch, M. (2001). Research report: Richness versus parsimony in modeling technology adoption decisions--Understanding merchant adoption of a smart card-based payment system. *Information Systems Research, 12*(2), 208–222. doi:10.1287/isre.12.2.208.9697

Schaupp, L., & Carter, L. (2008). The impact of trust, risk and optimism bias on E-file adoption. *Information Systems Frontiers.* doi:.doi:10.1007/s10796-008-9138-8

Shapiro, S. P. (1987). The social control of impersonal trust. *American Journal of Sociology, 93*(3), 623–658. doi:10.1086/228791

Sjoberg, L., & Fromm, J. (2001). Information Technology Risks as Seen by the Public. *Risk Analysis, 21*(3), 427–441. doi:10.1111/0272-4332.213123

Szajna, B. (1996). Empirical evaluation of the revised technology acceptance model. *Management Science, 42*(1), 85–92. doi:10.1287/mnsc.42.1.85

Thomas, J. C., & Streib, G. (2003). The new face of government: Citizen-initiated contacts in the era of E-government. *Journal of Public Administration: Research and Theory, 13*(1), 83–102. doi:10.1093/jpart/mug010

Thompson, R., Higgins, C., & Howell, J. (1991). Personal computing: toward a conceptual model of utilization. *Management Information Systems Quarterly, 15*(1), 124–143. doi:10.2307/249443

Thompson, R., Higgins, C., & Howell, J. (1994). Influence of experience on personal computer utilization: Testing a conceptual model. *Journal of Management Information Systems, 1*(11), 167–187.

Thong, J., Hong, W., & Tam, K. (2004). What Leads to User Acceptance of Digital Libraries? *Communications of the ACM, 47*(11), 78–83. doi:10.1145/1029496.1029498

Torkzadeh, G., Chang, J., & Demirhan, D. (2006). A contingency model of computer and Internet self-efficacy. *Information & Management, 43*(4), 541–550. doi:10.1016/j.im.2006.02.001

Van Slyke, C., Belanger, F., & Comunale, C. L. (2004). Factors Influencing the Adoption of Web-Based Shopping: The Impact of Trust. *The Data Base for Advances in Information Systems, 35*(2), 32–49.

Venkatesh, V., & Davis, F. D. (2000). Theoretical extension of the technology acceptance model: Four longitudinal field studies. *Management Science, 46*(2), 186–204. doi:10.1287/mnsc.46.2.186.11926

Venkatesh, V., Morris, M. G., & Davis, G. B. (2003). User Acceptance of Information Technology: Toward a Unified View. *Management Information Systems Quarterly, 27*(3), 425–478.

Wang, Y., & Emurian, H. (2005). Trust in E-Commerce: Consideration of Interface Design Factors. *Journal of Electronic Commerce in Organizations, 3*(4), 42–60.

Wang, Y., Lin, H., & Luarn, P. (2006). Predicting consumer intention to use mobile service. *Information Systems Journal, 16*(2), 157–179. doi:10.1111/j.1365-2575.2006.00213.x

Warkentin, M., & Gefen, D. (2002). Encouraging citizen adoption of e-government by building trust. *Electronic Markets, 12*(3), 157–162. doi:10.1080/101967802320245929

Weinstein, N. D. (1980a). Optimistic Bias about personal risks. *Science, 246*, 1232–1233. doi:10.1126/science.2686031

Weinstein, N. D. (1980b). Unrealistic optimism about future life events. *Journal of Personality and Social Psychology, 39*, 806–820. doi:10.1037/0022-3514.39.5.806

Welch, E. W., Hinnant, C. C., & Moon, M. J. (2005). Linking Citizen Satisfaction with E-Government and Trust in Government. *Journal of Public Administration: Research and Theory, 15*(3), 371–391. doi:10.1093/jopart/mui021

Zucker, L. G. (1986). Production of trust: Institutional sources of economic structure. *Research in Organizational Behavior, 8*(1), 53–111.

APPENDIX: E-FILE ADOPTION ITEMS

Table 5.

Effort Expectancy Learning to use an Internet tax-filing method would be easy for me. I would find an e-file system easy to use. It would not be easy for me to become skillful at using an e-file system. (REVERSE CODED) It would be easy for me to input and modify data when I use an e-file system Instructions for using an e-file system will be easy to follow. Using an e-file system would make filing my taxes clearer and more understandable.
Performance Expectancy Internet Tax filing will be of no benefit to me (Reverse Coded) Using Internet Tax filing will speed the tax filing process The advantages of Internet tax filing will outweigh the disadvantages. Overall, using Internet tax filing will be advantageous.
Social Influence People who influence my behavior think that I should use an e-file system. People who are important to me think that I should use an e-file system. I use an e-file system to file my taxes because of the number of people around me who use it also. People around me who use the e-file system to file their taxes have more prestige.
Trust of the E-filer I think I can trust TurboTax. TurboTax can be trusted to carry out online transactions faithfully. In my opinion, TurboTax is trustworthy. I think TurboTax will keep my electronic information private and secure. I believe TurboTax is a reliable e-filing system.
Optimism Bias Please rate your ability to perform the following tasks compared to the average Internet user: Submit personal information to the IRS using an e-file system. Complete a transaction with the IRS using an e-file system. Download forms and documents from the www.irs.gov website.
Use I predict that I will use an e-file system in the future. Filing taxes via e-file is something that I would do. I would use the Internet to file my taxes. I will experiment with an e-file service and then decide whether or not to use it in the future. I intended to use an Internet filing method for my income tax return next year.

This work was previously published in International Journal of Electronic Government Research (IJEGR), edited by Vishanth Weerakkody, pp. 62-74, copyright 2009 by Information Science Reference (an imprint of IGI Global)

Chapter 16

The Key Organisational Issues Affecting E-Government Adoption in Saudi Arabia

Abdullah AL Shehry
Prince Nayef College, Saudi Arabia

Simon Rogerson
De Montfort University, UK

N. Ben Fairweather
De Montfort University, UK

Mary Prior
De Montfort University, UK

ABSTRACT

The e-government paradigm refers to utilizing the potential of Information and Communication Technology (ICT) in the whole government body to meet citizens' expectations via multiple channels. It is, therefore, a radical change within the public sector and in the relationship between a government and its stakeholders. In the light of that, the Kingdom of Saudi Arabia has a keen interest in this issue and thus it has developed a national project to implement e-government systems. However, many technological, managerial, and organisational issues must be considered and treated carefully before and after going online. Based on an empirical study, this article highlights the key organisational issues that affect e-government adoption in the Kingdom of Saudi Arabia at both national and agency levels.

INTRODUCTION

Over the past two decades, a series of initiatives to transform government processes has led to the development of concepts such as "reinventing the government" and "new public management"

(NPM) to address shortcomings in public administration and make government "work better" (Kettl, 2005). In addition, ICT and Globalisation has encouraged the idea of public sector reform and had a significant influence on the relationships people keep with other individuals, with the business community and, more recently with

DOI: 10.4018/978-1-60960-162-1.ch016

the government (Castells, 1996; Howard, 2001). The adoption of e-government is the decision to utilize electronic services to share information with other government agencies and provide services to stakeholders and to make full use of information and communications technology as the best course of action available (AL-Shehry et al., 2006). E-government is more than a technological issue as it is influenced by many factors such as organisational, human, social and cultural issues which are related to the nature of the government in a particular country and its responsibility in the society (Prins, 2001; Howard, 2001). Thus, this article focuses more on identifying the elements that will promote the successful introduction of e-government in a developing country context, rather than assessing the success or failure of an implemented e-government system in a developed country environment from an organisational perspective.

THEORETICAL BACKGROUND

Many theories have attempted to explain the acceptance of technology such as Theory for Reason Action (TRA), Technology Acceptance Model (TAM) and Diffusion of Innovation Theory. For example TAM uses two perceptions: perceived ease of use and usefulness to determine an individual's intention to use a technology (Davis et al., 1992). Igbaria et al (1997) found that perceived ease of use is a major factor in explaining perceived usefulness and system usage. Molla and Licher (2005) identified the organisational factors that might affect e-commerce adoption in South Africa which are human, business, technology resources and awareness. Wang (2003) examined the factors affecting the adoption of electronic tax-filing systems by using the Technology Acceptance Model (TAM). Tung and Rieck (2005) indicated a significant positive relationship be-

tween perceived benefits, external pressure and social influence, and the firm's decision to adopt e-government services. Kim and Lee (2004) suggested that organisational culture, structure and Information Technology all exert significant forces on knowledge-sharing capabilities among South Korean government employees.

Carter and Belanger (2004) indicated that perceived usefulness, relative advantage and compatibility are significant indicators of citizens' intention to use the state government's services online. Al-Dosari and King (2004) have developed a framework to deal with e-government implementation stages at the national level. Ebrahim and Irani's study (2005) identified the benefits and the barriers for e-government adoption. Gil-Garcia and Pardo (2005) identified some challenges in e-government such as: project size, complexity, organisational diversity and the lack of alignment between organisational goals and IT projects. Al-Fakhri et al (2008) indicated to the importance of increasing the awareness among government employees and the public at-large during e-government implementation; making Internet access more available across the full spectrum of society and adopting a flexible approach to technological change and the IT environment more generally. Amoretti (2007) pointed to a four way typology of e-government regimes which are: reform-oriented e-government; authoritarian e-government; managerial e-government and open e-government. However, the findings of several studies indicate that despite high costs of e-government projects, both tangible and intangible, many e-government efforts are failing or are slowly diffusing (United Nations 2001; Pardo and Scholl 2002; Heeks 2003; OCED, 2004; Dawes et al., 2004). In addition, previous studies pay little attention to the organisational aspects of this transformation, particularly, the interrelation between current reality of government administration and implementation of e-government.

Organisational Issues and IS Adoption

The term 'organisational issue' is frequently used in the literature to describe a wide range of non-technical aspects (Clegg et al., 1997). Thus, it is defined as "any distinct area on the interface between a technical system and the characteristics and requirements either of the host organisation or its individual employees which can lead to operational problems within the organisation" (Doherty and King, 1998: l05). A report conducted by OASIG (1996), cited in Cabrara et al. (2001) summarized the experiences of 45 leading Information Technology (IT) researchers and consultants in the UK and indicated that 80-90% of IT projects fail to meet their performance goals. (Table 1)

This is in part due to the fact that organisations give inadequate attention to non-technical details. Heeks (2006) mentioned that many information systems (IS) in developing countries can be categorized as failing either totally or partially. Many researchers (such as Ewusi-Mensah and Prazasnyski, 1994; Flowers 1996; Whyte and Bytheway 1996 and Gouliemmos, 2005) have indicated a number of different reasons for the failure of IS projects which summarized in Table (1).

In developing IS many researchers have examined a variety of factors affecting technology adoption. For instance, Weber and Schweiger (1992) observed that resistance is caused by a culture-conflict between the presumed organisational culture and the actual organisational culture; this hinders the implementation of the technology. The organisational culture has the potential to affect the acceptance of technology (Viswanath et al, 2003). This article, therefore, explores the impact of organisational issues on the development of e-government systems in developing countries and how we could treat them in practice.

Case Study Background (Kingdom of Saudi Arabia)

The official name of the country is The Kingdom of Saudi Arabia (KSA). Even so, internationally, it is widely called Saudi Arabia. It is a monarchy as the nation is ruled by a royal family although there were some municipal council elections late in 2004 at a local level. Table 2 shows some facts about Saudi Arabia's profile.

The economy of Saudi Arabia is oil-based. This is because the Kingdom has the largest reserves of petroleum in the world (25% of the proved reserves), ranks as the largest exporter of petroleum, and plays a leading role in OPEC

Table 1. Reasons for IS failure

Organisational issues	Technological issues	Other issues
Lack of senior management support	Lack of technical knowledge and expertise	Budget and time overruns
Behavioural problems / resistance to change	Technological inadequacies and shortcomings	Lack of user participation
Inadequate project management	Satisfying existing or emerging technology	Over-ambitious scope
Organisational problems (culture, structure etc)	Technology focused development	Poor communication between users and development staff
Inappropriate implementation strategy	Inexperienced staff to develop systems	Political pressures, influential outsiders, external and internal power struggles
Staff turnover and competency		Under-estimating complexity
Poor consultation and poor user training		

Table 2. Saudi Arabia's National Profile (MCIT, 2007)

Population	Population: 24M (74% Saudis); Annual growth (GACR) 2.5% (26M+ in 2010); 51% less than 20 years of age
Area	2,240,000 square kilometres
Economy	Natural resources: petroleum, natural gas, iron ore, gold and copper ; GDP(Gross Domestic Product) > 370 B $; Heavy investment in creating a leading ICT economy
Gov. organisations	350
ICT Indicators	Penetration rate in KSA by end of 2006: Fixed Lines 16.4% (4 M lines); Mobile 81.7% (19 M Cellular); Internet ~ 19.6% (4.1 M users) and Broadband ~1%

(World Fact Book, 2005). Saudi Arabia has occupied a special place in the Islamic world because it is the homeland of the Prophet Muhammad and the two Holy Mosques for Muslims (Al-Farsy, 2003). Saudi Arabia's culture is very religious and Islam plays a central role in defining this culture, acting as a major force in determining the social norms, patterns, traditions, obligations, privileges and practices of society (Al-Saggaf, 2004). Saudi Arabia has conserved, though in a new form, many values of Arab and Islamic civilization and the traditional system of power and government while, at the same time, adopting Western technology, a market economy, a modern state education system, and health-care and other public sector services (Vassiliev, 2000).

Adoption of technology was given top priority from the government of the Kingdom of Saudi Arabia (KSA). For instance, IT applications have spread rapidly to cover many sectors to enhance productivity in the fields of finance, industry, commerce, education, government and health care (Al-Tawil et al. 2003; Al-Turki & Tang, 1998 and Al-Sudairy, 2000). The utilization of IT systems in some sectors in KSA, namely in the banking, oil and petrochemical sectors, is considered to be among the most advanced in the world (Abdul-Gader & Alangari, 1994). However, IT diffusion in a country like Saudi Arabia is a very complex process and is often associated with many problems such as cultural, educational, economic and social issues.

E-government initiatives were launched as part of the country's overall information technology plans in 2003 and this focused on ICT as a tool for reforming public organisations (Saudi Computer Society, 2004). In the light of this, in 2005, the Saudi government created "Yesser", the e-government programme designed to achieve continuous growth and development within government (MCIT, 2004). The main objective for this project (Yesser) is to enable and facilitate e-government implementation by supporting governmental organisations in terms of methodologies, data, standards and knowledge (Al-Sabti, 2005). Among organisations in the public sector, there are many projects that have implemented e-government activities in a variety of ways. However, these efforts are at the initial stage and are not, as yet, working together. Thus, the new project, Yesser, aims to make these projects work co-operatively (AL-Shehry et al, 2006).

Research Methodology

This study aims to identify the key challenges that effect e-government adoption within the public sector in Saudi Arabia. The qualitative approach is appropriate for this kind of study (Yin, 2003). Under the qualitative umbrella the case–study is a well known strategy for an exploratory study (Stake, 1995 and Lam, 2005). The case study involved 22 semi-structured interviews; a focus group; observations; 12 electronic reports; analysis of 22 documents; 12 websites and 19 newspaper

Figure 1. Organisational issues affect e-government adoption

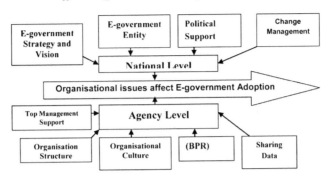

articles and press releases. Piloting revealed the need for interviews to last at least an hour and that interviewees were likely to dislike tape-recording. Thus, during the interviews, the researcher took notes and wrote down the interviewees' answers. The interviewees included stakeholders such as senior officials from the public sector; independent experts from the private sector; academics; the e-government project team and a number of IT managers from government ministries.

Data analysis means the separation of something into its component parts, which involves examining, categorising and tabulating the collected data with the aim of finding answers to the research questions (Yin, 2003; Denscombe 2007). Data analysis techniques, such as content analysis, cross-interview analysis, the interview guide approach, coding and classifications were manually applied in this study. The research followed broadly the three streams (i.e. data reduction, data display and data conclusion-drawing) that were suggested by Miles and Huberman (1994: 10-12). The data were coded according to the themes identified by the initial framework and the interview guide. Any new themes were separated, classified and added to their respective classes within specific categories, or sometimes by creating new classes within the same categories. Analytical techniques involved the classifications and categorisation of the data, noting regularities and patterns, deriving explanations, and reviewing and rechecking findings. Aldosri (2005) pointed

out that it was important that such processes were regularly related to the initial framework and, at the same time, enhanced the revision of the framework in the light of the analysed data.

E-GOVERNMENT ADOPTION AT NATIONAL LEVEL

The case study findings present the key organisational issues that influence e-government adoption at early stages of development. These essential factors need treatment before and after going online, as Figure 1 illustrates and the following sections explain.

E-Government Strategy and Vision

Developing a national e-government strategy is the cornerstone of adopting an e-government system. This strategy needs to offer a wide view of a multi-faceted system in order to reflect the needs of stakeholders. Also, the experiences of other countries must be considered in order to avoid mistakes and learn from their experience. At the same time, the conditions and circumstances of a particular country must be considered. Case study findings pointed out that Saudi Arabia has made good progress in drawing up its strategy, starting with putting in place a detailed national ICT plan, creating a specific Ministry dedicated to ICT, and having already established a dedi-

261

cated e-government facilitative body (Yesser). For example one of the e-government team said:

Yesser enables the implementation of e-government services by ministries and other public agencies. This can be achieved by building the national infrastructure and defining common standards; and by providing best practice examples of pilot services.

The vision for Saudi Arabia's e-government initiative points out that, by the end of 2010, e-government services will be accessible to all stakeholders through a variety of electronic means. However, many participants in the study had doubts about achieving this goal within this specific time limit. The alignment of e-government strategy with other government initiatives in the country is an important issue, yet the case study showed a lack of such alignment in the Saudi Arabian project. However, the e-government strategy needs to prioritise initiatives across government agencies; to be citizen-focused and to obtain the support of leadership. The literature review found many studies which outlined the need for an e-government strategy to deliver the idea of e-government as a reality (e.g. OECD, 2004; Heeks, 2006). Other research in the literature points out that a lack of vision and strategy or over-ambitious goals can be problematic in developing electronic services (e.g. United Nations, 2001; Heeks, 2003; Lam, 2005).

E-Government Entity

The e-government entity is one of the key issues raised by participants as an important indication that Saudi government is interested in e-government adoption. For instance one of the e-government team in Saudi Arabia stressed the significance of this issue by stating that:

Now we have an e-government body which is very important to manage this project effectively......

all regulations and standards will be put forward in order to achieve and develop successful e-government.

Forming an e-government entity is a crucial issue in the adoption of e-government in developing countries although this is no guarantee that the project will be successful. This entity (the Yesser programme) is responsible for coordinating and managing all e-government activities at a national level. At the same time, other ministries and public agencies are empowered to continue developing their online services, except those transactional services considered as a core part of the responsibilities of the e-government team. Many participants claimed that the whole project should be coordinated by a joint group and should be linked directly to the Cabinet because there is no follow up for e-government implementation. The literature (e.g. Al-Dosari and King, 2004; Vassilakis et al., 2005) similarly raises this issue as an essential factor that influences e-government systems in developing countries.

Political Support

An e-government system is a national project which requires clear political support. There is no doubt about the importance of this factor for the successful adoption of e-government projects (as noted in OECD, 2004; Heeks, 2006; Moon, 2002). In the case of Saudi Arabia, King Abdullah issued a royal decree backed by the decision of the Council of Ministers on the use of electronic services to improve existing ones, following the lead of other advanced countries. The case study emphasised the importance of following up such political interest on the ground; otherwise many obstacles are likely to slow up progress. The literature pointed out that the lack of support of political leaders will lead to barriers for implementing e-government systems (e.g. United Nations, 2001; Hackney and Jones, 2002; Molla and Licher, 2002; Heeks, 2003; Hahamis and Iles, 2005).

Change Management

The e-government system represents a radical change in the whole government body. At the same time, it is a tool for this change; hence, the most complex part of adopting an e-government system is how to embed and manage the change that will be introduced by that system. The public sector in Saudi Arabia is grounded in routines and patterns that are embedded in the cultural context. One of the e-government team stated

"Change management is a key challenge that faces e-government adoption in Saudi Arabia".

The case study findings show the vital importance of change management in the adoption of an e-government system. The radical changes that will be caused by the transaction and transformation phases may encounter significant resistance from, on the one hand, employees, where costs will be incurred to displace and retrain staff and, on the other, from ordinary citizens who must be satisfied with the e-government services. Many theories and studies explain the negative attitudes towards such a system (e.g. TRA: see Ajzen and Fishbein, 1980; TAM: see Davis, 1989). The case study demonstrates the lack of change management skills among ministries in the Saudi Arabian public sector. This requires the formation of a qualified team to work within each ministry to lead and direct the change initiative. There is much research which emphasises the need for change management to distribute the new idea (e.g. United Nations, 2001; Molla and Licher, 2002; Grant and Chau, 2005).

E-GOVERNMENT ADOPTION AT AGENCY LEVEL

An e-government system is not like other traditional IT projects that can be developed and managed by a single department in a single public organisa-tion. It is a multi-faceted system that takes place across many organisations. Thus, each ministry or organisation has its own project to automate its services in order to join the e-government system on a national project. The participants in this study identified many organisational factors that affect e-government adoption such as top management support, organisational culture, organisational structure, power distribution and reward system.

Top Management Support

E-government provides opportunities for more effective flow back of information to policy makers, assisting them to become better policy makers who are then more highly regarded by their area (UN, 2008). Top management support is one of the key factors that can facilitate the adoption of e-government initiatives in their respective agency. The case study highlighted the importance of creating awareness among top managers about the advantages of e-government while lack of leadership support is equivalent to an early failure of the project. The regulation and the political system's desire have influenced this level, according to the cabinet in Saudi Arabia *"All government organisations must use electronic services"*. One expert in this study commented:

The majority of top managers will follow the e-government system to show that they are of the modern style, rather than delay in this issue, especially after the political system desired movement towards using electronic services.

The literature in the area of e-government also supports the importance of leadership roles (e.g. Molla and Licher, 2002; Heeks, 2003; Hahamis and Iles, 2005).

Organisation Structure

The structure in any organisation plays an important role in defining and distributing responsibilities, coordinating routine activities and establishing channels of communication within the organisation and its context. The adoption of an e-government system at the first stage requires some changes in organisational structure; this becomes more necessary when moving to later stages. On the one hand, this necessitates involving all agency departments in e-government activities by engaging the managers and employees to participate in developing the e-government system. On the other hand, the organisation's hierarchical structure can help in enforcing the adoption of the e-government system. However, the case study indicated the importance of realising that the hierarchical structure does not reflect how decisions are made in the Saudi public sector and therefore leaders should select effective people and departments to be leaders in this initiative, encouraging others to participate in the successful implementation. One IT manager suggested that:

The e-government adoption must be within strategy planning for any target organisation and requiring adoption of the right organisational structure for that is a key for success.

This result is supported in the literature by studies such as those of (Heeks, 2006). Other studies have explained the effect of old structures and processes on the adoption of e-services (e.g. Tambouris, 2001; Hahamis and Iles, 2005).

Organisation Culture

Organisational culture is one of the main factors that were identified by the participants in this study as an important factor influencing the adoption of e-government. The cultural issues need careful treatment during the development of an e-government system at agency level. There are some unique characteristics of this culture in the public sector in SA according to the study findings. The concept of public service or professionalism is ambiguous in the public sector in Saudi Arabia;

many employees believe that the citizen should seek to obtain the service, and generally, the work style can be classified as work-oriented. Another important issue is the concept of time; the value of time does not exist within the mentality of many employees in the public sector in Saudi Arabia. However, many participants agreed that developing e-government system can reduce these problems effectively. Moreover, the way that many managers follow can be described as a bureaucratic rather than a democratic style. They focus on micro-managing everything while not spending any time on the big-picture or strategic thinking. So, the pattern of the decision-making is highly centralised in many of the public organisations. The cultural issues that might be overcome by well planned change management include, for example, retraining the staff and developing an effective reward system (Al-Fakhri et al, 2008).

Sharing Knowledge and Data

An aim of e-government is to make government services available from a one-stop portal at anytime, anywhere. Therefore its adoption requires cooperative efforts from various government agencies and functional units. However, the case study indicates there is a lack of sharing of knowledge and data among public sector agencies in Saudi Arabia. This serious problem threatens the implementation since the sharing of data; information, experience and knowledge, and the focus on citizens' needs as a primary aim are at the heart of the project. One expert's comment on this issue was

The culture of information and knowledge sharing does not exist in the Saudi public sector, so a great effort will be required to overcome this problem.

To overcome such a problem, Yesser aims to establish such collaboration and coordination among organisations in the public sector and that requires new legislation associated with e-government implementation. This will also help remove the fear factor connected with collaboration and will help establish a culture within the organisations that appreciates and promotes the concept of collaboration and information sharing. In the literature, the lack of organisational cooperation among administration departments was pointed out (e.g. Molla and Licher, 2002; Joia 2004; Grant and Chau, 2005; Jioa and Foundation, 2007 and Al-Nuaim, 2009).

Business Processes Reengineering (BPR)

E-government aims to reinvent organisation services and replicate them electronically to reduce the cost and to improve the quality of services. Reengineering work processes is an essential component of e-government adoption for any organisation. Most bureaucratic processes in the public sector are built around the assumption of a hierarchical structure and the existence of a paper trail. The participants in this study emphasised on the need for an effective program within each organisation for BPR associated with top management support in order to adopt e-government successfully. Reengineering work processes is essential prior to and during the implementation of an e-government project. Government agencies in Saudi Arabia need to change their activities to be citizen-orientated as automation alone will not improve results; instead, processes must be aligned to the demands of stakeholders. The case study findings indicated that some progress has been made in these areas in some public agencies in Saudi Arabia with the successful implementation of the smart card and e-Passport facilities. However, much remains to be done as many areas in other agencies still maintain traditional processes. There is much support in the literature for the study's findings. For example, several authors: Kawalek and Wastall (2005) and Barki et al (2007), have empirically shown, through case studies and simulations, that an absence of

a clear and well-executed process reengineering strategy significantly hinders e-government adoption and success.

CONCLUSION

The implementation of e-government is a key challenge for developing countries, including Saudi Arabia. Since there is no universal model that can be applied in all countries to guarantee success, organisational issues play an essential role for the success or failure of these projects. In this case, each country has its own circumstances which reflect its environment, including factors such as the economic, political, cultural and social systems which might influence the adoption of e-government systems. This article presents the importance of treating organisational issues, especially at early stages of development of an e-government project based on an empirical study. The awareness of these issues represents a milestone for overcoming the barriers to e-government implementation. Thus the policy makers and developers of e-government projects must take into consideration these issues which might facilitate or hinder e-government adoption. Overall, there is a general consensus among researchers and practitioners that e-government problems are not only about the complexity of technology, but also related to organisational, behavioural, institutional, socio-structural and cultural aspects. This study provides evidence of the influence of different organisational factors on the success of e-government, although the relative impact of these factors may be different according to specific initiatives and environmental conditions. Policy makers at all levels of government can increase their success rates by modifying certain organisational structures and processes, or engaging other stakeholders in organisational changes. Understanding the complexity of e-government can help to set more realistic goals and to develop better e-government.

REFERENCES

Abdul-Gader, A. H., & Alangari, K. (1994). *Information technology assimilation in the government sector: an empirical study*. Final report of funded research, King Abdul Aziz City of Science and Technology, Riyadh, Kingdom of Saudi Arabia.

Ajzen, I., & Fishbein, M. 1980. *Understanding attitudes and predicting social behaviour*. Englewood Cliffs, N.J.: Prentice-Hall.

Al-Dosari, R., & King, M. (2004). E-government lifecycle model. *Proceeding of UK Academy of Information Systems Conference*, Glasgow Caledonian University, UK.

Al-Fakhri, M., Cropf, R., Higgs, G., & Kelly, P. (2008). e-Government in Saudi Arabia: Between Promise and Reality. *International Journal of Electronic Government Research, 4*(2), 5–82.

Al-Farsy, F. (2003). *Modernity and traditional: the Saudi Equation*. Willing Clowes, Beccles, Suffolk, UK.

Al-Nuaim, H. (2009). How "E" are Arab Municipalities? An Evaluation of Arab Capital Municipal Web Sites. *International Journal Of Electronic Government Research,5*(1), 50, 14.

Al-Sabti, K. (2005). *The Saudi government in the information society*. Available online at www.yesser.gov.sa. Access on 20/1/2008.

Al-Saggaf, Y. (2004). The effect of online community on offline community in Saudi Arabia. *Electronic Journal Of Information Systems In Developing Countries, 16*(2), 1–16.

Al-Shehry, A., Rogerson, S., Fairweather, N. B., & Prior, M. (2006). The motivations to change towards e-government adoption. *E-government Workshop (eGOV06),* Brunel University, UK.

Al-Sudairy, M. (2000). *An empirical investigation of electronic data interchange (EDI) utilisation in the Saudi's Private organisations*. Unpublished PhD Thesis, University of Leicester, UK.

Al-Tawil, K., Sait, S., & Hussain, S. (2003). Use and effect of internet in Saudi Arabia. *Proceedings of the 6th World Multi-conference on Systemic,* Cybernetics and Informatics, Orlando, Florida, USA.

Al-Turki, S. M., & Tang, N. K. H. (1998). *Information technology environment in Saudi Arabia: a review*. Work report, Leicester University Management Centre. UK.

Amoretti, F. (2007). International organizations ICT policies: e-democracy and e-government for political development. *Review of Policy Research, 24*(4), 331–344. doi:10.1111/j.1541-1338.2007.00286.x

Barki, H., Titah, R., & Boffo, C. (2007). Information system use-related activity: an expanded behavioural conceptualization of individual-level information system use. *Information Systems Research, 18*(2), 173–192. doi:10.1287/isre.1070.0122

Carter, L., & Belanger, F. (2004). Citizen adoption of electronic government initiatives (ETEGM03). *Proceedings of the 37th Hawaii International Conference on System Sciences*, IEEE Computer Society, Washington, DC, USA.

Castells, M. (1996). *The Information age: economic, society and culture, 1*. The rise of the network society. Oxford: Blackwell.

Clegg, C., Axtell, C., Damodaran, L., Farbey, B., Hull, R., & Lloyd-Jones, R. (1997). Information technology: a study of performance and the role of human and organizational factors. *Ergonomics London, 40*(9), 851–871.

Davis, F. D. (1989). Perceived usefulness, perceived ease of use, and user acceptance of information technology. *MIS Quarterly, 13*(3), 319–340. doi:10.2307/249008

Davis, F. D., Bagozzi, R. P., & Warshaw, P. R. (1992). Extrinsic and intrinsic motivation to use computers in the workplace. *Journal of Applied Social Psychology*, *22*(14), 1–32. doi:10.1111/j.1559-1816.1992.tb00945.x

Dawes, S. S., Pardo, T. A., & Cresswell, A. M. (2004). Designing electronic government information access programs: a holistic approach. *Government Information Quarterly*, *21*(1), 3–23. doi:10.1016/j.giq.2003.11.001

Denscombe, M. 2007. *The good research guide: for small-scale social research projects*. Maidenhead: Open University Press.

Doherty, N. F., & King, M. (1998). The importance of organisational issues in systems development. *Information Technology & People*, *11*(2), 104–123. doi:10.1108/09593849810218300

Ebrahim, Z., & Irani, Z. (2005). E-government adoption: architecture and barriers. *Business Process Management Journal*, *11*(5), 589–611. doi:10.1108/14637150510619902

Ewusi-Mensah, K., & Przasnyski, Z. H. (1994). Factors contributing to the abandonment of information systems development projects. *Journal of Information Technology*, *9*(3), 185. doi:10.1057/jit.1994.19

Flowers, S. (1996). *Software failure, management failure: amazing stories and cautionary tales*. Chichester; New York: Wiley.

Gil-Garcia, J. R., & Pardo, T. A. (2005). E-government success factors: Mapping practical tools to theoretical foundations. *Government Information Quarterly*, *22*(2), 187–216. doi:10.1016/j.giq.2005.02.001

Grant, G., & Chau, D. (2005). Developing a generic framework for e-government. *Journal of Global Information Management*, *13*(1), 1–30.

Hackney, R., & Jones, S. (2002). *Towards e-government in the Welsh (UK) assembly: an information systems evaluation*. Work report, Manchester, Manchester metropolitan University, Business School.

Hahamis, P., & Iles, J. (2005). E-government in Greece: opportunities for improving the efficiency and effectiveness of local government. *Proceedings of European conference on e-government*, Antwerp, Belgium.

Heeks, R. (2003). *Most e-government for development projects fail: how can risks be reduced?* I-Government working paper, University of Manchester. Institute for Development, Policy and Management.

Heeks, R. (2006). *Implementing and managing e-government: an international text*. London; Thousand Oaks, Calif.: SAGE.

Howard, M. (2001). e-government across the globe: how will "e" change government? *Government Finance Review*, *17*(4), 6–9.

Igbaria, M., Zinatelli, N., Cragg, P., & Cavaye, A. L. M. (1997). Personal computing acceptance factors in small firms: a structural equation model. *Management Information Systems Quarterly*, *21*(3), 279–306. doi:10.2307/249498

Joia, J., & Foundation, G. (2007). A heuristic model to implement government-to-government projects. *International Journal of Electronic Government Research*, *3*(1), 49–67.

Joia, L. A. (2004). Developing government-to-government enterprises in Brazil: a heuristic model drawn from multiple case studies. *International Journal of Information Management*, *24*(2), 147–166. doi:10.1016/j.ijinfomgt.2003.12.013

Kawalek, P., & Wastall, D. (2005). Pursuing radical transformation in information age government: case studies using the SPRINT methodology. *Journal of Global Information Management*, *13*(1), 79–101.

Kettl, D. F. (2005). *The global public management revolution.* (2nd ed.), Washington, D.C.: Brookings Institution Press.

Kim, S., & Lee, H. (2004). Organizational factors affecting knowledge sharing capabilities in e-government: an empirical study. *Lecture Notes in Computer Science, 3035,* 265–277.

Lam, W. (2005). Barriers to e-Government. *Journal Of Enterprise Information Management, 18*(5), 511–530. doi:10.1108/17410390510623981

MCIT. (2004). *Ministry of Communication and Information Technology.* Online: www.mcit.gov.sa. Access on 20/2/2005.

MCIT. (2007). *Ministry Of Communication And Information Technology.* Online: www.mcit.gov.sa. Access on 20/2/2007.

Miles, M. B., & Huberman, A. M. 1994. *Qualitative data analysis: an expanded sourcebook.* 2nd edition, Thousand Oaks: Sage Publications.

Molla, A., & Licher, P. S. (2002). Information technology implementation in the public sector of a developing country: issues and challenges. *Proceedings of 3rd Annual global Information Technology Management World Conference,* New York, USA.

Molla, A., & Licher, P. S. (2005). E-commerce adoption in developing countries: a model and instrument. *Information & Management, 42*(6), 877–899. doi:10.1016/j.im.2004.09.002

Moon, M. J. (2002). The evolution of e-government among municipalities: rhetoric or reality? *Public Administration Review, 62*(4), 424–433. doi:10.1111/0033-3352.00196

OECD. (2004). *The E-Government imperative.* Organisation for Economic Co-operation and Development, Paris.

Pardo, T., & Scholl, H. J. (2002). Walking atop the cliffs: avoiding failure and reducing risk in large scale e-government projects. *Proceedings of the 35th Hawaii International Conference,* Hawaii.

Prins, C. (2001). *Designing e-government: on the crossroads of technological innovation and institutional change.* Hague; Boston: Kluwer Law International.

Saudi Computer Society. (2004). National Information Technology Plan. (NITP) (Draft version). S. C. Society. Riyadh, Saudi Arabia, Saudi . *Computers & Society,* 1–463.

Stake, R. E. (1995). *The art of case study research.* Thousand Oaks: Sage Publications.

Tambouris, E. (2001). An integrated platform for realising online one-stop government: the e-GOV project. *Proceedings of 12th international workshop on database and expert systems applications,* Munich, Germany, IEEE Computer Society.

Tung, L. L., & Rieck, O. (2005). Adoption of electronic government services among business organizations in Singapore. *The Journal of Strategic Information Systems, 14*(4), 417–440. doi:10.1016/j.jsis.2005.06.001

United Nations. (2001). *Benchmarking e-government: a global perspective, assessing the progress of the UN member states,* Work report, United Nations – DPEPA and ASP A. New York, USA.

Vassilakis, C., Lepouras, G., Fraser, J., Haston, S., & Georgiadis, P. (2005). Barriers to electronic service development. *E Service Journal, 4*(1), 41–64. doi:10.2979/ESJ.2005.4.1.41

Vassiliev, A. (2000). *The history of Saudi Arabia.* New York University press. NY, 10003.

Viswanath, V., Micha, M., Grdon, D., & Davis, F. (2003). User Acceptance of Information Technology: Toward a Unified View. *MIS Quarterly, 27*(3).

Wang, Y. S. (2003). The adoption of electronic tax filing systems: an empirical study. *Government Information Quarterly*, *20*(4), 333–352. doi:10.1016/j.giq.2003.08.005

Weber, Y., & Schweiger, D. (1992). Top management culture conflict in mergers and acquisitions: a lesson from anthropology. *The International Journal of Conflict Management*, *3*(4), 285–302. doi:10.1108/eb022716

Whyte, G., & Bytheway, A. (1996). Factors Affecting Information Systems' Success. *International Journal of Service Industry Management*, *7*(1), 74–93. doi:10.1108/09564239610109429

World Fact Book. (2005). Available online at http://www.cia.gov/cia/publ ications/factbook/index.html. Accessed in 6/10/2005.

Yesser (2006). *E-government project:-website of Saudi Arabian e-government*. Available online at:-http://www.yesser.gov.sa /english/default.asp.

Yin, R. K. (2003). *Case study research: design and methods* (3rd ed.). Thousand Oaks, CA: Sage Publications.

This work was previously published in International Journal of Electronic Government Research (IJEGR), edited by Vishanth Weerakkody, pp. 1-13, copyright 2009 by Information Science Reference (an imprint of IGI Global)

Chapter 17

Stages of Information Systems in E-Government for Knowledge Management:
The Case of Police Investigations

Petter Gottschalk
Norwegian School of Management, Norway

ABSTRACT

A stage model for knowledge management systems in policing financial crime is developed in this chapter. Stages of growth models enable identification of organizational maturity and direction. Information technology to support knowledge work of police officers is improving. For example, new information systems supporting police investigations are evolving. Police investigation is an information-rich and knowledge-intensive practice. Its success depends on turning information into evidence. This chapter presents an organizing framework for knowledge management systems in policing financial crime. Future case studies will empirically have to illustrate and validate the stage hypothesis developed in this paper.

INTRODUCTION

Knowledge management is concerned with simplifying and improving the process of sharing, distributing, creating, capturing, and understanding knowledge. Information and communication technology can play an important role in successful knowledge management initiatives. The extent of information technology can be defined in terms of growth stages for knowledge management systems. In this chapter, a model consisting of four stages is presented: investigator-to-

technology systems, investigator-to-investigator systems, investigator-to-information systems and investigator-to-application systems respectively.

Collier (2006) argues that technology is clearly a major impediment to progress in the intelligent application of knowledge in policing. Traditionally, inadequacies in computer systems have been evidenced in most countries, in terms of the lack of national police information strategy, the inability of systems in use by different forces to communicate with each other, and the lack of integration between computer systems in any one force. There has also been a lack of confidence held by internal investigators in intelligence systems, suggesting

DOI: 10.4018/978-1-60960-162-1.ch017

that tacit rather than explicit knowledge was still fundamental to how many internal police officers work. This is not surprising, as many police officers have considered their work a handicraft job rather than a knowledge job. The only way inexperienced officer could learn a new policing field was to observe and join experienced officer in their work as craftsmen.

KNOWLEDGE MANAGEMENT SYSTEMS

The potential of knowledge management systems to enable new organizational forms as well as inter-organizational relationships and partnerships useful in policing will be demonstrated in this chapter. Partnership working is becoming an increasingly common methodology in the public sector for addressing complex social issues such as poverty, economic development and crime. According to Wastell et al. (2004), information systems have a vital role to play in enabling such inter-organizational networks and in facilitating the multi-disciplinary collaboration that is essential to joint working when fighting financial crime.

Knowledge management systems refer to a class of information systems applied to manage organizational knowledge. These systems are IT applications to support and enhance the organizational processes of knowledge creation, storage and retrieval, transfer, and application. Knowledge management and collaboration systems are among the fastest-growing areas of corporate and government software investments (Laudon and Laudon, 2010).

Knowledge management and collaboration are closely related. Laudon and Laudon (2010) argue that knowledge that cannot be communicated and shared with others is nearly useless. Knowledge becomes useful and actionable when shared throughout an organization and between collaborating organizations.

The knowledge management technology stage model presented in this chapter is a multistage model proposed for organizational evolution over time. Stages of knowledge management technology are a relative concept concerned with information and communication technologies (ICT) ability to process information for knowledge work. The knowledge management technology stage model consists of four stages. When applied to law enforcement in this chapter, the stages are labeled investigator-to-technology, investigator-to-investigator, investigator-to-information, and investigator-to-application as illustrated in Figure 1.

Stages of knowledge management technology are such that ICT at later stages is more useful to knowledge work than ICT at earlier stages. The relative concept implies that ICT is more directly involved in knowledge work at higher stages, and that ICT is able to support more advanced knowledge work at higher stages.

STAGE 1: INVESTIGATOR TO TECHNOLOGY

Investigator-to-Technology Stage: *Tools for end users* are made available to knowledge workers. In the simplest stage, this means a capable networked PC on every desk or in every briefcase, with standardized personal productivity tools (word processing, presentation software) so that documents can be exchanged easily throughout a company. More complex and functional desktop infrastructures can also be the basis for the same types of knowledge support. Stage 1 is recognized by widespread dissemination and use of end-user tools among knowledge workers in the company. For example, lawyers in a law firm will in this stage use word processing, spreadsheets, legal databases, presentation software, and scheduling programs.

Related to the new changes in computer technology is the transformation that has occurred

Figure 1. The knowledge management systems stage model for policing

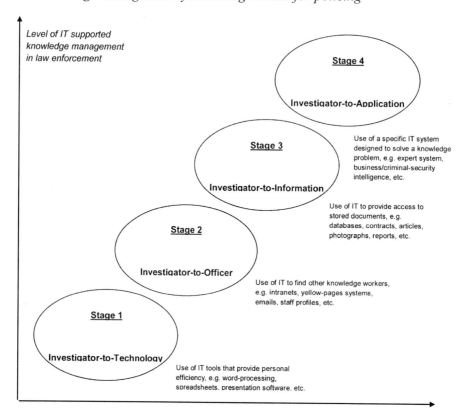

in report writing and recordkeeping in financial crime investigations. Every policing activity or crime incident demands a report on some kind of form. Most business organizations have computer information systems that can store reports electronically. Today, investigators can write reports on small notebook computers. Cursor keys and spell-check functions in these report programs are useful timesaving features.

An example of a specialized investigator-to-technology system is a fraud examination process tool that centers on the fraud hypothesis approach, which has four sequential steps (Ilter, 2009):

a. **Analyzing the available data:** An auditor gathers document-evidence depicting all of the business.

b. **Developing a fraud hypothesis:** Based on what is discovered during analysis, a fraud

examiner develops a hypothesis – always assuming a worst-case scenario – of what could have occurred. The hypothesis addresses one of the three major classifications of occupational (internal) fraud: asset misappropriations, corruption or fraudulent financial statements.

c. **Revising it as necessary:** If, for example, the facts do not point to a kickback scheme, the fraud examiner will look for the possibility of a billing scheme. Although the two schemes have several common elements, the latter raises own red flags.

d. **Confirming it:** Testing the hypothesis by combining theoretical elements with empirical evidence.

An example of a generalized investigator-to-technology system is the Major Incident Policy

Document in the UK. This document is maintained whenever a Major Incident Room using HOLMES system is in operation. Decisions, which should be recorded, are those which affect the practical or administrative features of the enquiry, and each entry has clearly to show the reasoning for the decision. When the HOLMES system is used, the SIO directs which policy decisions are recorded on the system.

The basic information entered into HOLMES is location of incident, data and time of incident, victim(s), senior investigating executive, and date enquiry commenced. During the enquiry, which has been run on the HOLMES system, a closing report is prepared and registered as another document linked to a category of Closing Report. The report will contain the following information: introduction, scene, the victim, and miscellaneous.

Most knowledge workers rely on office systems, such as word processors, voice mail, e-mail and presentation tools, which are designed to increase worker productivity. Some knowledge workers require highly specialized knowledge work systems with powerful graphics, analytical tools, and document management capabilities (Laudon and Laudon, 2010).

Stage 1 can be labeled *end-user-tools* or *people-to-technology* as information technology provide knowledge workers with tools that improve personal efficiency.

STAGE 2: INVESTIGATOR TO INVESTIGATOR

Investigator-to-Investigator Stage: Information about who knows what is made available to all people in the firm and to target outside partners. Search engines should enable work with a thesaurus, since the terminology in which expertise is sought may not always match the terms the expert uses to classify that expertise.

The creation of corporate directories, also referred to as the mapping of internal expertise, is a common application of knowledge management technology. Because much knowledge in an organization remains not codified, mapping the internal expertise is a potentially useful application of technology to enable easy identification of knowledgeable persons.

Here we find the cartographic school of knowledge management, which is concerned with mapping organizational knowledge. It aims to record and disclose who in the organization knows what by building knowledge directories. Often called Yellow Pages, the principal idea is to make sure knowledgeable people in the organization are accessible to others for advice, consultation, or knowledge exchange. Knowledge-oriented directories are not so much repositories of knowledge-based information as gateways to knowledge, and the knowledge is as likely to be tacit as explicit.

At Stage 2, firms apply the personalization strategy in knowledge management. The personalization strategy implies that knowledge is tied to the person who developed it and is shared mainly through direct person-to-person contact. This strategy focuses on dialogue between individuals: knowledge is transferred mainly in personal email, meetings and one-on-one conversations.

Electronic networks of practice are computer-mediated discussion forums focused on problems of practice that enable individuals to exchange advice and ideas with others based on common interests. Electronic networks have been found to support organizational knowledge flows between geographically dispersed co-workers and distributed research and development efforts. These networks also assist cooperative open-source software development and open congregation on the Internet for individuals interested in a specific practice. Electronic networks make it possible to share information quickly, globally, and with large numbers of individuals.

Information systems at stage 2 are knowledge networks systems, and they are also known as expertise location and management systems. These systems address the problem that arises when the appropriate knowledge is not represented as information in the form of a digital document at stages 1 or 3. Rather knowledge instead resides in the memory of expert individuals in the organization (Laudon and Laudon, 2010: 448):

Knowledge network systems provide an online directory of corporate experts in well-defined knowledge domains and use communication technologies to make it easy for employees to find the appropriate expert in the organization.

Some knowledge network systems go further than this by systematizing the solutions developed by experts and then storing the solutions in a knowledge support database as best practices or frequently asked questions (FAQ) repositories at stage 3.

An example of a stage 2 system in policing financial crime is a knowledge network system exchanging information about self-regulation in the private sector. Governance in the form of clear policies and procedures, formalized cross-company communication, along with performance-based salary for board members and employees reduces incidences of white-collar crime within corporations (Hansen, 2009).

The typical system at stage 2 of knowledge management technology in investigation and prevention of financial crime is the intranet. Intranets provide a rich set of tools for creating collaborative environments in which members of an organization can exchange ideas, share information, and work together on common projects and assignments regardless of their physical location. Information from many different sources and media, including text, graphics, video, audio, and even digital slides can be displayed, shared, and accessed across an enterprise through a simple common interface.

Stage 2 can be labeled *who-knows-what* or *people-to-people* as knowledge workers use information technology to find other knowledge workers.

STAGE 3: INVESTIGATOR TO INFORMATION

Investigator-to-Information Stage: Information from knowledge workers is stored and made available to everyone in the firm and to designated external partners. Data mining techniques can be applied here to find relevant information and combine information in data warehouses, as discussed earlier in this book.

On a broader basis, search engines are web browsers and server software that operate with a thesaurus, since the terminology in which expertise is sought may not always match the terms used by the expert to classify that expertise.

At Stage 3, firms apply the codification strategy in knowledge management. The codification strategy centers on information technology: knowledge is carefully codified and stored in knowledge databases and can be accessed and used by anyone. With a codification strategy, knowledge is extracted from the person who developed it, is made independent from the person and stored in form of interview guides, work schedules, benchmark data etc; and then searched and retrieved and used by many employees.

Two examples of knowledge management systems at stage 3 in law enforcement are COPLINK and geo-demographics. COPLINK has a relational database system for crime-specific cases such as gang-related incidents, and serious crimes such as homicide, aggravated assault, and sexual crimes. Deliberately targeting these criminal areas allows a manageable amount of information to be entered into a database. Geo-demographic profiles of the characteristics of individuals and small areas are central to efficient and effective deployment of law enforcement resources. Geo-computation is based on geographical information systems.

A third example of a stage 3 system in policing is business intelligence systems. Business intelligence (BI) systems provide the ability to analyze business information in order to support and improve management decision-making across a broad range of business activities (Elbashir et al., 2008). For example, Staffordshire Police in Great Britain uses a series of custom-made applications including crime recording, custody recording, file preparation, courts administration, and an intelligence system. A small number of criminals are committing most of the crime and detailed BI analysis of data reveal information about offenders and lead to their eventual prosecution. Functions in the system include general queries, property queries, statistical searches, crime profiling queries, and prolific-offender queries.

An important part of stage 3 systems are enterprise content management systems that belong to enterprise-wide knowledge management systems Enterprise-wide knowledge management systems are general-purpose business-wide efforts to collect, store, distribute, and apply digital content. Enterprise content management systems help organizations manage business-wide content. Such systems include agency repositories of documents, reports, presentations, and best practices, as well as capabilities for collecting and organizing semi-structured information such as e-mail (Laudon and Laudon, 2010). E-mail as a tool at stage 1 is developing into a content provider of information for knowledge sharing at stage 3.

An important source of information for stage 3 policing systems is private sector organization's contribution in the control of financial crime. Gilsinan et al. (2008) identified this contribution in terms of five distinct roles. Each role has its own dynamics and implications for successful suppression of unlawful conduct. The five roles are grudging informant, enthusiastic intelligence operative, agent provocateur, cop on the take, and investigator friendly:

- The *grudging informant* is an organization that complies with the requirement to produce information for public sector enforcement activity.

- The *enthusiastic intelligence operative* is an organization that finds it profitable to have a partnership with the government.

- The *agent provocateur* is an organization that adheres to the provision requiring the chief operating and financial executives to certify that the firm's internal controls provide transparency to the financial processes of the company.

- The *cop on the take* is an organization where corporate entities have the primary responsibility for policing their own ranks for compliance with regulatory strictures.

- The *investigator friendly* is an organization that acts from different motives than the government and with different abilities, and both end up engaging in a kind of token enforcement strategy.

Gilsinan et al. (2008) argue that a calculus of incentives and disincentives determines which role will be adopted by the private sector. For the grudging informant, a disincentive can be found in customers who demand secrecy in their financial dealings. For the enthusiastic intelligence operative, an incentive is technology that allows private brokers to gather large amounts of information and package these data as a commodity for sale to governments. For the agent provocateur an incentive is to better manage risk and thereby increasing their stock's market appeal. For the cop on the take, a disincentive is a potential stock market bubble that encourages risk taking with the lure of enormous, quickly realized profits. For the investigator friendly, an incentive is external business threats from fraudulent activity by competitors.

Gilsinan et al. (2008) argue further that the temptation towards malfeasance is high when the private sector is responsible for both the provision and production of industry regulation. Behaviors change when such provision and production is linked to fraudulent activity perpetrated by individuals external to the organization and when either government resources or likelihood of success are in short supply. In these kinds of situations, the government tends to be laid back while businesses tend to optimize their own utility.

Tellechea (2008) suggests the introduction of reverse corruption, where individuals and entities are attracted to incentives to uncover and report misconduct by quickly and efficiently giving them a share of any seized funds. There is precedent for this type of approach in the USA. If you provide sufficiently detailed information on a tax evader, then the Internal Revenue Service may award you up to 15 per cent of the amount recovered in taxes and penalties up to a maximum of $2 million. In 2003, whistleblowers in the USA received $4.1 million in rewards. In 2002, IRS paid $7.7 million and recovered $66.9 million in taxes, fines, penalties and interest. The record year was 2000, when $10.8 million was paid. Such incentives might stimulate the flow of information into stage 3 systems.

Stage 3 can be labeled *what-they-know* or *people-to-docs* as information technology provide knowledge workers with access to information that is typically stored in documents. Examples of documents are contracts and agreements, reports, manuals and handbooks, business forms, letters, memos, articles, drawings, blueprints, photographs, e-mail and voice mail messages, video clips, script and visuals from presentations, policy statements, computer printouts, and transcripts from meetings.

STAGE 4: INVESTIGATOR TO APPLICATION

Investigator-to-Application Stage: Information systems solving knowledge problems are made available to knowledge workers and solution seekers. Artificial intelligence is applied in these systems. For example, neural networks are statistically oriented tools that excel at using data to classify cases into one category or another.

Another example is expert systems that can enable the knowledge of one or a few experts to be used by a much broader group of workers requiring the knowledge. Investigator-to-application systems will only be successful if they are built on a thorough understanding of law enforcement.

An example of a stage 4 system not yet implemented is a system for evaluation of compliance levels according to recommendations by the Financial Action Task Force (FATF). The FATF was formed in 1989 by the G-7 group of countries, motivated by the General Assembly of the United Nations' adoption of a universal pledge to put a halt to money laundering, fuelled largely at that time by the laundering of illegal drug trade money. One of the FATF's first tasks was to develop measures to combat money laundering (Johnson, 2008).

A set of Forty Recommendations was issued by the FATF. They were designed to provide a comprehensive strategy for action against money laundering. FATF members have been evaluated over a number of years against these recommendations and more recently against the Nine Special Recommendations using self-assessment and/or mutual assessment procedures. Self-assessment is a questionnaire-based yearly exercise. Mutual evaluation involves an onsite visit by experts from other member countries in the areas of law, financial regulation, law enforcement, and international co-operation (Johnson, 2008).

The result of a mutual evaluation may be one of the following compliance levels (Johnson, 2008):

1. Non-Compliant (NC). There are major shortcomings, with a large majority of the essential criteria not being met.
2. Partially Compliant (PC). Some substantive action has been taken, and there is compliance with some of the essential criteria.
3. Largely Compliant (LC). Only minor shortcomings, with a large majority of the essential criteria being fully met.

4. Fully Compliant (FC). The recommendation is fully observed with respect to all essential criteria.

To be able to compare compliance across countries, each compliance level was assigned a numerical level: NC = 0, PC = 0.33, LC = 0.67 and FC = 1.0. The following countries achieved highest compliance scores (Johnson, 2008):

Belgium 0.77
UK 0.70
USA 0.69
Portugal 0.69
Norway 0.68
Switzerland 0.64
Ireland 0.63

Johnson (2008) argues that the results here should be used as a guide only to the ranking and compliance of countries rather than some exact measurement of compliance. This is because compliance levels are very broad, where substituting a single value for each compliance level provides only a crude measure of compliance for comparisons to be made. Only a future system based on artificial intelligence might provide an exact measure of compliance.

Another example of a stage-four system is based on artificial intelligence (AI), which is an area of computer science that endeavors to build machines exhibiting human-like cognitive capabilities. Most modern AI systems are founded on the realization that intelligence is tightly intertwined with knowledge. Knowledge is associated with the symbols we manipulate. The example is a system for auditing insurance fraud. The hybrid knowledge- and statistics-based system uses knowledge discovery techniques to: first, integrate expert knowledge with statistical information assessment to identify cases of unusual provider behavior; and second, the use of machine learning to develop new rules and

to improve identification processes (Yusuf and Babalola, 2009).

Artificial intelligence as well as database technology provides a number of intelligent techniques that organizations can use to capture individual and collective knowledge and to extend their knowledge base. Expert systems, case-based reasoning, and fuzzy logic are used for capturing knowledge from knowledge workers and making information representations and procedures available to other knowledge workers. Neural networks and data mining are used for knowledge discovery. They can discover underlying patterns, categories, and behaviors in large data sets that could not be discovered by intelligence officers alone or simply through experience (Laudon and Laudon, 2010).

Knowledge-based systems deal with solving problems by exercising knowledge. The most important parts of these systems are the knowledge base and the inference engine. The former holds the domain-specific knowledge whereas the latter contains the functions to exercise the knowledge in the knowledge base. Knowledge can be represented as either rules or frames. Rules are a natural choice for representing conditional knowledge, which is in the form of if-when statements. Inference engines supply the motive power to the knowledge. There are several ways to exercise knowledge, depending on the nature of the knowledge. For example, backward-chaining systems work backward from the conclusions to the inputs. These systems attempt to validate the conclusions by finding evidence to support them. In law enforcement this is an important system feature, as evidence determines whether a person is charged or not for a crime.

Case-based reasoning systems are a different way to represent knowledge through explicit historical cases. This approach differs from the rule-based approach because the knowledge is not complied and interpreted by an expert. Instead, the experiences that possibly shaped the expert's knowledge are directly used to make decisions.

Learning is an important issue in case-based reasoning, because with the mere addition of new cases to the library, the system learns. In law enforcement police officers are looking for similar cases to learn how they were handled in the past, making case-based reasoning systems an attractive application in policing.

An example of a stage IV system in policing is dynamic emergency response information system (DERMIS) conceptually introduced by Turoff et al. (2006). They developed a set of general and supporting design principles and specifications for DERMIS by identifying design premises resulting from the use of indexes. The principles are based on the assumption that implicit in crises of varying scopes and proportions are communication and information needs that can be addressed by today's information and communication technologies. What is required, however, is organizing the premises and concepts that can be mapped into a set of generic design principles.

Turoff et al. (2006) identified the following eight design premises for DERMIS design:

1. System training and simulation. An emergency system that is not in use on a regular basis before an emergency will never be useful in an actual emergency.
2. Information focus. People responding to an emergency are working 14-18 hour days and have no tolerance or time for things unrelated to dealing with the crisis.
3. Crisis memory. Learning and understanding what actually happened before, during, and after the crisis is extremely important for the improvement of the response process.
4. Exceptions as norms. Almost everything in a crisis is an exception to the norm.
5. Scope and nature of crisis. The critical problem of the moment is the nature of the crisis, a primary factor requiring people, authority, and resources to be brought together at a specific period of time for a specific purpose.

6. Role transferability. It is impossible to predict who will undertake what specific role in a crisis situation. The actions and privileges of the role need to be well defined in the software of the system and people must be trained for the possibility of assuming multiple or changing roles.
7. Information validity and timeliness. Establishing and supporting confidence in a decision by supplying the best possible up-to-date information is critical to those whose actions may risk lives and resources.
8. Free exchange of information. Crises involve the necessity for many hundreds of individuals from different organizations to be able to exchange information freely, delegate authority, and conduct oversight, without the side effect of information overload.

Some of the premises remind us of stage 2, where communication and information exchange between people is the most important feature. Thus, a DERMIS may be developed according to the stage model by first including communication aspects at stage II, then move into information bases at stage 3, and finally combine systems users and information sources into information services in emergency situation.

Stage Four can be labeled *how-they-think* or *people-to-systems* where the system is intended to help solve a knowledge problem.

KNOWLEDGE WORK

Information technology to support knowledge work of internal investigators is improving. For example, new information systems supporting investigation processes are evolving. A criminal investigation is an information-rich and knowledge-intensive practice. Its success depends on turning information into evidence. However, the process of turning information into evidence is neither simple nor straightforward. The raw information that is gathered through the investigative process is often required to be transformed into usable knowledge before its value as potential evidence can be realized. Hence, in an investigative context, knowledge acts as an intervening variable in this transformative process of converting information via knowledge into evidence.

The extent to which knowledge management systems as described above are used by internal investigators is dependent on a number of factors. One important factor frequently discussed in the research literature, is the task technology fit. Task technology theory argues that the use of a technology may result in different outcomes, depending upon its configuration and the task for which it is used. Four elements are part of the theory: task characteristics, technology characteristics, which combine to affect the fit, and which affects the outcome in terms of performance or utilization. Tasks are broadly defined as the actions carried out in turning inputs to outputs in order to satisfy information needs. Perceived technology fit depends on the agreement between the perceived capabilities of the technology, the needs of the task, and the competence of the users. (Lin and Huang, 2008).

Kappos and Rivard (2008) argue that culture plays an increasingly important role in information systems initiatives. Depending on cultural values, information systems initiatives will be stimulated or prevented. For example if legality is more important than effectiveness, and if formal is more important than informal, then initiatives will emerge more easily.

Knowledge management systems have created incentives for promoting knowledge sharing among organizational members and for fostering innovation within public and private institutions. Knowledge management systems can support four knowledge management processes (Hsiao, 2008): (i) knowledge creation is a process of proactively determining what knowledge is desired and needed, (ii) knowledge development is the process of establishing valuable knowledge, (iii) knowledge

reuse is the process of putting knowledge in a reusable form, and (iv) knowledge transfer is the process of disseminating knowledge effectively.

It is important to stress here that stages of growth models are very different from life cycle models. While stage models define and describe accumulated improvements in knowledge management technology to support policing the business organization, life cycle models represent a cycle of birth, growth, decline, and eventually death of information technology.

In future research there is a need to validate the stage model both theoretically and empirically. Furthermore, there is a need for benchmark variables that will have different content for different stages. In the current presentation of our model, the stages are lacking both theoretical background and practical situations. While stage 4 may seem understandable and viable, the remaining stages are in need of further conceptual work. Core questions in future research will be whether officer-to-technology, officer-to-officer, officer-to-information, and officer-to-application are valid, practicable, and reliable concepts.

In future research, pros (strengths) and cons (weaknesses) to the suggested model have to be taken into account. We need to provide a more critical analysis of a stage model such as the one suggested. It is not at all intuitively obvious that the progression over time is from end-user-tools, via who-knows-what and what-they-know, to how-they-think. Why not what-they-know, via who-knows-what and end-user-tools, to how-they-think; or end-user-tools via what-they-know and who-knows-what to how-they-think? The conceptual research presented here is lacking empirical evidence. Only a questionnaire based on Guttman scaling rather than Likert scaling can verify the suggested sequence or alternatively identify another sequence.

The important contribution of this chapter is the introduction of the stage hypothesis to knowledge management technology in policing the business organization. Rather than thinking of knowledge management technology in terms of alternative strategies, we suggest an evolutionary approach where the future is building on the past, rather than the future being a divergent path from the past. Rather than thinking that what was done in the past is wrong, past actions are the only available foundation for future actions. If past actions are not on the path to success, direction is changed without history being reversed.

REFERENCES

Collier, P.M. (2006). Policing and the Intelligent Application of Knowledge. *Public Money & Management*, April, 109-116.

Elbashir, M. Z., Collier, P. A., & Davern, M. J. (2008). Measuring the effects of business intelligence systems: The relationship between business process and organizational performance. *International Journal of Accounting Information Systems*, *9*, 135–153. doi:10.1016/j.accinf.2008.03.001

Gilsinan, J. F., Millar, J., Seitz, N., Fisher, J., Harshman, E., Islam, M., & Yeager, F. (2008). The role of private sector organizations in the control and policing of serious financial crime and abuse. *Journal of Financial Crime*, *15*(2), 111–123. doi:10.1108/13590790810866854

Hansen, L. L. (2009). Corporate financial crime: social diagnosis and treatment. *Journal of Financial Crime*, *16*(1), 28–40. doi:10.1108/13590790910924948

Hsiao, R. L. (2008). Knowledge Sharing in a Global Professional Service Firm. *MIS Quarterly Executive*, *7*(3), 123–137.

Ilter, C. (2009). Fraudulent money transfers: a case from Turkey. *Journal of Financial Crime*, *16*(2), 125–136. doi:10.1108/13590790910951803

Johnson, R. R. (2008). Officer Firearms Assaults at Domestic Violence Calls: A Descriptive Analysis. *The Police Journal*, *81*(1), 25–45. doi:10.1350/pojo.2008.81.1.407

Kappos, A., & Rivard, S. (2008). A Three-Perspective Model of Culture, Information Systems, and Their Development and Use. *Management Information Systems Quarterly*, *32*(3), 601–634.

Laudon, K. C., & Laudon, J. P. (2010). *Management Information Systems: Managing the Digital Firm* (11th ed.). London, UK: Pearson Education.

Lin, T. C., & Huang, C. C. (2008). Understanding knowledge management system usage antecedents: An integration of social cognitive theory and task technology fit. *Information & Management*, *45*, 410–417. doi:10.1016/j.im.2008.06.004

Tellechea, A. F. (2008). Economic crimes in the capital markets. *Journal of Financial Crime*, *15*(2), 214–222. doi:10.1108/13590790810866908

Turoff, M., Walle, B. V. d., Chumer, M., & Yao, X. (2006). The Design of a Dynamic Emergency Response Management Information System (DERMIS). *Annual Review of Network Management and Security*, *1*, 101–121.

Wastell, D., Kawalek, P., Langmead-Jones, P., & Ormerod, R. (2004). Information systems and partnership in multi-agency networks: an action research project in crime reduction. *Information and Organization*, *14*, 189–210. doi:10.1016/j.infoandorg.2004.01.001

Yusuf, T. O., & Babalola, A. R. (2009). Control of insurance fraud in Nigeria: an exploratory study. *Journal of Financial Crime*, *16*(4), 418–435. doi:10.1108/13590790910993744

Chapter 18

Leaders as Mediators of Global Megatrends:
A Diagnostic Framework in Praxis

Katarina Giritli-Nygren
Midsweden University, Sweden

Katarina Lindblad-Gidlund
Midsweden University, Sweden

ABSTRACT

The idea of eGovernment is moving rapidly within supra-national and national and local institutions. At every level leaders are interpreting the idea, attempting to grasp either the next step or indeed the very essence of the idea itself. This chapter outlines a diagnostic framework, resting on three different dimensions; translation, interpretative frames and sensemaking, to create knowledge about the translation processes and by doing so, emphasize enactment rather than vision. The diagnostic framework is then empirically examined to explore its possible contribution to the understanding of the complexity of leader's translating and mediating the idea of eGovernment in their local context. In conclusion it is noted that the diagnostic framework reveals a logic of appropriateness between local mediators, eGovernment, different areas of interest and appropriate organisational practices.

INTRODUCTION

When overarching ideas such as that of eGovernment begin to travel (through different types of policy documents), this causes the operative leaders to play a specific role in translating the ideas into their own division's "language" (Røvik, 2000; Latour, 1996; Czarniawska & Joerges, 1996). Local authorities are placed in the role of institutional mediators between global mega-

trends and local conditions (Anttiroikko, 2002; O'Toole, 2007). In the perspective of translation institutional entrepreneurs has an important role in organisational transformation processes (see Di Maggio, 1988; Lawrence & Phillips, 2004). Institutional entrepreneurs are described as "organised actors with sufficient resources" (Di Maggio, 1988, p.14) whom, by attributing their subjective and intersubjective meaning to the idea, construct meaningful means of using it, which also can be seen as an activity of sense making (Weick, 1995; Orlikowski & Gash, 1994).

DOI: 10.4018/978-1-60960-162-1.ch018

This activity is often referred to as an enactment process, the ideas become *enacted,* and it is in this first/early enactment process that leaders play an important role (see for example Fountain, 2001; Weick, 2001; and Yildiz, 2007). Additionally, the strategic management level of eGovernment is quite unarticulated and unproblematised and the correct way forward becomes difficult to decipher (Andersen, 2004; Andersen & Henriksen, 2005).

In this paper we develop an analytical framework to analyse how leaders, in their role of institutional entrepreneurs or mediators, make sense of eGovernment goals. The objective of the article is then to analyse leaders as mediators of megatrends with the help of a theoretical framework resting on the concept of translation by Shein, interpretative frames by Orlikowski and sensemaking by Weick. The aim is therefore one of theory testing with the aspiration of gaining deeper insight into the local interpretation processes of global megatrends. In addition the value of the framework as a practical tool has been further analysed through two different workshops and the results of those are presented and discussed.

The main argument put forward in this paper is that it is important to examine and analyse the actions of operative leaders and how they make sense of overarching eGovernment goals. Analysing leaders' translations of the concept of eGovernment becomes interesting as leaders not only shape their own image of information technology's potential (Moon & Norris, 2005; Ogawa & Scribner, 2002; Avolio et al., 2001) but are also important actors in shaping how the organisation makes sense of technological implementations i.e. how leaders influence the way technology is assimilated and adapted by organisations.

After this introduction the paper has the following disposition: in section two the discussion focuses on eGovernment as a megatrend. In section three, a theoretical framework in order to analyse the enactment of eGovernment is put forward. The framework consists of three analytically different dimensions (i) translation of the goals

and their reasons, (ii) how they make sense and are legitimated according to the organisational understanding, and (iii) interpretative frames related to position and responsibility. To provide an enriched picture an explorative examination of the diagnostic framework is provided in section five. In section six the value of the framework as a practical tool is further discussed in view of two different workshop settings. The implications relating to unreflected and unarticulated management strategies that are enacted are then discussed and the manner in which this proposed framework might contribute to the strategic management level of eGovernment.

THE MEGATREND OF EGOVERNMENT

We focus on eGovernment as a travelling megatrend acquired by its own rhetoric and special attributes. For an idea to achieve status as a "megatrend" it has to be in unison with the, at the time, dominating discourse. A homogenisation is taking place when an organisational concept becomes a symbol for the modern society. By becoming a symbol the idea obtains a penetrating power and is legitimated as the "best" or "only" way to reform and modernise an organisation. A common aspect relating to overarching megatrends (Røvik, 2002; Anttiroikko, 2002; Czarniawska & Joerges, 1996) is their global sphere of action and that they are associated with foundational modern values such as rationality, efficiency, science, nous and progress.

The idea of eGovernment is often described (Löfgren, 2007; Moon & Norris, 2005; Yildiz, 2007; Bekker & Homberg, 2007; Kraemer & King, 2006) as an object or the "the principal tool" by which public administrations can improve their government activities both internally, for improved efficiency and effectiveness, and externally for improved relations with stakeholders. The idea of eGovernment therefore consists of different processes which, with the assistance of ICT, are supposed to increase: accountability, quality delivery of service, efficiency delivery of service, transparency, access, political participation of citizens and data transfer services (information exchange) (Yildiz, 2007; Bekker & Homberg, 2007; Löfgren, 2007; Grönlund, 2002, among others).

In accordance with Jaeger (2003) the idea of eGovernment could be described as an idea embraced with great enthusiasm by many governments which has now advanced into a global phenomenon. However, as there is no standard definition of what exactly eGovernment is (Yildiz, 2007; Bekker & Homberg, 2007) the idea appears to be being accepted as a common set of beliefs.

The goals of eGovernment tidily combine quality with efficiency since both are seemingly required in order to provide service. However, in a typical manner the values, efficiency and quality, are combined, without reference to any possible contradiction, and with a rhetoric that implies that both can be attained simultaneously. In order to, at the same time rationalise the public administration, and increase the service towards citizens, this requires a combination of elements from two different management strategies. In government areas these two management strategies used to be refereed to as *mass customised bureaucracy* and *customer oriented bureaucracy* (Frenkel et al, 1998; Korczynski, 2002).

From a transformational perspective, similar technology and organisational structure can serve different socio-economic aims and it might be the case, for good or bad, that the technological progression is impeded by existing socio-economic modes of production. We want to draw attention to the important process through which the myth is translated by institutional mediators into different divisions of practice. It is necessary to analyse how different socio-economic modes of production are intermingled without taken into consideration if they are of contradictory nature or not.

A DIAGNOSTIC FRAMEWORK FOR EGOVERNMENT MANAGEMENT

As has been shown above, our focus relies on a presumption that leaders, as individual agents, use existing institutional settings to make sense of the eGovernment concept. Karl Weick (1995) has been one of the main advocates of the sensemaking perspective within organisation theory. Weick (2001) describes sensemaking in terms of a number of properties that stress its dynamic, social and retrospective nature. Sensemaking consists of the three elements; a frame, a cue and the connection between them. It also involves the way that individual leaders attempt to cope with ambiguity in their interpretation of different events. This means that the concept of 'sensemaking' could be used to explain the degree to which individuals may make sensible interpretations of equivocal information in organisational events. Along with the recommendations offered by Weick (ibid.) we will study how different events are structured into meaningful actions. Notably, with regard to our study, is the point that Weick argues that sensemaking is a part of the process through which managers understand the organisation they are supposed to manage. Management could therefore be seen as a process of sensemaking (ibid.) activated when managers attempt to find meaning in the actions that their organisations have performed or should perform. This implies that the manner in which leaders understand their organisation (potentials and limitations) is important with regards to how they make sense of the eGovernment concept. Sensemaking, as an organisational understanding, is therefore to be seen as a key to understanding why the eGovernment idea is enacted to a local context in a particular way.

How technology or technologically driven transformations becomes enacted according to different interpretations is further explained by Wanda J Orlikowski (1992, 1994). She uses Giddens' structuration theory to explore the dualistic relation between technology and organization and to show how they influence each other reciprocally. She (Orlikowski, 1994) uses the term interpretative flexibility to grasp the complexity of how different people interpret and create the meaning of technology and how the interpretations determine how it is used and how it can contribute to the organization.

Orlikowski (1992) uses Giddens' structuration theory but, instead of Giddens' concept 'self reflexivity' Orlikowski uses the term interpretative flexibility, which can be seen as a consequence of self reflexivity and institutional knowledge. Different interpretations are created by self reflexive processes in different institutional contexts. The term "interpretative flexibility" (Orlikowski & Gash, 1994) is aimed at an attempt to grasp the complexity of how different people interpret and create the meaning of technology and how the interpretations determine how it is used and how it can contribute to the organisation. This implies that different leaders will translate and use the idea differently due to their interpretative frame. In order to be able to understand their viewpoint of the transformation it is necessary to understand both their position and their responsibility.

Hitherto, we have discussed how ideas becomes enacted into an organisational setting form three different perspectives; translation, sense making and interpretative frames. These concepts are used in this context to analytically differentiate between the dimensions embedded in the process of enactment. We understand local translations of eGovernment in relation to how the idea makes sense in relation to their organisational understanding and we understand their organisational understanding in relation to their interpretative frames. By doing so, we can draw attention to how our respondents individual self reflexive processes are related to their institutionalised forms of understanding. In other words, how organisational members will use existing knowledge to interpret and make sense of the concept of eGovernment. How they (i) translate the goals and the reasons behind the policy, (ii)

how they make sense and legitimise the change according to the organisational understanding, and (iii) their interpretative frames connection with position and responsibility.

The organisational theorist Schein (1994) has developed a diagnostic framework to enable an analysis regarding how leaders describe their perceptions of leadership which we find rewarding. According to him, the leaders' critical role in each developmental stage of an organisation has to do with the fact that they not only construct their own view onto the task but are also in charge of shaping the organisation. Schein (1992) describes their work as a process in order to:

> …perpetually diagnose the particular assumption of the culture and figure out how to use those assumptions constructively or to change them if they are constraints (Schein, 1992, p.381)

As we are interested in how a changing process is understood and how management strategies and intentional actions are developed we find a framework to diagnose management rewarding. The model makes it possible to focus on strategies but also on the underlying assumptions behind them and the sensemaking process by which they are legitimised. This means that we understand management as a process of sense making (Weick, 1995) which reflects the way that leaders and managers view their organisation's potentials and limitations. Though our aim is to examine and analyse leaders' perceptions in relation to a very specific subject i.e. the sense making process by which they enact the overarching eGovernment vision to their practical circumstances and Schein's model is for diagnosing leadership in general, we modified Schein's general model to make it more compatible with our aim. Following Schein (1994) the concepts of translation, sense making and interpretative frames corresponds with the method by which Schein distinguishes between perceptions of *self, task and subordinates* in the process of transforming the local municipality toward an electronic government. Together these perceptions express how they diagnose the necessary eGovernment action and how they will deliver it further.

In order to be able to focus on sensemaking, we must know how our respondents understand their organisations and their subordinates and the component perceptions of subordinates, does in fact provide that information. Together these perceptions express how they diagnose the necessary eGovernment action and how they will deliver it further. Our model of analysis could be concluded in this way:

RESEARCH DESIGN

The objective of the article is then to analyse leaders as mediators of megatrends with the help of a theoretical framework (see Figure 1; A diagnostic framework) resting on the concept of translation by Shein, interpretative frames by Orlikowski and sensemaking by Weick. The aim is therefore one of theory testing with the aspiration of gaining deeper insight into the local interpretation processes of global megatrends. There are several methodological implications associated with such an intention and we will briefly touch upon some of them since they have guided us in our research design.

First, we rest upon the interpretative tradition in IS research (Walsham, 1995; Orlikowski & Baroudi, 1991; Klein & Myers, 1999) focusing on the complexity of human sensemaking. Complexity in terms of acknowledging conflicting interpretations among stakeholders (see e.g. Klein & Myers, 1999) but also the need of being sensitive to rich, in-depth and idiographic meanings that the participants assign to them. The intent is to increase the understanding of the interpretation process in its natural setting (Orlikowski & Baroudi, 1991). The ambition is therefore not to gain repeatability or generalisations in a positivistic sense, the value of the results is rather judged in terms of the extent to which it allows others to understand the phenomenon (Walsham, 1995).

Figure 1. A theoretical and empirical diagnostic framework for analyzing leaders' interpretations of an eGovernment transformation

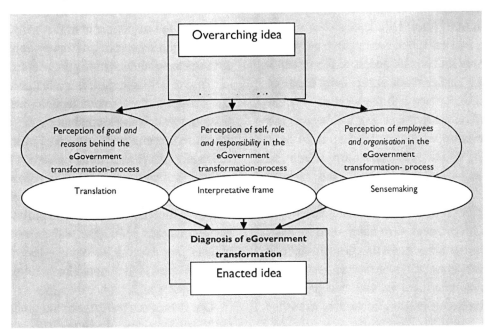

Second, which is closely related to the above, this understanding is believed to be deepened by the proposed theoretical framework and is as such using the theoretical framework as a set of sensitizing concepts to guide the analysis. Notable, the analysis is not predicted by any propositions (in terms of Cavaye, 1996) but is strictly explorative and aims at testing the value of the framework for a richer understanding the interpretation process. The anticipation is therefore not to create answers to a set of questions but to analyse whether the theoretical framework could generate a deeper understanding of the phenomenon in its context which is believed can then be used to inform other settings (Reason & Rowan, 1981; Denzin, 1970; Orlikowski & Baroudi, 1991; Weick, 1995; Giddens, 1984).

Third, the empirical material is gathered through loosely structured deep interviews to be able to allow the explorative nature at the same time as linking the assigned perceptions to the structure of the theoretical framework. And the material is treated as 'displays of perceptions' (Silverman, 1993) or manifestations and not valued in terms of true or false representations. This also gives that dissensus of interpretations signifies "not that we are indifferent to them but that they matter" (ibid, p. 222).

Fourth, we conducted the research as part of a larger study and gained access to the research site through earlier work at the same location. However, the contact with the location was initially based on the fact that they had begun their journey towards what they labelled a 24 hour municipality and strived to take the next step and formulate action strategies. And as such was very suitable for our study since our intentions were to study the process of how overarching and global ideas of eGovernment were interpreted by local leaders. Furthermore, as we do consider the idea of eGovernment as still in an early and formative stage, to stay with the same case and go deeper into the translation process, instead of choosing

a new case to make comparisons, seemed to be more adequate in relation to our research design.

Last, the three respondents in the study were chosen in line with the concept of 'institutional entrepreneurs' (Di Maggio, 1988; Lawrence & Phillips, 2004; Maguire et al., 2004) where Di Maggio defines institutional entrepreneurs as organised actors holding both sufficient resources and opportunity to realise their interest. By acknowledging that the institutionalisation of new practices in a local organisation will be driven by the institutional entrepreneurs the focus is shifted from external fields to internal organisational actors who may frame the new practice. As such, the institutional entrepreneurs are seen as situated actors and institutionally embedded in different settings which are intimately linked to the suggested theoretical framework of translation. Traditionally this concept has been used to describe individuals who attempt to change a whole field; here we see it as an institutional mediator comparable with Latour's 'Mr Manybodies' (1987). A 'Mr Manybody' is a person/actor who has a numerous of allies and following, a suggested interpretation are stronger and more likely to be accepted as the truth if it can refer to support of a strong ally e.g. leader. And in line with such an argument the respondents in this study held the position, the resources, the opportunity and the network to be labelled institutional entrepreneurs.

To sum it up, the choices made throughout the study rested on these five methodological implications and we decided to conduct a series of loosely structured deep interviews with a limited number of leaders (considered to be important actors i.e. institutional entrepreneurs) in an organisation in public sector that was in the process of enacting the idea of eGovernment. And, as mentioned above, since we already had contact with the municipality at hand[1], we asked if we could return with these questions and if they could be helpful in single out interviewees.

Our first contact with the municipality was via the IT-strategy department and the result of this contact was an initial meeting, together with a sample of representatives from the executive level. Following that meeting, these representatives were asked to choose the most prominent figures in the eGovernment transformation process who should participate in the study. We wanted the participants who considered themselves (as representatives for the organisation) to be in charge in order to shape the future implementation process. They were not steered by us in any direction as it was very important that they themselves made this selection so as to avoid the scenario that the selection process reflected the researchers' analytical presumptions. This resulted in a group of three leading figures; (a) the head of the IT-strategy department, (b) the personnel manager and (c) the head of administration. From our research perspective this non-interference on our part proved to be a finding in itself as it highlighted the fact that these were the persons perceived to be the most important.

After finding our respondents we performed the three deep-interviews. The deep-interviews were loosely structured (following on the model of analysis inspired by Schein) into three main areas: the leader's perception of the task, self (role and responsibility) and subordinates. Our guiding questions were (1) for the perception of task: - *how would s/he describe the goal and reasons behind eGovernment transformation*, (2) for perception of self, role and responsibility: - *what is s/he supposed to do,* and (3) for perception of subordinates: - *in relation to the eGovernment transformation how does s/he view the organisation and the employees within?* Each interview lasted between 60-90 minutes and was recorded and transcribed into text. The respondents were guaranteed anonymity and it was promised that the research site i.e. the municipality, would be presented in such a way that it would be difficult to identify a respondent. Thus the respondent was able to answer in as safe an environment as was possible.

We then returned to the data in several steps and thematically analysed them in order to develop initial categories, grouping data and identifying patterns (Strauss, 1987; Doolin, 2004). As such, the empirical material was analysed in the light of "thematic dimensions" derived from theory: *the idea* (translating goal and reason), *interpretative frames* (role and responsibility) and *organisational understanding* (a way to make sense). During this process the transcripts were read and re-read, to identify statements that reflected assumptions, knowledge and experiences of eGovernment.

EMPIRICAL FINDINGS AND DISCUSSION

The next step was to analyse the material more closely in relation to 'the diagnostic framework model' to explore its possible contribution to the understanding of the complexity of leader's translating and mediating the idea of eGovernment in their local context. Given that the diagnostic framework model is not used in this case as a comparative or quantitative tool to illustrate dissimilarities and/or numbers regarding how many possible translations exist for each kind, the result will consequently not be presented in such a manner. The focus, in this case, is rather on how the interviewees (the leaders) translate and make sense of the overarching idea and the insights their translations might provide. It should also be mentioned that the translations could be regarded both independently and collectively which could illustrate itself differently in whatever case chosen.

In the following the data is presented and analysed in accordance with the structure of the diagnostic framework model: (i) the idea (translating goal and reason), (ii) interpretative frames (role and responsibility), and (iii) organisational understanding (a way to make sense).

The Idea – What is Supposed to be Enacted – Translating Goal and Reason

In the first step we focus on eGovernment as a travelling megatrend, it means that it is acquired by its own rhetoric and special attributes. By becoming a symbol the idea obtains a penetrating power and is legitimated as the "best" or "only" way to reform and modernise an organisation. Through the lens of the diagnostic framework the eGovernment rhetoric and attributes are translated by our respondents into a local context as a matter of goal and reason behind the eGovernment transformation. Considering this, they are the local mediators of eGovernment in this particular context. The global idea of eGovernment is most easily described as being a common set of beliefs and we have seen that the dominant rhetoric of the overarching eGovernment idea was repeated in our interviews.

The head of Personnel (HP) identified the eGovernment idea as primarily being a separation between simple and complex processes, which should be supported, and not dominated, by technological solutions. The simple processes could be automated (and simple and repetitive work processes could be transformed into self-service and case handling systems) and in so doing time is created for handling the more complex processes.

"Well, the introduction of electronic administration ought to mean a simplified administration and more core business. You've got to go back and think about the mission when you're allocating resources, especially things like such-and-such a section will always have too few resources". (HP)

From that perspective the idea of eGovernment is interpreted as a money or time saving tool. At a later stage, the HP also stressed the fact that it is important to reflect upon the original assignment of the municipality as otherwise it is easy to be caught up in the wrong issues when attempting to grasp how to transform the municipality.

"... if you think about this thoroughly, we exist for the citizens, and I hope that they will become more demanding in the future" (HP)

When the head of administration (HA) described his perception of the goals and reasons behind the eGovernment he said that it concerned an "evolution from using websites for information to now being able to act like a citizen on the web". This evolution was described to create opportunities for a more rational management. He thought of it as mostly a question of business development. Additionally, words such as 'increased quality' and 'effectiveness' were used to describe the goals and reasons behind eGovernment. According to both the head of personnel (HP) and the HA it is important that the organisational 'needs' are prior to technological 'needs' since it is necessary to stay strategic. As the quotations below shows it might otherwise be easy to be caught up in the wrong issues.

"It is important that you, I think it's important that you, that it isn't IT but it's the operations that determine things, what we want to achieve" (HA)

"Well, we have to decide how we are going to use the resources to change operations; get rid of the focus on IT". (HP)

The head of the IT-strategic department (HI) seamed to be sharing the same depiction that it is important to remember that technology is not the objective; it is rather the thought processes on how to run the organisation that should be put in the centre. When the head of the IT-strategic department (HI) described his perception of the goals and reasons behind the eGovernment he pointed out that:

"the idea has more to do with changed organisational processes and ways of thinking than it has to do with information technology" (HI)

The aim of reducing costs, enhancing quality and effectiveness through automatic processes were mentioned at a later stage.

The HI also stressed the fact that it is important not to present the idea of eGovernment as a case handling system (i.e. a back office tool) but instead constantly work from a citizen/user perspective. According to the HI it is too often the case that the development runs in the same structures as before without careful analysis. However, the HI mentioned that avoiding predestined development paths is important. For example, the HI mentions that it is now possible (through available technology) to create a decentralised decision-making at the same time as a centralised administrative support system. Furthermore, the image of implementation and diffusion is perceived of as somewhat undeveloped, there is more to be gained in this than often put forward (for example intuitive systems to avoid costly educations and familiar interfaces and structures to avoid confusion). Put together, the image of the transformation towards eGovernment according to the HI contains an inherent tension since one is supposed to:

"...enhance quality inside existing financial frames and at the same time preferably reduce costs" (HI)

This way of reasoning is also repeated in the other two interviews. The head of administration talks about reducing costs for the administration:

"Of course, we have to reduce costs for administrative routines; you can't justify having such an expensive administrative apparatus" (HA)

While the head of personnel talk about redirecting resources:

"It's a matter, as I see it, then it's a matter of being able to redirect resources, and by that I mean money and so on, to what is our core business" (HP)

Additionally, according to the HP the fundamentals of a municipality are that its existence is legitimised by the citizens (which the HP was under the impression was often forgotten). The main argument has to be the usefulness for the citizen and to ease the interaction between the citizen and their authorities. From such a perspective, the HP also addressed the question with regards to the structure of the organisation, since the resources are, in a sense, always considered to be too limited. This also corresponds to the statement that the municipality is a quite inert organisation by nature, and adjustments and changes are carried out slowly. Another, complementary, driving force which he saw as essential for the transformation was;

"...not to be old fashioned" (HP)

This again stressed that it might be necessary to rethink the local municipalities' original assignment.

This part of the diagnostic framework shows that words such as reducing costs, enhanced service, effectiveness and accessibility were a common way to understand the goals and reason behind eGovernment. Another, complementary, driving force mentioned was "image" and "not to be old fashioned". This strengthens the fact that eGovernment is, in many respects, perceived as being a megatrend which it is necessary to follow in order to not be described as out of date. This is also supported when eGovernment is described as an unavoidable change, even if they highlight three different aspects of this unavoidable change; business development, organisational change and a rethinking of the assignment.

Interpretative Frames - Own Role and Responsibility

The second step in the diagnostic framework is to view and express the mediators "interpretative frames" (Orlikowski & Gash, 1994). To analyse their interpretative frames is an attempt to grasp the complexity of how different people interpret and use the idea differently due to their interpretative frame. In order to be able to understand their viewpoint of the transformation it is necessary to understand both their position and their responsibility. Interpretative frames could be understood as a combination of self-reflexivity and institutional context here referred to as own role and responsibility.

When the mediators were asked to describe their own role and responsibility in the eGovernment transformation process three different interpretative frames appeared. These three different frames illustrate how the translations due to the respondent's different formal positions i.e. how they are trying to practically transfer their perceptions into management strategies. Even if they, on an abstract level, agree upon the overarching idea their perception of focus in the following step differs. The head of personnel described his responsibility as:

"...on a general level everything about the employees" (HP)

This was later specified as being a matter of creating good relations with the union, allocating resources, and a management strategy resting on creating understanding.

"...ultimately it is about allocating resources... and to lead to create understanding among the employees" (HP)

The importance of the employees are also mentioned by the head of administration and in the descriptions of his own role and responsibility the HA several times mentioned that he (i.e. the organisation) should be the one to define the organisational need and that the IT department had a responsibility to provide the necessary technology. This was understood by us as being an emphasized statement that organisational requirements

must be considered prior to any technological development. Another important function he saw for himself in the transformation process was to "get everyone on the train" and that the coming generation could be viewed as an opportunity for a shift in strategic recruitment policy.

In comparison the descriptions of the HI's own role and responsibility he emphasized that he did not have any personnel responsibility and said that:

"My responsibility is the technical improvement, to support the organisation and to be sensitive for the needs they communicate" (HI)

Instead he saw as an important function for himself to play a visionary role and to:

"...send signals out in the organisation about what information technology could contribute to" (HI)

To summarise, the diagnostic framework so far has showed that the overarching ideas of the eGovernment concept structured the mediators understanding of the vision in a similar way. But when the concept of eGovernment was connected to own role and responsibility the mediators have to couple the idea to a practical situation and the circumstances became more problematic. From the analysis three interpretative frames emerged, that of (a) a personnel perspective with a leading responsibility of creating understanding among the employees that the municipality's existence is totally dependent on the citizen, (b) a focus on the need for strategic management with a leading responsibility to "get everyone on the train" and make sure that the organisation has employees who can adjust and find meaning in the transformation, and finally (c) a focus on technological support with a leading responsibility to be visionary and send signals about possible changes.

These three different frames also illustrate the translations due to the respondent's different formal positions i.e. how they are attempting to transfer their perceptions in a practical manner into management strategies. Even if they, at an abstract level, agree upon the overarching idea, their perception of the focus in the following step differs.

Organisational Understanding: A Way to Make Sense

The third part of the diagnostic framework focuses upon the sense making process by which leaders provide a practical view regarding the guiding eGovernment vision. As we have seen their perceptions of role and responsibility affects their suggestions with regard to different strategies for change. In this section the concept of 'sensemaking' are used as a matter of organisational understanding in pointing out the appropriate practices for eGovernment.

When the head of personnel were describing the organisational view he mentioned that the organisation still has a few managers who are still unable to use their own email, a problem he also refeered to as a broad lack of IT-skills in the organisation. Thus, from the HP´s perspective, this transformation is a very slow process. Another problem mentioned was:

"...too many managers in an organisation together with a never ending lack of resources" (HP)

The HA understood it as a necessity to:

"...create such a situation where everybody is directed in the same way" (HA)

It was described as being a difficult issue to manage an organisation which had a high average age, low computer maturity and weak central management.

In the HI´s view with regards to the organisation and employees, he stressed the necessity to define:

"...how we should use our resources to develop the business" (HI)

An important issue was thus to consider that in this municipality:

"...it is very much about certain individuals and each development is tightly connected to specific persons which is not that good" (HI)

The transformation process was said to need:

"...a clear goal picture" (HI)

When the head of personnel emphasized the necessity to firstly define or redefine what the municipality assignment is or should be, the head of administration saw the employees as the key and emphasized the necessity of a strategic recruitment. However, the head of IT strategic department saw the most important aspect as being to create intuitive systems and avoid static development processes.

If we understand this along with the recommendations offered by Weick (1995) it means that the way that different events are structured into meaningful actions by the mediators is a way to make sense. The ways the mediators understand their organisation (potentials and limitations) are important with regards to how they make sense of the eGovernment strategies. For example, it is a means of sense making to suggest a strategic recruitment when the organisational problem is considered to be low IT-skill and an average age which is too high. It also makes sense to suggest a front office focus if the problems are considered to be a consequence of the back office perspective and so on.

Diagnosis of eGovernment Transformation: Emphasizing Enactment

Hitherto, we have discussed the diagnostic framework according to how an idea becomes enacted into an organisational setting from three different perspectives; translation, sense making and interpretative frames. These concepts have served as diagnostic tools to analytically differentiate between three different dimensions embedded in the process of enactment. The model has until now examined local translations of eGovernment in relation to how the idea makes sense in relation to the organisational understanding and the organisational understanding in relation to interpretative frames. By doing so, the diagnostic framework offers a possibility to draw attention to how different mediators self reflexive processes are related to their institutionalised forms of understanding. In other words, how organisational members will use existing knowledge to interpret and make sense of the concept of eGovernment. How they (i) translate the goals and the reasons behind the policy, (ii) their interpretative frames in connection with position and responsibility, and (iii) how they make sense and legitimise the change according to the organisational understanding.

By putting these dimensions together three different diagnosis emerges. In the first diagnose the central image regarding how to succeed in the eGovernment transformation process, according to the HP, involved ever more careful structuring of the organisation. In a new and more suitable structure (among other things, by separating the simple and complex processes) he saw the potential of better allocating resources (financial and human). To do so, the HP stressed a leadership based on mutual understanding.

In the second diagnose, the central theme of the eGovernment transformation process was more of an enhanced strategic management process, with recruitment being one of the dominant tools. Low computer literacy and attitudes toward information systems cannot be addressed by education only but has to be partly dealt with by a generational shift. According to the HA it is important that the organisational 'needs' are dealt with prior to technological 'needs' since it is necessary to remain in strategic mode. Otherwise, it becomes easy to become caught up in the wrong issues. The HA emphasised that it is important to put the organisational development in front of the electronic transformation. A subject which in a self-reflexive statement is stated as:

"... my impression is that one has not sufficiently dealt with this subject" (HA)

And finally in the third diagnose, the head of the IT-strategic department (HI) said that it is important to remember that technology is not the objective; it is rather the thought processes with regards to how to run the organisation that should be the central issue. The HI also stressed the fact that it is important not to present the idea of eGovernment as a case handling system (i.e. a back office tool) but instead to constantly work from a citizen/user perspective. Until the present time, there was felt to be too much focus placed on the case handlers as users, which from his view was disadvantageous and caused a static development. Instead, it was suggested that the starting point should be focussed on the citizen. The implementation was seen as the key, but it was required to be more front office oriented.

According to the HI, it is too often the case that the development runs in the same structures as before without careful analysis. However, the HI mentioned that avoiding predestined development paths is important. Put together, the image of the transformation towards eGovernment according to the HI contains an inherent tension since one is supposed to:

"...enhance quality inside existing financial frames and at the same time preferably reduce costs" (HI)

When analysing these three diagnoses as a whole, in order to provide a deeper understanding regarding the diagnostic framework it becomes obvious that the way that an idea is made sense of differs based on different interpretative frames and the central image of how to succeed in the eGovernment transformation process are intimately related to interpretative frames and ways of making sense. For the HP, who saw his role as being to create understanding it was important to rethink and define the municipality assignment. To be able to create understanding it can be useful to have a clear picture of the assignment. The HI, who saw his role as being a visionary, emphasised a shift in focus, from back to front office but the HA who saw his role as being to "get everyone on the train" highlighted the importance of strategic management and recruitment. In this matter the diagnostic framework reveals the logic of appropriateness between eGovernment, different areas of interest and appropriate organisational practices – the logic between translation, interpretative frame and way of making sense.

A DIAGNOSTIC FRAMEWORK IN PRAXIS: TWO WORKSHOP SETTINGS

In order to communicate, access and analyse our research findings further we have arranged two different workshops, one on the subject of "Managing e-government transformations in local state agencies" and another one on the subject of "IT-visions in rural areas" with a number of managers within the public administration. Managers from different authorities and local governments were invited to the workshops. Both of the workshops took place at the Mid Sweden University. They lasted from ten am to three pm and was structured into three different themes: (1) What is supposed to be done and why, (2) How do I want to do it and (3) How do I look upon the organisation and the employees? The managers were structured into working groups. After completing the first group discussions around the first theme they were asked to write down their reflections which we then together reflected upon. This repeated itself three times, one for each theme.

And once again the method and model seemed to be rewarding. As in the first study the over-

arching idea of eGovernment showed itself to be loosely coupled to their respective practical situations. When asked to reflect upon the relation between the idea of eGovernment and possible practical implications, several managers seemed surprised about the fact that they actually never really had given it any careful examination. Few of them had come across any national policies and they had definitely not seen the policies produced on a European level.

Still they shared a similar, but from obvious reasons not that detailed, understanding about the idea of eGovernment. In the next phase they tried to analyse how they wanted it to be done and then they presented to some extent more elaborated strategies. The strategies differed between the managers and between the working groups but they existed, and by sharing them between themselves they seemed to get hold of valuable information about possible perspectives on how to transform the public administration. It also became clear that some perspectives were more common than others, for example, to make back-office more efficient from an economical perspective seemed to be the prominent idea. But they also reflected upon the fact that the idea of eGovernment often is communicated as a way to enhance service for the citizens and that there might exist a tension between economic rationalisation and higher quality of service and that such a possible tension seldom were touched upon.

Furthermore, their discussions around how they looked upon the organisation and the employees exposed several interesting choice of strategies. For example, some of the managers revealed an increasing need to be in charge of the process, to be able to steer their subordinates in a more direct manner. As they experience it, the major obstacle in the implementation process is the employees who from different reasons (low computer skills, inability to adjust to change etc.) make the daily work integration process impossible. Different strategies were suggested and some of the managers even came to the conclusion that they actually needed to remove some of the most hesitant employees. On the other hand, they also discussed the possibility to do this in a different mode. To actually acknowledge the difficulties instead of simply wishing them away, and by doing so constructively deal with them, but they lacked methods to do so. Many implementation methods for increased digitalisation of government do not consider these issues but instead dismiss the problem as a generation gap which left the managers without support.

Even though the workshop was thought of as mostly an opportunity to communicate and test the model, than actually performing a scientific study. It again manifested the impression that the model highlights the fact that a deeper analysis of the eGovernment transformation makes them question what is and what should be the fundamental assignment of the public administration (government, local authorities etc.).

To highlight this insight, in for example strategic workshops, gives the local managers a possibility to reflect upon their role as eGovernment managers in order to their task, responsibility and their subordinates. If the local public administration managers come together and try to diagnose their eGovernment management strategies they will be able to identify the core processes and its internal obstacles.

To sum it up, the way that local managers understand their organisation (assignment, potentials and limitations) are of utmost importance for how they make sense of the eGovernment concept. We think that a diagnostic framework for eGovernment management is a useful method to identify both obstacles and facilitators embedded in the local context of the public administration.

CONCLUSION

In the beginning of this paper we argued that it is important to consider leaders as local mediators of mega trends and to create knowledge about the diverse translation processes, with the main focus on enactment rather than on vision. In order to do so, we developed a theoretical and empirical diagnostic framework for analyzing leaders' interpretations of an eGovernment vision. The diagnostic framework put forward in this paper provides a dynamic three step analysis of the mediating process which reveals how an idea becomes enacted into an organisational setting from three different perspectives; translation, interpretative frames and sense making. Put together these three steps serve both as diagnostic tools to analytically differentiate between dimensions embedded in the process of enactment and to diagnose the necessary transformation.

However, in this case, one of the interesting outcomes of the diagnostic framework is that on one hand the goals and reasons of eGovernment was translated by the respondents in a harmonizing way but on the other hand the individual diagnosis showed an incongruence in the way they understood how this should be executed. In the light of the suggested diagnostic framework it becomes obvious how different key processes and management strategies are made invisible by an "imaginary" feeling of agreement concerning the overarching idea (what is supposed to be enacted) among the local mediators. As the diagnostic framework shows, managers' perceptions of both their subordinates and their own responsibility, affects how they suggest different strategies for change. The vagueness and loosely coupled vision of eGovernment upon which they agreed did not provide them with any directions as to how it was to be enacted. Thus, managers are placed in a situation of ambiguity, which they have to manage. When they attempt to cope with this ambiguity, they give the vision a practical sense, but, as Schein (1992) points out, they are not always aware of how they are doing it. In summation, the way they understand their organisations (assignment, potentials and limitations) is of utmost importance with regard to how they make sense of the eGovernment concept. Therefore the model provides the tools to disclose their understandings and enhance the possibility of managing the transformation since the enactment process, as a matter of translation, interpretative frames and sense making, holds invaluable clues for the eGovernment transformation.

In conclusion, the explorative examination of the diagnostic framework reveals the logic of appropriateness between local mediators, eGovernment, different areas of interest and appropriate organisational practices – the logic between translation, interpretative frame and way of making sense. The diagnostic framework makes it possible to trace the administrative culture (and institutional memories) in the different frames that can emerge. It makes it possible to analyse how local managers translate task, as well as responsibility and view on subordinates, due to their organisational culture and their different formal positions. In other words, how they make sense of the transformation are shown by the means they use in transferring their perceptions into actual management strategies. We would like to claim that inside these perceptions the diagnosis holds important knowledge for the eGovernment management. When analysing these three diagnoses as a whole, in order to provide a deeper understanding regarding the diagnostic framework, it becomes obvious that the way that an idea is made sense of differs based on different interpretative frames. And the central image of how to succeed in the eGovernment transformation process is intimately related to these interpretative frames and different ways of making sense. From their different formal positions they identify different administrative core processes which carry, as mentioned above, obstacles for the eGovernment transformation. The means by which local leaders and managers translate the overarching idea of eGovernment is significant for the overall implementation process. We therefore argue that even though the framework is theoretically based it has a stronger value in praxis as a tool for reflection, interpretation and cooperation.

REFERENCES

Andersen, K. V. (2004). *E-Government and Public Sector Process Rebuilding: Dilettantes, Wheel Barrows, and Diamonds*. New York: Springer.

Andersen, K. V., & Henriksen, H. Z. (2005). The First Leg of E-Government Research: Domains and Application Areas 1998-2003. *International Journal of Electronic Government Research, 3*(4), 26–44.

Anttiroiko, A. (2002). Strategic Knowledge Management in Local eGovernment. In Grönlund, Å. (Ed.), *Electronic Government – Design, Applications, and Management*. Hershey, PA: Idea Group Publishing.

Avolio, B. J., Kahai, S., & Dodge, G. E. (2001). E-leadership – Implications for theory, research, and practice. *The Leadership Quarterly, 11*(4), 615. doi:10.1016/S1048-9843(00)00062-X

Beaumaster, S. (2002). Local Government IT implementation Issues: A challenge for Public Administration. *Proceedings of the 35th Hawaii International Conference on Systems Science*

Bekkers, V., & Homberg, V., (2007). The myths of E-Government: Looking beyond the Assumptions of A New and better Government in *The Information Society, 2* p. 373-382

Cavaye, A. L. M. (1996). Case study research: a multifaceted approach for IS in *Information Systems Journal, 6*, pp. 227-242

Cooper, R. B., & Zmud, R. W. (1990). Information technology Implementation research: A technological Diffusion Approach. *Management Science, 36*(2), 123–139. doi:10.1287/mnsc.36.2.123

Czarniawska, B., & Joerges, B. (1996). Travels of ideas. In Czarniawska, B., & Sevon, G. (Eds.), *Translating organizational change*. New York: Walter De Gryter.

Denzin, N. K. (1970). *The research act in Sociology: A theoretical introduction to sociological methods*. London: Butterworths.

Di Maggio, P. J. (1988). Interest and Agency in Institutional Theory. In Zucker, L. G. (Ed.), *Patterns and Organizations: Culture and environment*. Cambridge: Ballinger Publishing Company.

Doolin, B. (2004). Power and resistance in the implementation of a medical management information system. *Information Systems Journal, 14*(4), 343–351. doi:10.1111/j.1365-2575.2004.00176.x

Fountain, J. (2001). *Building the virtual state: Information technology and institutional change. Brookings Institution: Washington, DC: Giddens, A., (1984). The Constitution of Society*. Berkeley: University of California Press.

Frenkel. et al. (1998). Beyond bureaucracy? Work organization in call centres in *The international Journal of Human resource Management, 9*(6). pp.957-979

Grönlund, Å. (2002). *Electronic Government – Design, Application and Management*. Hershey, PA: Idea Group Publishing.

Jaeger, P. T. (2003). The endless wire: E-government as global phenomenon in *Government Information Quarterly, 20*, p.323-331

Klein, H., & Meyers, M., (1999). A set of principles for conducting and evaluating interpretive field studies in *MIS Quarterly, 23*, pp. 67-93.

Kraemer, K., & King, J. L. (2006). Information Technology and Administrative Reform: Will E-Government Be Different? *International Journal of Electronic Government Research, 2*(1), 1–20.

Latour, B. (1987). *Science in action: How to follow Scientists and engineers through society*. Cambridge, MA: Harvard University Press.

Latour, B. (1996). *Aramis or the Love of Technology*. Cambridge, MA: Harvard University Press. Translated by Catherine Porter.

Lawrence, T. B., & Phillips, N. (2004). From Moby Dick to Free Willy: Macro cultural Discourse and institutional Entrepreneurship in Emerging Institutional Fields. *Organizations, 11*(5), 689–711. doi:10.1177/1350508404046457

Lenk, K., & Traunmuller, R. (2000). Presentation at the IFIP WG 8.5 Working Conference on *"Advances in Electronic Government"*, Zaragoza, 10-11 February.

Löfgren, K., (2007). The Governance of E-government A governance Perspective on the Swedish E-government Strategy in *Public, Policy and administration* 22 (3). p. 335-352

Maguire, S., Hardy, C., & Lawrence, T. B. (2004). Institutional entrepreneurship in emerging fields: HIV/AIDS treatment Advocacy in Canada. *Academy of Management Journal, 47*(5), 657–679. doi:10.2307/20159610

Moon, M. J., & Norris, D. F. (2005). Does managerial orientation matter? The adoption of reinventing government and e-government at the municipal level. *Information Systems Journal, 15*, 43–60. doi:10.1111/j.1365-2575.2005.00185.x

O'Toole, K. (2007). E-Governance in Australian Local Government: Spinning a Web Around Community. *International Journal of Electronic Government Research, 3*(4), 58–83.

Ogawa, R., & Scribner, W. (2002). Leadership: spanning the technical and institutional dimensions of organizations. *Journal of Educational Administration, 40*(6), 576. doi:10.1108/09578230210446054

Orlikowski, W. (1992). The duality of technology: Rethinking the concepts of technology in organizations. *Organization Science, 3*(3), 398–427. doi:10.1287/orsc.3.3.398

Orlikowski, W., & Baroudi, J. (1991). Studying Information Technology in Organizations: Research Approaches and Assumptions. *Information Systems Research, 2*(1). doi:10.1287/isre.2.1.1

Orlikowski, W., & Gash, D. C. (1994). Technological Frames: making Sense of Information Technology in Organizations. *ACM Transactions on Information Systems*, 12()2 April, pp 174-20758-83 pages

Reason, P., & Rowan, J. (1981). *Human Inquiry: A sourcebook of new Paradigm research.* Chichester, UK: John Wiley.

Røvik, K. A. (2000). *Moderna organisationer: trender inom organisationstänkande vid millennieskiftet.* Stockholm: Liber.

Røvik, K. A. (2002). The secret of the winners: Management ideas that flow in Sahlin-Andersson, K & Engwall (ed.), L *The expansion of management knowledge. Carriers, flows and sources.* Stanford, CA: Stanford University Press

Schein, E. H. (1992). *Organizational culture and leadership: A dynamic view.* San Francisco, CA: Jossey Bass.

Schein, E. H. (1994). *Organizational Psychology.* London: Prentice Hall.

Senyucel, Z. (2005). *Towards successful eGovernment facilitation in UK local authorities presented at eGovernment Workshop '05 (eGOV2005), September 13 2005.* Brunel University, West London.

Silverman, D. (1993). *Interpreting Qualitative data methods for Analysing Talk, text and Interaction.* London: SAGE Publications Ltd.

Strauss, A. (1987). *Qualitative Analysis for Social Scientists.* Cambridge, MA: Cambridge University Press. doi:10.1017/CBO9780511557842

Walsham, G. (1995). Interpretative case in IS research: nature and method. *European Journal of Information Systems, 4,* 74–81. doi:10.1057/ejis.1995.9

Weick, K. E. (1995). *Sensemaking in organisations.* London: SAGE Publications.

Weick, K. E. (2001). *Making Sense of the Organization.* Malden, MA: Blackwell publishing

Yang, K., (2003). Neoinstitutionalism and E-government: Beyond Jane Fountain in *Social Science Computer review.* 4(21), p.432-442

Yildiz, M., (2007). E-government research: Reviewing the literature, limitations and ways forward in *Government Information Quarterly,* 24(3), p. 646-665

ENDNOTE

[1]   The research site is a medium sized Swedish municipality with approximately 95 000 inhabitants.

Chapter 19
Metagoverning Policy Networks in E-Government

Karl Löfgren
Roskilde University, Denmark

Eva Sørensen
Roskilde University, Denmark

ABSTRACT

Since the late 1990s, an explicit goal of most industrialized states has been to integrate electronic access to government information and service delivery, examples being 'the 24/7 agency' or 'Joined-up governance'. This aim, which goes beyond the establishment of 'single' governmental websites, calls for both horizontal, as well as vertical integration of otherwise separate public agencies and authorities who are supposed to collaborate towards 'joint' and 'needs-based' electronic solutions to the benefit of citizens. While many authors have described this implementation of a policy aim in purely technical interoperability terms, the authors frame this development as a policy process of metagoverning self-regulating networks. This chapter is primarily a theoretical think piece in which we will present a systematic framework for the analysis of meta-governing the policy process of electronic government. In addition to the value of framing the process as a metagovernance process, they wish to discuss how the metagovernance approach also sheds light on whether or not the on-going process of vertical and horizontal integration leads to more or less centralization, and whether it may contribute to a more democratic process. Their arguments will be supported by empirical illustrations mainly adopted from Scandinavian research.

INTRODUCTION

This article analyzes the current processes of integrating different governmental on-line information and service delivery initiatives which go toward providing single entry points for citizens and businesses. Based on notions of 24/7 Agency, Gateways, Single-windows, One-stop-shops, Supersites, and Joined-up government, the underlying vision is to make electronic government information and services more accessible and interactive, and functionally needs-based (for

DOI: 10.4018/978-1-60960-162-1.ch019

example, based on 'life-situations' such as e.g. birth, marriage etc). In order to fulfill this vision, integration of information domains are needed (Bekkers, 2007). Information domains can be seen as distinctive fields of influence, ownership and control over information; and how information is being used (Bellamy & Taylor, 1998). Naturally, the integration of otherwise autonomous and self-regulatin information domains creates a number of problems. In particular since several actors need to work together.

This challenge of integrating and coordinating a number of concerned actors calls for a meta-governor, usually the responsible political leadership, who by discursive and organizational means, can manage a (policy) network of otherwise self-governing and self-regulating actors, and mobilize, and 'guide' them towards a certain policy goal (Sørensen and Torfing, 2007; Triantifillou, 2007). As in other policy fields in modern societies, current electronic government policies are based around a keystone idea that there has been a dislocation of the traditional hierarchical ('silo') concept of governing in which a strong and unitary state is at the centre of the polity. It is therefore our contention that the policy processes of fully integrated, electronic, single entry points are taking place in a political setting more characterized by governance than government. In other words direct commands, and legal provisions, have been replaced by institutionalized negotiations between otherwise autonomous actors. The theorization of this transition also includes the prospect of metagoverning these networks, which provides us with a completely different approach thus filling out some of the gaps we are able to see in the current stock of literature on e-government.

First, there has so far been a strong technical bias in the literature on e-government towards the design of integrating different systems, whereas the public administration and policy research of the processes has, with some notable exceptions, been almost completely absent from the field (Dunleavy et al., 2006, p. 469). The traditional information

systems literature has some blind spots in terms of the developments in the field. It is too focused on information and process integration in terms of the technical interoperability and interconnectivity, such as e.g. semantic standards (cf. Traumüller and Wimmer, 2004; Klischweski, 2004; Bajaj & Ram, 2007; Guijarro, 2007), whereas the political management and organization of integration has been notably overlooked. Second, there is still a tendency to envisage implementation processes of governmental information systems as vertical processes (albeit acknowledging that they can be either top-down or bottom-up) in which the individual public agency is at the center of the study (cf. Heeks, 2006). Consequently, the managerial and organizational aspects of horizontal on-line integration processes that take place between several interdependent actors are somehow missed out. Third, the question whether the integration processes should be conceived as a response to the fragmentation of public management caused by either old functional and geographical borders, or by New Public Management (NPM) reforms (cf. Eggers, 2005; Dunleavy et al., 2006), is actually an issue that can be addressed through a metagovernance approach. The process of integrating various agencies' electronic information and service facilities does not necessarily mean centralization (Bekkers, 2007). A metagovernance process can actually, as we will discuss below, perpetuate the existing autonomy of various public agents.

Although the approach presented here does not dismiss previous research, we find it essential to expand the domain of inquiry in order to give modern political management processes a more prominent position. The meta-governance approach is not by default a universal framework for describing all the intricacies involved in the process of integrating governmental information systems, but it provides a novel perspective to understanding the policy complexities involved.

The vision of integrated on-line information and services goes beyond the normal internal use of information and communication technologies

(ICTs) in public management. It also goes beyond various public organizations' exclusive websites where the electronic services usually are no more than complementary to standard administrative routines. A basic challenge in this process is that the use of strong policy instruments, such as hierarchical commands and legal provisions, is not always 'appropriate', or possible, when we span functional and geographical borders. This is due to either constitutional vertical restrictions (such as in, e.g. federal political systems, or where sub-national authorities are autonomous *vis-à-vis* central governments), or horizontal constraints (such as in e.g. systems with ministerial government, or strong autonomous agencies, or *quangos*, within bureaucracy). Although these constraints are well-known problems of coordinating policy (Thomas, 1997; Peters, 2006), they simply become more manifest in the case of electronic government. As stated by Robert Denhardt in a comment on the future of public management:

[Future] [p]ublic management will be turned upside-down as the traditional top-down orientation of the field is replaced – not necessarily by a bottom-up approach, but by a system of shared leadership (Denhart, 1999, p. 285)

The journey to the ultimate goal of electronic integrated single entry points is accompanied with structural, political, legal, managerial and cultural challenges (Kernaghan, 2007, p. 112). Consequently, the formations of networks, bringing together the 'stakeholders', have become an increasingly widespread mechanism in electronic government strategies across industrialized democracies, although the organizational design may vary (cf. Fountain, 2001; Bellamy, 2002; Acaud & Lakel, 2003; Eifert & Püschel (eds.), 2004, Pratchett, 2004; Jensen & Kähler, 2006; Lim et al., 2007; Löfgren, 2007). These networks of stakeholders in the field of electronic government usually include representatives of those various governmental agencies (on different levels) which

are supposed to integrate their electronic information and service delivery, but can sometimes also include private, and voluntary actors, or organized interests (such as e.g. national associations of local governments). By integrating all concerned actors, the idea is to make the policy process more inclusive, coordinated, and transparent, avoid duplication, pool resources, and not least, to engender a more 'successful' implementation of the policy vision.

The objective of this article is to apply meta-governance theories to the specific field of electronic government in general. More specifically, the objective is to present a framework for the formation of governmental electronic single entry points in order to produce a more policy-oriented understanding of the on-going e-government integration processes, and thereby generate some claims which we hope can be used for future comparative case studies. In particular, we wish to shed light on whether the integration of access to electronic information and service leads to more or less centralization, and whether it can be used to enhance the democratic aspects of e-government design.

Section two, the theoretical part, will outline the concept of meta-governance. The following four sections will present four different forms of mechanisms for governing self-regulated networks along the dimensions of 'hands-on/ hands-off' and 'limited/strong intervention'. These sections will also a systematic list of meta-governance mechanisms in the field of integrated electronic information and service delivery of governments. The theoretical discussion will be supported by, primarily Scandinavian, empirical illustrations. This focus on Scandinavian countries stems from the fact that there are certain vertical and horizontal demarcations in these countries that underscore problems of integration, although they do occur elsewhere as well. Also, these illustrations are to be conceived as empirical illustrations of our theoretical discussions, rather than as empirical evidence. Finally, the concluding section will discuss how meta-governance in this field affects the production of outputs and outcomes of network governance, and whether metagovernance should be perceived as either centralization or decentralization.

META-GOVERNANCE AS A THEORETICAL APPROACH

Seen from the perspective of the large body of governance theory that has evolved since the 1990s, current developments can be perceived as a part of a general transition within public governance from sovereign forms of bureaucratic rule to meta-governance of self-regulating actors (Kooiman, 2000; Jessop, 2003; Scharpf, 1997; Sørensen & Torfing, 2007). Hence, governance theorists argue that the increasing functional and organizational complexity, dynamism, and fragmentation of public governance processes have spurred on the search for new forms of governance that combine de-centered self-regulation and centralized strategic leadership. In other words, increased fragmentation calls for increased coordination.

The many reform programmes that have been implemented over the last 25 years can be seen as an effort to transform political systems aiming to perform sovereign rule into meta-governing systems in which public authorities seek to regulate self-regulating actors. The New Public Management (NPM) reform programme, may be perceived as a specific meta-governance strategy that aims to meta-govern self-regulating actors through the establishment of market-based competition between public and private actors.

However, in recent years this competition based meta-governance strategy has been modified, and supplemented, by a network oriented strategy aiming to enhance coordination and cooperation between fragmented actors through the meta-governance of self-regulating intra- and inter-organizational governance networks. In an increasingly more complex, functionally divided, and organizationally fragmented world of public governance, such networks have proven to be crucial for the promotion of vertical and horizontal coordination. Governance theorists argue that self-regulating governance networks are valuable because they are able to ensure a highly flexible form of coordination, reduce resistance through the enhancement of ownership, promote resource pooling among stakeholders, and make these resources i.e. knowledge, engagement, and manpower an asset in the promotion of public values (Kooiman, 2002; Jessop, 2003; Kickert & Koppenjan, 2004; Peters & Pierre, 2000). A governance network is defined as a cluster of interdependent actors who coordinate their actions on the basis of negotiated agreements that are reached with reference to a self-constituted regulatory, normative, cognitive and imaginary framework, and by doing so contribute to the production of public values (Torfing, 2005).

The energies and capacities of governance networks, which also exist in self-regulating markets, are closely related to their relative autonomy vis-à-vis public authorities. As such, the ability to harvest the potential benefits of governance networks depends on the degree to which public authorities are able to influence the actions of self-regulating networks without undermining their autonomy. This is exactly what meta-governance is about: the regulation of self-regulation.

Different Forms of Meta-Governance

A review of the theoretical literature on governance networks (Sørensen & Torfing, 2007), and an analysis of studies describing the empirical developments in contemporary liberal democracies (Rhodes & Marsh, 1992; Markussen & Torfing, 2007; Bogason & Zølner, 2007; v. Heffen, *et al.*, 2000; Bogason *et al.*, 2004), points to the presence of four main categories of meta-governance that are available for public authorities in their efforts to meta-govern self-regulating networks and other self-regulating actors. These are: policy and resource framing, institutional design, network facilitation and network participation. As envisaged in Table 1, these four forms of meta-governance techniques vary according to the level and form of intervention exercised by the meta-governor and according to whether meta-governance is performed hands-off at a distance, or hands-on through close interaction between the meta-governor and the self-regulating actors.

Below we will present these four forms of meta-governance one at a time in order to identify the different ways in which meta-governance of self-regulating networks are, or can be, carried out in policy studies of integrating access to public information and services.

META-GOVERNANCE THROUGH POLICY AND RESOURCE FRAMING

First, meta-governance can be carried out through the demarcation of the political and financial conditions under which networks are granted autonomy to govern themselves. Political framing is exercised through the formulation of some overall political goals and governance objectives that the networks must meet. This form of meta-governance is identical with what the NPM-terminology denotes 'management by objectives'. Resource framing takes place through the allocation of a specific amount of fiscal or administrative resources that the self-regulating networks are authorized to use in their self-regulated effort to reach the overall objectives set out in the political framing. As such, policy and resource framing are closely interrelated. As long as the networks encapsulate these general political goals, and do

so without exceeding the resources delegated by the meta-governor, the network maintains a high level of autonomy. If not, however, the level of autonomy is likely to be reduced. As such, meta-governance through policy and resource framing is performed in what Scharpf (1994, p. 40) denotes a 'shadow of hierarchy' that puts pressure on the networks to fulfill their part of the job, and thus earn their autonomy. Policy and resource framing is exercised hands-off in the sense that it does not necessarily call for direct interaction between the meta-governor and the self-regulating networks. The framing establishes a distribution of labor between what is governed by the meta-governor and what is governed by the networks. A part of the bargain behind this distribution of labor is a low level of intervention on the part of the meta-governor vis-à-vis the self-regulating network.

In terms of the electronic government field, this soft mechanism of meta-governance is about communicating some boundaries for the otherwise self-governing networks, and allocating resources to the activities of the same. As a point of departure, the policy domain of electronic government is not 'given', but constructed. In our framework we identify the policy framing by the meta-governor in a) the received significance among the actors for the policy, b) the network actors' responsibility, and c) the evolutionary understanding of the implementation process.

First, in most countries, the e-government domain has been framed by the central actors in government (i.e. the metagovernor), as one of the core pillars of the future public administration or even as a central element in the national future information society strategies as well (cf. Muir & Oppenheim, 2002). Consequently, its imperative

Table 1.

Forms of meta-governance:	Limited intervention	Strong intervention
Hands-off	Policy and resource framing	Institutional design
Hands-on	Network facilitation	Network participation

status is reflected, and repeated, in most official documents regarding the future of public administration. Also, the issue is officially framed as an organizational issue, rather than merely a technical or a public procurement matter.

Secondly, the different actors, on various levels of the public sector, are assigned *responsibility* for the fulfillment of the vision of integrated electronic services. One can just briefly take two Scandinavian (Denmark and Sweden) examples of governmental strategies.

In general terms, Project eGovernment will create a common framework and support cross-cutting co-operation, but the realization of specific gains will require the involvement and commitment of individual public authorities across the boundaries of sectors and levels of authority throughout the public sector (The Danish Government 2004, p.10)

[The Government's] assessment is that the 24/7 Agency must, through its choice and implementation of service channels and electronic services, become part of the larger context that is central e-government. This calls for voluntary collaboration between agencies or Government-led development and strategy throughout the central public administration (SAFAD 2000, p. 11).

Consequently, the issue is framed as a division of labor between the meta-governor and the actors, but where the responsibility, by and large, is placed with a larger group of implementers, i.e. the public agencies, and thus, the members of the networks. Implicitly, this encourages collaboration between a number of actors in one form or another.

Thirdly, we can also conclude, as mentioned above, that the policy programmes for e-government development across industrialized countries, in which fully integrated service solutions is the main aim, is currently mentioned as an overall policy objective. What unites many of these programmes is the focus on evolutionary development paths, usually in the shape of 'ladders' in which the

service development, and the integration between public agents' different services, will go through a series of stages (Layne & Lee, 2001; Goldkuhl & Persson, 2006; Choudrie & Weerrakody, 2007). Examples of this can be found in the Swedish '24/7 Agency model' (SAFAD, 2000), which in turn is based on the Australian national audit office plan for electronic service delivery (ANAO, 1999). As a result, the final policy aim is already set out in advance, whereas deviating policy paths become less likely to materialize.

META-GOVERNANCE THROUGH INSTITUTIONAL DESIGN

Meta-governance can also be exercised through the strategic design of the institutional conditions under which networks govern themselves. By strategically designing institutional structures, meta-governors are able to enhance the propensity of self-regulating networks to act as desired by the meta-governor. The diverse understandings of how institutions structure action in the large complex of contemporary institutional theory produce different tool kits for meta-governing networks (Sørensen and Torfing 2000, pp. 25*ff*). Traditional institutionalism, which focuses on the structuring effects of the formal institutional set up, points to how governance networks can be influenced through some formal guidelines regulating the composition of a governance network, e.g. what stakeholder groups are to participate? What formal competencies does it have? What formal procedures should be followed? Rational choice institutionalism (Scharpf 1997, p. 45; Kooiman 1993, p. 251) points to how the actions of self-regulating networks can be influenced through a strategic design of incentive structures directed towards both the network actors and the governance networks. Strong networks are promoted through the construction of plus-sum games while competition between networks and other actors enhanced through the construction of zero-sum games. Finally, sociological neo-institutionalism (March & Olsen, 1995; Hajer, 1995) shows how self-regulating actors can be meta-governed through the construction and institutionalization of specific discursive story-lines that shape the perceptions of purpose, interests and collective points of identification of governance networks and other self-regulating actors. The strategic launching of such story-lines can promote a sense of shared destiny and meaning in governance networks that spur action in line with the wishes of the story-telling authorities.

The different forms of meta-governance inspired by the different institutionalisms are not alternatives but should be seen as complementary. As such, meta-governance through institutional design may take many forms just as the different forms can be combined in multiple ways that either reinforce or weaken each other. Like policy and resource framing, meta-governance through institutional design is exercised hands-off and at a distance, since designing institutions can take place without involving the implicated networks. However, in contrast to policy and resource framing, meta-governance through institutional design is highly interventionist. The meta-governor's aim is not only to demarcate an autonomous space within which governance networks are allowed to regulate themselves. With institutional design follows influencing the content of the self-regulation through the composition of governance networks, through the strategic construction of incentive steering, and through the internalization of a specific collective points of identification and meaning.

If we turn the attention to the field of electronic government, we can discuss four aspects of meta-governance and institutional design: a) production of discourses, b) financial incentives, c) audit and control, and d) selection of participants. In most of the e-government plans there are signs of a top-down process in which the overarching vision of an 'information society' has been a sign to follow by the networks (cf. Hall, 2005; Hall & Löfgren,

2006). By inspiring a group of otherwise autonomous actors to witness a mental picture in which the classical dilemmas of accessibility, service orientation, and cost effectiveness can be solved through the integration of information systems, the meta-governor(s) can produce a 'story-line'. This language of the new age is not immediately interventionist, although it as a discourse systematically arranges representations of reality with the purpose of shaping the very same (Foucault, 1991). By keeping up a high level of production of visions one ultimately arrives at a shared view on what direction we are moving towards and why. This is also an attempt to shape a common identity around an objective definition of the future among those who for a foreseeable future will carry out the policy. In the electronic government field this can, in particular, be witnessed in the rich publication of policy documents from governments wherein certain buzz-words such as, 'modernization', 'change', 'needs-based service', 'network society', and 'citizen-orientation' are repeated, and replicated, through the whole publication series of the official publications. In addition, the plethora of public management conferences, and similar conventions of the community of civil servants and politicians, fills the function of diffusing the vision. As demonstrated in Hall and Löfgren (2004), there are several examples of how the discourse on electronic government, as expressed in different policy documents, and diffused through conferences, is paraphrased by various actors during interview studies. Likewise, a quote from a Danish study on meta-governance of electronic government by Jensen and Kähler (2007) describes the power of the discourse:

"What we do here", says a Local Government Executive "is a product of thinking in terms of the information society. And it seems to work even if I cannot prove it scientifically. I am convinced that there is something fundamentally right..." (quoted from Jensen and Kähler 2007, p.185).

While the overall visions might be intangible, it is worth remembering that the basic struggles embedded in the discourse of electronic government are well known and far from esoteric.

In addition, there are examples of a more solid and stronger intervention, through financial support to research and development activities in which governments seek to encourage various actors to participate in, for example, technical system development. This is particular true in Sweden where certain funding schemes, organized by the Swedish agency of innovation offer some financial incentives for both private and public actors to form research and development networks in the field of electronic government.

Finally, a rather strong interventionist tool - at a distance - is of course the wide-spread use of auditing techniques, benchmarking and best practices in the field of e-government. By regularly requiring reports on the progress of the on-line integration, the meta-governor is capable of ensuring that the actors in the networks are moving in the right direction. Equally, benchmarking techniques have become a strong interventionist instrument of steering the otherwise autonomous networks. Even though there is evidence that these benchmarking studies are not always consistent, or even relevant in terms of the policy aims (Jansen et al., 2004), they do play an imperative role for governing the actors (Hall & Löfgren, 2006). As Rose points out 'rendering something auditable shapes the processes that are to be audited' (Rose 1996, p. 351). Indeed, this has been a significant element in the strategy of reaching the objectives of fully integrated on-line service and service delivery in many countries. By means of audit and best practice reports, benchmarking exercises, and even straight-forward 'competitions', the actors need to adjust their work to what is demanded in the audit exercises. Examples of this can be found in much of the Swedish e-government policy in which one of the main aims of the policy is to identify readily measurable indicators for the progress of the on-line integration, to regularly

require reports from the networks, and even inspire the actors of the networks through competitions (Hall & Löfgren, 2006).

NETWORK FACILITATION

But meta-governance cannot only be performed hands-off. As suggested by a number of governance network theorists (Kickert et al., 1997; Rhodes, 1997), the hands-on facilitation of governance networks plays an important part in promoting successful network cooperation. Due to the general instability of governance networks, which derives from the fact that they are based on negotiated cooperation between autonomous actors, they are in constant danger of failing in their efforts to regulate themselves. If governance networks are to function successfully, that is to coordinate action among autonomous actors, it is essential that they are able to surmount internal distrust and destructive conflicts between the network actors. The ability of governance networks to develop mutual trust and to turn destructive conflicts into constructive negotiated agreements can be increased considerably through skilful hands-on facilitation of network cooperation. What is called for is a facilitating meta-governor who takes part in the daily activities of the governance network, and supporting the ability of the network to define and solve the overall governance tasks and public values it has set out to fulfill. This network facilitation can take many forms: initiating contacts between potential network actors, giving administrative support to existing networks and hence reducing transaction costs of network participation, mediating conflicts that occur in the negotiation processes, functioning as an ambassador for network actors with few resources, and processing two-way information and communication between a network and its meta-governor that might enhance mutual understanding and recognition.

Even though this form of meta-governance is exercised hands-on, and takes place within the realm of self-regulating networks, it is characterized by a *low level of intervention* in the content of the network governance. Hence, the major objective of the facilitating meta-governor is to enhance the ability of the network to define common goals and to coordinate their actions in the pursuit of these goals as successfully as possible.

As described above in section two, network meta-governance also entails the facilitation of networks in which the meta-governor initiates contacts between the actors who are supposed to participate in the network, gives administrative support, mediates in conflicts, and acts as an ambassador for the network. Even though this is a more interventionist strategy, it is still a rather subtle and 'soft' way of governing the networks whereas the promotion of networking *per se* is the objective. In terms of on-line integration, we will here discuss the following mechanisms: a) initiating, sponsoring and composing networks, b) supporting knowledge sharing, and c) trust building. To begin with initiating networks, they are probably the most common mechanisms of meta-governance in the field of electronic government. There is a general tendency among the networks which integrate on-line information and service delivery that they originate from centrally located actors. A comparative study of the EU countries shows that all the national information society (IS) strategies have been accompanied by the formation of inter-ministerial committees, boards of stakeholders, task forces, advisory boards, public-private forums etc, initiated from above (Chatrie & Wraight, 2000, p. 12; see also, Accenture, 2006). Partly this is the result of the underlying rationale in the IS strategies of governments trying to limit their own roles. But this low-key strategy has also, as mentioned above, to do with the problems of coordinating autonomous actors. Consequently, we can witness how the first steps of designing a network are usually taken by the meta-governor who invites the con-

cerned stakeholder to the network, and caters for possible meetings. Following these initial steps, the meta-governor pulls out or tries to remain on the sideline of the network. This was at least the case of the Swedish 'E-forum' (Hall & Löfgren, 2006), and the Danish 'Digital Task-force' (Jensen & Kähler, 2007). With this follows also the prerogative of selecting the 'right' members of the networks. Here it is important that the meta-governor is capable of identifying the right blend of actors dependent on which public values the single entry point should entail. A new electronic service website, which relates to, for example, the industry's needs (e.g. taxes, VAT, etc), should ideally include members of the business community, or trade associations. Later on in the process, the network can be opened to new participants, but at that stage there is probably already a high degree of path-dependency which sets the limits for new issues, or diverging strategies.

In addition to initiating the networks, the meta-governor can also supply the network with knowledge. While many of the networks working for the integration of on-line information and services lack financial means to achieve their objectives, or for various reasons are unable to obtain additional resources from their own organizations, the meta-governor can act as a supplier of knowledge. By supplying indicators, statistics, and other forms of resources, the meta-governor can distribute the information s/he wants to disseminate thereby presenting both the problems and the solutions to the policy problem of the network. This has, for example, been the case in Sweden where the government, through its agency for administrative development, has been the main producer of statistics, user surveys, and other forms of reports which they then have distributed to the electronic government networks (Hall & Löfgren, 2006). A similar example of knowledge dissemination is the Finnish electronic portal 'Kärkiverkosto' which operates as a knowledge sharing electronic database between practitioners and researchers, and which is maintained by the Finnish national fund for research and development (SITRA) (Hyyryläinen, 2004).

Finally, the meta-governor is important for the trust building of the networks. By granting a certain authority to the works of the networks, thereby giving certain seriousness to the same, the meta-governor implicitly enhances the understanding and trust between the actors, and also supports an interactive process between the actors.

NETWORK PARTICIPATION

Finally, meta-governance can be exercised through participation in governance networks (Dunsire 1993, p. 34; Mayntz 1991, p. 18). Network participation represents yet another *hands-on* form of meta-governance that grants the meta-governor a direct platform for interacting with the network actors, and for participating in the debates and negotiations within governance networks. This direct participation and interaction in governance networks gives the meta-governor an important insight into the effects that the hands-off forms of meta-governance have on the self regulation processes within a governance network. In addition, direct participation in governance networks provides meta-governors with a platform for story-telling and for explaining the reasons for the policy and resource framing that defines the autonomy of the network. As such, network participation enhances the vertical coordination, trust and shared understanding between meta-governors and network actors.

Network participation is a far more interventionist form of meta-governance than network facilitation because the meta-governor, like the rest of the network participators, actively takes part in network negotiations in order to gain influence on the shared goals and strategies of the network. Since public authorities tend to have more resources than most of the other network participants, they are in most instances able to dominate the negotiation process and get their

way. The potential capacity of public authorities to obtain influence through the participation in governance networks places them in a difficult position. If they make full use of this capacity, they undermine the horizontal interaction, negotiation, and cooperation logic that constitute governance networks. An asymmetrical distribution of power between the network actors is a condition of being for most governance networks, but if such asymmetries result in hierarchical patterns of interaction the networks fall apart. As such, public authorities and other strong meta-governors who participate in governance networks, must constantly balance their efforts to gain influence against the need to maintain and promote the horizontal patterns of interaction that is the glue that keeps governance networks together. The difficult act of participating in network governance, without undermining the self-regulating capacities of the governance network, points to a general consideration for meta-governors: how to avoid an overregulation of governance networks that will undermine the constituting autonomy of the governance network, and how to avoid under-regulation that leaves a governance network to regulate itself without any overall direction vis-à-vis the surrounding society.

In terms of the integration process of electronic information and service delivery we wish to point to the role of the meta-governor participating as an active member of a working electronic government network. While the theoretical metagovernance literature usually presumes networks with a considerable high level of participation of private, or voluntary, sector actors, the non-public sector is usually not very well represented in the networks of electronic government. And even if they are, they usually play a limited role such as that of supplier, external partner or consultant, where the interaction is regulated through a contract. However, network participation makes sense if we envisage metagovernance as policy coordination between central – local actors where both sides are interdependent. This was actually the case in the Danish 'Project eGovernment' where the leading central actor in the policy field, the Ministry of Finance, took part on equal terms with Local Government Denmark and Danish Regions. Here, the Ministry of Finance acted as a metagovernor who actively took part in all parts of the process, and tried to influence the process. However, the dilemmas between over- and under regulation, mentioned above, became visible over the years as discussions on more strategic issues, including major economic and organizational repercussions, caused recoil away from the more consensual network mode of governance (Jensen & Kähler, 2007).

CONCLUDING REMARKS

So, how can theories of metagovernance contribute to the on-going discussion on the implementation of single entry points across industrialized democracies? The current process of integrating various authorities and agencies' electronic information provision and service delivery is a revolutionary attempt in the history of public administration which, provided it succeeds, holds the capacity to recast the previous organization of discrete, and often autonomous, public agencies. By applying a metagovernance perspective, we can generate a theoretical framework which can be employed for both prescriptive purposes ('how to go about'), as well as a more analytical point of departure for empirical studies. In relation to the latter, and despite our Scandinavian empirical examples, this framework still needs illuminating by further empirical and comparative case studies. There is all reason to believe that empirical studies will demonstrate a rich institutional variety.

Still, to metagovern networks also demands some further considerations. First, as we have tried to demonstrate above, self-regulating networks can be meta-governed in a number of ways. An effective and successful meta-governance must seek to combine all four forms of meta-governance, but the choice of the most suitable combination

between them depends on the precise character of the governance network in question. As governance networks materialize differently, it makes little sense to search for a general model for the meta-governance of governance networks. Rhodes & Marsh (1992) place networks on a running scale from policy communities to issue networks. Whilst some of the networks involved in the process of creating single entry-points can be classified as policy communities based on previous long-term, and bottom-up based, collaboration between agencies with 'natural' interfaces (such as e.g. tax and welfare benefits), others are the result of top-down processes in which reluctant actors are more or less 'forced' into collaboration. In this context, the former seem to be more averse towards too interventionist forms of metagovernance (as that inevitably means loss of autonomy), whereas the former somehow presuppose stronger interventionist forms of metagovernance.

Regarding one of our overall questions concerning whether the integration processes of electronic single entry points leads to more or less centralization, we envisage two different paths.

First, there is risk of failure if the meta-governor does not intervene at all, or demonstrate ambiguity, or opaqueness, in terms of the final policy objective of electronic government integration. This was for example revealed in an audit report of the Swedish government's policy on 'the 24/7 Agency' in which the integration process of the central government's vision of a fully integrated 'network administration' very quickly came to a halt, or rather, never really started (SNAO, 2003). The vision behind the Cabinet's aspiration of integrating electronic services never became clear to the actors, and by employing a hands-off design, only entailing policy framing to a group of actors the integration process became a failure. Even though this is an example *par excellence* of classical implementation problems (Pressman & Wildawsky, 1973), ambiguity of policy objectives is devastating for meta-governance.

Thirdly, and quite the opposite, too much emphasis on control and hands on might also jeopardize the prospects of the fulfilling the vision. The network actors must be able to see some incentives, and acceptance of the meta-governor. If the network just becomes a realm for idle talk without any prospects that the members will gain anything, the chances of success are limited. The examples presented by Jensen and Kähler (2007) demonstrate that the actors in the Danish 'Project eGovernment' started to retract and withdraw once they could no longer see any beneficial effects, and that the whole set up of meta-governing just was part of a general governmental cut-back on public expenditures.

Finally, it should be emphasized that the metagovernance approach can actually be used to enhance the democratic influence in the integration, and thereby the design, of new on-line services. While there has been a strong movement towards more 'participatory design' from the 1970s and on-wards, this movement has mainly concerned two groups of actors (the designers and the end-users), and has chiefly discussed the democratic aspects in the context of industrial relations. The overall issue whether on-line integration processes actually should be subject to democratic control, and become more than merely a 'technical-administrative' matter is rarely discussed in the literature. Sørensen & Torfing argue in an article that the metagovernance perspective heralds a new form of 'democratic anchorage' in political decision-making, given that we go beyond the representative (liberal) understanding of democracy (Sørensen & Torfing, 2005). However, democratic metagovernance requires that we seek to approximate four different democratic anchorage points. First, there needs to be some form of control by elected politicians; at least at a distance. Second, the involved actors must be anchored in the membership basis of the participating groups and organizations. Third, there should be a territorial anchorage in terms of the citizenry. That is, there should be some

mechanisms for holding networks accountable for the actions. Finally, networks need to comply with a certain set of democratic rules and norms which provides the network with a stable set of democratically established frames. Naturally, there is no guarantee that the network will develop in a more democratic direction. However, the meta-governance perspective at least provides us with some frames for a different and less technocratic approach to on-line integration.

REFERENCES

Acaud, D., & Lakel, A. (2003). Electronic government and the French state: A negotiated and gradual reform. *Information Polity*, 3-4(8), 117–132.

Accenture (2006). *Leadership in customer service: building the trust*. Retrieved October, 21, 2008 from: http://www.accenture.com/xdoc/en/industries /government/acn_2006_govt_report_FINAL2.pdf]

Andersen, K. V. (2004). *E-government and public sector process rebuilding: dilettantes, wheel barrows, and diamonds*. Boston: Kluwer International. Bajaj, A., & Ram, S. A comprehensive framework towards information sharing between government agencies. *International Journal of Electronic Government Research*, 3(2), 29–44.

Bekkers, V. (2007). The governance of back-office integration. *Public Management Review*, 9(3), 377–400. doi:10.1080/14719030701425761

Bellamy, C. (2002). From automation to knowledge management: modernizing British government with ICTs. *International Review of Administrative Sciences*, 68(2), 213–230. doi:10.1177/0020852302682004

Bogason, P., & Zølner, M. (2007). Methods for network governance research: an introduction . In Bogason, P., & Zølner, M. (Eds.), *Methods in Democratic Network Governance* (pp. 1–20). Basingstoke: Macmillan.

Chatrie, I., & Wright, P. (2000). *Public strategies for the information society in the member states of the European Union*. Brussels: Information Society Activity Centre, DG Information Society.

Choudrie, J., & Weerrakody, V. (2007). Horizontal process integration in e-government: the perspective of a UK local authority. *International Journal of Electronic Government, 3*(3), 22–39.

Denhardt, R. B. (1999). The future of public administration. *Public Administration and Management, 4*(2), 279–292.

Dunleavy, P., Margetts, H., Bastow, S., & Tinkler, J. (2006). New public management is dead – Long live digital era governance. *Journal of Public Administration: Research and Theory, 16*(3), 467–494. doi:10.1093/jopart/mui057

Dunsire, A. (1993). Modes of governance . In Kooiman, J. (Ed.), *Modern governance: New Government-Society Interactions* (pp. 21–34). London: Sage.

Egger, W. D. (2005). *Government 2.0: Using technology to improve education, cut red tape, reduce gridlock & enhance democracy*. New York: Rowman & Littlefield Publishers.

Eifert, M., & Püschel, J. O. (Eds.). (2004). *National electronic government: comparing governance structures in multi-layer administrations*. London: Routledge.

Foucault, M. (1991). Politics and the study of discourse . In Burchell, G., Foucault, M., & Gordon, C. (Eds.), *The Foucault effect. Hertfordshire: Harvester Wheatsheaf* (pp. 53–72).

Fountain, J. E. (2001). *Building the virtual state. Information technology and institutional change*. Washington, D.C.: Brookings Institution Press.

Goldkuhl, G., & Persson, A. (2006). *From E-ladder to E-diamond – reconceptualising models for public e-services*. Paper for the 14th European Conference on Information Systems (ECIS2006), June 12-14, Gothenburg, Sweden.

Guijarro, L. (2007). Interoperability frameworks and enterprise architectures in e-government initiatives in Europe and the United States. *Government Information Quarterly, 24*(1), 89–101. doi:10.1016/j.giq.2006.05.003

Hajer, M. (1995). *The politics of environmental discourse: ecological modernization and policy process*. Oxford, UK: Clarendon Press.

Hall, P., & Löfgren, K. (2004). The rise and decline of a visionary policy: Swedish ICT-policy in retrospect. *Information Polity, 9*(3-4), 149–165.

Hall, P., & Löfgren, K. (2006). *Politisk styrning i praktiken* [Political Governance in Practice]. Malmö: Liber.

Heeks, R. (2006). *Implementing and managing eGovernment. An international text*. London: Sage.

vHeffen, O., Kickert, W. J. M., & Thomassen, J. A. (2000). *Governance in modern society. effects, change and formation of government institutions*. Dordrecht: Kluwer Academic Publishers.

Janssen, D., Rotthier, S., & Snijkers, K. (2004). If you measure it, they will score: an assessment of international eGovernment benchmarking. *Information Polity, 9*(3-4), 121–130.

Jensen, L., & Kähler, H. (2007). The Danish ministry of finance as meta-governor – the case of public sector digitalisation . In Marcussen, M., & Torfing, J. (Eds.), *Democratic network governance in Europe* (pp. 174–191). Basingstoke: Macmillan.

Jessop, B. (2002). *The Future of the capitalist State*. Cambridge, UK: Polity Press.

Kernagaghan, K. (2007). Beyond bubble gum and goodwill: integrating service Delivery. In Sandford, B., Kernaghan, K., Brown, D. Bontis, N., Perri 6, Thompson, F.(Eds.), *Digital state at the leading edge*, Toronto: University of Toronto Press. (Ch. 4).

Kickert, W. J. M., Klijn, E.-H., & Koppenjan, J. F. M. (Eds.). (1997). *Managing complex networks: strategies for the public sector*. London: Sage.

Klichewski, R. (2004). Information integration or process integration? How to achieve interoperability in administration . In Traumüller, R. (Ed.), *EGOV2004, Berlin*. Heidelberg: Springer Verlag.

Kooiman, J. (Ed.). (1993). *Modern governance. New government-society interactions*. London: Sage.

Kooiman, J. (2003). *Governing as governance*. London: Sage.

Koppenjan, J. F. M., & Klijn, E.-H. (2004). *Managing uncertainties in networks – a network approach to problem-solving and decision-making*. London: Routledge.

Layne, K., & Lee, J. (2001). Developing fully function e-government: a four stage model. *Government Information Quarterly*, *18*(2), 122–136. doi:10.1016/S0740-624X(01)00066-1

Lim, E. T. K., Tan, C.-W., & Pan, S.-L. (2007). E-government implementation: balancing collaboration and control in stakeholder management. *International Journal of Electronic Government Research*, *3*(2), 1–28.

Löfgren, K. (2007). The governance of e-government. A governance perspective on the Swedish e-government strategy. *Public Policy and Administration*, *3*(22), 335–352.

March, J. G., & Olsen, J. P. (1995). *Democratic governance*. New York: The Free Press.

Mayntz, R. (1991). *Modernization and the logic of interorganizational networks*. MPFIG Discussion Paper 8. Max Planck Institut für Gesellschaftsforschung.

Mayntz, R. (1993). Governing failure and the problem of governability: some comments on a theoretical paradigm . In Kooiman, J. (Ed.), *Modern governance: new government-society interactions*. London: Sage.

Muir, A., & Oppenheim, C. (2002). National information policy developments worldwide I: electronic government. *Journal of Information Science*, *3*(28), 173–186. doi:10.1177/016555150202800301

Peters, G. P. (2006). Concepts and theories of horizontal policy management . In Peters, G. B., & Pierre, J. (Eds.), *Handbook of public policy*. London: Sage.

Pierre, J., & Peters, G. P. (2000). *Governance, politics and the state*. London: St. Martin's Press.

Pratchett, L. (2004). Electronic government in Britain . In Eifert, M., & Püschel, J. O. (Eds.), *National electronic government: comparing governance structures in multi-layer administrations*. London: Routledge.

Pressman, J. L., & Wildavsky, A. (1973). *Implementation*. Berkeley: University of California Press.

Rhodes, R. A. W. (1997). *Understanding governance. Policy Networks, governance, reflexivity and accountability*. Buckingham: Open University Press.

Rhodes, R. A. W., & Marsh, D. (Eds.). (1992). *Policy networks in British government*. Oxford, UK: OUP.

Rose, N. (1999). *Powers of freedom: reframing political thought*. Cambridge, UK: Cambridge University Press. doi:10.1017/CBO9780511488856

Scandinavian Political Studies, *28*(3), 195–218. doi:10.1111/j.1467-9477.2005.00129.x

Scharpf, F. W. (1994). Games real actors could play: positive and negative coordination in embedded negotiations . *Journal of Theoretical Politics*, *6*(1), 27–53. doi:10.1177/0951692894006001002

Scharpf, F. W. (1997). *Games real actors play. Actor-centred institutionalism in policy research*. Oxford, UK: West View Point.

Sørensen, E., & Torfing, J. (2000). *Skanderborg på landkortet – et studie af lokale styringsnetværk og politisk handlekraft* [Skanderborg on the map: a study of local governance networks and political effectivness]. Copenhagen: Jurist og Økonomforbundets Forlag.

Sørensen, E., & Torfing, J. (2003). Network politics, political capital, and democracy. *International Journal of Public Administration, 26*(6), 609–634. doi:10.1081/PAD-120019238

Sørensen, E. & Torfing, J. (2005). The democratic anchorage of governance networks,

Sørensen, E., & Torfing, J. (2007). Theoretical approaches to meta-governance . In Sørensen, E., & Torfing, J. (Eds.), *Theories of democratic network governance* (pp. 169–182). Basingstoke: Macmillan.

Swedish National Audit Office (SNAO) [Riksrevisionen]. 2003. *Vem styr den elektroniska förvaltningen?* [Who Governs the Electronic Government?], Report 2003:19. Stockholm.

Thomas, C. W. (1997). Public management as interagency cooperation: testing epistemic community theory at the domestic level. *Journal of Public Administration: Research and Theory, 7*(2), 221–246.

Torfing, J. (2005). Governance network theory: towards a second generation. *European Political Science, 24*(4), 305–315. doi:10.1057/palgrave.eps.2210031

Traumüller, R., & Wimmer, M. (2004). E-Government: the challenges ahead . In Traumüller, R. (Ed.), *EGOV2004, Berlin*. Heidelberg: Springer Verlag.

Triantifillou, P. (2007). Governing the formation and mobilization of governance networks . In Sørensen, E., & Torfing, J. (Eds.), *Theories of democratic network governance* (pp. 183–198). Basingstoke: Macmillan.

Chapter 20

New Directions for IT Governance in the Brazilian Government

Fabio Perez Marzullo
Federal University of Rio de Janeiro, Brazil

Jano Moreira de Souza
Federal University of Rio de Janeiro, Brazil

ABSTRACT

This chapter presents an IT Governance Framework and a Competency Model that was developed to identify the intellectual capital and the strategic actions needed to implement an efficient IT Governance programme in Brazilian Government Offices. This work is driven by the premise that the human assets of an organization should adhere to a set of core competencies in order to prioritize and achieve business goals that, when seen from a government perspective are related to public resources management. IT Governance may help the organization to succeed in its business domain; consequently, through effective investment policies and strategic decisions on IT assets, the organization can come up with a business-IT alignment proposal, capable of enabling and achieving highly integrated business services.

INTRODUCTION

Research made in recent years has shown that, to efficiently apply IT resources in a well-designed and responsible fashion it is important to implement an IT Governance programme in a way that it might allow the organization to maximize its actions in its business context, therefore, gaining competitive advantage (CARR, 2004). In many countries, the Government economy share might range from 20% to 50% (Weill, 2004).

The Brazilian economy is not different and also has a significant impact on several sectors of the economy, and, as exemplified by the Brazilian Growth Acceleration Programme (PAC), the need to control huge investment amounts imposes new strategic approaches and requires new support endeavours (PAC, 2007).

What must be understood is that, with massive Government investments in all economic sectors, how does one establish an efficient IT infrastructure that must be pervasive and reach every government sector in order to avoid effort and public resource waste? And mostly, how can

DOI: 10.4018/978-1-60960-162-1.ch020

IT become a strategic asset for the Brazilian government (or any Government), helping with the investment process, and consequently improving the quality of its services? To reach a plausible answer, we first need to understand the Government IT Governance scenario and then identify ways in which government organizations might benefit from more efficient IT Governance programmes, therefore, improving the infrastructure needed to support government services.

Many are the factors that influence this goal. Organizations continuously review their IT investments, seeking new integration models and strategic use of IT services in order to come up with better strategies for their business domain problems (Fernandes, 2006). For public organizations such effort directs towards the creation and aggregation of 'public value', meaning better use of the 'Machinery of Government', always focused on the public well-being. Adding to what is explained by Weill (2004), public organizations should provide services that perform in an effective and efficient way, promote such public well-being. Public services may be perceived as the real assets the governmental apparatus represents to the population. Consequently, the incremental optimization of these services allows a sustained evaluation of these public values.

IT Governance today is a change driver and, as such, must be considered in high levels of power as it will affect every investment policy the organization might come up with, and this is one of many reasons that motivate us to create this chapter. The pursuit of this vision is presented in the following sections as we describe what has been done so far, and what is still under research.

MOTIVATIONS

We have reached a new era of competition. Organizations can no longer afford to delegate IT decisions to IT officers. What we see now is an increasing need for business integration and such integration can only be achieved through strategic business alignment with IT services. Also, the objective should be the creation of an organizational strategic 'thinking' which would be responsible for driving IT efforts towards the achievement of business goals (Marzullo, 2009).

At the same time, the organization should be aware that IT investments are to be controlled and prioritized as the word of order is to create value while relentlessly reducing costs. In fact, IT elements have become important business assets that not only contribute to achieving business goals but also revolutionize the organization as a whole. IT is an excellent innovation driver and, as such, should be properly used and controlled by the organization (Weill, 1998).

This vision is shared amongst non-profit organizations as well. Considering Government organizations, extreme care should be exercised when using public resources to sponsor public programmes. Challenges faced by Government strategists resemble those of private organizations: costs must be minimal and results maximized. However, it is even more complicated as they are dealing with public resources, and therefore controlling mechanisms should be applied as fiercely as possible.

Business and Government officers ask themselves constantly how resources can be invested without risking failing business goals, or, more specifically, how governments can invest public resources to obtain the best results for the taxpayer. And one of the answers, in today's 'Information Age' is: an efficient Information Technology strategy. It not only contributes to Government actions, but also helps coordinate better ways to attend to the needs of the country, whether by means of new services or through the improvement of old ones. In fact it can be used to reinvent the way in which the Government reaches its citizens, and regarding the Brazilian Government, the so called Electronic Government (e-Government) programmes are at the top of such actions (MP, 2009).

Information technology is the single largest capital expense in many organizations that, when correctly managed, allows to efficiently achieve business goals (Weill, 1998). Seeking that efficiency, the Brazilian Department of Planning (MP, 2009), in conjunction with the Department of Defence (MD, 2009) and the Federal University of Rio de Janeiro, strived to create IT Governance directives that were capable of guiding efforts towards the standardization of IT in support to the needs of government and the population.

Given the above analysis the following structure has been conceived to (1) understand IT Governance and its current scenario in public organizations; and (2) propose the competency model and governance framework that is being used in Governance Models at SERPRO (Serpro, 2010)

IT GOVERNANCE FOUNDATIONS

Today, as explained in (WEILL, 1998), we experience a competitive business environment in which organizations are forced to manage numerous assets – human, financial, physical – and have to invest heavily on technology, trying to achieve four main objectives:

a. Transactional;
b. Structural;
c. Informational;
d. Strategic.

These objectives, on one hand, help establishing the ground for sponsoring the right initiatives to manage all the information regarded to be important and necessary to comply with business needs. In order for the organization to obtain proper benefits from these actions, i.e., actually achieve a competitive advantage, it is necessary to build a technological infrastructure capable of helping it manage its intellectual capital. More precisely, to help the organization create a learning

environment with the conception of "knowledge isles" that would help fomenting organizational initiatives to meet its strategic goals.

On the other hand, bad information management makes the organization vulnerable to ill-conceived actions and strategies that are not properly aligned with organizational goals and business needs. This scenario would eventually leverage undesirable management behaviour, making strategic decisions less reliable and risky (Marzullo, 2008).

This can be avoided by restructuring the organizational processes, strengthening the use of IT Governance in order to control IT assets properly. However, the way IT Governance is approached today, is limited to the use of good practices embodied by methodologies like CobIT (Cobit, 2007) and ITIL (ITIL, 2002), and this chapter was designed to show the reader that IT Governance can be much more useful, spatially helping organizations build their IT-business alignment plans (FERNANDES, 2006). So, what is proposed here is to answer questions like:

1. What is IT Governance?
2. Why use IT Governance?
3. Who is responsible for assuring IT Governance? and
4. How to establish IT governance efficiently throughout the organization, in order to fulfil the needs of the organizations?

Such questions are approached here because they are receiving so much attention from CEOs worldwide, and this is a reflection from the fact that IT Governance can leverage actions towards gaining competitive advantage (Broadbent, 2005). Experienced organizations are now engaged in revising their IT investments seeking new models and integration techniques to accurately use IT assets in the business environment, hoping that at some level, innovations and, consequently, larger benefits would result on higher return of investments (WEILL, 2006).

WHAT IS IT GOVERNANCE?

Before understanding the concepts that surround IT Governance it is necessary to analyze a few Corporate Governance concepts. Corporate Governance defines the idea of how and why an enterprise should be governed. It states roles and responsibilities of those who are in charge of the organization.

With the emergence of large corporations, the responsibilities and decisions were executed and assumed by Boards of Directors, where the group itself had the power to define the strategies and business objectives. The owners assumed a role of shareholders and managers came to be hired in the market.

This transition establishes that employees should be hired directly from the market to fill positions that would directly influence the success or failure of the organization. This generated a series of new questions that, in practice, raised doubts on the guarantees that those who were involved in executive control would be able to hold strategic positions envisioning the organization's best interests (KAEN, 2003).

The control of the organization belongs to shareholders, but strategic decisions are now taken by those who are part of the Board of Directors, including the Figure of the CEO. It is up to the board and the CEO to identify business needs and think strategies that would ensure the survival of the organization.

Therefore, good corporate governance, as published in 1999 by the Organization for Economic Cooperation and Development (www.oecd.org) means rethinking the organization to create an organizational structure that will target business goals and monitor performance to ensure the achievement of these goals.

This structure should identify and communicate which are the desired behaviours, as well as the roles and responsibilities of each stakeholder, and clearly state that misconduct from those re-

sponsible for the organization is liable to punishment (Weill, 2004).

For example, in July 2002, amidst the biggest corporate scandals amongst FORTUNE 100 North American corporations, new legislation was created to achieve a new control standard for corporate finance, which would directly affect the segment of Information Technology (Ruzbacki, 2005). A new law called Sarbanes-Oxley was enacted in November of 2004, aimed at preventing future scandals by forcing organizations to follow a set of reporting rules that would allow better control mechanisms (Peterson, 2004).

As shown in studies presented by (Ross, 2004) the key issue in the management of effective IT governance is the ability to identify those who are responsible for making strategic decisions for the organization as well as those who will respond on these same decisions.

The study showed that only 38% of organizations knew how to identify the way IT was governed in their structure and whether it was profitable. Understanding how IT is governed, and how to align and integrate their services in meeting business goals is the main indicator of success, chosen by 60% of executives in large corporations when asked if their organizations failed to reach their market goals.

IT Governance, in its broader sense, corresponds not only to the acronyms commonly seen in the market, but also to the decision-making rights and responsibilities that encourage appropriate behaviour in the use of IT (Ross, 2004). IT Governance reflects the general principles of corporate governance while focusing on managing IT resources to achieve business goals. IT should not be considered in isolation but as a fundamental part in delivering the service to end users. Its integration with the processes in the organization is essential for one to achieve higher quality levels than those associated with the competitive environment in which the organization operates.

For example, Figure 2 presents the corporate hierarchy structure that should be considered

Figure 1.Broad outline of how modern corporations are structurally organized (KAEN, 2003)

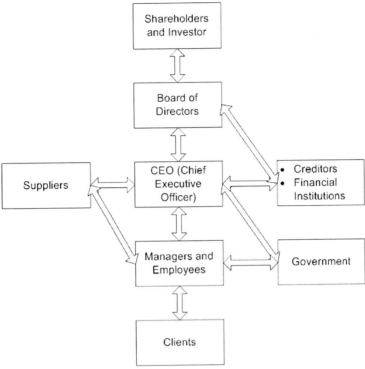

Figure 2. General modern corporation structure associated with IT elements (adapted from KAEN, 2003)

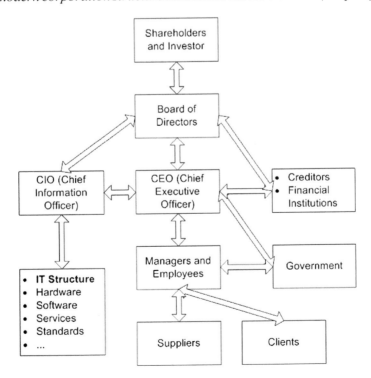

when IT governance is seriously considered and sponsored by the organization.

WHY USE IT GOVERNANCE?

Effective IT management requires a careful understanding of the organization's decision-making arrangements. For example, IT Governance is concerned with questions such as: How decisions should be made? Who should make them? Who has the knowledge to ensure consistency with organizational needs? How to monitor strategic decisions? And how should stakeholders account for good and bad decisions?

Apart from dealing with bureaucratic issues such as rules and roles, it is an IT Governance attribute to determine how business executives (CEOs, CIOs, CFOs, etc.) should be accountable in the decision-making process, and also create an infrastructure capable of improving the speed to align organization needs with IT directives (Broadbent, 2005).

To better understand this aspect, we need to assess how IT governance can be formulated in any organization. Through the governance arrangements matrix, as shown by (Weill, 2006), it is possible to determine which key decisions need to be taken and who should take them. The five key decisions are listed below:

1. **IT Principles:** which clarifies the role of IT in the organization;
2. **IT Architecture:** which defines the requirements for integration and standardization;
3. **IT Infrastructure:** which defines the services and physical assets that will support the business;
4. **The need for business applications:** which specifies the commercial need of IT applications, and also if they will be acquired outside the company or developed internally;
5. **IT investment and prioritization:** which chooses the IT initiatives that will be financed

and how it will affect the organizational processes.

Each decision represents an important aspect that must be addressed to determine the actual IT assets that will be needed in the organization. Each decision points to an organizational structure that must be built so that a comprehensive decision-making process aligned with corporate strategic goals can be conceived.

Some may say that this is not a simple task, however, the archetypes presented below define how management groups can be arranged to simplify the decision making process. They can be seen as six organizational decision groups:

1. **Business monarchy:** senior managers make up the council where all decisions are taken, including those related to IT;
2. **IT monarchy:** the IT specialists are responsible for defining key decisions;
3. **Feudalism:** each business unit has its own decision-making committee and are independent from each other;
4. **Federalism:** a combination of the corporate centre and business units, with or without the involvement of the IT staff;
5. **IT duopoly:** the IT group and some other group (i.e., top management or leaders of business units);
6. **Anarchy:** making decisions individually or in small groups;

By combining the key decisions with the archetypes, it is possible to create an institutional decision arrangement, shown in Tables 1 and 2, which help formalize the organizational decision arrangement. This arrangement is important because it helps establishing the correct functional structure for each decision.

An interesting example can be seen in the case of Motorola. It demonstrates how the arrangement matrix may be useful in identifying decision-making structure in the organization. As shown

Table 1. Principles, architecture and IT strategy decision arrangement

Decision x Archetypes	IT Principles	IT Architecture	Strategies for IT Infrastructure
Business Monarchy			
IT monarchy			
Feudalism			
Federalism			
Duopoly			
Anarchy			

in (WEILL, 2006), Motorola has relied on close relations between IT and business needs. At the end of the restructuring process, it identified that its decision-making arrangement should have the configuration shown in Tables 3 and 4.

- Arrangement that contributes to decision making.
- Arrangement that decides.

This example illustrates how the matrix can be used to formalize the decision-making arrangements. One should bear in mind that every organization should, based on its business needs, assess what the best arrangement to achieve the desired alignment is.

Table 2. Decisions regarding application needs in business and IT investments

Decision x Archetypes	The need for business applications	IT Investments
Business Monarchy		
IT monarchy		
Feudalism		
Federalism		
Duopoly		
Anarchy		

Table 3. Decisions regarding principles, architecture and IT strategy at Motorola

Decision x Archetypes	IT Principles	IT Architecture	Strategies for IT Infrastructure
Business Monarchy			
IT monarchy		IT Leaders IT (II)	IT Leaders (II)
Feudalism			
Federalism	Business Leaders (I)		
Duopoly	Administrative Council and IT Leaders (II)	CIO and IT Leaders (I)	CIO and IT Leaders (I)
Anarchy			

Table 4. Decisions on application needs of business and IT investments

Decision x Archetypes	The need for business applications	IT Investments
Business Monarchy		
IT monarchy		
Feudalism		
Federalism	IT Sector and Business Leaders (I)	IT Sector and Business Leaders (I)
Duopoly	Staff of the CIO (II) Business Leaders and IT Leaders (II)	Business Leaders and IT Leaders (II)
Anarchy		

WHO IS RESPONSIBLE FOR ENSURING IT GOVERNANCE?

Considering what was presented in the previous sections, we took care to identify skill dimensions required within organizations that want to sponsor IT governance programmes that are engaged in guaranteeing that the decision-making process is carried out by the right people, and meet the previously set archetypes.

For this purpose, from research work carried out by the IT Governance group at COPPE/UFRJ we identified the most common fields that should be considered by Brazilian public organizations when engaged in processes to implement IT Governance (Marzullo, 2008).

IT GOVERNANCE VALUE FOR PUBLIC ORGANIZATIONS

The intensive and extensive use of IT in public administration is affecting, on a daily basis, Government actions aimed at its citizens. The successive growth in tax collection and new economies brought up by electronic Government programmes, are examples of the pervasive application of IT resources (MP, 2009)

Despite the increasing rate in which, nowadays, IT services contribute to governmental actions, and assumes a critical position as a new Government demand, IT elements have not always been considered as an essential and strategic asset. What we have seen (especially in Brazil) was dependence on private organizations. As a sub-product of this line of action, outsourcing has become a common practice where corrupt politicians linger on defrauding the Brazilian Government. These actions not only were used to obtain financial advantage, but have also cast most of the Government infrastructure and services control into obsolescence.

This situation compelled the present Government to develop a new strategy that would help regaining control over critical IT resources and as such, impose security reliability to the public administration.

As stated by (Fernandes, 2006), IT has become an indispensable part of modern organizations; it has gained value and attention of the top management in private organizations, and as a natural consequence, has also become critical to the high levels of this new public administration.

Understanding the Government Scenario

The conception of an unified IT Governance competency model was related to two questions: (1) how does one acquire proper knowledge to understand the currently employed IT approach; and (2) how does one use the knowledge to model the unified competency and IT Governance framework without imposing unnecessary organizational and cultural changes. The first question was answered by running a series of interviews with IT officers (CIOs) of most of the major IT sectors of Government organizations (see the attached questionnaire exhibit).

After years of Government impositions and budget restrictions, the IT areas presented a scenario of service quality impoverishment. By not understanding that IT services were no longer fit for outsourcing (at least the strategic ones), without risking missing Government goals and incurring into quality depreciation, the impacts were crippling the Government IT infrastructure. Decisions were made with little concern for proper impact analysis and, most of the time, using criteria that were aligned with political interest and as a function of the cost and speed in which the actions resulted for the group in charge. Governance arrangements were typically feudalistic, and all IT principles lacked proper definition and integration.

This old perspective was critical to promote a shift in the way IT assets should be managed. The new approach should be directed to Government areas that were highly dependent on IT services and the output of this new course of action would be a series of statements that would incorporate new directions to sustain the necessary organizational changes (MP, 2009), such as:

1. ***Human Resources Policies.*** In recent years, and adhering to the new vision, the Federal Government has been valuing the so called 'State Careers' which denoted major concerns regarding essential Government activities. This new value imposes new management methods that should be integrated through IT services.
2. ***Government IT Organizations.*** Created with the purpose of assisting the Government. Public IT organizations such as (SERPRO, 2010), needed to align their internal strategies with the new Government strategies. The old scenario, where budget restrictions imposed systematic outsourcing and

degradation of services had to be reviewed and replaced. Suffering from an identity crisis, these organizations needed to regain strength by rethinking and defining new governance strategies that would account for business and service integration, standardizing management models, specific to public organizations, and guaranteeing the quality of services to their customers.

3. ***New Success Strategies.*** Given the foregoing, the new Government strategies for IT assets were essential to improve service quality. New perspectives meant that new solutions, either through the adoption of new policies or through the implementation of new management models, should be defined to improve IT governance inside public organizations. Organizations that managed to succeed, despite the chaotic scenario, must be taken as examples of successful initiatives (Fazenda, 2009) and (TCU, 2009).

As to how efficiently establish IT governance programmes, in order to fulfil organizations needs, the following describes the IT Governance Framework and the competency domain model needed to implement the framework.

THE FRAMEWORK

Elaborating and implementing an appropriate IT Governance programme is not an easy task. It is necessary to understand several organization aspects such as Government vision, mission, business strategies, goals, functional structure, human assets and, along the process, to evaluate the level of maturity attained by the organization. Having identified the current status of the organization, it is necessary to plan all the steps needed to improve the IT Governance structure, and how it can be implemented and managed (Grembergen, 2004).

The public sector faces the increasing burden of compliance with legislative and regulatory demands. The belief is that much of this increase is due to regulation alignment with the information economy, and the need to apply tighter controls, not only in relation to the mobility of information itself, but mainly to guarantee proper resource investment (Jennings, 2004), (MP, 2009) and (MD, 2009).

The framework defined below tries to align aspects oriented by business goals and IT goals without ignoring public administration aspects such as political views, responsible investments, and the best interests of the population, and Ethics. Basically, business strategies and IT strategies should be carefully aligned and integrated to promote the best results in terms of management, decision-making and IT portfolio (which involves IT services, infrastructure, architecture, and standards). There should be considerations regarding outsourcing as well as this is common practice in public organizations, information security techniques, and investment priorities.

The IT Governance Framework uses three distinct aspects to define the Governance Cycle. Initially, all goals and policies should be carefully understood in order to plan Government strategy; this process is commonly achieved by understanding the organization mission and vision, defining strategic business (government) goals and assessment indicators (possibly with the use of balanced score cards). By doing this, IT strategies can be set in order to provide ways to align, integrate and service Government strategies; along this process, performance measurement programmes should be planed, aiming at evaluating business and IT indicators, and also at guaranteeing integration and service quality, so that investments are not wasted in futile efforts. Also, administrative responsibility and Ethics should be employed to avoid the waste of public resources and corruption, which is a serious problem in the practice of many Brazilian politicians.

Outputs from the Governance Cycle are rules and principle definitions for all IT aspects, from infrastructure, architecture and services, to out-

sourcing standards and performance indicators. This approach allows a specific design for each organization, meaning that each organization will have its organizational culture preserved as long as it does not influence the outputs. The final stage is to put the IT-Business alignment and integration programme designed for it into practice.

Finally, an iterative process of performance evaluation is conducted to guarantee that the proposed framework and competency dimensions will converge to an efficient IT Governance model. The model validated was used by the Brazilian Planning Department as baseline guidance to conduct future implementations of IT Governance programmes.

The Dimensions of IT Competency

In general, individual and organizational knowledge are the main aspects that should be considered when analyzing intellectual capital, and it is important to know that knowledge transfer activities should be planned in order to create effective learning processes. Through these learning processes, organizations create background to develop new experts and, consequently, improve intellectual capital (Liebowitz, 2001).

In practice, accurately choosing and implementing the best governance model, designed specifically for each public organization requires, from the top management and particularly top IT managers (CIOs), a great diversity of knowledge elements and skills. In the creation of a systematic vision of such characteristics we proposed a general competency domain that groups different competency aspects needed to implement IT Governance programmes, as shown below.

Each competency dimension described below represents a multidisciplinary subject that adds significant value to the competency domain model and helps establishing a general Government Knowledge Requirements all IT Governance programmes might rely on as an organizational standard.

Figure 3. The government IT governance framework was specifically designed to accommodate actual needs and constraints identified in the Brazilian IT Organizations evaluated. Its elements, however, were extracted and adapted from (Weill, 2004) and (Fernandes, 2006).

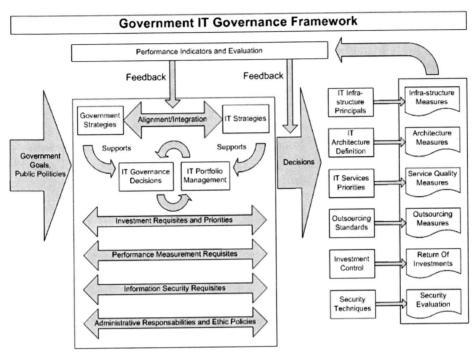

Figure 4. Dimensions of IT governance competency, adapted from (PMCDF, 2001)

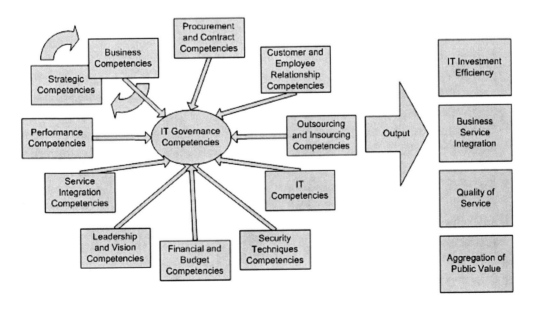

1. ***Business and Strategic Competencies.***
   Business and Strategic competencies are pivotal elements for a successful IT Governance programme. This competency dimension falls outside classic technical skills and leverage the ability to acquire and manage the needs of the organization through IT services (Schubert, 2004). The following sources help creating the background needed to evaluate business skills that should be retained when creating an IT Governance programme for the Brazilian Government.

   a. **Brazilian 1988 Federal Constitution:** Defines elements for national regulations;

   b. **Multi-Annual Plan (*Plano Plurianual – PPA*):** This is a Government instrument to plan a new Brazil. It establishes directives, objectives, and goals for public administration, for a stated period of no less than four years. Planning a PPA is to decide and to prioritize investments inside social development projects (PPA, 2009);

   c. **Budget Directives Act (*Lei de Diretrizes Orçamentárias - LDO*):** Should be defined prior to the LOA. It shows which programmes should be prioritized in the coming year and provides guidance that should be followed by those responsible to make all the investments; it should also detail Government cash flow, stating whether the budget will be sufficient to sponsor the necessary activities (TCU, 2009);

   d. **Annual Budget Act (*Lei Orçamentária Anual - LOA*):** Annual programming of government actions regarding investments described by the PPA. It must identify the origin of the resources (sponsor) and the destination (pursuant to the PPA) (TCU, 2009).

   e. **Law 8112:** Institutes the legal regime public servants have to comply with.

   f. **Law 4320:** Sets legal regulations to prepare and control public budget.

   g. **Tax Responsibility Act:** (*Lei de Responsablidade Fiscal*): Sets regulations to control public budget; it is concerned with tax accountability in order to prevent risks of resource wastage and corruption (TN, 2009).

2. ***Customer and Employee Relationship Competencies.*** This is a knowledge dimension that should be highly considered. Mostly, Government clients are its population and, eventually, other governments and the private enterprise, so relationship competency is at the centre when aiming at successfully achieving business goals. In parallel, as in any organization, the Government has to recruit and retain employees who are capable of coping with their daily jobs with responsibility and quality. It is important to form a workforce of the right competency and size to achieve Government goals.

3. ***Outsourcing and Insourcing Competencies.*** This competency dimension is critical when deciding whether a service can or should be done within or outside the organization. Governments tend to contract external services to comply with business demands but they should also know when to 'bring it home'. It is important to know that outsourcing is a strategic decision that affects the organization budget, goals and structure. Extreme caution should be exercised not only to know when to outsource but also what should be outsourced, and why. In the political environment one needs to carefully analyze and be aware of whether the outsourcing strategy is relevant for the organization and hence decisive to achieve business goals, or if it is only a strategy to retain or gain political power.

4. ***IT or Technical Competencies.*** This competency dimension involves IT aspects, such as infrastructure, architectures, frameworks,

patterns, maturity models, and programming standardizations. The following examples, extracted from (Schubert, 2004), help evaluating the technical skills that should be retained when creating an IT Governance programme.

    a.   Computer System Architectures
- Shared x Partitioned memories;
- Internal x External disk systems;
- Single processors x multiprocessors;
- Host-based x client/server-based;
- Centralized x Distributed.

    b.   Computer System Implementations
- Batch x Interactive;
- Laptop x Desktop;
- Object-Oriented x procedural programming;
- Database implementations.

5.   ***Service Integration Competencies.*** This competency dimension requires strategies to model business and IT services together so that they might be able to assist business needs in an efficient manner. Integration foments system standardization, cost cutting, alignment of IT and Business strategies (Peterson, 2004) and is now commonly achieved through the use of Service-Oriented Architectures (SOAs) as presented in (Marzullo, 2009).

6.   ***Performance Competencies.*** Requires knowledge and skills to analyze and measure whether the aspects presented in the Governance Cycle are complying with business needs and public expectations.

7.   ***Procurement and Contract Competencies.*** Despite some people advocating that this competency falls outside IT Governance aspects, we decided to include it in our Competency Domain because a Government environment will always have needs related to technical aspects, such as equipment, and therefore IT officers should be capable of managing service and supply contracts (Schubert, 2004).

8.   ***Security Technique Competencies.*** Information and supporting processes, systems, and networks are important business assets. Defining, achieving, maintaining, and improving information security may be essential to maintain competitive advantage, cash flow, profitability, legal compliance, and commercial image (ISO/IEC 17799, 2005). As for public organizations, information security is extremely important in order to guarantee national security (MD, 2009).

9.   ***Financial and Budget Competencies.*** Government interests in financial and budget management are vital for political base stability. Dealing with public resources requires ethics and responsibility, otherwise corruption might prevail, and consequently government stability may suffer. Proper financial planning directly affects the lives of the ordinary citizen. With that in mind, the Government should adhere to performance measurement and public, accountable, control systems as the primary means in resource management. Managing budgetary processes, including preparing and justifying a budget and operating a budget under strict rules imposes a level of expertise that should be more efficient than those applied in private organizations.

10.   ***Leadership and Vision Competencies.*** The modern leader requires qualities that are innate or acquired. Besides having technical competencies, personnel skills and conceptual capacity, a strong personality is one of the most important criteria in choosing an organization leader. A leader should show direction, and understand and be able to explain the organization's vision, and have integrity and optimism (Warren, 2001). For public organizations all of these competencies should be empowered by Ethics

and responsibility towards the population. In a leadership position, its competencies determine the success of the organization, as well as whether the organization vision will be achieved or not. His or her behaviour should inspire one's staff, giving directions and mostly hope and meaning to their activities. One should communicate effectively, understand business processes and operations, have strategic thinking and negotiation skills, important items to succeed in contract disputes, and an ability to influence and demonstrate extreme technical proficiency (Schubert, 2004).

This competency dimension domain is a general overview of the elements that make the IT Governance Cycle. IT Governance effectiveness is only partially dependent on IT Officers, more specifically, the competency domain contains competencies that are pervasive to all organizations business aspects, and as such, depends on all top managers to be successfully implemented.

Also, for a complete IT Governance programme, there should be room for the top management officers in the organization to communicate and deliver their message properly. In that case, another scenario should be evaluated: How can IT best support a complex organization consisting of diverse, globally operating business units? And, how do IT Officers guarantee proper commitment from all levels of powers in the organizations?

To successfully answer these questions, the organization should create a learning environment in which every 'member of power' (in that case the CEO, CFO, CTO, etc., in the organization) can communicate and understand how IT influences and helps business needs. According to this research, two levels of competency are necessary to leverage IT Governance in Government organizations: (1) the public Top IT Manager (equivalent to CIO) level of competency; and (2) the other public Top Manager (equivalent to all the CxOs) levels of competency.

Focusing on the Top IT Manager Level, one should have the highest degree of proficiency in each of the competency areas previously described, according to their decision level. It means that a CIO is only really effective if one can think and act according to strategic business plans. As for all the other officers, their competency level should be higher for business-related areas and sufficient to cope with IT terms and theories.

It is important to understand that 'Top Business Managers' in Brazilian Government organizations are more commonly addressed as the 'Authorization Bosses'. They are the ones with the political power to supply the necessary funds to sponsor all IT initiatives. They are the 'decision-making' figures and therefore should have a certain level of IT competence to effectively allow the work to be done. Normally they do not have any competency regarding IT aspects, nonetheless it must be highlighted that it should become a requirement to public managers nationwide.

GOVERNMENT INICIATIVES

We understood, along this chapter, that the use of IT to support government services should be seen as a basic condition to improve the quality of life of the society. Some say that the use of IT should be seen as a prerequisite for the development of social and economic models, especially when the goal is to force the Government to deliver services more efficiently, at low cost, without affecting quality. Governments that promote the use of IT in their bodies improve the levels of integration between the different power domains, promoting accountability in the management of public assets, avoiding ambiguities, and simplifying the Machinery of Government.

As in any business environment, success depends on ethical and secure information management. At a technological level, this requires the use of management techniques and patterns that help the organization define and develop management tools capable of enhancing the way they operate in their business processes. The efficient handling of information, from simple solutions to more complexes often brings considerable benefits to the population as it exposes the corrupt politician that seeks, above all, personal gain and undue empowerment, which is a sickness that affects our society.

Hopefully, today, it is already possible to point out a few ongoing initiatives aimed at creating government rules and standards which might produce good results in the short term. By being aligned with the recent movement of building a stronger Government IT infrastructure, started from the need to establish a general IT Policy, the following topics present a few successful initiatives and show the reader that it is possible to use the Machinery of Government efficiently.

e-PING Standard

One of the initiatives is called e-PING [80]. It defines a set of IT rules and governing policies for interoperability between all spheres of power and society. This work is coordinated by three government agencies:

- The Planning Department's Office of Logistics and Information Technology (SLTI-MP);
- The National Information Technology Institute (ITI, PR);
- SERPRO, Brazilian Finance Department Data Processing Company.

It is focused on defining and developing modern technical procedures, related to:

- **Connectivity:** Determines connectivity conditions between government agencies and society. Focuses on defining standards for message exchanging, infrastructure, and network services.

- **Safety:** Determines the vision for government security aspects such as communication protocols - IP, email messages, encryption criteria, systems development, and network services.
- **Accessibility:** Establishes the electronic devices that can be used to access government services. Define policies for the use of the following devices:
  - **Workstations:** determines the use of Internet browsers, character sets and alphabets allowed in each station, hypertext exchange formats, types of documents allowed for each area, databases (DBMSs) to be used as repositories, and audio and video technologies.
  - Smart Cards, Tokens and others: determining how data should be handled by electronic components that will be used, their communication protocols, security criteria, and physical interfaces.
  - **Mobility:** determines allowed communication protocols, browsers, hypertext standards, message exchanging, and communication technologies for video, audio, and image.
  - **Digital TV:** sets the rules that will be adopted by the Brazilian Technical Standards Association (ABNT).
  - **Information organization and exchange:** sets standards for information exchange between government electronic services. Specifies language standards for building messages, the definition of which data can be transmitted and government metadata standards (e-PMG).
  - **Areas of integration for e-government:** determines the use or construction of XML specifications, giving more robustness to the process of communication and message

exchanging between services. Some patterns like data catalogue (CPD), the XML schemas catalogue and services catalogues (Web Services).

The first version of the e-PING standard was launched in 2005 and today is already in version 4.0 (www.governoeletronico.gov.br). Every year, the document is updated after consultation with the society. As its reference document exemplifies, the e-PING board is focused on developing more formal interactions between the public and private sectors, seeking to improve service delivery. This determines a change of stance from the government regarding how IT solutions are offered to, and by, the private sector.

It establishes a new model of partnership whose goal is to prevent the enslavement of government to proprietary solutions, which is exactly the problem presented at the beginning of this chapter and is very common nowadays. As regards citizen concerns, it brings a light at the end of the tunnel, as it helps strengthen the alignment between government services and IT, which favours the development of better mechanisms to meet the needs of the population in a transparent and accessible way.

The Government Open-Source Portal

Another good initiative that creates a new licensing model and management solution developed by (and for) the Government. Although not directly associated with a technological model for service implementation, the software portal demonstrates how the Government can apply IT solutions to their strategic objectives aiming to minimize the waste of public resources. This portal allows, for example, that problems like the development of similar systems in different government agencies (which is extremely common) can be minimized. Its focus is primarily on free software initiatives, which helps reducing costs, improving technical

solutions and ultimately stimulates the modernization of Government services.

The Comprasnet Electronic Trading Portal

The electronic trading portal, last year alone, brought an economy of hundreds of millions of dollars, lowering the costs for the Government in direct purchases from the private sector. Initiatives as this show us how our money is misused and helps explaining the stratospheric load on public taxes.

Comprasnet, besides being a mechanism to democratize the relationship between business and government, brings more transparency on how things are conducted, exposing purchasing and contracting deals, giving the citizens tools for monitoring and auditing.

Although there are a few more initiatives we could discuss here, there is one that we should emphasize as a great application of the Service-Oriented theory and which is producing amazing benefits to the society. It is presented in the following case study.

Minimizing Bureaucracy at Government-Enterprises Level

'Brazil is the champion of bureaucracy'. This is a statement that symbolizes the precariousness of public services in the country. This problem is pointed out by most organizations as the main obstacle for the creation and expansion of businesses, which ultimately means fewer jobs and fewer opportunities, and that affects the development of the country.

Here is an example: in a report published by the newspaper A Gazeta in the city of Vitória, state of Espírito Santo, on January 10th, 2009, a professional wishing to start a company faced an 11-step bureaucratic process, which are:

1. **Prior consultation on the feasibility of the operation:** To start the process of opening a business, the professional has to go to the City Hall and request clearance on the feasibility of operation, which aims to identify if the entrepreneur can establish its company at the specified address. This prevents, for example, a chemical plant to be built next to a hospital.

2. **Name Search Certificate:** the applicant has to provide a set of three names one understands to be fit for one's business and send it to the State Board of Trade. There, a search will be conducted in order to identify that at least one of the names is not being used by another company.

3. **Articles of Incorporation:** the applicant has to file its Articles of Incorporation with the State Board of Trade.

4. **Corporate Taxpayer registration Number (CNPJ):** after filing the Articles of Incorporation, the entrepreneur receives from the IRS its CNPJ. This registration was created to replace the General Taxpayer Register, the CGC, and aims to unify the identity of companies in all spheres of power, local, state and federal.

5. **State Registration:** In this fifth step, the applicant, when necessary, has to go to the state Department of Finance and request the State Registration number. The number of such registration serves to register the company in the VAT registry and, therefore, it is replaced by formal registration with the State Revenue Department.

6. **Fire Brigade Permit:** the future entrepreneur should ask the Fire Brigade Department a permit for the business. This permit states that an inspection and technical survey was conducted by the Fire Brigade Department to clear the applicant to open the business at the required location.

7. **License to Operate:** Issued by the City Urban Development Department. This document approves the business, taking into account the address, economic activity, environment, safety, morality, public peace, etc.

8. **Request for Document Printing Authorization:** upon approval, the entrepreneur can go to a printer's and have its receipt books printed. This step is done at the Federal Revenue Agency (RFA).

9. **Social Security Registration:** is the act whereby the citizen is registered with the Department of Social Welfare.

10. **Registration with the Employer's Association:** every company is required to join an Union in its class and pay an annual fee.

11. **Licensing of other public bodies:** finally, when necessary, the applicant has to obtain other certifications, such as a health surveillance certificate, in order to start the business.

Simple? Well, more bureaucratic, impossible. In Brazil, these eleven steps impose the entrepreneur a 152-day process (on average). That's right! Five months, a few dozen documents, endorsements and many fees to start a company. Amazing! Anyone would read this and ask how this is possible. Is there anything that can be done to improve this situation? Fortunately, there is, and, more importantly, this is already happening.

Imagine the following scenario: you decide to start an IT consulting firm. You know that consultancy on service-oriented solutions is 'hot' right now and intend to use the knowledge gained by reading a great SOA book (Marzullo, 2009) to help both private and public organizations deploy more efficient business strategies.

You then begin the adventure of opening your business, heading to the City Hall. On arriving there, they hand you a document with some instructions and ask you to go to a room with half a dozen computers. You sit at a terminal and, as described in the document, access the system. You fill up an electronic form with the necessary information, and at the end, after finishing the registration, you receive a submittal number and a message saying that within 48 hours you can access the results on the possibility of starting your business.

You feel in doubt: "Should I go to the Board of Trade to continue the process?" Not knowing what to do, you go to the City Hall Information Officer to find out what to do next. He answers that you should go home and return after 48 hours to check if there is any pending information that will keep your request from being completed. Despite being a little incredulous, you decide to accept the advice and return after the specified period.

Just to be sure, you decide to return three days later, after all, it is always good to give a margin of error. Once there, you address the Information Officer and he requests that you go to the terminal, enter the user name (your submittal number) and the password you received three days before and access any issues that might have been generated in you procedure. After doing all these steps, you enter the report and read: "Congratulations, your application has successfully passed. Please pay the necessary fees and return to complete the procedure."

Unbelievable, huh? Everyone saying that you would take at least five months to open your company and yet, it only took three days. How is this possible? Well, let us explain. It happens that the Mayor of your town probably decided to work for the people and implemented an integrated registration system that streamlines communication between all the departments needed to start a company. The city IT officer decided that by creating a system based on a service-oriented architecture paradigm it would be possible to help the citizens shorten the company start procedure. It runs as follows: the system installed in City Hall (Figure 5) aggregates data input, so the applicant needs only to go to one place to enter all the necessary information (personal and of the company). The system then identifies and separates the information set, necessary for all the departments involved,

splitting it into distinct messages, and activates a set of services installed as Web Services. All departments have a counterpart Web Service where they receive their due messages, and a system connected to each Web Service can validate the data. From that point, all the communication can now be done remotely, which eliminates the 'city tour' to hand the documentation in person at each department.

By using these simple ideas it was possible to speed the process is such a way that it is certainly worth any government investment, but better yet, the solution is as simple as economic, so it only needs minimal political effort to become a national process. It is an efficient solution that adds enormous value as a government service.

Luckily, this initiative is already available and being implemented, in a few states. For example, this solution can now be used in states such as Santa Catarina and Espírito Santo. Of course, not all types of business will have the benefits highlighted here. For example, those who need federal intervention, such as pharmaceutical industries, will still have to go through a longer process, but well below the average of 152 days.

FUTURE DIRECTIONS

Many are the possibilities that arise from good Government initiatives. Mainly, when such initiatives focus on improving the quality of life of the citizens. This paper presented only a few actions that are being conducted to achieve such improvements. In the future, we expect to increase the coverage of the results achieved so far and broaden the number of standards and services that are created and controlled by good IT policies.

Besides presenting a competency domain and an IT Governance framework, future work also involves the optimization of internal processes, identification of new competency domains, the evaluation of change impacts in every targeted

Figure 5. From a single location - the City Hall - the taxpayer can start a company in less than 48 hours

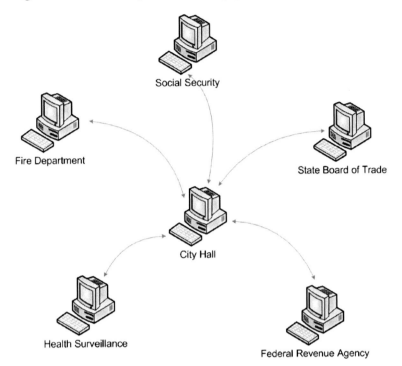

organization, and the development of supporting tools.

CONCLUSION

This chapter presented not only a competency domain model and an IT Governance framework, but an overall analysis of the Brazilian Government IT scenario. It is clear that this research succeeded in creating both the domain competency and IT Governance framework, but the intended outcome was to foment the discussion of applying good IT policies in the Government. After studying the IT assets and policies of current organizations we proposed a governance standard that, according to the Brazilian Planning Department, should be considered as a baseline implementation.

Our contribution is only a small step towards a much broader restructuring of IT services that will follow this sequence of events:

1. To guarantee a baseline level of competency in all aspects represented in the competency domain;
2. To keep implementing the framework in public organization on a nationwide basis;
3. To analyze and correct flaws identified along the implementation process;
4. To validate and finalize the IT Governance Model through feedback collected in performance measurements;
5. To make the model available to other Government sectors, to train and restructure their IT initiatives by designing IT Governance specific programmes and implementing then.

The IT Governance Cycle created for the Brazilian government helps understanding how an IT governance should be conducted, to ensure the alignment of IT and business needs. By using the competency model and the IT Governance Framework, it is possible to come up with the necessary background to effectively govern IT issues without imposing drastic changes to the organization.

Finally, this research also concluded that, without a proper alignment between business and IT in the competitive market, the chances of failure are significant, as it increases the risk of investment loss; and despite the concept of 'business' being different from commercial to government organizations, we encourage the use of the framework to ensure that the IT initiatives can improve the quality of services rendered to users.

EXHIBIT

Field Research Form

Organization:
    City and State:
    Company Sector or Area:
    Occupation:
    Do you influence the decisions in the organization? () Yes () No

This research is focused on data, as related to the IT structure in Brazilian Government organizations. All data is considered confidential and will be used anonymously for academic research purposes.

Part 1. Organizational Structure

Q1) What is the purpose of the organization and area of activity/concern?

Q2) How many people does the organization have?

Q3) How many staff does the IT sector have?

Q4) What is the organization functional hierarchy model?

Q5) What is the IT functional hierarchy model?

Part 2. Business and IT Strategies

Q6) How does the organization plan business strategies? (Describe the process and people involved)

Q7) How does the organization plan IT strategies? (Describe the process and people involved)

Q8) Does the organization outsource its IT assets? (Describe the process and people involved)

Q9) How much is the organization budget and how much is annually invested in IT?

Q10) How are IT investments justified?

Q11) How is IT planned to support business strategies?

Q12) How is IT effectively used to support business strategies?

Q13) Is there any defined IT architectural model? (Describe process, patterns and decisions involved)

Q14) Describe the IT infrastructure in the organization. How are IT services coordinated?

Q15) Is there any ongoing study evaluating business needs and IT support?

Q16) Are the IT areas independent or are they subordinated to higher areas?

Q17) Regarding IT investments, how does project selection occur?

Q18) Fill the table below with your understanding of the proficiency degree a CIO should have in order to achieve business goals (mark with an X the expected level of proficiency):

Table 5.

Competencies	Low	Medium	High
Strategic and Business			
Customer and Employee Relationship			
Outsourcing and Insourcing			
IT or Technical			
Service Integration			
Performance Measurement			
Procurement and Contract			
Security Technique			
Financial and Budget			
Leadership and Vision			

Part 3. IT Governance

Q19) What do you understand by IT Governance?

Q20) How are IT principles established for decision support?

Q21) Is there any ongoing activity to formalize best practices and to identify business opportunities?

Q22) Describe you perception towards cost-benefit in overall IT investments and achieved results.

REFERENCES

Broadbent, M. (2005). *Why Governance Matters.* CIO Insight Magazine. http://www.cioinsight. com/ c/a/ Past-Opinions/Why-Governance-Matters/. Last access: August, 2010.

Carr, N. G. (2004). *Does IT Matter? – Information Technology and the Corrosion of Competitive Advantage.* First Edition, Pub. Boston:Harvard Business School.

COBIT 4.1, (2007), IT Governance Institute. *Framework, Control Objectives, Management Guidelines, Maturity Models.* IT Governance Institute.

Fernandes, A. A., & Abreu, V. F. (2006). *Implantando a Governança de TI – da Estratégia à Gestão dos Processos e Serviços.* Brazil: Brasoft.

Grembergen, W. V. (2004). *Strategies for Information Technology Governance.* Idea Group Publishing.

Haes, S. D., Gremberger, W. V., & Guldentops, E. (2004). *Structures, Processes and Relational Mechanisms for IG Governance.* Hershey, PA: Idea Group, Inc. "ISO/IEC 17799, (2005). *Information Technology – Security Techniques – Code of Practice for Information Security Management.* International Standard Organization.

HP. (2002). *ITIL Practitioner Change Management.* Mountain View, CA: Hewlett-Packard.HP Education Services.

Jennings, T. (2004). *Change Management – An Essential Tool for IT Governance.* Butler Direct Limited.

KAEN. Fred R., (2003). *A Blueprint for Corporate Governance.* First Edition. Amacon, New York.

Liebowitz, J. (1999). *Building Organizational Intelligence: A Knowledge Management Primer-Transforming Organizational Learning into Organizational Learning.* Rockville, MD:Johns Hopkins University.

Marzullo, F. P. (2009). *SOA na Prática: inovando seu negócio por meio de soluções orientadas a serviços.* 1. ed. São Paulo: Novatec, 1. 392 p.

Marzullo, F. P., & Souza, J. M. (2008). *A Competence Based Approach to IT Governance in the Brazilian.* WORLDCOMP'08 - The 2008 World Congress in Computer Science, Computer Engineering and Applied Computing, 2008, Las Vegas, Nevada.

Marzullo, F. P., & Souza, J. M. (2009). New Directions for IT Governance in the Brazilian Government. *International Journal of Electronic Government Research, 5,* 57–69.

Ministério da Defesa, www.defesa.gov.br Last Access: June, 2010.

Ministério da Fazenda - Receita Federal, http://www.receita.fazenda.gov.br/ Last access: June, 2010.

Ministério do Planejamento. www.planejamento. gov.br. Last Access: June, 2010.

PAC – Brazilian Growth Acceleration Programme http://www.fazenda.gov.br/ portug ues/ releases/2007/r220107-PAC.pdf. Last access: May, 2007.

Peterson, R. R. (2004). *Integration strategies and Tactics form Information Technology Governance.* Hershey, PA: Idea Group Inc.

PMCDF - Project Management Competency Development Framework. http://www.pmi.org/ 2001. Last access: May, 2010.

PPA - Plano Plurianual. http://www.planobrasil. gov.br, Last access: May, 2007.

Ross, J., & Weill, P. (2004). *IT Governance on One Page.* MIT Sloan School of Management.

Ruzbacki, T. (2005). *Sarbanes-Oxley, and Enterprise Change Management.* MKS White Paper.

Schubert, K. D. (2004). *CIO Survival Guide – The Roles and Responsibilities of the Chief Information Officer.* Wiley.

SERPRO. http://www.serpro.gov.br/. Last access: June, 2009.

Tesouro Nacional. http://www.tesouro.fazenda. gov.br 2009. Last access: June, 2010.

Tribunal de Contas da União. http://www.tcu.gov. br 2009. Last access: June, 2010.

Warren, B. (2001). *Uma força Irresistível.* HSM Management.

Weill, P., & Broadbent, M. (1998). *Leveraging the New Infrastructure – How Market Leaders Capitalize on Information Technology.* Boston: Harvard Business School Press.

Weill, P., & Ross, W. J. (2004). *IT Governance: How Top Performers Manage IT Decision Rights for Superior Results.* Boston: Harvard Business Press.

Compilation of References

Abdul-Gader, A. H., & Alangari, K. (1994). *Information technology assimilation in the government sector: an empirical study*. Final report of funded research, King Abdul Aziz City of Science and Technology, Riyadh, Kingdom of Saudi Arabia.

Abramson, M. A., & Means, G. E. (2001). *E-Government 2001*. New York: Rowman & Littlefield Publishers, Inc.

Acaud, D., & Lakel, A. (2003). Electronic government and the French state: A negotiated and gradual reform. *Information Polity, 3-4*(8), 117–132.

Accenture (2004). *e-government Leadership: High Performance, Maximum Value*. Retrieved May 7, 2005 from: http://www.accenture.com/xdoc/en/ind ustries/government/ gove_egov_value.pdf

Accenture (2006). *Leadership in customer service: building the trust*. Retrieved October, 21, 2008 from: http://www.accenture.com/xdoc/en/industries /government/acn_2006_govt_report_FINAL2.pdf]

Acosta, C., & Siu, N. (1993). Dynamic event trees in accident sequence analysis: application to steam generator tube rupture. *Reliability Engineering & System Safety, 41*, 135–154. doi:10.1016/0951-8320(93)90027-V

Adar, E. (2002). *End-to-End Security Assessment*. Paper presented at the Analysis & Assessment for Critical Infrastructure Protection, Brussels.

Adar, E., & Wuchnet, A. (2005). *Risk Management for Critical Infrastructure Protection (CIP) Challenges, Best Practices & Tools*. Paper presented at the First IEEE International Workshop on Critical Infrastructure Protection (IWCIP'05), Darmstadt, Germany.

Agarwal, R., & Prasad, J. (1998). A conceptual and operational definition of personal innovativeness in the domain of information technology. *Information Systems Research, 9*(2), 204–215. doi:10.1287/isre.9.2.204

Agarwal, R., & Prasad, J. (1999). Are individual differences germane to the acceptance of new information technologies? *Decision Sciences, 30*(2), 361–391. doi:10.1111/j.1540-5915.1999.tb01614.x

Agarwal, R., Sambamurthy, V., & Stair, R. (2000). The evolving relationship between general and specific computer self-efficacy: An empirical investigation. *Information Systems Research, 11*(4), 418–430. doi:10.1287/isre.11.4.418.11876

Ahlstrom, V., & Longo, K. (Eds.). (2003). *Human factors design standard for acquisition of commercial-off-the-shelf subsystems, non-developmental items, and developmental systems (DOT/FAA/CT-03/05/HF-STD-001)*. *Atlantic City International Airport*. NJ: FAA William J. Hughes Technical Center.

Ajzen, I., & Fishbein, M. 1980. *Understanding attitudes and predicting social behaviour*. Englewood Cliffs, N.J.: Prentice-Hall.

Akbulut, A. Y. (2002). An Investigation of the Factors that Influence e-Information Sharing Between State and Local Agencies, *Proceedings of the 8th Americas Conference on Information Systems*, USA, (pp. 2454-2460.

Akinci, H. (2004), *Geospatial Web Services for e-Municipality*, XX ISPRS Congress. Istanbul, Turkey. Retrieved from http://www.isprs.org/proceedings/XXXV/congress/comm2/papers/210.pdf

Akman, I., Yazici, A., Mishra, A., & Arifoglu, A. (2005). E-Government: A Global View and an Empirical Evaluation of some Attributes of Citizens. *Government Information Quarterly*, *22*(2), 239–257. doi:10.1016/j.giq.2004.12.001

Al-Dosari, R., & King, M. (2004). E-government lifecycle model. *Proceeding of UK Academy of Information Systems Conference*, Glasgow Caledonian University, UK.

Alesina, A., Baqir, W., & Easerly, W. (1999). Public Goods and Ethnic Divisions. *The Quarterly Journal of Economics*, *114*(4), 1243–1284. doi:10.1162/003355399556269

Alesina, A., Devleeschauwer, A., Easterly, W., Kurlat, S., & Wacziarg, R. (2003). Fractionalization. *Journal of Economic Growth*, *8*, 155–194. doi:10.1023/A:1024471506938

Alesina, A., & La Ferrara, E. (2002). Who Trusts Others? *Journal of Public Economics*, *85*, 207–234. doi:10.1016/S0047-2727(01)00084-6

Al-Fakhri, M. O., Cropf, R. A., Kelly, P., & Higgs, G. (2008). E-Government in Saudi Arabia: Between Promise and Reality. *International Journal of Electronic Government Research*, *4*(2), 59–85.

Al-Fakhri, M., Cropf, R., Higgs, G., & Kelly, P. (2008). e-Government in Saudi Arabia: Between Promise and Reality. *International Journal of Electronic Government Research*, *4*(2), 5–82.

Al-Farsy, F. (2003). *Modernity and traditional: the Saudi Equation*. Willing Clowes, Beccles, Suffolk, UK.

Al-Gahtani, S., & King, M. (1999). Attitudes, satisfaction and usage: factors contributing to each in the acceptance of information technology. *Behaviour & Information Technology*, *18*(4), 277–297. doi:10.1080/014492999119020

Allendoerfer, K., Friedman-Berg, F., & Pai, S. (2007). *Usability assessment of the* fly.faa.gov *Web site* (DOT/FAA/TC-07/10). Atlantic City International Airport, NJ: Federal Aviation Administration William J. Hughes Technical Center.

Al-Nuaim, H. A. (2009). How "E" are Arab Municipalities? An Evaluation of Arab Capital Municipal Web Sites. [IJEGR]. *International Journal of Electronic Government Research*, *5*(1), 50–63.

Al-Sabti, K. (2005). *The Saudi government in the information society*. Available online at www.yesser.gov.sa. Access on 20/1/2008.

Al-Saggaf, Y. (2004). The effect of online community on offline community in Saudi Arabia. *Electronic Journal Of Information Systems In Developing Countries*, *16*(2), 1–16.

Al-Sebie, M., & Irani, Z. (2005). Technical and organisational challenges facing transactional e-government systems: an empirical study. *Electronic Government. International Journal (Toronto, Ont.)*, *2*(3), 247–276.

Al-Shafi, S. (2008). Free Wireless Internet park Services: An Investigation of Technology Adoption in Qatar from a Citizens' Perspective. *Journal of Cases on Information Technology*, *10*, 21–34.

Al-Shafi, S., & Weerakkody, V. (2007). Implementing and managing e-government in the State of Qatar: a citizens' perspective. *Electronic Government, an International Journal*, *4*(4), 436-450.

Al-Shafi, S., & Weerakkody, V. (2009). Examining the Unified Theory of Acceptance and Use Of Technology (UTAUT) Of E-Government Services Within The State Of Qatar. In *Handbook of Research on ICT-Enabled Transformational Government: A Global Perspective*. Hershey, PA: Information Science Reference.

Al-Shehry, A., Rogerson, S., Fairweather, N. B., & Prior, M. (2006). The motivations to change towards e-government adoption. *E-government Workshop (eGOV06)*, Brunel University, UK.

Al-Sudairy, M. (2000). *An empirical investigation of electronic data interchange (EDI) utilisation in the Saudi's Private organisations*. Unpublished PhD Thesis, University of Leicester, UK.

Al-Tawil, K., Sait, S., & Hussain, S. (2003). Use and effect of internet in Saudi Arabia. *Proceedings of the 6th World Multi-conference on Systemic*, Cybernetics and Informatics, Orlando, Florida, USA.

Al-Turki, S. M., & Tang, N. K. H. (1998). *Information technology environment in Saudi Arabia: a review.* Work report, Leicester University Management Centre. UK.

Amoretti, F. (2007). International organizations ICT policies: e-democracy and e-government for political development. *Review of Policy Research, 24*(4), 331–344. doi:10.1111/j.1541-1338.2007.00286.x

Anand, P. (1993). *Foundations of Rational Choice Under Risk.* Oxford University Press.

Andersen, K. V. (2004). *E-government and public sector process rebuilding: Dilettantes, wheel barrows, and diamonds.* Boston: Kluwer Academic Publishers.

Andersen, K. V. (2004). *E-Government and Public Sector Process Rebuilding: Dilettantes, Wheel Barrows, and Diamonds.* New York: Springer.

Andersen, K. V., & Henriksen, H. Z. (2005). The First Leg of E-Government Research: Domains and Application Areas 1998-2003. *International Journal of Electronic Government Research, 3*(4), 26–44.

Andersen, K. V. (2004). *E-government and public sector process rebuilding: dilettantes, wheel barrows, and diamonds.* Boston: Kluwer International. Bajaj, A., & Ram, S. A comprehensive framework towards information sharing between government agencies. *International Journal of Electronic Government Research, 3*(2), 29–44.

Anderson, C., & Paskeviciute, A. (2006). How Ethnic and Linguistic Heterogeneity Influence the Prospects for Civil Society: A Comparative Study of Citizenship Behavior. *The Journal of Politics, 68*(4), 783–802. doi:10.1111/j.1468-2508.2006.00470.x

Andrews, J. D., & Moss, T. R. (1993). *Reliability and Risk Assessment* (1st Ed. ed.). Longman Group UK.

Annett, A. (2001). Social Fractionalization, Political Instability, and the Size of the Government. *IMF Staff Papers, 48*(3), 561–592.

Anttiroiko, A. (2002). Strategic Knowledge Management in Local eGovernment. In Grönlund, Å. (Ed.), *Electronic Government – Design, Applications, and Management.* Hershey, PA: Idea Group Publishing.

Ateetanan, P. (2001). *Country Report Thailand.* Retrieved November 14, 2006 from: http:// unpan1.un.org/intradoc/ groups/public/ documents/APCITY/UNPAN012806.pdf

Australian Government Information Management Office (2005). *Australian's use of and satisfaction with e-government services.* Department of Finance and Administration, Australian Government Information Management Office. Barton: Commonwealth of Australia

Aven, T. (1992). *Reliability and Risk Analysis* (1st Ed. ed.): Elsevier Applied Science.

Avolio, B. J., Kahai, S., & Dodge, G. E. (2001). E-leadership – Implications for theory, research, and practice. *The Leadership Quarterly, 11*(4), 615. doi:10.1016/S1048-9843(00)00062-X

Awan, M. A. (2007). Dubai e-Government: An Evaluation of G2B Websites. *Journal of Internet Commerce, 6*(3), 115–129. doi:10.1300/J179v06n03_06

Ba, S. and Pavlou, P.A., (2002). Evidence of the Effect of Trust Building Technology in Electronic Markets: Price Premiums and Buyer Behavior, MIS Quarterly, (26:3), 2002, pp. 243-268.

Babbie, E. R. (2007). *The practice of social research* (11th ed.). Belmont, CA: Thomson Wadsworth.

Baier, A. C. (1986). Trust and AntiTrust. *Ethics, 96,* 231–260. doi:10.1086/292745

Bailey, J. E., & Pearson, S. W. (1983). Developing a tool for measuring and analyzing computer user satisfaction. *Management Science, 29*(5), 530–545. doi:10.1287/mnsc.29.5.530

Bailey, J. P., & Bakos, J. Y. (1997). An Exploratory Study of the Emerging Role of Electronic Intermediaries. *International Journal of Electronic Commerce, 1*(3), 7–20.

Bandura, A. (1986). *Social foundations of thought and action: a social cognitive theory*. Englewood Cliffs, NJ: Prentice Hall.

Bandura, A. (1982). Self-efficacy mechanism in human agency. *The American Psychologist*, *37*(2), 122–147. doi:10.1037/0003-066X.37.2.122

Bandura, A. (1997). *Self-Efficacy: The Exercise of Control*. New York: Freeman.

Banerjee, A., Iyer, L., & Somanathan, R. (2005). History, Social Divisions, and Public Goods in Rural India. *Journal of the European Economic Association*, *3*, 639–647.

Bannister, F. (2007). The curse of the benchmark: An assessment of the validity and value of e-government comparisons. *International Review of Administrative Services*, *73*, 171–188. doi:10.1177/0020852307077959

Barber, B. (1984). *Strong democracy: Participatory politics for a new age*. Berkely, CA: University of California Press.

Barki, H., Titah, R., & Boffo, C. (2007). Information system use-related activity: an expanded behavioural conceptualization of individual-level information system use. *Information Systems Research*, *18*(2), 173–192. doi:10.1287/isre.1070.0122

Barne, S. J., & Vidgen, R. T. (2006). Data triangulation and Web quality metrics: A case study in e-government source. *Information & Management*, *4*, 767–777. doi:10.1016/j.im.2006.06.001

Barnett, G. A. (2001). A longitudinal analysis of the international telecommunication network, 1978-1996. *The American Behavioral Scientist*, *44*, 1638–1655.

Barney, D. (2004). *The network society*. Malden, MA: Polity Press.

Baroudi, J. J., Olson, M. H., & Ives, B. (1986). An empirical study of the impact of user involvement on system usage and information satisfaction. *Communications of the ACM*, *29*(3), 232–238. doi:10.1145/5666.5669

Barro, R. J. (1999). Determinants of Democracy. *The Journal of Political Economy*, *107*(2), 158–183. doi:10.1086/250107

Bass, C. & Lee, J.M. (2002). Building a Business Case for EAI, *EAI Journal*, 18-20.

Beath, C. M. (1991). Supporting the Information Technology Champion. *Management Information Systems Quarterly*, *15*(3), 355–372. doi:10.2307/249647

Beatty, J. R. (2000). *Statistical Methods* (6th ed.). New York: McGraw-Hill.

Beaumaster, S. (2002). Local Government IT implementation Issues: A challenge for Public Administration. *Proceedings of the 35th Hawaii International Conference on Systems Science*

Becerra, M., & Gupta, A. (1999). Trust Within the Organization: Integrating the Trust Literature with Agency Theory and Transaction Costs Economics. *Public Administration Quarterly*, *23*(2), 177–203.

Becerra, H. (2006). Welcome to Maywood, Where Roads Open Up for Immigrants. Los Angeles Times, March 21, 2006, p. A-1.

Bekkers, V. (2007). The governance of back-office integration. *Public Management Review*, *9*(3), 377–400. doi:10.1080/14719030701425761

Bekkers, V., & Homberg, V., (2007). The myths of E-Government: Looking beyond the Assumptions of A New and better Government in *The Information Society*, 2 p. 373-382

Belanger, F., & Carter, L. (2009). The Impact of the Digital Divide on E-Government Use. *Communications of the ACM*, *52*(4), 132–135. doi:10.1145/1498765.1498801

Bélanger, F., Hiller, J., & Smith, W. (2002). Trustworthiness in electronic commerce: the role of privacy, security, and site attributes. *The Journal of Strategic Information Systems*, *11*, 245–270. doi:10.1016/S0963-8687(02)00018-5

Bélanger, F., & Carter, L. (2008). Trust and Risk in E-government Adoption. *The Journal of Strategic Information Systems, 17*(2), 165–178. doi:10.1016/j.jsis.2007.12.002

Bélanger, F., & Hiller, J. (2006). A Framework for E-Government: Privacy Implications. *Business Process Management Journal, 12*(1), 48–60. doi:10.1108/14637150610643751

Bell, D., Cox, L., Jackson, S., & Schaefer, P. (1992). *Using Causal Reasoning for Automated Failure & Effects Analysis (FMEA).* Paper presented at the Annual Reliability and Maintainability Symposium.

Bellamy, C. (2002). From automation to knowledge management: modernizing British government with ICTs. *International Review of Administrative Sciences, 68*(2), 213–230. doi:10.1177/0020852302682004

Bellman, B., & Rausch, F. (2004). Enterprise Architecture for eGovernment. In Traunmüller, R. (Ed) *Electronic Government*, conference proceedings,(LNCS # 3183, Springer Verlag Heidelberg et al,pp. 48 – 56).

Benamou, N. (2006). Bringing eGovernment Interoperability to Local Governments in Europe. *egovInterop'06 Conference*, Bordeaux, France.

Benbasat, I., & Barki, H. (2007). Quo Vadis TAM. *Journal of the Association for Information Systems, 8*(4), 211–218.

Bennett, W. L., & Entman, R. (2001). *Mediated politics: Communication in the future of democracy.* New York: Cambridge University Press.

Bentivegna, S. (2002). Politics and new media. In Lievrouw, L. A., & Livingstone, S. (Eds.), *Handbook of new media* (pp. 50–61). London: Sage Publications.

Bertot, J. C., & Jaeger, P. T. (2006). User-centered e-government: Challenges and benefits for government Web sites. *Government Information Quarterly, 23*, 163–168. doi:10.1016/j.giq.2006.02.001

Bertot, J. G., Jaeger, P. T., & McClure, C. R. (2008). Citizen-centered e-government services: benefits, costs, and research needs. *ACM International Conference Proceeding Series, 289*, 137-142 (Proceedings of the 2008 International Conference on Digital Government Research).

Bevan, N. (1999). Quality in use: Meeting user needs for quality. *Journal of Systems and Software, 49*(1), 89–96. doi:10.1016/S0164-1212(99)00070-9

Beynon-Davies, P. (2005). Constructing Electronic Government: The Case of the UK Inland Revenue. *International Journal of Information Management, 25*(1), 3–20. doi:10.1016/j.ijinfomgt.2004.08.002

Beynon-Davies, P., & Williams, M. D. (2003). Evaluating Electronic Local Government in the UK. *Journal of Information Technology, 18*, 137–149. doi:10.1080/0268396032000101180

Bharat, K. and Broder, A. (1998). A technique for measuring the relative size and overlap of public Web search engines. In 7th WWW,.

Bhattacherjee, A. (2001). Understanding information systems continuance: An expectation-confirmation model. *Management Information Systems Quarterly, 25*(3), 351–370. doi:10.2307/3250921

Bhattacherjee, A. (2002). Individual Trust in Online Firms: Scale Development and Initial Trust. *Journal of Management Information Systems, 19*(1), 211–241.

Bimber, B., & Davis, R. (2005). *Campaigning online.* New York: Oxford University Press.

Bitner, M. J., Booms, B. H., & Mohr, L. A. (1994). Critical service encounters: The employee's viewpoint. *Journal of Marketing, 58*(4), 95–106. doi:10.2307/1251919

Blackstone, E., Bognanno, M., & Hakim, S. (2005). Electronic government: Review, evaluation, and anticipated impact. In E. Blackstone, M. Bognanno & S. Hakim (Eds.), *Innovations in E-Government: The Thoughts of Governors and Mayors.* New York, Boulder, Oxford, Lanham: Rowman & Littlefield Publishers, Inc.

Boehm, B. W. (1991). Software risk management: principles and practices. *IEEE Software, 8*, 32–41. doi:10.1109/52.62930

Bogason, P., & Zølner, M. (2007). Methods for network governance research: an introduction. In Bogason, P., & Zølner, M. (Eds.), *Methods in Democratic Network Governance* (pp. 1–20). Basingstoke: Macmillan.

Bonham, G., Seifert, J., & Thorson, S. (2001). *The Transformational Potential of e-Government: The Role Of Political Leadership*. Paper presented at the 4th Pan European International Relations Conference of the European Consortium for Political Research, University of Kent, Canterbury, UK.

Bonoma, T. (1985). Case Research in Marketing: Opportunities, Problems and a Process. *JMR, Journal of Marketing Research, 22*, 199–208. doi:10.2307/3151365

Bordewijk, J. L., & van Kaam, B. (1986). Towards a new classification of teleinformation services. *Intermedia, 14*, 16–21.

Bouti, A., & Kadi, D. A. (1994). A state-of-the-art review of FMEA/FMECA. *International Journal of Reliability Quality and Safety Engineering, 1*(4), 515–543. doi:10.1142/S0218539394000362

Bovens, M., & Zouridis, S. (2002). From Street-Level to System-Level Bureaucracies: How Information and Communication Technology is Transforming Administrative Discretion and Constitutional Control. *Public Administration Review, 62*(2), 174–184. doi:10.1111/0033-3352.00168

Bozeman, B., & Kingsley, G. (1998). Risk culture in public and private organizations. *Public Administration Review, 58*(2), 109–118. doi:10.2307/976358

Bradford, M., & Florin, J. (2003). Examining the Role of Innovation Diffusion Factors on the Implementation Success of ERP Systems. *International Journal of Accounting Information Systems, 4*(3), 205–225. doi:10.1016/S1467-0895(03)00026-5

Bridges.org. (2003/2004). *Our perspective on the digital divide*. Retrieved August 7, 2005 from http://www.bridges.org/perspectives/digitaldivide.html

Broadbent, M. (2005). *Why Governance Matters*. CIO Insight Magazine. http://www.cioinsight.com/ c/a/ Past-Opinions/Why-Governance-Matters/. Last access: August, 2010.

Burn, J., & Robins, G. (2003). Moving Towards E-government: A Case Study of Organizational Change Processes. *Logistics Information Management, 16*(1), 25–35. doi:10.1108/09576050310453714

Bush, G. W. (2007, November 15). President Bush discusses aviation congestion. *Office of the Press Secretary* [Press release]. Retrieved February 22, 2008, from http://www.whitehouse.gov/news/ releases/2007/11/20071115-6.html

Cabinet Office - Office of the E-Envoy. (2004). *E-Government Interoperability Framework (e-GIF)*, London, UK. Version 6.0, URL: http://www.alis-etsi.org/IMG/pdf/ UK _e-Gov_V6_April_04.pdf, p. 4.

California Labor Code. Section 3550(d). Accessed on March 24, 2006 from http://www.leginfo.ca.gov/ cgi-bin/displaycode ?section=lab&group=03001-04000&file=3550-3553.

Carlsson, C., Carlsson, J., Hyvonen, K., Puhakainen, J., & Walden, P. (2006). Adoption of mobile devices/services-searching for answers with the UTAUT. Proceedings of the 36th Hawaii International Conference on System Sciences (2006) IEEE.

Carr, N. G. (2004). *Does IT Matter? – Information Technology and the Corrosion of Competitive Advantage*. First Edition, Pub. Boston:Harvard Business School.

Carter, L., & Belanger, F. (2005). The Utilization of E-government Services: Citizen Trust, Innovation and Acceptance Factors. *Information Systems Journal, 15*(1), 5–25. doi:10.1111/j.1365-2575.2005.00183.x

Carter, B., Hancock, T., Morin, J. M., & Robins, M. (2001). *Introducing Riskman Methodology – The European Project Risk Management Methodology*. Oxford, UK: NCC Blackwell Ltd.

Carter, L., & Belanger, F. (2005). The Utilization of E-government Services: Citizen Trust, Innovation and Acceptance Factors. *Information Systems Journal, 15*(1), 2–25. doi:10.1111/j.1365-2575.2005.00183.x

Carter, L., & Belanger, F. (2004). Citizen adoption of electronic government initiatives (ETEGM03). *Proceedings of the 37th Hawaii International Conference on System Sciences*, IEEE Computer Society, Washington, DC, USA.

Carter, L., & Bélanger, F. (2003). *Diffusion of Innovation and Citizen Adoption of E-government Services.* The Proceedings of the 1st International E-Services Workshop, 57–63.

Castelli, W. A., Huelke, D. F., & Celis, A. (1969). Some basic anatomic features in paralingual space. *Oral Surgery, Oral Medicine, and Oral Pathology, 27*(5), 613–621. doi:10.1016/0030-4220(69)90093-0

Castells, M. (2000). Toward a sociology of the network society. *Contemporary Sociology, 29*, 693–699. doi:10.2307/2655234

Castells, M. (2001). *The Internet galaxy: Reflections on the Internet, business, and society.* New York: Oxford University Press.

Castells, M. (1996). *The Information age: economic, society and culture, 1.* The rise of the network society. Oxford: Blackwell.

Cavaye, A. L. M. (1996). Case study research: a multifaceted research approach for IS. *Information Systems Journal, 6*(3), 227–242. doi:10.1111/j.1365-2575.1996.tb00015.x

Cavaye, A. L. M. (1996). Case study research: a multi-faceted approach for IS in *Information Systems Journal, 6*, pp. 227-242

Center for Democracy and Technology and InfoDev. (2002). *The e-Government Handbook for Developing Countries.* Washington, D.C.: The World Bank, 2002. Retrieved from http://www.cdt.org/egov/handbook/2002-11-14egovhandbook.pdf

Chamlertwat, K. (2001). Current status and issues of e-government in Thailand. *15ᵗʰ Asian Forum for the Standardization of Information Technology*, Kathmandu, Nepal.

Chapman, R. J. (1998). The effectiveness of working group risk identification and assessment techniques. *International Journal of Project Management, 16*(6), 333–343. doi:10.1016/S0263-7863(98)00015-5

Chase-Dunn, C., & Grimes, P. (1995). World-systems analysis. *Annual Review of Sociology, 21*, 387–417. doi:10.1146/annurev.so.21.080195.002131

Chatrie, I., & Wright, P. (2000). *Public strategies for the information society in the member states of the European Union.* Brussels: Information Society Activity Centre, DG Information Society.

Chau, P. Y. K. (1996). An empirical assessment of a modified technology acceptance model. *Journal of Management Information Systems, 13*(2), 185–204.

Chau, P. Y. K., & Hu, P. J. H. (2001). Information technology acceptance by individual professionals: a model comparison approach. *Decision Sciences, 32*(4), 699–719. doi:10.1111/j.1540-5915.2001.tb00978.x

Chau, P. Y. K. (1996). An Empirical Assessment of a Modified Technology Acceptance Model. *Journal of Management Information Systems, 13*(2), 185–204.

Chau, Y. K., & Hu, J. H. (2001). Information technology acceptance by individual professionals: a model comparison approach. *Decision Sciences, 32*(4), 699–718. doi:10.1111/j.1540-5915.2001.tb00978.x

Chen, D.-Y., Huang, T.-Y., & Hsiao, N. (2006). Reinventing Government through On-Line Citizen Involvement in the Developing World: A Case Study of Taipei City Major's E-Mail Box in Taiwan. *Public Administration and Development, 26*(5), 409–423. doi:10.1002/pad.415

Chen, Y., & Dimitrova, D. (2006). Electronic government and online engagement: Citizen interaction with government via web portals. *International Journal of Electronic Government Research, 2*(1), 54–76.

Chen, Y. C., & Gant, J. (2001). Transforming Local E-Government Services: The Use of Application Services Providers. *Government Information Quarterly, 18*(4), 343–355. doi:10.1016/S0740-624X(01)00090-9

Chen, C., Tseng, S., & Huang, H. (2006). A comprehensive study of the digital divide phenomenon in Taiwanese government agencies. *International Journal Of Internet And Enterprise Management, 4*(3), 244–256. doi:10.1504/IJIEM.2006.010917

Chen, H. (2005). *Adopting Emerging Integration Technologies in Organisations*, PhD Thesis, Department of Information Systems and Computing, West London, UK, Brunel University.

Cheng, T., Lam, D., & Yeung, A. (2006). Adoption of internet banking: An empirical study in Hong Kong. *Decision Support Systems, 42*, 1558–1572. doi:10.1016/j.dss.2006.01.002

Cheng-Yi Wu, R. (2007). Enterprise Integration in e-Government. *Transforming Government: People. Process and Policy, 1*(1), 89–99.

Chircu, A. M., & Kauffman, R. J. (1999). Strategies of Internet Middlemen in the Intermediation/Disintermedation/Reintermediation cycle. *Electronic Markets, 9*(2), 109–117. doi:10.1080/101967899359337

Chircu, A.M., & Lee, D.H-D. (2005). E-government: Key Success Factors For Value Discovery And Realisation. *E- Government, an International Journal, 2*(1), 11-25.

Choudrie, J., & Dwivedi, Y. (2005). A Survey of Citizens Adoption and Awareness of E-Government Initiatives. In *The Government Gateway: A United Kingdom Perspective*. E-Government Workshop. Brunel University, West London.

Choudrie, J., & Weerrakody, V. (2007). Horizontal process integration in e-government: the perspective of a UK local authority. *International Journal of Electronic Government, 3*(3), 22–39.

CIA. (2005). *The world factbook 2005*. Retrieved November 3, 2005, from Central Intelligence Agency's Website https://www.cia.gov/cia/publication s/factbook/index.html

CIA. (2008). *The 2008 world factbook*. Retrieved September 29, 2008, from Central Intelligence Agency's Website https://www.cia.gov/library/publications/ the-world-factbook/index.html

Ciborra, C., & Navarra, D. D. (2003). "Good Governance and Development Aid: Risks and Challenges of E-government in Jordan". Paper presented at the IFIP WG 8.2 - WG 9.4, Athens, Greece.

CIO Council (2001). *Practical Guide to Federal Enterprise Architecture*. Chief Information Officer Council, Version 1.0.

Clegg, C., Axtell, C., Damodaran, L., Farbey, B., Hull, R., & Lloyd-Jones, R. (1997). Information technology: a study of performance and the role of human and organizational factors. *Ergonomics London, 40*(9), 851–871.

Clemons, E. K., & Row, M. C. (1992). Information Technology and Industrial Cooperation: The Changing Economics of Coordination and Ownership. *Journal of Management Information Systems, 9*(2), 9–28.

Coase, R. (1937). The Nature of the Firm. *De Economía, 4*, 386–405. doi:10.1111/j.1468-0335.1937.tb00002.x

COBIT 4.1, (2007), IT Governance Institute. *Framework, Control Objectives, Management Guidelines, Maturity Models*. IT Governance Institute.

Cohen, J. E. (2006). Citizen Satisfaction with Contacting Government on the Internet. *Information Policy, 11*(1), 51–65.

Cojazzi, G., & Cacciabue, P. C. (1994). *The DYLAM Approach for the Reliability Analysis of Dynamic System*. Berlin, Heidelberg: Springer-Verlag.

Coleman, S., & Gøtze, J. (2001). Bowling Together: Online Public Engagement in Policy Deliberation. Hansard Society, London. Retrieved October 4, 2005 from http://bowlingtogether.net.

Collier, P. (1998). *The Political Economy of Ethnicity.* Washington, DC: World Bank.

Collier, P. (2002). Social Capital and Poverty: A Microeconomic Perspective. In Van Bastelaer, T. (Ed.), *The Role of Social Capital in Development* (pp. 19–41). Melbourne: Cambridge University Press. doi:10.1017/CBO9780511492600.003

Collier, P.M. (2006). Policing and the Intelligent Application of Knowledge. *Public Money & Management,* April, 109-116.

Compeau, D., & Higgins, C. (1995). Computer Self-Efficacy: Development Of A Measure And Initial Test. *Management Information Systems Quarterly, 19*(2), 189–211. doi:10.2307/249688

Compeau, D., Higgins, C., & Huff, S. (1999). Social Cognitive Theory and Reactions to Computing Technology: A Longitudinal Study. *Management Information Systems Quarterly, 23*(2), 145–158. doi:10.2307/249749

Compeau, D. R., & Higgins, C. A. (1995). Computer self efficacy: development of a measure and initial Test. *Management Information Systems Quarterly, 19*(2), 189–211. doi:10.2307/249688

Compeau, D., Higgins, C. A., & Huff, S. (1999). Social Cognitive Theory and Individual Reactions to Computing Technology: A Longitudinal Study. *Management Information Systems Quarterly, 23*(2), 145–158. doi:10.2307/249749

Compeau, D. R., & Higgins, C. A. (1995). Computer Self-Efficacy: Development of a Measure and Initial Test. *Management Information Systems Quarterly, 19*(2), 189–211. doi:10.2307/249688

Connors, H., Koretz, P., Knowle, S., & Thibodeau, M. (1999). Municipal Web Sites in Onondaga County: A Study Comparing Selected Characteristics. *Community Benchmarks Program. Maxwell School of Citizenship and Public Affairs.* Syracuse University. Retrieved from http://www.maxwell.syr.edu/benchm arks/newsite/reports/web_down.html

Contractor, N., & Monge, P. (2003). *Theories of communication networks.* New York: Oxford.

Cook, K., & Cooper, R. (2003). Experimental Studies of Cooperation, Trust, and Social Exchange. In Ostrom, E., & Walker, J. (Eds.), *Trust and Reciprocity, Interdisciplinary Lessons from Experimental Research* (pp. 209–244). New York: Russell Sage Foundation.

Cook, M. E., La Vigne, M. F., Pagano, C. M., Dawes, S. S., & Pardo, T. A. (2002). *Making a case for local E-government.* Retrieved September 29, 2008, from Center for Technology in Government, University at Albany, SUNY Website http://www.ctg.albany.edu/publications/guides/making_a_case/making_a_case.pdf

Cooper, R. B., & Zmud, R. W. (1990). Information technology Implementation research: A technological Diffusion Approach. *Management Science, 36*(2), 123–139. doi:10.1287/mnsc.36.2.123

Cornford, T., & Smithson, S. (1997). *Project Research in Information Systems: A Student's Guide.* London: Macmillan Press.

Couldry, N. (2003). Digital Divide or discursive design: On the emerging ethics of information space. *Ethics and Information Technology, 5,* 89–97. doi:10.1023/A:1024916618904

Coursey, D., & Norris, D. F. (2008). Models of E-Government: Are They Correct? An Empirical Assessment. *Public Administration Review, 68*(3), 523–536. doi:10.1111/j.1540-6210.2008.00888.x

Cronin, J. J. Jr, & Taylor, S. A. (1992). Measuring service quality: A reexamination and extension. *Journal of Marketing, 56*(3), 55–68. doi:10.2307/1252296

Cross, M. (2007). £5m e-government awareness campaign flops. *The Guardian.* http://www.guardian.co.uk/technology/ 2006/oct/12/marketingandpr.newmedia

Çukurçayır, M., & Eroğlu, H. (2010). E-Cities: A Content Analysis of the Web Pages of Heidelberg and Konya Metropolitan Municipalities. *Current Rserach Journal of Social Sciences, 2*(1), 7–12.

Cullen, R., & Hernon, P. (2004). *Wired for Well-being Citizens' Response to E-government.* Retrieved March 15, 2005 from: http://www.e-government.govt.n z/docs/vuw-report-200406/

Cunliffe, D., Jones, H., Jarvis, M., Egan, K., Huws, R., & Munro, S. (2002). Information Architecture for Bilingual Web Sites. *Journal of the American Society for Information Science and Technology, 53*(10), 866. doi:10.1002/asi.10091

Cuntz, N., & Kindler, E. (2006): *EPC Tools.* http://wwwcs.upb.de/cs/ kindler/Forschung/EPCTools/, last accessed: March 2009.

Currall, S., & Judge, T. (1995). Measuring Trust Between Organizational Boundary Role Persons. *Organizational Behavior and Human Decision Processes, 1995*(64), 151–170. doi:10.1006/obhd.1995.1097

Curthoys, N., & Crabtree, J. (2003). SmartGov: Renewing Electronic Government for Improved Service Delivery, ISociety Report, Available from http://www.pwc.com/uk/eng/ about/ind/gov/smargovfinal.pdf

Czarniawska, B., & Joerges, B. (1996). Travels of ideas. In Czarniawska, B., & Sevon, G. (Eds.), *Translating organizational change.* New York: Walter De Gryter.

Dadidrajuh, R. (2007). Towards measuring true e-readiness of a Third-World country: A case study on Sri Lanka. In Al-Hakim, L. (Ed.), *Global E-Government: Theory, Applications and Benchmarking. (*(pp. 185–199). London: Idea Group.

Daemen, H., & Schaap, L. (2000). Developments in Local Democracies: An Introduction. Daemen. In H & Schaap, L. (eds.). *Citizen and City: Developments in fifteen local democracies in Europe.* Delft: Eburon.

Davis, F. D. (1989). Perceived usefulness, perceived ease of use, and user acceptance of information technology. *Management Information Systems Quarterly, 13*(3), 319–340. doi:10.2307/249008

Davis, F. (1989). Perceived Usefulness, Perceived Ease Of Use, And User Acceptance Of Information Technology. *Management Information Systems Quarterly, 13,* 319–339. doi:10.2307/249008

Davis, F., Bagozzi, R., & Warshaw, P. (1992). Extrinsic and Extrinsic Motivation to Use Computers in the Workplace. *Journal of Applied Social Psychology, 22*(14), 1111–1132. doi:10.1111/j.1559-1816.1992.tb00945.x

Davis, F. D. (1989). Perceived usefulness, perceived ease of use, and user acceptance of information technology. *Management Information Systems Quarterly, 13*(3), 319–339. doi:10.2307/249008

Davis, F. D. (1993). User acceptance of information technology: system characteristics, user perceptions and behavioral impacts. *International Journal of Man-Machine Studies, 38,* 475–487. doi:10.1006/imms.1993.1022

Davis, F. D., Bagozzi, R. P., & Warshaw, P. R. (1982). Extrinsic and Intrinsic Motivation to Use Computers in the Workplace. *Journal of Applied Social Psychology, 22*(14), 1111–1132. doi:10.1111/j.1559-1816.1992.tb00945.x

Davis, F. D., Bagozzi, R. P., & Warshaw, P. R. (1989). User acceptance of computer technology: a comparison of two theoretical models. *Management Science, 35*(8), 982–1002. doi:10.1287/mnsc.35.8.982

Davis, F. D., & Venkatesh, V. (1996). A critical assessment of potential measurement biases in the technology acceptance model: three experiments. *Internet Journal of Human-computer Studies, 45*(1), 19–45. doi:10.1006/ijhc.1996.0040

Davis, F. D. (1989). Perceived usefulness, perceived ease of use, and user acceptance of information technology. *Management Information Systems Quarterly, 13,* 319–340. doi:10.2307/249008

Davis, F. D., Bagozzi, R. P., & Warshaw, P. R. (1989). User Acceptance of Computer Technology: A Comparison of Two Theoretical Models. *Management Science, 35*(8), 982–1003. doi:10.1287/mnsc.35.8.982

Davis, F. D. (1989). Perceived usefulness, perceived ease of use, and user acceptance of information technology. *Management Information Systems Quarterly, 13*(3), 319–340. doi:10.2307/249008

Davis, F. D. (1989). Perceived usefulness, perceived ease of use, and user acceptance of information technology. *MIS Quarterly, 13*(3), 319–340. doi:10.2307/249008

Davis, F. D., Bagozzi, R. P., & Warshaw, P. R. (1992). Extrinsic and intrinsic motivation to use computers in the workplace. *Journal of Applied Social Psychology, 22*(14), 1–32. doi:10.1111/j.1559-1816.1992.tb00945.x

Davison, R. M., Wagner, C., & Ma, L. C. (2005). From government to e-government: a transition model. *Information Technology & People, 18*(3), 280–299. doi:10.1108/09593840510615888

Dawes, S. S., Pardo, T. A., & Cresswell, A. M. (2004). Designing electronic government information access programs: a holistic approach. *Government Information Quarterly, 21*(1), 3–23. doi:10.1016/j.giq.2003.11.001

De Long, M., & Lentz, L. (2006). Scenario evaluation of municipal Web sites: Development and Use of an Expert-Focused Evaluation Tool. *Government Information Quarterly, 23*, 191–206. doi:10.1016/j.giq.2005.11.007

DeBenedictis, A., Howell, W., Figueroa, R., & Boggs, R. A. (2002). *E-government defined: An overview of the next big information technology challenge.* Retrieved September 29, 2008, from http://www.zeang.com/RobertFig/egov.pdf

Dejoy, D. (1989). The Optimism Bias and Traffic Accident Risk Perception. *Accident; Analysis and Prevention, 21*(4), 333–340. doi:10.1016/0001-4575(89)90024-9

Delbecq, A. L., Van de Ven, A. H., & Gustafson, D. H. (1975). *Group Techniques for Program Planning: A Guide to Nominal Group and Delphi Processes.* Glenview, Illinois: Scott, Foresman and Company.

Delhey, J., & Newton, K. (2004). "Social Trust: Global Pattern or Nordic Exceptionalism?" *Wissenschaftszentrum Berlin für Sozialforschung (WZB) Discussion Paper* SP I 2004-202.

DeLone, W. H., & McLean, E. R. (1992). Information systems success: The quest for the dependent variable. *Information Systems Research, 3*(1), 60–95. doi:10.1287/isre.3.1.60

DeLone, W. H., & McLean, E. R. (2003). The DeLone and McLean model of information systems success: A ten year update. *Journal of Management Information Systems, 19*(4), 9–30.

DeLone, W. H., & McLean, E. R. (2004). Measuring eCommerce success: Applying the DeLone and McLean information system success model. *International Journal of Electronic Commerce, 9*(1), 31–47.

DeLone, W. H., & McLean, E. R. (1992). Information Systems Success: The Quest for the Dependent Variable. *Information Systems Research, 3*, 60–95. doi:10.1287/isre.3.1.60

DeLone, W. H., & McLean, E. R. (2003). The DeLone and McLean Model of Information Systems Success: A Ten-Year Update. *Journal of Management Information Systems, 19*(4), 9–30.

DeLone, W. H., & McLean, E. R. (2002). Information systems success revisited. *Proceedings of the 35th Hawaii International Conference on System Science, 3*(1), 2966–2976.

Denhardt, R. B. (1999). The future of public administration. *Public Administration and Management, 4*(2), 279–292.

Denscombe, M. 2007. *The good research guide: for small-scale social research projects.* Maidenhead: Open University Press.

Denzin, N. Y. K., & Lincoln, Y. (1994). *Handbook of Qualitative Research.* London: SAGE Publications.

Denzin, N. K. (1970). *The research act in Sociology: A theoretical introduction to sociological methods.* London: Butterworths.

Devadoss, P. R., Pan, S. L., & Huang, J. C. (2002). Structurational Analysis of e-government Initiatives: A Case Study of SCO. *Decision Support Systems, 34*(3), 253–269. doi:10.1016/S0167-9236(02)00120-3

Devadoss, P. R., Pan, S. L., & Huang, J. C. (2002). Structurational analysis of e-government initiatives: a case study of SCO. *Decision Support Systems, 34*(3), 253–269. doi:10.1016/S0167-9236(02)00120-3

Dhar, S. (2003). Introduction to smart card. Data Security Management <http://sumitdhar.blogspot.com/2004 /11/introduction-to-smart-cards.html> (Accessed March 1, 2007).

Di Maggio, P. J. (1988). Interest and Agency in Institutional Theory. In Zucker, L. G. (Ed.), *Patterns and Organizations: Culture and environment*. Cambridge: Ballinger Publishing Company.

Digital Futures Project. (2008). Available: http://www.digitalcenter.org/pages/current_report.asp?intGlobalId=19

Dillon, A., & Morris, M. (1996). User acceptance of new information technology: theories and models. In Williams, M. (Ed.), *Annual Review of Information Science and Technology*. Medford, NJ: Information Today.

Doherty, N. F., & King, M. (1998). The importance of organisational issues in systems development. *Information Technology & People, 11*(2), 104–123. doi:10.1108/09593849810218300

Doll, W. J., & Torkzadeh, G. (1988). The measurement of end-user computing satisfaction. *Management Information Systems Quarterly, 12*(2), 259–274. doi:10.2307/248851

Doolin, B. (2004). Power and resistance in the implementation of a medical management information system. *Information Systems Journal, 14*(4), 343–351. doi:10.1111/j.1365-2575.2004.00176.x

Doyle, E., Stamouli, I., & Huggard, M. (2005). Computer anxiety, self-efficacy, computer experience: an investigation throughout a computer science degree. October 19-22, 2005, Indianapolis, IN 35th ASEE/IEEE Frontiers in Education Conference S2H-3.

Dryzek, J. S. (1990). *Discursive democracy*. New York: Cambridge University Press.

Dugdale, A., Daly, A., Papandrea, F., & Maley, M. (2005). Accessing E-Government: Challenges for Citizens and Organizations. *International Review of Administrative Sciences, 71*(1), 109–118. doi:10.1177/0020852305051687

Dunleavy, P., Margetts, H., Bastow, S., & Tinkler, J. (2006). New public management is dead – Long live digital era governance. *Journal of Public Administration: Research and Theory, 16*(3), 467–494. doi:10.1093/jopart/mui057

Dunsire, A. (1993). Modes of governance. In Kooiman, J. (Ed.), *Modern governance: New Government-Society Interactions* (pp. 21–34). London: Sage.

Durham, C. C. (2002). *Implementing Electronic Government Statement*. Retrieved. from http://www.durham.gov.uk/.

Durofee, A. J., Walker, J. A., Alberts, C. J., Higuera, R. P., Murphy, R. L., & Williams, R. J. (1996). *Continuous Risk Management Guidebook*. Pittsburg, PA: Carnegie Mellon University.

Dwivedi, Y., Papazafeiropoulou, A., & Gharavi, H. (2006). Socio-Economic Determinants of Adoption of the Government Gateway Initiative in the UK. *Electronic Government, 3*(4), 404–419. doi:10.1504/EG.2006.010801

Dwivedi, Y., Papazafeiropoulou, A., & Gharavi, H. (2006). Socio-Economic Determinants of Adoption of the Government Gateway Initiative in the UK. *Electronic Government, 3*(4), 404–419. doi:10.1504/EG.2006.010801

Dwivedi, Y., & Weerakkody, V. (2007). Examining the factors affecting the adoption of broadband in the Kingdom of Saudi Arabia. *Electronic Government, an International Journal, 4*(1), 43-58.

Easterly, W., & Levine, R. (2001). What have we learned from a decade of empirical research on growth? It's not Factor accumulation: Stylized facts and growth models. *The World Bank Economic Review, 15*(2), 177–219. doi:10.1093/wber/15.2.177

Eastin, M. A., & LaRose, R. L. (2000). Internet self-efficacy and the psychology of the digital divide. *Journal of Computer-Mediated Communication, 6*(1).

Ebbers, W., Pieterson, W., & Noordman, H. N. (2008). Electronic Government: Rethinking Channel Management Strategies. *Government Information Quarterly, 25*(2), 181–201. doi:10.1016/j.giq.2006.11.003

Ebrahim, Z., & Irani, Z. (2005). E-Government Adoption: Architecture and Barriers. *Business Process Management Journal, 11*(5), 589–611. doi:10.1108/14637150510619902

Ebrahim, Z., Irani, Z., & Sarmad, S. (2004). Factors Influencing the Adoption of E-Government in Public Sector. *European & Mediterranean Conference on Information Systems*, Tunis Tunisia.

Economist Intelligence Units's. *E readiness Rankings* (EIU). Available online http://globaltechforum.eiu.com/in dex.asp?layout=rich_story&doc_id=6427 Accessed January 10, 2007.

Economist Intelligence Unit. (2006). The 2006 E-Readiness Rankings. *The Economist*. Retrieved from http://graphics.eiu.com/files/ad_ pdfs/2006Ereadiness_Ranking_WP.pdf

eEurope (2004) *Top of the web: User satisfaction and usage survey of eGovernment services*. Retrieved October 25, 2005, from http://www.europa.eu.int/egovernment_research (also available at http://www.cisco.at/pdfs/publicse ctor/egov_service-survey_02-05.pdf accessed on July 30, 2008)

Egger, W. D. (2005). *Government 2.0: Using technology to improve education, cut red tape, reduce gridlock & enhance democracy*. New York: Rowman & Littlefield Publishers.

E-government Handbook. (2007). Available online http://www.cdt.org/egov/ handbook/trust.shtml, Accessed March 15, 2007.

Eifert, M., & Püschel, J. O. (Eds.). (2004). *National electronic government: comparing governance structures in multi-layer administrations*. London: Routledge.

Elbashir, M. Z., Collier, P. A., & Davern, M. J. (2008). Measuring the effects of business intelligence systems: The relationship between business process and organizational performance. *International Journal of Accounting Information Systems, 9*, 135–153. doi:10.1016/j.accinf.2008.03.001

Elieson, B. D. (2006). Construction of an IT Risk Framework, Available from http://www.isaca.org/ContentManagement /ContentDisplay.cfm?ContentID=33595

Estabrook, L., Witt, E., & Rainie, L. (2007). *Information Searches that Solve Problems*. Retrieved July 31, 2009 from http://www.pewinternet.org/Reports/2007/Information-Searches-That-Solve-Problems.aspx.

European Commission. (2003). *Linking up Europe: The importance of interoperability for e-government services*. Staff Working Document 2003, URL: http://europa.eu.int/information_s ociety/activities/eGovernment_research/archives/ events/egovconf/doc/interoperability.pdf, p. 6.

European Communities IDABC – EIF. (2004). *European Interoperability Framework for Pan-European E-Government Services*. URL: http://europa.eu.int/id abc/en/document/3761, p. 3.

Evangelidis, A. (2007). FRAMES – A Risk Assessment Framework for e-Services. *Electronic. Journal of E-Government, 2*(1), 21–30.

Evangelidis, A., Akomode, J., Taleb-Bendiab, A., & Taylor, M. (2002). *Risk Assessment & Success Factors for e-Government in a UK Establishment*. Paper presented at the Electronic Government, First International Conference, Aix-en-Provence France.

Evans, D., & Yen, D. C. (2005). E-government: An analysis for implementation: Framework for understanding cultural and social impact. *Government Information Quarterly, 22*(3), 354–373. doi:10.1016/j.giq.2005.05.007

Evans, K. (2006, December). *Expanding E-government: Making a Difference for the American People Using Information Technology*. Executive Office of the President, Office of Management and Budget.

Evaristo, R., & Kim, B. (2005). A strategic framework for a G2G e-government excellence center. In Huang, W., Siau, K., & Wei, K. K. (Eds.), *Electronic government strategies and implementation* (pp. 68–83). London: Idea Group.

Ewusi-Mensah, K., & Przasnyski, Z. H. (1994). Factors contributing to the abandonment of information systems development projects. *Journal of Information Technology*, *9*(3), 185. doi:10.1057/jit.1994.19

Fairley, R. (1994). Risk management for software projects. *IEEE Software*, 57–64. doi:10.1109/52.281716

Fallows, D. (2007). *Chinese online population explosion: What it may mean for the Internet globally…and for US users.* Pew Internet & American Life Project. Retrieved November 11, 2007 from http://www.pewinternet.org/pdfs/China_Internet_July_2007.pdf

Fang, Z. (2002). E-government in digital era: Concept, practice, and development. *International Journal of the Computer. The Internet and Management*, *10*(2), 1–22.

Fernandes, A. A., & Abreu, V. F. (2006). *Implantando a Governança de TI – da Estratégia à Gestão dos Processos e Serviços.* Brazil: Brasoft.

Fichman, R. (1992). Information Technology Diffusion A Review of Empirical Research, *13th International Conference on Information Systems*, (pp. 195-206).

Fishbein, M., & Ajzen, I. (1975). *Belief, Attitude, Intention And Behavior: An Introduction To Theory And Research.* Reading, MA: Addison-Wesley.

Fishbein, M., & Ajzen, I. (1975). *Belief, attitude, intention and behavior: an Introduction to theory and research.* Reading, MA: Addison-Wesley.

Fishbein, M., & Ajzen, I. (1975). *Belief, Attitude, Intention and Behaviour: An Introduction to Theory and Research.* Reading, MA, USA: Addison-Wesley.

Floridi, L. (2001). Information ethics: An environmental approach to the digital divide. *Philosophy in the Contemporary World*, *9*(1), 1–7.

Flowers, S. (1996). *Software failure, management failure: amazing stories and cautionary tales.* Chichester; New York: Wiley.

Foley, P. (2005). The real benefits, beneficiaries and value of e-government. *Public Money & Management*, *25*, 4–6.

Foucault, M. (1991). Politics and the study of discourse. In Burchell, G., Foucault, M., & Gordon, C. (Eds.), *The Foucault effect. Hertfordshire: Harvester Wheatsheaf* (pp. 53–72).

Fountain, J. (2001). *Building the virtual state: Information technology and institutional change. Brookings Institution: Washington, DC: Giddens, A., (1984). The Constitution of Society.* Berkeley: University of California Press.

Fountain, J. E. (2001). *Building the virtual state. Information technology and institutional change.* Washington, D.C.: Brookings Institution Press.

Freedom House. (2006). *Freedom of the World.* Available online http://www.freedomhouse.org/ accessed January 10, 2007.

Frei, R., Kingston, J., Koornneef, F., & Schallier, P. (2002). NRI MORT User's Manual, Available from http://www.nri.eu.com/NRI1.pdf

Freiheit, J., Mondorf, A. (2007*): Formal Analysis of web service standards.* Deliverable WP4-D4, R4eGov - Towards e-Administration in the large, project number IST-2004-026650.

Frenkel. et al. (1998). Beyond bureaucracy? Work organization in call centres in *The international Journal of Human resource Management*, *9*(6). pp.957-979

Fu, J.-R., Farn, C.-K., & Chao, W.-P. (2006). Acceptance of electronic tax filing: A study of taxpayer intentions. *Information & Management*, *43*(1), 109–126. doi:10.1016/j.im.2005.04.001

Fukuyama, F. (1995). *Trust: The Social Virtues and the Creation of Prosperity.* New York: Free Press.

Fukuyama, F. (2001). Social Capital, Civil Society and Development. *Third World Quarterly*, *22*(1), 7–20. doi:10.1080/713701144

Fullwood, R. R., & Hall, R. E. (1988). *Probabilistic Risk Assessment in the Nuclear Power Industry* (1st ed.). Pergamon Press.

Gail, E. T., Marshall, I. M., & Jone, S. (1995). One in the eye to plastic card fraud. *International Journal of Retail and Distribution Management, 23*, 3–11. doi:10.1108/09590559510089195

Ganesan, S., & Hess, R. (1997). Dimensions and Levels of Trust: Implications for Commitment to a Relationship. *Marketing Letters, 8*(4), 439–448. doi:10.1023/A:1007955514781

Gardner, C., & Amoroso, D. (2004). *Development of an Instrument to Measure the Acceptance of Internet Technology by Consumers.* Paper presented at the Proceedings of the 37th Hawaii International Conference on System Sciences, Hawaii, USA.

Garson, G. D. (2004). The promise of digital government. In Pavlichev, A., & Garson, G. D. (Eds.), *Digital government: Principles and best practices* (pp. 2–15). London: Idea Group Publishing.

Gassert, H. (2004). "How to Make Citizenz Trust E-Government," University of Fribourg, E-Government Seminar, Information Systems Research Group. Available from http://edu.mediagonal.ch/unifr/ egov-trust/slides/html/title.html accessed November 29, 2005.

Gefen, D., Rose, G., Warkentin, M., & Pavlou, P. A. (2005). Cultural Diversity and Trust in IT Adoption: A Comparison of Potential e-Voters in the USA and South Africa. *Journal of Global Information Management, 13*(1), 54–78.

Gefen, D. (2000). E-commerce: The Role Of Familiarity and Trust. *Omega: The International Journal of Management Science, 28*(6), 725–737. doi:10.1016/S0305-0483(00)00021-9

Gefen, D., Karahanna, E., & Straub, D. W. (2003). Inexperience and experience with online stores: The importance of TAM and trust. *IEEE Transactions on Engineering Management, 50*(3), 307–321. doi:10.1109/TEM.2003.817277

Gefen, D., Rose, G., Warkentin, M., & Pavlou, P. (2005). Cultural Diversity and Trust in IT Adoption: A Comparison of USA and South African e-Voters. *Journal of Global Information Management, 13*(1), 54–78.

Gefen, D., Warkentin, M., Pavlou, P. A., & Rose, G. M. (2002). EGovernment Adoption. Eighth Americas Conference on Information Systems, Association for Information Systems.

Gellman, R. (1996). Disintermediation and the Internet. *Government Information Quarterly, 13*(1), 1–8. doi:10.1016/S0740-624X(96)90002-7

Giaglis, G. M., Klein, S., & O'Keefe, R. M. (2002). The role of intermediaries in electronic marketplaces: developing a contigency model. *Information Systems Journal, 12*(3), 231–246. doi:10.1046/j.1365-2575.2002.00123.x

Gichoya, D. (2005). Factors Affecting the Successful Implementation of ICT Projects in Government. *The Electronic. Journal of E-Government, 3*(4), 175–184.

Gilbert, D., Kelly, L. L., & Barton, M. (2003). Technophobia, gender influences and consumer decision-making for technology-related products. *European Journal of Innovation Management, 6*(4), 253–263. doi:10.1108/14601060310500968

Gilbert, D., Balestrini, P., & Littleboy, D. (2004). Barriers and Benefits in the Adopiton of E-government. *International Journal of Public Sector Management, 17*(4/5), 286–301. doi:10.1108/09513550410539794

Gil-Garcia, J. R., & Pardo, T. A. (2005). E-government success factors: mapping practical tools to theoretical foundations. *Government Information Quarterly, 22*, 187–216. doi:10.1016/j.giq.2005.02.001

Gilsinan, J. F., Millar, J., Seitz, N., Fisher, J., Harshman, E., Islam, M., & Yeager, F. (2008). The role of private sector organizations in the control and policing of serious financial crime and abuse. *Journal of Financial Crime, 15*(2), 111–123. doi:10.1108/13590790810866854

Giorgini, P., et al. (2004). Requirements Engineering meets Trust Management: Model, Methodology, and Reasoning, In *Proc. of iTrust 2004*, (LNCS, vol. 2995, pp. 176–190). Springer-Verlag Heidelberg.

Girion, L. (2006). Language Becoming an Issue for Health Insurers. Los Angeles Times, March 20, 2006, p. C-1.

Glaeser, E. L., Laibson, D., Scheinkman, J., & Soutter, C. (2000). Measuring Trust. *The Quarterly Journal of Economics, 115*(3), 811–846. doi:10.1162/003355300554926

Goldberg, J. (2009). *State of Texas Municipal Web Sites: A Description of Website Attributes and Features of Municipalities with Populations Between 50,000-125,000.* Texas State University-San Marcos. Retrieved from http://ecommons.txstate.edu/arp/307

Goldkuhl, G., & Persson, A. (2006). *From E-ladder to E-diamond – reconceptualising models for public e-services.* Paper for the 14th European Conference on Information Systems (ECIS2006), June 12-14, Gothenburg, Sweden.

Goulding, A. (2001). Information Poverty or Overload? *Journal of Librarianship and Information Science September, 33*(3), 109-111.

Granovetter, M. (1973). The Strength of Weak Ties. *American Journal of Sociology, 78*(6), 1360–1380. doi:10.1086/225469

Grant, G., & Chau, D. (2005). Developing a generic framework for e-government. *Journal of Global Information Management, 13*(1), 1–30.

Green, S. B. (1991). How many subjects does it take to do a regression analysis? *Multivariate Behavioral Research, 26*(3), 499–510. doi:10.1207/s15327906mbr2603_7

Grefenstette, G., & Nioche, J. (2000). Estimation of English and non-English Language Use on the WWW. Xerox Research Centre Europe. Available from http://arxiv.org/pdf/cs.CL/0006032 accessed March 19, 2006.

Grembergen, W. V. (2004). *Strategies for Information Technology Governance.* Idea Group Publishing.

Gritzalis, D., & Katsikas, S. (2004). *Autonomy and political disobedience in cyberspace.* Athens: Papasotiriou.

Grönlund, A., & Horan, T. A. (2004). Introducing E-Gov: History, Definitions, and Issues. *Communications of the Association for Information Systems, 15*, 713–729.

Grönlund, Å. (2002). *Electronic Government – Design, Application and Management.* Hershey, PA: Idea Group Publishing.

Guijarro, L. (2007). Interoperability frameworks and enterprise architectures in e-government initiatives in Europe and the United States. *Government Information Quarterly, 24*(1), 89–101. doi:10.1016/j.giq.2006.05.003

Guijarro, L. (2004). Analysis of the Interoperability Frameworks in eGovernment Initiatives, in Traunmüller, R. (Ed.) *Electronic Government*, conference proceedings (LNCS # 3183, pp. 36-39) Springer Verlag Heidelberg et al.

Gulli, A., & Signorini, A. (2005). The Indexable Web is More than 11.5 Billion Pages. WWW 2005, May 10–14, 2005, Chiba, Japan.

Gupta, V., & Gupta, S. (2005). *Experiments in Wireless Internet Security.* Statistical Methods in Computer Security.

Hacker, K. (2002a). Network democracy and the fourth world. *European Journal of Communication Research, 27*, 235–260.

Hacker, K. (2002b). *Network democracy, political will and the fourth world: Theoretical and empirical issues regarding computer-mediated communication (CMC) and democracy. Keynote address to EURICOM.* The Netherlands: Nigmegan.

Hacker, K., & Mason, S. (2003). Ethics gaps in studies of the digital divide. *Ethics and Information Technology, 5*, 99–115. doi:10.1023/A:1024968602974

Hacker, K. (2004). The potential of computer-mediated communication (CMC) for political structuration. *Javnost/ The Public, 11*, 5-26.

Hackney, R., & Jones, S. (2002, April). Towards E-government in the Welsh (UK) Assembly: an Information Systems Evaluation. *Proceedings of the ISOneWorld Conference and Convention*, Las Vegas, USA.

Hackney, R., & Jones, S. (2002). *Towards e-government in the Welsh (UK) assembly: an information systems evaluation*. Work report, Manchester, Manchester metropolitan University, Business School.

Haes, S. D., Gremberger, W. V., & Guldentops, E. (2004). *Structures, Processes and Relational Mechanisms for IG Governance*. Hershey, PA: Idea Group, Inc. "ISO/IEC 17799, (2005). *Information Technology – Security Techniques – Code of Practice for Information Security Management*. International Standard Organization.

Hahamis, P., & Iles, J. (2005). E-government in Greece: opportunities for improving the efficiency and effectiveness of local government. *Proceedings of European conference on e-government,* Antwerp, Belgium.

Hair, J. F. Jr, Anderson, R. E., Tatham, R. L., & Black, W. C. (1992). *Multivariate Data Analysis*. New York: Macmillian.

Hair, J. F. Jr, Anderson, R. E., Tatham, R. L., & Black, W. C. (1998). *Multivariate Data Analysis* (5th ed.). Upper Saddle River, NJ: Prentice-Hall.

Hajer, M. (1995). *The politics of environmental discourse: ecological modernization and policy process*. Oxford, UK: Clarendon Press.

Hall, P., & Löfgren, K. (2004). The rise and decline of a visionary policy: Swedish ICT-policy in retrospect. *Information Polity, 9*(3-4), 149–165.

Hall, P., & Löfgren, K. (2006). *Politisk styrning i praktiken* [Political Governance in Practice]. Malmö: Liber.

Hampshire, C. C. (2006). *Section 5: Risk Assessment*. Retrieved. from http://www.hants.gov.uk/ egovernment/ IEG2-sec5.html.

Hansen, L. L. (2009). Corporate financial crime: social diagnosis and treatment. *Journal of Financial Crime, 16*(1), 28–40. doi:10.1108/13590790910924948

Hansson, S. O. (1994). *Decision Theory, A Brief Introduction,* Available from http://www.infra.kth.se/~soh/decisiontheory.pdf

Hargittai, E. (1999). Weaving the Western Web: Explaining Differences in Internet Connectivity Among OECD Countries. *Telecommunications Policy, 23*(10/11).

Harris, J. F., & Schwartz, J. (2000. June 22). Anti drug website tracks visitors. *Washington Post*, (p. 23).

Hart, P., & Saunders, C. (1997). Power and Trust: Critical factors in the adoption and use of electronic data interchange. *Organization Science, 8*(1), 23–42. doi:10.1287/orsc.8.1.23

Hart-Teeter. (2003). The New E-government Equation: Ease, Engagement, Privacy and Protection. A report prepared for the Council for Excellence in Government, Retrieved November 27, 2005 from http://www.excelgov.org/usermedia/ images/uploads/PDFs/egovpoll2003.pdf.

He, D., & Lu, Y. (2007). Consumers' perceptions and acceptances towards mobile advertising: an empirical study in China. IEEE. 3770-3773.

Heeks, R., & Bailur, S. (2007). Analysing eGovernment research. *Government Information Quarterly, 22*, 243–265. doi:10.1016/j.giq.2006.06.005

Heeks, R. (Ed.). (2001). *Reinventing government in the information age: International practice in IT-enabled public sector reform*. London: Routledge.

Heeks, R. (2006). *Implementing and managing eGovernment. An international text*. London: Sage.

Heeks, R. (2002). E-Government in Africa: promise and practice. *Information Polity, 7*(2, 3), 97-114.

Heeks, R. (2003). *Most e-government for development projects fail: how can risks be reduced?* I-Government working paper, University of Manchester. Institute for Development, Policy and Management.

vHeffen, O., Kickert, W. J. M., & Thomassen, J. A. (2000). *Governance in modern society. effects, change and formation of government institutions*. Dordrecht: Kluwer Academic Publishers.

Helbig, N., Gil-Garcia, J. R., & Ferro, E. (2009). Understanding the complexity of electronic government: Implications from the digital divide literature. *Government Information Quarterly, 26*(1), 89–97. doi:10.1016/j.giq.2008.05.004

Herriot, R. E., & Firestone, W. A. (1983). Multi-site Qualitative Policy Research: Optimising Description and Generalisability. *Educational Researcher, 12*, 14–19.

Hibbing, J. R., & Theiss-Morse, E. (2002). *Stealth Democracy: Americans' Beliefs About How Government Should Work*. Cambridge: Cambridge University Press. doi:10.1017/CBO9780511613722

Hiller, J., & Belanger, F. (2001). *Privacy strategies for electronic government. E-Government series*. Arlington, VA: Pricewaterhouse Coopers Endowment for the Business of Government.

Hinton, P. R., Brownlow, C., McMurvay, I., & Cozens, B. (2004). *SPSS explained*. East Sussex, England: Routledge Inc.

Ho, A. (2002). Reinventing local government and the e-government initiative. *Public Administration Review, 62*, 434–444. doi:10.1111/0033-3352.00197

Hofer, A. (2005). Architektur zur Prozessinnovation in Wertschöpfungsnetzwerken. In Scheer, A.-W. (Ed.), *Veröffentlichungen des Instituts für Wirtschaftsinformatik* (*Vol. 181*).

Holderness, M. (1998). Who are the world's information poor? In Loader, B. (Ed.), *Cyberspace Divide* (pp. 35–56). London: Routledge.

Holliday, I. (2002). Building e-government in East and Southeast Asia: Regional rhetoric and national (in)action. *Public Administration and Development, 22*, 323–335. doi:10.1002/pad.239

Hollingsworth, D. (1995). *Workflow Management Coalition (WfMC) - The Workflow Reference Model*. URL: http://www.wfmc.org/stand ards/docs/tc003v11.pdf.

Holsapple, C. W., & Sasidharan, S. (2005). The dynamics of trust in B2C e-commerce: a research model and agenda. *Information System E-Business Management, 3*(4), 377–403. doi:10.1007/s10257-005-0022-5

Holzer, M., & Kim, S. (2005). Digital Governance in Municipalities Worldwide: A Longitudinal Assessment of Municipal Websites Throughout the World. *The E-Governance Institute. National Center for Public Productivity*. Rutgers, The State University of New Jersey. Newark. Retrieved from http://unpan1.un.org/intradoc/groups/ public/documents/ASPA/UNPAN022839.pdf

Horrigan, J. B. (2004). *How Americans get in Touch with Government*. Retrieved July 31, 2009 from http://www.pewinternet.org/Reports/2004/ How-Americans-Get-in-Touch-With-Government.aspx

Horst, M., Kuttschreutter, M., & Gutteling, J. M. (2007). Perceived usefulness, personal experiences, risk perception, and trust as determinants of adoption of e-government services in the Netherlands. *Computers in Human Behavior, 23*, 1838–1852. doi:10.1016/j.chb.2005.11.003

Horton, R. P., Buck, T., Waterson, P. E., & Clegg, C. W. (2001). Explaining intranet use with the technology acceptance model. *Journal of Information Technology, 16*(2), 237–249. doi:10.1080/02683960110102407

Howard, M. (2001). e-government across the globe: how will "e" change government? *Government Finance Review, 17*(4), 6–9.

Howard, P., Rainie, L., & Jones, S. (2002). Days and nights on the Internet. In Wellman, B., & Haythornwaite, C. (Eds.), *The Internet in everday life* (pp. 45–73). Oxford, UK: Blackwell Publishers. doi:10.1002/9780470774298.ch1

Howell, D. C. (2007). *Chi-square with ordinal data*. Retrieved January 22, 2007, from

HP. (2002). *ITIL Practitioner Change Management*. Mountain View, CA: Hewlett-Packard.HP Education Services.

Hsiao, R. L. (2008). Knowledge Sharing in a Global Professional Service Firm. *MIS Quarterly Executive*, *7*(3), 123–137.

Hsu, M., & Chiu, C. (2004). Internet self-efficacy and electronic service acceptance. *Decision Support Systems*, *38*(3), 369–381. doi:10.1016/j.dss.2003.08.001

http://www.uvm.edu/~dhowell/StatPages/ More_Stuff/OrdinalChisq/OrdinalChiSq.html

Hu, P. J., Chau, P. Y. K., Sheng, O. R. L., & Tam, K. Y. (1999). Examining the technology acceptance model using physician acceptance of telemedicine technology. *Journal of Management Information Systems*, *16*(2), 91–112.

Huang, Z. (2007). A comprehensive analysis of U.S. counties' e-Government portals: development status and functionalities. *European Journal of Information Systems*, *16*(2), 149–164. doi:10.1057/palgrave.ejis.3000675

Hull, E., Jackson, K., & Dick, J. (2005). *Requirements Engineering*. London: Springer-Verlag.

Hung, S. Y., Chang, C. M., & Yu, T. J. (2006). Determinants of user acceptance of the e-government services: The case of online tax filing and payment system. *Government Information Quarterly*, *23*, 97–122. doi:10.1016/j.giq.2005.11.005

Iacovou, C., Benbasat, I., & Dexter, A. (1995). Electronic Data Interchange and Small Organisations – Adopting and Impact of Technology. *Management Information Systems Quarterly*, *19*(4), 465–485. doi:10.2307/249629

IctQATAR. (2007). *Free wireless internet in Qatar's public parks*. Retrieved from http://www.ict.gov.qa/output/page422.asp

Ifinedo, P. (2007). Moving towards e-government in a developing society: Glimpses of the problems, progress, and prospects in Nigeria. In Al-Hakim, L. (Ed.), *Global E-Government*.

Igbaria, M., Guimaraes, T., & Davis, G. B. (1995). Testing the determinants of microcomputer usage via a structural equation model. *Journal of Management Information Systems*, *11*(4), 87–114.

Igbaria, M., Zinatelli, N., Cragg, P., & Cavaye, A. L. M. (1997). Personal computing acceptance factors in small firms: a structural equation model. *Management Information Systems Quarterly*, *21*(3), 279–306. doi:10.2307/249498

Iivari, J. (2005). An empirical test of the DeLone-McLean model of information system success. *The Data Base for Advances in Information Systems*, *36*(2), 8–27.

Ilter, C. (2009). Fraudulent money transfers: a case from Turkey. *Journal of Financial Crime*, *16*(2), 125–136. doi:10.1108/13590790910951803

infoDev and CDT (Center for Democracy and Technology) (2002). *The E-government Handbook for Developing Countries*. Retrieved November 14, 2006 from: http://www.cdt.org/egov/handbook/2002-11-14egovhandbook.pdf

InfoDev. (2002). *The e-Government Handbook for Developing Countries*. Retrieved from http://www.cdt.org/egov/handbook.

Inglehart, R., & Baker, W. (2000). Modernization, Cultural Change, and the Persistence of Traditional Values. *American Sociological Review*, *65*(1), 19–51. doi:10.2307/2657288

International City/County Management Association (ICMA). (2004). Electronic Government Survey 2004. accessed February 22, 2008 from: http://icma.org/upload/bc/attach/%7B 9BA2A963-DDCC-40B7-836D-F1CFC17DCD98%7Degov2004Web.pdf

International Telecommunication Union. (2008). ITU/ICT Statistics. Retrieved September 29, 2008 from ITU website http://www.itu.int/ITU-D/ict/statistics/

International Telecommunications Union. (2003). World Summit. Retrieved November 1, 2007 at http://www.itu.int/wsis/index.html

International Telecommunications Union. (2007). *World Information Society Report 2007*. Retrieved November 2, 2007 at http://www.itu.int/osg/spu/publications/worldinformationsociety/2007/report.html

Internet World Stats. (2007). Internet Usage Statistics. *The Internet Big Picture World Internet Users and Population Stats. Miniwatts* Marketing Group. Retrieved from http://www.internetworldstats. com/stats.htm

Internet World Stats. (2008): *Usage and population statistics*. Retrieved September 29, 2008, from http://www.internetworldstats.com/

Irani, Z., Love, P. E. D., & Montazemi, A. (2007). E-Government: Past, present, and future. *European Journal of Information Systems*, *16*, 103–105. doi:10.1057/palgrave.ejis.3000678

Irani, Z., Al-Sebie, M., & Elliman, T. (2006). Transaction stage of E-Government systems: Identification of its location & importance. *Proceedings of the 39th Hawaii International Conference on System Sciences, IEEE*, (pp. 1-9).

Irani, Z., Al-Sebie, M., & Elliman, T. (2006). Transaction Stage of e-Government Systems: Identification of its Location & Importance. *Proceedings of the 39th Hawaii International Conference on System Science*.

IRS. (2004). *Visie op dienstverlening 2010*. Utrecht: Dutch Inland Revenue Service.

IRS. (2004). *IRS e-Strategy for Growth*. http://www.irs.gov/pub/irs-pdf/p3187.pdf

Ishahak, J. (2006). Smart card 'success' - gauged by number of cards issued or card usage? Available at <http://www.frost.com/prod/servlet/market-insight-top.pag?docid=4813276> (Accessed November 30, 2006).

ISO/IEC. (2005a). *27001:2005 Information security management systems - Requirements*, Current Stage 90.92 Available from http://www.iso.org/iso/iso_catalogue/catalogue_tc/catalogue_detail.htm?csnumber=42103

ISO/IEC. (2005b). *27002:2005 Code of practice for information security management*, Current Stage 90.92 Available from http://www.iso.org/iso/iso_catalogue/catalogue_ics/catalogue_detail_ics.htm?csnumber=50297

ISO/IEC. (2008). *27005:2008 Information security risk management*, Current Stage 90.92 Available from http://www.iso.org/iso/iso_catalogue/catalogue_tc/catalogue_detail.htm?csnumber=42107

ISO/IEC. (2009). *27004:2009 Information security management -- Measurement*, Current Stage 60.60 Available from http://www.iso.org/iso/iso_catalogue/catalogue_tc/catalogue_detail.htm?csnumber=42106

ISO/IEC. (2010). *27003:2010 Information security management system implementation guidance*, Current Stage 60.60 Available from http://www.iso.org/iso/iso_catalogue/catalogue_tc/catalogue_detail.htm?csnumber=42105

Israel, G. D. (1992). Sampling the evidence of extension program impact. Program Evaluation and Organizational Development, IFAS, University of Florida. PEOD-5. October.

Jaccard, J., & Becker, M. A. (1990). *Statistics for the Behavioral Sciences*. Belmont, CA, USA: Wadsworth Publishing Company.

Jadu, (2005). Enterprise Content Management for Public and Private Sector. Available from http://www.jadu.co.uk/ego v/jadu_egov_econsultation.php accessed November 26, 2005.

Jaeger, P. T. (2003). The endless wire: E-government as global phenomenon. *Government Information Quarterly*, *20*, 323–331. doi:10.1016/j.giq.2003.08.003

Jaeger, P. T. (2003). The endless wire: E-government as global phenomenon in *Government Information Quarterly*, 20, p.323-331

Janssen, M., Gortmaker, J., & Wagenaar, R. W. (2006, Spring). Web Service Orchestration in Public Administration: Challenges, Roles, and Growth Stages. *Information Systems Management*, 44–55. doi:10.1201/1078.10580530/45925.23.2.20060301/92673.6

Janssen, M., & Klievink, B. (2009). The Role of Intermediaries in Multi-channel Service Delivery Strategies. [IJEGR]. *International Journal of E-Government Research*, *5*(3), 36–46.

Janssen, M., & Sol, H. G. (2000). Evaluating the role of intermediaries in the electronic value chain. *Internet Research. Electronic Networking Applications and Policy, 19*(5), 406–417. doi:10.1108/10662240010349417

Janssen, M., & Verbraeck, A. (2005). Evaluating the Information Architecture of an Electronic Intermediary. *Journal of Organizational Computing and Electronic Commerce, 15*(1), 35–60. doi:10.1207/s15327744joce1501_3

Janssen, D., Rotthier, S., & Snijkers, K. (2004). If you measure it, they will score: an assessment of international eGovernment benchmarking. *Information Polity, 9*(3-4), 121–130.

Janssen, M., & Cresswell, A. (2005). Enterprise Architecture Integration in E-Government. *38ᵗʰ Hawaii International Conference on System Sciences,* Hawaii, (pp. 1-10).

Jarvenpaa, S. L., & Tractinsky, N. (2000). Consumer Trust in an Internet Store. *Information Technology Management, 1*(1-2), 45–70. doi:10.1023/A:1019104520776

Jennings, T. (2004). *Change Management – An Essential Tool for IT Governance.* Butler Direct Limited.

Jensen, L., & Kähler, H. (2007). The Danish ministry of finance as meta-governor – the case of public sector digitalisation. In Marcussen, M., & Torfing, J. (Eds.), *Democratic network governance in Europe* (pp. 174–191). Basingstoke: Macmillan.

Jeong, M., & Lambert, C. U. (2001). Adaptation of an information quality framework to measure customers' behavioral intentions to use lodging web sites. *Hospital Management, 20*(2), 129–146. doi:10.1016/S0278-4319(00)00041-4

Jessop, B. (2002). *The Future of the capitalist State.* Cambridge, UK: Polity Press.

Jeyaraj, A., Rottman, J., & Lacity, M. (2006). A review of the predictors, linkages, and biases in IT innovation adoption research. *Journal of Information Technology, 21*(1), 1–23. doi:10.1057/palgrave.jit.2000056

Jiang, J. J., Hsu, M. K., Klein, G., & Lin, B. (2000). E-commerce user behavior model: an empirical study. *Human Systems Management, 19*(4), 265–276.

Jiwire (2006). *JiWire Launches Worldwide Point-of-Connection Wi-Fi Hotspot Advertising Network.* Retrieved from http://www.jiwire.com/about/announ cements/press-advertising-network.htm

Johansson-Stenman, O., Mahmud, M., & Martinsson, P. (2006). *Trust and Religion: Experimental evidence from Bangladesh.* Department of Economics, Göteborg University, Mimeo.

Johnson, R. R. (2008). Officer Firearms Assaults at Domestic Violence Calls: A Descriptive Analysis. *The Police Journal, 81*(1), 25–45. doi:10.1350/pojo.2008.81.1.407

Joia, J., & Foundation, G. (2007). A heuristic model to implement government-to-government projects. *International Journal of Electronic Government Research, 3*(1), 49–67.

Joia, L. A. (2004). Developing government-to-government enterprises in Brazil: a heuristic model drawn from multiple case studies. *International Journal of Information Management, 24*(2), 147–166. doi:10.1016/j.ijinfomgt.2003.12.013

Jones, S., Hackney, R., & Irani, Z. (2007). Towards E-government Transformation: Conceptualising "Citizen Engage". *Transforming Government: People. Process and Policy, 1*(2), 145–152.

Jouko, S., & Rouhiainen, V. (1993). *Quality Management of Safety and Risk Analysis.* New York: Elsevier Science Publishers B.V.

June, L., Chun-Sheng, Y., Chang, L., & James, E. (2003). Technology Acceptance Model for Wireless Internet. *Internet Research: Electronic Networking Application And Policy, 13*(3), 206–222. doi:10.1108/10662240310478222

Kaaya, J. (2004). Implementing e-government services in East Africa: Assessing status through content analysis of government websites. *Electronic Journal of E-Government, 2(1),* 39-54. Retrieved September 29, 2008, from EJEG Website http://www.ejeg.com/volume-2/volume2-issue-1/v2-i1-art5-kaaya.pdf

KAEN. Fred R., (2003). *A Blueprint for Corporate Governance.* First Edition. Amacon, New York.

Kalu, K. N. (2007). Capacity building and IT diffusion: A comparative assessment of e-government environment in Africa. *Social Science Computer Review, 25*(3), 358–371. doi:10.1177/0894439307296917

Kamal, M. M. (2006). IT Innovation Adoption in the Government Sector: Identifying the Critical Success Factors. *Journal of Enterprise Information Management, 19*(2), 192–222. doi:10.1108/17410390610645085

Kamal, M. M., Themistocleous, M., & Morabito, V. (2008). *Evaluating Information Systems: Public and Private Sector*. Published by Butterworth-Heinemann.

Kamal, M. M., & Themistocleous, M. (2007). Investigating EAI Adoption in LGAs: A Case Study Based Analysis. *Proceedings of the 13ᵗʰ Americas Conference on Information Systems*, Keystone, Colorado, USA, (pp. 1-13).

Kamal, M. M., Themistocleous, M., & Elliman, T. (2008). Extending IT Infrastructures in LGAs through EAI, *Proceedings of the 14ᵗʰ Americas Conference on Information Systems*, Toronto, Canada, (pp. 1-12).

Kamal, M. M., Themistocleous, M., & Morabito, V. (2009). Justifying the Decisions for EAI Adoption in LGAs: A Validated Proposition of Factors, Adoption Lifecycle Phases, Mapping and Prioritisation of Factors. *Proceedings of the 42ⁿᵈ Hawaii International Conference on System Sciences*, Hilton Kaikoloa Village Resort, Hawaii, (pp. 1-10).

Kannabiran, G., Xavier, M. J., & Banumathi, T. (2008). E-Governance and ICT Enabled Rural Development in Developing Countries: Critical Lessons from RASI Project in India. *International Journal of Electronic Government Research, 4*(3), 1–19.

Kappos, A., & Rivard, S. (2008). A Three-Perspective Model of Culture, Information Systems, and Their Development and Use. *Management Information Systems Quarterly, 32*(3), 601–634.

Kara-Zaitri, C., Keller, A. Z., Barody, I., & Fleming, P. V. (1991). *An Improved FMEA methodology*. Paper presented at the Annual Reliability and Maintainability Symposium.

Kara-Zaitri, C., Keller, A. Z., & Fleming, P. V. (1992). *A Smart Failure Mode and Effect Analysis Package*. Paper presented at the Annual Reliability and Maintainability Symposium.

Karim, M. R. A., & Khalid, N. M. (2003). *E-government in Malaysia: Improving responsiveness and capacity to serve*. Selangor D.E. Malaysia: Pelanduk Publications.

Kaufmann, D. (2004). "*Corruption, Government and Security: Challenges for the Rich country and the World*" Chapter in the Global Competitiveness Report 2004/2005. Available online http://siteresources.worldbank.org/INTWBIGOVANTCOR/Resources/ETHICS.xls Accessed January 11, 2007.

Kawalek, P., & Wastell, D. (2005). Pursuing radical transformation in information age government: Case studies using the SPRINT methodology. *Journal of Global Information Management, 13*, 79–101.

Kawalek, P., & Wastall, D. (2005). Pursuing radical transformation in information age government: case studies using the SPRINT methodology. *Journal of Global Information Management, 13*(1), 79–101.

Ke, W., & Wei, K. K. (2004). Successful e-government in Singapore. *Communications of the ACM, 47*(6), 95–99. doi:10.1145/990680.990687

Kenny (2006). *Overselling the web? Development and the Internet*. London: Lynne Rienner Publishers, Inc.

Kernagaghan, K. (2007). Beyond bubble gum and goodwill: integrating service Delivery. In Sandford, B., Kernaghan, K., Brown, D. Bontis, N., Perri 6, Thompson, F.(Eds.), *Digital state at the leading edge*, Toronto: University of Toronto Press. (Ch. 4).

Kettl, D. F. (2005). *The global public management revolution*. (2nd ed.), Washington, D.C.: Brookings Institution Press.

Khalil-babnet, M. WSIS Prepcom-2: Cybersecurity an Issue for All. Available from http://www.babnet.net/en_detail.asp?id=935 accessed November 29, 2005.

Khoumbati, K., Themistocleous, M., & Irani, Z. (2006). Evaluating the Adoption of Enterprise Application Integration in Healthcare Organisations. *Journal of Management Information Systems, 22*(4), 69–108. doi:10.2753/MIS0742-1222220404

Kickert, W. J. M., Klijn, E.-H., & Koppenjan, J. F. M. (Eds.). (1997). *Managing complex networks: strategies for the public sector.* London: Sage.

Kim, S., & Lee, H. (2004). Organizational factors affecting knowledge sharing capabilities in e-government: an empirical study. *Lecture Notes in Computer Science, 3035,* 265–277.

Kim, H. J., & Bretschneider, S. (2004). Local Government Information Technology capacity: An Exploratory Theory. *Proceedings of the 37th Annual Hawaii International Conference on System Sciences,* (pp. 121-130).

Kim, K., & Prabhakar, B. (200) "Initial Trust, Perceived Risk, and the Adoption of Internet Banking," *Proceedings of ICIS 2000,* Brisbane, Australia, Dec. 10-13, 2000.

Klein, J. H., & Cork, R. B. (1998). An approach to technical risk assessment. *International Journal of Project Management, 16*(6), 345–351. doi:10.1016/S0263-7863(98)00006-4

Klein, H., & Meyers, M., (1999). A set of principles for conducting and evaluating interpretive field studies in *MIS Quarterly,* 23, pp. 67-93.

Klichewski, R. (2004). Information integration or process integration? How to achieve interoperability in administration. In Traumüller, R. (Ed.), *EGOV2004, Berlin.* Heidelberg: Springer Verlag.

Klievink, B., & Janssen, M. (2008). Improving Government Service Delivery with Private Sector Intermediaries. *European Journal of ePractice, 1*(5), 17-25.

Klischewski, R., & Scholl, H. J. (2006). *Information quality as a common ground for key players in e-government integration and interoperability.* In Proceedings of HICSS'06.

Klitgaard, R., Justesen, M.K., & Klemmensen R. (2005). "The Political Economy of Freedom, Democracy and Terrorism", Dept. of Political Science and Public Management University of Southern Denmark, mimeo.

Knox, N. W., & Eicher, R. W. (1992). *MORT User's Manual, rev. 3: US Department of Energy.* System Safety Development Center EG&G Idaho Inc.

Kooiman, J. (Ed.). (1993). *Modern governance. New government-society interactions.* London: Sage.

Kooiman, J. (2003). *Governing as governance.* London: Sage.

Koppenjan, J. F. M., & Klijn, E.-H. (2004). *Managing uncertainties in networks – a network approach to problem-solving and decision-making.* London: Routledge.

Kotamraju, N. (2004). Art vs. code. In Howard, P., & Jones, S. (Eds.), *Society online* (pp. 189–200). London: Sage Publications.

Kovačić, Z. J. (2005). The Impact of National cultures on Worldwide E government Readiness. *Informing Science Journal, 8,* 143–159.

Kraemer, K., & King, J. L. (2006). Information Technology and Administrative Reform: Will E-Government Be Different? *International Journal of Electronic Government Research, 2*(1), 1–20.

La Porta, R., Lopez-de-Silanes, F., Shleifer, A., & Vishny, R. (1997). Trust in Large Organizations. *American Economic Review Papers and Proceedings, LXXXVII,* 333–338.

La Porte, T. M., Demchak, C. C., & de Jong, M. (2002). Democracy and bureaucracy in the age of the web. *Administration & Society, 34,* 411–446. doi:10.1177/0095399702034004004

La Prensa-San Diego. Available from http://laprensa-sandiego. org/rates/rates.html accessed November 27, 2005.

Lam, W. (2005). Barriers to E-Government Integration. *Journal of Enterprise Information Management, 18*(5), 511–530. doi:10.1108/17410390510623981

Lam, W. (2005). Investigating Success Factors in Enterprise Application Integration: A Case Driven Analysis. *European Journal of Information Systems, 14*(2), 175–187. doi:10.1057/palgrave.ejis.3000530

Lam, W. (2005). Barriers to e-Government. *Journal Of Enterprise Information Management, 18*(5), 511–530. doi:10.1108/17410390510623981

Larsen, E., & Rainie, L. (2002). "The rise of the e-citizen: How people use government agencies' web sites". *Pew Internet and American Life Project*. Available Online http://www.pewinternet.org/ reports/toc.asp?Report=57 accessed January 13, 2007.

Larsen, E., & Rainie, L. (2002). *The Rise of the E-Citizen: How People Use Government Agencies' Web Site*. Retrieved April 12, 2006 from: Available: http:// www.pewinternet.org/pdfs/ PIP_Govt_Web site_Rpt.pdf

Latour, B. (1987). *Science in action: How to follow Scientists and engineers through society*. Cambridge, MA: Harvard University Press.

Latour, B. (1996). *Aramis or the Love of Technology*. Cambridge, MA: Harvard University Press. Translated by Catherine Porter.

Lau, T. Y., Aboulhoson, M., Lin, C., & Atkin, D. J. (2008). Adoption of e-government in three Latin American countries: Argentina, Brazil and Mexico. *Telecommunications Policy, 32*(2), 88–100. doi:10.1016/j.telpol.2007.07.007

Laudon, K. C., & Laudon, J. P. (2010). *Management Information Systems: Managing the Digital Firm* (11th ed.). London, UK: Pearson Education.

LaVigne, M., Simon, S., Dawes, S., Pardo, T., & Berlin, D. (2001). *Untangle the Web: Delivering Municipal Services Through the Internet. Center for Technology in Government. University at Albany*. SUNY.

LaVoy, D. J. (2001, Fall). Trust and Reliability. *Public Management, 30*(3), 8.

Lawrence, T. B., & Phillips, N. (2004). From Moby Dick to Free Willy: Macro cultural Discourse and institutional Entrepreneurship in Emerging Institutional Fields. *Organizations, 11*(5), 689–711. doi:10.1177/1350508404046457

Layne, K., & Lee, K. (2001). Developing fully functional E-government: a four stage model. *Government Information Quarterly, 18*, 122–136. doi:10.1016/S0740-624X(01)00066-1

Lean, O. K., Zailani, S., Ramayah, T., & Fernando, Y. (2009). Factors influencing intention to use e-government services among citizens in Malaysia. *International Journal of Information Management, 29*(6), 458–475. doi:10.1016/j.ijinfomgt.2009.03.012

Lee, M., & Turban, E. (2001). A Trust Model for Internet Shopping. *International Journal of Electronic Commerce, 6*, 75–91.

Lee, Y., & Kozar, K. A. (2006). Investigating the effect of Web site quality on E-Business success: An analytic hierarchy process (AHP) approach. *Decision Support Systems, 42*(3), 1383–1401. doi:10.1016/j.dss.2005.11.005

Lee, S., & Treacy, M. E. (1998). Information technology impacts on innovation. *R & D Management, 18*(3), 257–271. doi:10.1111/j.1467-9310.1988.tb00592.x

Lee, S. M., Tan, X., & Trimi, S. (2005). Current practices of leading e-government countries. *Communications of the ACM, 48*(10), 99–104. doi:10.1145/1089107.1089112

Lee, M. K. O., & Turban, E. (2001). A Trust Model for Consumer Internet Shopping. *International Journal of Electronic Commerce, 6*(1), 75–91.

Lee, C. S., Chandrasekaran, R., & Thomas, D. (2003). Examining IT Usage across Different Hierarchical Levels in Organisations: A Study of Organizational, Environmental, and IT Factors. *9th Americas Conference on Information Systems*, (pp. 1259-1269).

Lee, H., & Luedemann, H. (2007). A Lightweight Decentralized Authorization Model for Inter-domain Collaborations. *ACM Workshop on Secure Web-Services*, Fairfax, Virginia, USA, SWS '07. ACM, New York, pp. 83-89.

Lehr, W., & McKnight, L. W. (2003). Wireless Internet access: 3G vs. WiFi? *Telecommunications Policy, 27*, 351–370. doi:10.1016/S0308-5961(03)00004-1

Leigh, A. (2006). Trust Inequality and Ethnic Heterogeneity. *The Economic Record, 82*(258), 268–280. doi:10.1111/j.1475-4932.2006.00339.x

Lenk, K., & Traunmuller, R. (2000). Presentation at the IFIP WG 8.5 Working Conference on *"Advances in Electronic Government"*, Zaragoza, 10-11 February.

Levi, M., & Stoker, L. (2000). Political trust and trustworthiness. *Annual Review of Political Science, 3*, 475–507. doi:10.1146/annurev.polisci.3.1.475

Liebowitz, J. (1999). *Building Organizational Intelligence: A Knowledge Management Primer- Transforming Organizational Learning into Organizational Learning.* Rockville, MD:Johns Hopkins University.

Lim, E. T. K., Tan, C. H., & Pan, S. L. (2007). E-Government Implementation: Balancing Collaboration and Control in Stakeholder Management. *International Journal of Electronic Government Research, 3*(2), 1–28.

Lim, E. T. K., Tan, C.-W., & Pan, S.-L. (2007). E-government implementation: balancing collaboration and control in stakeholder management. *International Journal of Electronic Government Research, 3*(2), 1–28.

Limayem, M., Hirt, S. G., & Cheung, C. M. K. (2003). Habit in the context of IS continuance: Theory extension and scale development. *Proceedings of the 11ᵗʰ European Conference on Information Systems (ECIS 2003).* Retrieved April 12, 2006 from: http://is2.lse.ac.uk/asp/aspecis/20030087.pdf

Lin, T. C., & Huang, C. C. (2008). Understanding knowledge management system usage antecedents: An integration of social cognitive theory and task technology fit. *Information & Management, 45*, 410–417. doi:10.1016/j.im.2008.06.004

Linthicum, D. (2000). *Enterprise Application Integration.* Massachusetts, USA: Addison-Wesley.

Localization Industry Standards Association. LISA, (2008). Localization. Accessed February 23, 2008 from http://www.lisa.org/Localization.61.0.html.

Löfgren, K. (2007). The governance of e-government. A governance perspective on the Swedish e-government strategy. *Public Policy and Administration, 3*(22), 335–352.

Löfgren, K., (2007). The Governance of E-government A governance Perspective on the Swedish E-government Strategy in *Public, Policy and administration* 22 (3). p. 335-352

Löfstedt, U. (2005). E-Government – Assessment of Current Research and Proposals for Future Directions, Available from http://www.hia.no/iris28/Docs/ IRIS2028-1008.pdf

Lohmann, N., Gierds, C., & Znamirowski, M. (2007). *BPEL2oWFN - Translating BPEL Process to Open Workflow Nets.* Version 2.0.3. Http://www.gnu.org/software/bpel2owfn/index.html, last accessed: November 2008.

Lu, J., Yu, C. S., Liu, C. and Yao, J. E. (2003). Technology acceptance model for wireless internet. Internet research: electronic networking application and policy 13(3), 206-222.

Macias, E., & Temkin, E. (2005). Trends And Impact Of Broadband In The Latino Community. Tomás Rivera Policy Institute, 2005. Available from http://www.trpi.org/PDFs/broadband.pdf accessed March 21, 2006.

Maguire, S., Hardy, C., & Lawrence, T. B. (2004). Institutional entrepreneurship in emerging fields: HIV/AIDS treatment Advocacy in Canada. *Academy of Management Journal, 47*(5), 657–679. doi:10.2307/20159610

Malaysia, M. S. C. MSC flagship Applications Updates. Available at: <http://www.msc.com.my/updates/flagships.asp> (Accessed December 2, 2006).

Malone, T. W., Yates, J., & Benjamin, R. I. (1987). Electronic Markets and Electronic Hierarchies. *Communications of the ACM, 30*(6), 484–497. doi:10.1145/214762.214766

Mannes, G. (2003). Bahrain's proposed smart ID cards. The Risk Digest: Forum on risks to the public and computers and related systems, 22 (89). Available at: <http://catless.ncl.ac.uk/Risks/22.89.html#subj8.1> (Accessed June 10, 2008).

Mantzana, V., Themistocleous, M., Irani, Z., & Morabito, V. (2007). Identifying Healthcare Actors Involved in the Adoption of Information Systems. *European Journal of Information Systems*, *16*(1), 90–102. doi:10.1057/palgrave.ejis.3000660

Marakas, G. M., Yi, M. Y., & Johnson, R. D. (1998). The multilevel and multifaceted character of computer self-efficacy: Toward clarification of the construct and an integrative framework for research. *Information Systems Research*, *9*(2), 126–163. doi:10.1287/isre.9.2.126

March, J. G., & Olsen, J. P. (1995). *Democratic governance*. New York: The Free Press.

Margetts, H., & Dunleavy, P. (2002). *Cultural Barriers to E-Government* (Working Paper). University Collage of London and London School of Economics for National Audit Office.

Martin, B., & Byrne, J. (2003). Implementing e-government: Widening the lens. *Electronic Journal of E-Government*, *1*(1), 11–22.

Martin, N. (2005). Why Australia needs a SAGE: a security architecture for the Australian government environment. *Government Information Quarterly*, *22*, 96–107. doi:10.1016/j.giq.2004.10.007

Marzullo, F. P., & Souza, J. M. (2009). New Directions for IT Governance in the Brazilian Government. *International Journal of Electronic Government Research*, *5*, 57–69.

Marzullo, F. P. (2009). *SOA na Prática: inovando seu negócio por meio de soluções orientadas a serviços*. 1. ed. São Paulo: Novatec, 1. 392 p.

Marzullo, F. P., & Souza, J. M. (2008). *A Competence Based Approach to IT Governance in the Brazilian*. WORLDCOMP'08 - The 2008 World Congress in Computer Science, Computer Engineering and Applied Computing, 2008, Las Vegas, Nevada.

Maskell, P. (2000). Social Capital, Innovation, and Competitiveness. In Baron, S., Field, J., & Schuller, T. (Eds.), *Critical Perspectives*. New York: Oxford University Press.

Massachusetts. (2004) *A Recipe for Success Building a Citizen-Centric Website: Commonwealth of Massachusetts*. December. Retrieved from www.mass.gov/Aitd/docs/ Cookbook_ver2.pdf

Massuthe, P., Reisig, W., & Schmidt, K. (2005). An Operating Guideline Approach to the SOA. Annals of Mathematics. *Computing & Teleinformatics*, *1*(3), 35–43.

Massuthe, P., & Weinberg, D. (2007). *Functional Interaction Analysis for open Workflow Nets*. Version 2.0. Http://www2.informatik.hu-berlin.de/top/tools4bpel/fiona/, last accessed: November 2008.

Matheis, T. et.al. (2004). Methodical interoperability requirements for eGovernment, *Deliverable D8.1, R4eGov – Towards e-Administration in the large*, IST-2004-026650.

Matheis, T., & Loos, P. (2008): Monitoring cross-organizational business processes. *International Conference on E-Learning, E-Business, Enterprise Information Systems, and E-Government*, (EEE 2008), Las Vegas, USA.

Matheis, T., Ziemann, J., & Loos, P. (2006). A Methodical Interoperability Framework for Collaborative Business Process Management in the Public Sector. In *Proceedings of the Mediterranean Conference on Information Systems*, Venice, Italy.

Matheis, T., Ziemann, J., Schmidt, D., Freiheit, J. (2008). *R4eGov IOP tool suite for modelling cross-organizational processes and data*. Deliverable WP4-D10, R4eGov - Towards e-Administration in the large, project number IST-2004-026650.

Mathieson, K., Peacock, E., & Chin, W. W. (2001). Extending the Technology Acceptance Model: the influence of perceive user resources. *The Data Base for Advances in Information Systems*, *32*(3), 86–112.

Mauro, P. (1995). Corruption and Growth. *The Quarterly Journal of Economics*, *110*(2), 681–712. doi:10.2307/2946696

Mayer, R. C., Davis, J. H., & Schoorman, F. D. (1995). An integrative model of organizational trust. *Academy of Management Review*, *20*(3), 709–734. doi:10.2307/258792

Mayntz, R. (1993). Governing failure and the problem of governability: some comments on a theoretical paradigm. In Kooiman, J. (Ed.), *Modern governance: new government-society interactions*. London: Sage.

Mayntz, R. (1991). *Modernization and the logic of interorganizational networks*. MPFIG Discussion Paper 8. Max Planck Institut für Gesellschaftsforschung.

McClave, J. T., Benson, P. G., & Sincich, T. (2008). *Statistics for Business & Economics, 10/E*. Upper Saddle River, New Jersey, USA: Prentice Hall.

MCIT. (2004). *Ministry of Communication and Information Technology*. Online: www.mcit.gov.sa. Access on 20/2/2005.

MCIT. (2007). *Ministry Of Communication And Information Technology*. Online: www.mcit.gov.sa. Access on 20/2/2007.

McIvor, R., McHugh, M., & Cadden, C. (2002). Internet Technologies Supporting Transparency in the Public Sector. *International Journal of Public Sector Management, 15*(3), 170–187. doi:10.1108/09513550210423352

McKnight, D. H., Choudhury, V., & Kacmar, C. (2002). Developing and Validating Trust Measures for E-Commerce: An Integrative Approach. *Information Systems Research, 13*(3), 334–359. doi:10.1287/isre.13.3.334.81

McKnight, D. H., Cummings, L. L., & Chervany, N. L. (1998). Initial Trust Formation in New Organizational Relationships. *Academy of Management Review, 23*(3), 473–490. doi:10.2307/259290

Mendling, J., & Nüttgens, M. (2004): Transformation of ARIS Markup Language to EPML. In Nüttgens, M., Rump, F. (eds.). *Proceedings of the 3rd GI Workshop on Event-Driven Process Chains (EPK 2004)*. Luxembourg, Luxembourg.

Mendling, J., & Nüttgens, M. (2005). *EPC Markup Language (EPML) - An XML-Based Interchange Format for Event-Driven Process Chains (EPC)*. Technical Report JM-2005-03-10. Vienna University of Economics and Business Administration. MODINIS program (2006). *Study on interoperability at local and regional level*. URL: http://www.egov-iop.ifib.de/index.html.

Menou, M. J. (2002, March). *Digital and social equity? Opportunities and threats on the road to empowerment*. Paper prepared for The Digital Divide from an Ethical Viewpoint, International Center for Information Ethics Symposium, Ausberg, Germany.

Mercuri, R. T. (2005). Trusting in Transparency. Association for Computing Machinery. *Communications of the ACM, 48*(5), 15. doi:10.1145/1060710.1060726

Metzger, M. J., Flanagin, A. J., & Zwarun, L. (2003). College student web use, perceptions of information credibility, and verification behavior. *Computers & Education, 41*, 271–290. doi:10.1016/S0360-1315(03)00049-6

Michael, B., & Bates, M. (2005). Implementing and assessing transparency in digital government: Some issues in project management. In Huang, W., Siau, K., & Wei, K. K. (Eds.), *Electronic Government Strategies and Implementation* (pp. 20–43). London: Idea Group.

Michael, G. (2005, August 6). Gov[ernmen]t uses US software to control use of revenue. *The Guardian* (Tanzania). Retrieved August 6, 2005, from http://www.ippmedia.com

Miles, M., & Huberman, A. (1994). *Qualitative Data Analysis: An Expanded Sourcebook*. Newbury Park, California: Sage.

Miles, M. B., & Huberman, A. M. 1994. *Qualitative data analysis: an expanded sourcebook*. 2nd edition, Thousand Oaks: Sage Publications.

Ministério da Defesa, www.defesa.gov.br Last Access: June, 2010.

Ministério da Fazenda - Receita Federal, http://www.receita.fazenda.gov.br/ Last access: June, 2010.

Ministério do Planejamento. www.planejamento.gov.br. Last Access: June, 2010.

Mkonya, J. (2007, March 21). Deputy PS roots for e-government agenda. *The Guardian* (Tanzania). Retrieved March 21, 2007, from http://www.ippmedia.com

Molla, A., & Licker, P. S. (2001). E-Commerce systems success: An attempt to extend and respecify the Delone and Maclean model of IS success. *Journal of Electronic Commerce Research, 2*(4), 131–141.

Molla, A., & Licher, P. S. (2005). E-commerce adoption in developing countries: a model and instrument. *Information & Management, 42*(6), 877–899. doi:10.1016/j.im.2004.09.002

Molla, A., & Licher, P. S. (2002). Information technology implementation in the public sector of a developing country: issues and challenges. *Proceedings of 3rd Annual global Information Technology Management World Conference*, New York, USA.

Moller, R. M. (2000). Profile of California Computer and Internet Users. California Research Bureau. California State Library. Available from http://www.library.ca.gov/crb/00/01/00-002.pdf accessed April 9, 2006.

Montagnier, P., Muller, E., & Vickery, G. (2002, August). *The digital divide: Diffusion and use of ICTs*. Paper presented at the IAOS Conference, London, U.K.

Moon, M. (2002). The Evolution of E-Government among Municipalities: Rhetoric or Reality? *Public Administration Review, 62*(4), 424–434. doi:10.1111/0033-3352.00196

Moon, M. J., & Welch, E. W. (2005). Same Bed, Different Dreams? A Comparative Analysis of Citizen and Bureaucratic Perspectives on E-Government. *Review of Public Personnel Administration, 25*(3), 243–264. doi:10.1177/0734371X05275508

Moon, M. J., & Norris, D. F. (2005). Does managerial orientation matter? The adoption of reinventing government and e-government at the municipal level. *Information Systems Journal, 15*, 43–60. doi:10.1111/j.1365-2575.2005.00185.x

Moon, J. M., Welch, W. E., and Wong, W. (2005). "What Drives Global E-governance? An Exploratory Study at a Macro Level" Proceedings of the 38th Hawaii International Conference on System Sciences, pp 1-10.

Moore, M. (1999). Truth, Trust and Market Transactions: What Do We Know? *The Journal of Development Studies, 36*(1), 74–88. doi:10.1080/00220389908422612

Moore, G. C., & Benbasat, I. (1991). Development of instrument to measure the perception of adopting an information technology innovation. *Information Systems Research, 2*(3), 192–222. doi:10.1287/isre.2.3.192

Moore, G., & Benbasat, I. (1991). Development of an instrument to measure the perceptions of adopting an information technology innovation. *Information Systems Research, 2*(3), 192–222. doi:10.1287/isre.2.3.192

Moorman, C., Deshpand, R., & Zaltman, G. (1993). Factors Affecting Trust in Market Research Relationships. *Journal of Marketing, 57*, 81–101. doi:10.2307/1252059

Morgan, R. M., & Hunt, S. (1994). The Commitment-Trust Theory of Relationship Marketing. *Journal of Marketing, 58*(3), 20–38. doi:10.2307/1252308

Morris, M. G., & Venkatesh, V. (2000). Age differences in technology adoption decisions: implications for a changing work force. *Personnel Psychology, 53*(2), 375–403. doi:10.1111/j.1744-6570.2000.tb00206.x

Mossberger, K., Tolbert, C., & Stansbury, M. (2003). *Virtual inequality: Beyond the digital divide*. Washington, DC: Georgetown University Press.

Muhlberger, P. (2002, October). *Political values and attitudes in Internet political discussion: Political transformation or politics as usual?* Paper presented at the Euricom Colloquium: Electronic Networks & Democracy, Nijmegen, Netherlands.

Muir, A., & Oppenheim, C. (2002). National information policy developments worldwide I: electronic government. *Journal of Information Science, 3*(28), 173–186. doi:10.1177/016555150202800301

Mulder, E. (2004). A Strategy for E-government in Dutch Municipalities. *Proceedings. 2004 International Conference on Information and Communication Technologies: From Theory to Application, 19*(23), 5 – 6.

Muninetguide (2006). *Striving for Online Excellence*. Retrieved from http://www.muninetguide.com/article s/Striving-for-Online-Excellence---161.php

Mutula, S. M. (2002). Africa's web content: Current status. *Malaysian Journal of Library & Information Science, 7*(2), 35–55.

National City. (2008). National City Web Site. Available from http://www.ci.national-city.ca.us/main.asp accessed February 22, 2008.

National Performance Review. (1993). *From Red Tape To Results: Creating A Government That Works Better And Costs Less*. Washington, D.C.: Government Printing Office.

Navarra, D. D., & Cornford, T. (2003). A Policy Making View of E-Government Innovations in Public Governance. *Proceedings of the Ninth Americas Conference on Information Systems.*

NECCC. (2000). *Risk Assessment Guidebook for e-Commerce/e-Government*. Available from http://www.ec3.org/Downloads/2000/Risk_Assessment_Guidebook.pdf

NECTEC (National Electronics and Computer Technology Center). (2003). *Thailand Information and Communications Technology (ICT) Master Plan (2002-2006)*. Retrieved November 6, 2006 from: http://www.nectec.or.th/pld/masterplan/ document /ICT_Masterplan_Eng.pdf

NECTEC (National Electronics and Computer Technology Center). (2005a). *Thailand ICT Indicators 2005*. Retrieved November 6, 2006 from: http://www.nectec.or.th/ pub/book/ ICTIndicators.pdf

NECTEC (National Electronics and Computer Technology Center). (2005b). *Internet User Profile of Thailand 2005*. Retrieved November 6, 2006 from: http://www.nectec.or.th/ pld/internetuser/Internet %20User%20 Profile%202005.pdf

Negash, S., Ryan, T., & Igbaria, M. (2003). Quality and effectiveness in web-based customer support systems. *Information & Management, 40*(8), 757–768. doi:10.1016/S0378-7206(02)00101-5

Newton, K., & Delhey, J. (2005). Predicting Cross-National Levels of Social Trust: Global Pattern or Nordic Exceptionalism? *European Sociological Review, 21*(4), 311–327. doi:10.1093/esr/jci022

Nielsen, J. (2003). *Severity ratings for usability problems*. Retrieved April 15, 2006, from http://www.useit.com/papers/ heuristic/severityrating.html

Noce, A. A., & McKeown, L. (2008). A New Benchmark for Internet Use: A Logistic Modeling of Factors Influencing Internet use in Canada, 2005. *Government Information Quarterly, 25*(3), 462–476. doi:10.1016/j.giq.2007.04.006

Norris, P. (2001). *Digital divide: Civic engagement, information poverty, and the Internet worldwide*. New York: Cambridge University Press.

Norris, D. F., & Moon, M. J. (2005). Advancing e-government at the grassroots: Tortoise or hare? *Public Administration Review, 65*(1), 64–75. doi:10.1111/j.1540-6210.2005.00431.x

Norris, P. (2001). *Digital divide: Civic engagement, information poverty, and the Internet worldwide*. Cambridge, NY: Cambridge University Press.

Norris, P., & Curtice, J. (2006). If you build a political website, will they come?: The Internet and political activism in Britain. *International Journal of Electronic Government Research, 2*(2), 1–21.

Norris, D. F. (1999). *Leading Edge Information Technologies and their Adoption: Lessons from US Cities*. Hershey, PA: Idea Group Publishing.

Nunnaly, J. C. (1978). *Psychometric Theory* (2nd ed.). New York, NY: McGraw-Hill.

O'Donnell, S. (2002, October). *Internet use and policy in European union and implications for e-democracy*. Paper prepared for the European Colloquium, Nijmegen, Netherlands.

O'Toole, K. (2007). E-governance in Australian local government: spinning a web around community? *International Journal of Electronic Government Research, 3*(4), 58–75.

OECD. (2001). *The Hidden Threat to E-Government: Avoiding large government IT failures*. Retrieved. from http://www.oecd.org/datao ecd/19/12/1901677.pdf.

OECD. (2004). *The E-Government imperative*. Organisation for Economic Co-operation and Development, Paris.

OECD. (2007). *e-Government for better government. OECD e-Government Studies*. Paris: OECD Publishing.

Ogawa, R., & Scribner, W. (2002). Leadership: spanning the technical and institutional dimensions of organizations. *Journal of Educational Administration, 40*(6), 576. doi:10.1108/09578230210446054

Okello, D. (1999). *Towards sustainable regional integration in East Africa: Voices and visions*. Konrad Adenauer Stiftung Occasional Papers: East Africa, 1/1999. Nairobi: Konrad Adenauer Foundation.

Ong, C.-S., & Wang, S.-W. (2009). Managing Citizen-Initiated Email Contacts. *Government Information Quarterly, 26*(3), 498–504. doi:10.1016/j.giq.2008.07.005

Ongaro, E. (2004). Process management in the public sector: the experience of one-stop shops in Italy. *International Journal of Public Sector Management, 17*(1), 81–107. doi:10.1108/09513550410515592

Organisation for Economic Cooperation and Development. (2004). *Regulatory reform as a tool for bridging the digital divide*. Retrieved June 26, 2005 from http://www.oecd.org/topic/0,2686,en_2649_37441_1_1_1_1_37441,00.html

Organisation for Economic Cooperation and Development. (2007). *OECD Communications Outlook 2007*. Retrieved March 11, 2008 from http://213.253.134.43/oecd/pdfs/browseit/9307021E.PDF

Organization for Economic Co-operation and Development (OECD). (2001). *Citizens as Partners*. Information, Consultation and Public Participation in Policy-Making.

Orlikowski, W. J., & Baroudi, J. J. (1991). Studying Information Technology in Organizations: Research Approaches and Assumptions. *Information Systems Research, 2*, 1–28. doi:10.1287/isre.2.1.1

Orlikowski, W. (1992). The duality of technology: Rethinking the concepts of technology in organizations. *Organization Science, 3*(3), 398–427. doi:10.1287/orsc.3.3.398

Orlikowski, W., & Baroudi, J. (1991). Studying Information Technology in Organizations: Research Approaches and Assumptions. *Information Systems Research, 2*(1). doi:10.1287/isre.2.1.1

Orlikowski, W., & Gash, D. C. (1994). Technological Frames: making Sense of Information Technology in Organizations. *ACM Transactions on Information Systems, 12*()2 April, pp 174-20758-83 pages

Osborne, D., & Gaebler, T. (1992). *Reinventing Government: How The Entrepreneurial Spirit Is Transforming The Public Sector*. Reading, MA: Addison-Wesley.

OSHA4LESS.COM, "Sales Book - lew 2.pdf", 2006. Received as attachment to electronic mail message, March 24, 2006.

O'Toole, K. (2007). E-Governance in Australian Local Government: Spinning a Web Around Community. *International Journal of Electronic Government Research, 3*(4), 58–83.

PAC – Brazilian Growth Acceleration Programme http://www.fazenda.gov.br/ portug ues/releases/2007/r220107-PAC.pdf. Last access: May, 2007.

Panagopoulos, C. (2004). Consequences of the cyberstate: The political implications of digital government in international context. In Pavlichev, A., & Garson, G. D. (Eds.), *Digital government: Principles and best practices* (pp. 116–132). London: Idea Group Publishing.

Paralinguism in the Theatres and the International Theatre Festivals. Sasho Ognenovski. Intercultural Communication, ISSN 1404-1634, 1999, August, issue 1. Available from http://www.immi.se/intercultural/ accessed November 19, 2005.

Parasuraman, A., Zeithaml, V. A., & Berry, L. L. (1985). A conceptual model of service quality and its implications for future research. *Journal of Marketing, 49*(4), 41–50. doi:10.2307/1251430

Parasuraman, A., Zeithaml, V. A., & Berry, L. L. (1988). SERVQUAL: A multiple-item scale for measuring consumer perceptions of service quality. *Journal of Retailing, 64*(1), 12–40.

Pardo, T. A. (2000). *Realizing the Promise of e-Government. Center for Technology in Government. University at Albany.* SUNY.

Pardo, T., & Scholl, H. J. (2002). Walking atop the cliffs: avoiding failure and reducing risk in large scale e-government projects. *Proceedings of the 35th Hawaii International Conference,* Hawaii.

Parent, M., Vandebeek, C. A., & Gemino, A. C. (2005). Building Citizen Trust through E-Government. *Government Information Quarterly, 22*(4), 720–736. doi:10.1016/j.giq.2005.10.001

Pate-Cornell, M. E. (1984). Fault Tree vs. Event Trees in Reliability Analysis. *Risk Analysis, 4*(3), 177–186. doi:10.1111/j.1539-6924.1984.tb00137.x

Pate-Cornell, M. E. (1993). Risk Analysis and Risk Management for Offshore Platforms: Lessons from the Piper Alpha Accident. *Journal of Offshore Mechanics and Arctic Engineering, 115,* 179–190. doi:10.1115/1.2920110

Pavlou, P. (2003). Consumer Acceptance of Electronic Commerce: Integrating Trust and Risk with the Technology Acceptance Model. *International Journal of Electronic Commerce, 7*(3), 69–103.

Pedhazur, E. J., & Schmelkin, L. P. (1991). *Measurement, Design, and Analysis: An Integrated Approach.* Hillsdale, NJ: Lawrence Erlbaum Associates, Inc.

Peikari, C., & Fogie, S. (2003). *Maximum Wireless Security.* Retrieved from http://www.berr.gov.uk/files/file9972.pdf.

Pelaez, C. E., & Bowles, J. B. (1995). *Applying Fuzzy Cognitive-Maps Knowledge- Representation to Failure Modes Effects Analysis.* Paper presented at the Annual Reliability and Maintainability Symposium.

Persell, C., Green, A., & Gurevich, L. (2001). Civil Society, Economic Distress, and Social Tolerance. *Sociological Forum, 16*(2), 203–230. doi:10.1023/A:1011048600902

Peters, G. P. (2006). Concepts and theories of horizontal policy management. In Peters, G. B., & Pierre, J. (Eds.), *Handbook of public policy.* London: Sage.

Peterson, E., & Seifert, J. (2002). Expectation and challenges of emergent electronic government: The promise of all things E? *Perspectives on Global Development and Technology, 1*(2), 193–212. doi:10.1163/156915002100419808

Peterson, R. R. (2004). *Integration strategies and Tactics form Information Technology Governance.* Hershey, PA: Idea Group Inc.

Pew Internet & American Life Project. (2006). *Home Broadband Adoption.* Retrieved from http://www.pewinternet.org/pdfs /PIP_Broadband_trends2006.pdf

Pew Research Center. (2008). *The Internet's broader role in campaign 2008: Social networking and online videos take off.* Retrieved February 19, 2008 from http://people-press.org/report/384/internets-broader-role-in-campaign-2008

Pierre, J., & Peters, G. P. (2000). *Governance, politics and the state.* London: St. Martin's Press.

Pieterson, W., & Ebbers, W. (2008). The Use of Service Channels by Citizens in the Netherlands: Implications for Multi-Channel Management. *International Review of Administrative Sciences, 74*(1), 95–110. doi:10.1177/0020852307085736

Pieterson, W. (2009). *Channel Choice: Citizens' Channel Behavior and Public Service Channel Strategy.* Netherlands: Thesis, University of Twente.

Pitt, L. F., Watson, R. T., & Kavan, C. B. (1995). Service quality: A measure of information systems effectiveness. *Management Information Systems Quarterly, 19*(2), 173–185. doi:10.2307/249687

Plouffe, D. R., Hulland, J. S., & Vandenbosch, M. (2001). Research Report: Richness versus parsimony in modeling technology adoption decisions-understanding merchant adoption of a smart card-Based Payment System. *Information Systems Research, 12*(2), 208–222. doi:10.1287/isre.12.2.208.9697

Plouffe, C. R., Hulland, S. J., & Vandenbosch, M. (2001). Research report: Richness versus parsimony in modeling technology adoption decisions--Understanding merchant adoption of a smart card-based payment system. *Information Systems Research, 12*(2), 208–222. doi:10.1287/isre.12.2.208.9697

PMCDF - Project Management Competency Development Framework. http://www.pmi.org/ 2001. Last access: May, 2010.

Poston, R. S., Akbulut, A. Y., & Looney, C. A. (2007). Online advice taking: examining the effects of self-efficacy, computerized sources, and perceived credibility, Available at: <http://sigs.aisnet.org/SIGHCI/Research/ICIS_workshop_2005.html> (Accessed March 4, 2007).

PPA - Plano Plurianual. http://www.planobrasil.gov.br, Last access: May, 2007.

Pratchett, L., Wingfield, M., & Polat, R. K. (2006). Local democracy online: An analysis of local government web sites in England and Wales. *International Journal of Electronic Government Research, 2*(3), 75–92.

Pratchett, L. (2004). Electronic government in Britain. In Eifert, M., & Püschel, J. O. (Eds.), *National electronic government: comparing governance structures in multilayer administrations.* London: Routledge.

Press, S. J. (1989). *Bayesian Statistics: Principles, Models and Applications.* New York: Wiley.

Press, L. (1993). *Relcom: An appropriate technology network.* Retrieved August 21, 2002 from ibiblio database, http://www.ibiblio.org/pub/academic/russian-studies/Networks/Relcom/relcom.history

Pressman, J. L., & Wildavsky, A. (1973). *Implementation.* Berkeley: University of California Press.

Price, C. J., Hunt, J. E., Lee, M. H., & Ormsby, R. T. (1992). A Model-based Approach to the Automation of Failure Mode Effects Analysis for Design. *IMechE, Part D: the Journal of Automobile Engineering, 206,* 285-291.

Prins, C. (2001). *Designing e-government: on the crossroads of technological innovation and institutional change.* Hague; Boston: Kluwer Law International.

Prychodko, N. (2001). *Municipalities and Citizen-Centred Service.* Report to the Public Sector Service Delivery Council December 4, 2001. Retrieved from http://www.iccs-isac.org/eng /pubs/iccs_muni.pdf

Puschmann, T., & Alt, R. (2001). Enterprise Application Integration – The Case of the Robert Bosch Group. *Proceedings of the 34th Hawaii International Conference on System Sciences* (pp. 1-10) Maui, Hawaii, IEEE.

Putnam, R. D. (1995). Bowling Alone: America's Declining Social Capital. *Journal of Democracy, 6*(1), 65–78. doi:10.1353/jod.1995.0002

Rai, A., Lang, S., & Welker, R. (2002). Assessing the Validity of IS Success Models: An Empirical Test and Theoretical Analysis. *Information Systems Research, 13*(1), 50–69. doi:10.1287/isre.13.1.50.96

Rangkaian Segar Sdn Bhd (RSSB). Facts and figures. Available at: <http://www.touchngo.com.my/MediaCentre_FF.html> (Accessed November 30, 2006)

Reason, P., & Rowan, J. (1981). *Human Inquiry: A sourcebook of new Paradigm research.* Chichester, UK: John Wiley.

Reddick, C. G. (2005a). Citizen-Initiated Contacts with Government: Comparing Phones and Websites. *Journal of E-Government, 2*(1), 27–51. doi:10.1300/J399v02n01_03

Reddick, C. G. (2005b). Citizen Interaction with E-Government: From the Streets to Servers? *Government Information Quarterly, 22*(1), 38–57. doi:10.1016/j.giq.2004.10.003

Reddick, C. G. (2004). A two-stage model of e-government growth: Theories and empirical evidence for U.S. cities. *Government Information Quarterly*, *21*(1), 51–64. doi:10.1016/j.giq.2003.11.004

Reddick, C. G. (2008). Perceived Effectiveness of E-Government and its Usage in City Governments: Survey Evidence from Information Technology Directors. *International Journal of Electronic Government Research*, *4*(4), 89–104.

Reddick, C. G. (2009). Factors that Explain the Perceived Effectiveness of E-Government: A Survey of United States City Government Information Technology Directors. *International Journal of Electronic Government Research*, *5*(2), 1–15.

Reffat, R. (2003). *Developing A Successful E-Government* (Working Paper). University Of Sydney, Australia.

Relyea, H. C., & Nunno, R. M. (2000). *Electronic government and electronic signatures*. Huntington, NY: Novinka Books.

Relyea, H. C. (2002). E-gov: Introduction and overview. *Government Information Quarterly*, *19*(1), 9–35. doi:10.1016/S0740-624X(01)00096-X

Resnick, P., Zeckhauser, R., & Avery, C. (1995). Roles for Electronic Brokers. In G. W. Brock (Ed.), *Towards a Competitive Telecommunication Industry* (pp. 289-306). New Jersey: Mahwah.

Reynolds, M. M., & Regio-Micro, M. (2001). The Purpose Of Transforming Government-E-Government as a Catalyst In The Information Age. *Microsoft E-Government Initiatives*. Retrieved from http://www.netcaucus.org/books/eg ov2001/pdf/EGovIntr.pdf

Rhodes, R. A. W. (1997). *Understanding governance. Policy Networks, governance, reflexivity and accountability*. Buckingham: Open University Press.

Rhodes, R. A. W., & Marsh, D. (Eds.). (1992). *Policy networks in British government*. Oxford, UK: OUP.

Riley, T. B. (2000). *Electronic Governance and Electronic Democracy: Living and Working in the Wired World*. London: Commonwealth Secretariat.

Ritzen, J., Easterly, W., & Woolcock, M. (2001). "On 'Good' Politicians and 'Bad' Policies: Social Cohesion, Institutions and Growth", *Working Paper, Washington DC. World Bank*

Rodrick, D. (1999). Where did all the growth go?' External Shocks, Social Conflict, and Growth Collapses. *Journal of Economic Growth*, *1*, 149–187.

Rodriguez, J. R (2005). IRIS project: Promoting civic attitudes in Barcelona through a customer service request platform. *The 2005 Ministerial e-Government Conference on transforming Public Services*. Retrieved from http://archive.cabinetoffice.gov. uk/egov2005conference/

Rodriguez, R., Estevez, E., Giulianelli, D., & Vera, P. (2009). Assessing E-Governance Maturity through Municipal Websites Measurement Framework and Survey Results. *Proceedings of the 6th Workshop on Software Engineering, Argentinean Computer Science Conference*. San Salvador de Jujuy, Argentina, October 5-9, 2009. Retrieved from http://egov.iist.unu.edu/cegov/ OUTPUTS/ PUBLICATIONS

Rogers, E. M. (1995). *Diffusion of innovations*. New York.

Rose, R. (2005). A Global diffusion model of e-Governance. *Journal of Public Policy*, *25*(1), 5–27. doi:10.1017/S0143814X05000279

Rose, N. (1999). *Powers of freedom: reframing political thought*. Cambridge, UK: Cambridge University Press. doi:10.1017/CBO9780511488856

Rosell, S., Gantwerk, H., & Furth, I. (2005). Listening To Californians: Bridging The Disconnect. Viewpoint Learning, Inc. Retrieved January 15, 2006 from http://www.viewpointlearni ng.com/pdf/HI_Report_FINAL.pdf.

Ross, J., & Weill, P. (2004). *IT Governance on One Page*. MIT Sloan School of Management.

Røvik, K. A. (2000). *Moderna organisationer: trender inom organisationstänkande vid millennieskiftet.* Stockholm: Liber.

Røvik, K. A. (2002). The secret of the winners: Management ideas that flow in Sahlin-Andersson, K & Engwall (ed.), L *The expansion of management knowledge. Carriers, flows and sources.* Stanford, CA: Stanford University Press

Roy, J. (2003). E-government. *Social Science Computer Review, 21*(1), 3–5. doi:10.1177/0894439302238966

Ruzbacki, T. (2005). *Sarbanes-Oxley, and Enterprise Change Management.* MKS White Paper.

Rycroft, R. W., & Kash, D. E. (2004). Self-organizing innovation networks: implications for globalization. *Technovation, 24*, 187–197. doi:10.1016/S0166-4972(03)00092-0

Saaty, T. L. (2001). *Decision Making for Leaders – The Analytical Hierarchy Process for Decisions in a Complex World.* Pittsburgh, PA: RWS Publications.

Salem, J. A. (2003). Public and private sector interests in e-government: A look at the DOE's PubSCIENCE. *Government Information Quarterly, 20*, 13–27. doi:10.1016/S0740-624X(02)00133-8

Salem, J. A. Jr. (2003). Public and private sector interests in e-government: a look at the DOE's PubSCIENCE. *Government Information Quarterly, 20*, 13–27. doi:10.1016/S0740-624X(02)00133-8

Sam, H. K., Othman, A. E. A., & Nordin, Z. S. (2005). Computer Self-Efficacy, Computer Anxiety, and Attitudes toward the Internet: A Study among Undergraduates in Unimas. *Journal of Educational Technology & Society, 8*(4), 205–219.

Sandoval, V. A., & Adams, S. H. (2001). Subtle Skills for Building Rapport: Using Neuro-Linguistic Programming in the Interview Room. Available from http://www.fbi.gov/publications /leb/2001/august2001/aug01p1.htm accessed November 19, 2005.

Sarkar, M. B., Butler, B., & Steinfield, C. (1995). Intermediaries and Cybermediaries: A Continuing Role for Mediating Players in the Electronic Marketplace. *Journal of Computer-Mediated Communication, 1*(3).

Saudi Computer Society. (2004). National Information Technology Plan. (NITP) (Draft version). S. C. Society. Riyadh, Saudi Arabia, Saudi. *Computers & Society*, 1–463.

Saunders, M., Lewis, P., & Thornhill, A. (2002). *Research methods for business students* (3rd ed.). Harlow: Prentice Hall.

Saunders. M., Lewis, P., & Thornhill, A. (2000). *Research Methods for Business Students'.* Essex, Pearson Education Ltd.

Scandinavian Political Studies, 28(3), 195–218. doi:10.1111/j.1467-9477.2005.00129.x

Scharpf, F. W. (1994). Games real actors could play: positive and negative coordination in embedded negotiations. *Journal of Theoretical Politics, 6*(1), 27–53. doi:10.1177/0951692894006001002

Scharpf, F. W. (1997). *Games real actors play. Actor-centred institutionalism in policy research.* Oxford, UK: West View Point.

Schaupp, L., & Carter, L. (2008). The impact of trust, risk and optimism bias on E-file adoption. *Information Systems Frontiers.* doi:.doi:10.1007/s10796-008-9138-8

Schein, E. H. (1992). *Organizational culture and leadership: A dynamic view.* San Francisco, CA: Jossey Bass.

Schein, E. H. (1994). *Organizational Psychology.* London: Prentice Hall.

Schelin, S. H. (2003). E-government: An overview. In Garson, G. D. (Ed.), *Public information technology: Policy and management issues* (pp. 120–137). Hershey, PA: Idea Group Publishing.

Schellong, A. (2008). *Citizen Relationship Management: A Study of CRM in Government.* Berlin: Peter Lang Publishing Group.

Schmidt, R. C., Lyytinen, K., Keil, M., & Cule, P., P. (2001). Identifying software project risks: an international Delphi study. *Journal of Management Information Systems, 17*(4), 5–36.

Schneider, G. P. (2003). *Electronic Commerce, Fourth Annual Edition.* Boston: Thomson Course Technology.

Schubert, K. D. (2004). *CIO Survival Guide – The Roles and Responsibilities of the Chief Information Officer.* Wiley.

Schutz, A. (1967). *The phenomenology of the social world.* Evanston, IL: Northwestern University Press.

Scruton, R. (1982). *A dictionary of political thought.* New York: Hill & Hwang.

Secretariat, Treasury Board of Canada, (2006). Communications Policy of the Government of Canada. Available from http://www.tbs-sct.gc.ca/pubs_ pol/sipubs/comm/comm1_e.asp#04 accessed March 21, 2006.

Seddon, P. B., & Kiew, M.-Y. (1994). A partial test and development of the DeLone and McLean's model of IS success. *Proceedings of the International Conference on Information Systems (ICIS 94)*, 99–110.

Seifert, J., & Petersen, E. (2002). The Promise Of All Things E? Expectations and Challenges of Emergent E- Government. *Perspectives on Global Development and Technology, 1*(2), 193–213. doi:10.1163/156915002100419808

Senyucel, Z. (2005). *Towards successful eGovernment facilitation in UK local authorities presented at eGovernment Workshop '05 (eGOV2005), September 13 2005.* Brunel University, West London.

Senyucel, Z. (2005). Towards Successful e-Government Facilitation in UK Local Authorities. *e-Government Workshop* (pp. 1-16).

SERPRO. http://www.serpro.gov.br/. Last access: June, 2009.

Shapiro, S. P. (1987). The social control of impersonal trust. *American Journal of Sociology, 93*(3), 623–658. doi:10.1086/228791

Shaw, N. G. (2000). Capturing the Technological Dimensions of IT Infrastructure Change. *Journal of the Association for Information Systems, 2*(8).

Shleifer, A., & Vishny, R. (1993). Corruption. *The Quarterly Journal of Economics, 108*(3), 599–618. doi:10.2307/2118402

Siegel, S. (1956). *Nonparametric Statistics for the Behavioral Sciences.* New York: McGraw-Hill Book Co.

Signore, O., Chesi, F., & Pallotti, M. (2005). E-Government: Challenges and Opportunities. *CMG Italy – XIX Annual Conference*, Florence, Italy.

Silcock, R. (2001). What is e-government? *Parliamentary Affairs, 54*, 88–101. doi:10.1093/pa/54.1.88

Silverman, D. (1993). *Interpreting Qualitative data methods for Analysing Talk, text and Interaction.* London: SAGE Publications Ltd.

Singh, A., & Sahu, R. (2008). Integrating Internet, Telephones, and Call Centers for Delivering Better Quality E-Governance to all Citizens. *Government Information Quarterly, 25*(3), 477–490. doi:10.1016/j.giq.2007.01.001

Singh, S., & Naidoo, G. (2005). Towards an e-government solution: A South African perspective. In Huang, W., Siau, K., & Wei, K. K. (Eds.), *Electronic Government Strategies and Implementation* (pp. 325–353). London: Idea Group.

Singh, H., Das, A., & Joseph, D. (2004). "Country-level determinants of e-government maturity". Nanyang Technological University, Working Paper.

Singh, S. (2006). "RM500m letdown," New Strait Times (December 25, 2006).

Siu, N. (1994). Risk Assessment for dynamic systems: An overview. *Reliability Engineering & System Safety, 43*, 43–73. doi:10.1016/0951-8320(94)90095-7

Sjoberg, L., & Fromm, J. (2001). Information Technology Risks as Seen by the Public. *Risk Analysis, 21*(3), 427–441. doi:10.1111/0272-4332.213123

Snellen, I. (2002). Electronic governance: Implications for citizens, politicians and public servants. *International Review of Administrative Sciences, 68*(2), 183–198. doi:10.1177/0020852302682002

Solomon, M. R., Surprenant, C. F., Czepiel, J. A., & Gutman, E. G. (1985). A role theory perspective on dyadic interactions: the service encounter. *Journal of Marketing, 49*(1), 99–111. doi:10.2307/1251180

Sommerville, J., & Sawyer, P. (2003). *Requirments Engineering – A good practice guide*. New York: Wiley.

Sørensen, E., & Torfing, J. (2000). *Skanderborg på landkortet – et studie af lokale styringsnetværk og politisk handlekraft* [Skanderborg on the map: a study of local governance networks and political effectivness]. Copenhagen: Jurist og Økonomforbundets Forlag.

Sørensen, E., & Torfing, J. (2003). Network politics, political capital, and democracy. *International Journal of Public Administration, 26*(6), 609–634. doi:10.1081/PAD-120019238

Sørensen, E., & Torfing, J. (2007). Theoretical approaches to meta-governance. In Sørensen, E., & Torfing, J. (Eds.), *Theories of democratic network governance* (pp. 169–182). Basingstoke: Macmillan.

Sørensen, E. & Torfing, J. (2005). The democratic anchorage of governance networks,

Spreng, R. A., MacKenzie, S. B., & Olshavsky, R. W. (1996). A reexamination of the determinants of consumer satisfaction. *Journal of Marketing, 60*(3), 15–32. doi:10.2307/1251839

Spulber, D. F. (1996). Market Microstructure and Intermediation. *The Journal of Economic Perspectives, 10*(3), 135–152.

Stahl, B. C. (2005). The paradigm of e-commerce in e-government and e-democracy. In Huang, W., Siau, K., & Wei, K. K. (Eds.), *Electronic Government Strategies and Implementation* (pp. 1–19). London: Idea Group.

Stake, R. E. (1995). *The art of case study research*. Thousand Oaks: Sage Publications.

Stal, M. (2002). Web Services: Beyond Component – Based Computing. *Communications of the ACM, 45*(10), 71–76. doi:10.1145/570907.570934

Stalder, F. (2006). *Manuel Castells: The theory of the network society*. Malden, MA: Polity Press.

Stamatis, D. H. (1995). *Failure Mode and Effect Analysis - FMEA from Theory to Execution*. ASQC Quality Press.

Steyaert, J. C. (2004). Measuring the Performance of Electronic Government Services. *Information & Management, 41*, 369–375. doi:10.1016/S0378-7206(03)00025-9

Stowers, G. N. L. (2004). Issues in e-commerce and e-government service delivery. In Pavlichev, A., & Garson, G. D. (Eds.), *Digital government: Principles and best practices* (pp. 169–185). London: Idea Group Publishing.

Strauss, A. (1987). *Qualitative Analysis for Social Scientists*. Cambridge, MA: Cambridge University Press. doi:10.1017/CBO9780511557842

Streib, G., & Navarro, I. (2006). Citizen Demand for Interactive E-Government: The Case of Georgia Consumer Services. *American Review of Public Administration, 36*(3), 288–300. doi:10.1177/0275074005283371

Studenmund, A. H. (1992). *Using Econometrics: A Practical Guide* (2nd ed.). New York, NY: HarperCollins.

Sturm, J. (2007). Interoperability and standards. In Zechner, A. (Ed.), *E-Government Guide Germany: Strategies, solutions and efficiency* (pp. 31–37). Stuttgart, Germany: Fraunhofer IRB Verlag.

Sun, J. (August 2005). User readiness to interact with information systems- a human activity perspective. Doctoral Dissertation, Texas A&M University

Sutton, I. S. (1992). *Process Reliability and Risk Management* (1st Ed. ed.): Van Nostrand Reinhold.

Svensson, J. (1998). Investment, Property rights and Political Instability: Theory and Evidence. *European Economic Review, 42*(7), 1317–1342. doi:10.1016/S0014-2921(97)00081-0

Swedish National Audit Office (SNAO) [Riksrevisionen]. 2003. *Vem styr den elektroniska förvaltningen?* [Who Governs the Electronic Government?], Report 2003:19. Stockholm.

Szajna, B. (1996). Empirical evaluation of the revised technology acceptance model. *Management Science*, *42*(1), 85–92. doi:10.1287/mnsc.42.1.85

Szymanski, D. M., & Hise, R. T. (2000). E-Satisfaction: An initial examination. *Journal of Retailing*, *76*(3), 309–322. doi:10.1016/S0022-4359(00)00035-X

Tambouris, E. (2001). An integrated platform for realising online one-stop government: the e-GOV project. *Proceedings of 12th international workshop on database and expert systems applications*, Munich, Germany, IEEE Computer Society.

Tambouris, E., & Tarabanis, K. (2004). Overview of DC-Based eGovernment Metadata Standards and Initiatives. In Traunmüller, R. (Ed.), *Electronic Government*, (LNCS # 3183, pp.40-47), Springer Verlag Heidelberg et al.

Tan, C. W., Pan, S. L., & Lim, E. T. K. (2007). Managing Stakeholder Interests in E-Government Implementation: Lessons Learned from a Singapore E-Government Project. *International Journal of Electronic Government Research*, *3*(1), 61–84.

Tan, C., Pan, S., & Lim, E. (2005). Managing stakeholder interests in e-government implementation: Lessons learned from a Singapore e-government project. *Journal of Global Information Management*, *13*(1), 31–53.

Tan, C. W., Pan, S. L., & Lim, E. T. K. (2005). Towards the Restoration of Public Trust in Electronic Governments: A Case Study of the E-Filing System in Singapore", Proceedings of the 38th Annual Hawaii International Conference on System Sciences, IEEE Computer Society Press

Tan, C.-W., Benbasat, I., & Cenfetelli, R. T. (2008). Building Citizen Trust towards e-Government Services: Do High Quality Web sites Matter? Proceedings of the 41st Hawaii International Conference on System Sciences, IEEE Computer Society Press.

Tasmania. (2005). *Risk Management Resource Kit.* from http://www.egovernment.tas.gov.au/themes/ project_management/risk_management_resource_kit

Taylor, S., & Todd, P. A. (1995). Understanding information technology usage: a test of competing models. *Information Systems Research*, *6*(2), 144–176. doi:10.1287/isre.6.2.144

Tellechea, A. F. (2008). Economic crimes in the capital markets. *Journal of Financial Crime*, *15*(2), 214–222. doi:10.1108/13590790810866908

Teo, T., Srivastava, S. C., & Jiang, L. (2008). Trust and Electronic Government Success: An Empirical Study. *Journal of Management Information Systems*, *25*(3), 99–131. doi:10.2753/MIS0742-1222250303

Tesouro Nacional. http://www.tesouro.fazenda.gov.br 2009. Last access: June, 2010.

The Heritage Foundation. (2005). *Index of Economic Freedom.* Retrieved Dec 20, 2006 from http://www. heritage.org/re search/features/index/.

The Heritage Foundation. *Index of Economic Freedom* (2005), Available online http://www.heritage.org/index/, accessed January 15, 2007.

The Peninsula Newspaper. (2008). *Bursting at the seams.* Retrieved from http://www.thepeninsulaqatar.com/Display_ne ws.asp?section=Local_News&month=January 2008&file=Local_News200801296298.xml.

The World Bank Group. (*GDPPC* 2004). Available online http://www.worldbank.org/data, accessed January 15, 2007.

The Year of Languages. (2005). Special project of the American Council on the Teaching of Foreign Languages (ACTFL). Available from http://www.yearoflanguages. org/i4 a/pages/index.cfm?pageid=3419 accessed March 21, 2006.

Themistocleous, M. (2004). Justifying the decisions for EAI implementations: A validated proposition of influential factors. *Journal of Enterprise Information Management*, *17*(2), 85–104. doi:10.1108/17410390410518745

Themistocleous, M., & Irani, Z. (2001). Benchmarking the Benefits and Barriers of Application Integration, *Benchmarking. International Journal (Toronto, Ont.)*, *8*(4), 317–331.

Themistocleous, M., Irani, Z., & Love, P. E. D. (2005). Developing E-Government Integrated Infrastructures: A Case Study. *Proceedings of the 38ᵗʰ Annual Hawaii International Conference on System Sciences* (pp. 1-10). Big Island, Hawaii: IEEE.

Theory, Applications and Benchmarking. (pp 143-166. London: Idea Group.

Thomas, J. C., & Streib, G. (2003). The New Face of Government: Citizen-Initiated Contacts in the Era of E-government. *Journal of Public Administration: Research and Theory, 13*(1), 83–102. doi:10.1093/jpart/mug010

Thomas, J. C., & Streib, G. (2005). E-Democracy, E-Commerce, and E-Research: Examining the Electronic Ties between Citizens and Governments. *Administration & Society, 37*(3), 259–280. doi:10.1177/0095399704273212

Thomas, C. W. (1997). Public management as interagency cooperation: testing epistemic community theory at the domestic level. *Journal of Public Administration: Research and Theory, 7*(2), 221–246.

Thomas, M. (2004). Is Malaysia's MyKad the 'One Card to Rule Them All'? The Urgent Need To Develop a Proper of Legal Framework for the Protection of Personal Information in Malaysia, Melbourne University Law Review.

Thompson, R. L., Higgins, C. A., & Howell, J. M. (1991). Personal Computing: Toward a Conceptual Model of Utilization. *Management Information Systems Quarterly, 15*(1), 124–143. doi:10.2307/249443

Thompson, R., Higgins, C., & Howell, J. (1994). Influence of experience on personal computer utilization: Testing a conceptual model. *Journal of Management Information Systems, 1*(11), 167–187.

Thong, J., Hong, W., & Tam, K. (2004). What Leads to User Acceptance of Digital Libraries? *Communications of the ACM, 47*(11), 78–83. doi:10.1145/1029496.1029498

Titah, R., & Barki, H. (2006). E-Government Adoption and Acceptance: A Literature Review. *International Journal of Electronic Government Research, 2*(3), 23–57.

Tolbert, C. J., & Mossberger, K. (2006). The Effects of E-Government on Trust and Confidence in Government. *Public Administration Review, 66*(3), 354–369. doi:10.1111/j.1540-6210.2006.00594.x

Torfing, J. (2005). Governance network theory: towards a second generation. *European Political Science, 24*(4), 305–315. doi:10.1057/palgrave.eps.2210031

Torkzadeh, G., Chang, J., & Demirhan, D. (2006). A contingency model of computer and Internet self-efficacy. *Information & Management, 43*(4), 541–550. doi:10.1016/j.im.2006.02.001

Town of Freetown. MA. (2007). *Top Ten Reasons to Have a Municipal Website.* Retrieved from http://town.freetown. ma.us/dept/faq _detail.asp?DeptID=WEB&FAQID=76

Traumüller, R., & Wimmer, M. (2004). E-Government: the challenges ahead. In Traumüller, R. (Ed.), *EGOV2004, Berlin*. Heidelberg: Springer Verlag.

Triantifillou, P. (2007). Governing the formation and mobilization of governance networks. In Sørensen, E., & Torfing, J. (Eds.), *Theories of democratic network governance* (pp. 183–198). Basingstoke: Macmillan.

Tribunal de Contas da União. http://www.tcu.gov.br 2009. Last access: June, 2010.

Truehits (2006). *Truehits 2005 Awards.* Retrieved November 16, 2006 from: http://truehits.net/awards2005

Tschannen-Moran, M., & Hoy, W. K. (2000). A Multidisciplinary Analysis of the Nature, Meaning, and Measurement of Trust. *Review of Educational Research, 70*(4), 547–593.

Tseng, M. M., Kyellberg, T., & Lu, S. C. Y. (2003). Design in the new e-manufacturing era. *Annals of the CIRP, 52*(2). doi:10.1016/S0007-8506(07)60201-7

Tung, L. L., & Rieck, O. (2005). Adoption of electronic government services among business organizations in Singapore. *The Journal of Strategic Information Systems, 14*(4), 417–440. doi:10.1016/j.jsis.2005.06.001

Turoff, M., Walle, B. V. d., Chumer, M., & Yao, X. (2006). The Design of a Dynamic Emergency Response Management Information System (DERMIS). *Annual Review of Network Management and Security, 1,* 101–121.

UN. (2008). *World public sector report: UN E-Government survey.* New York: From E-Government To Connected Governance.

Undheim, T. A., & Blakemore, M. (Eds.). (2007). *A Handbook for Citizen-centric eGovernment:* cc:eGov.

UNDP. (2008). *Human Development Report 2007/8—Fighting climate change: Human solidarity in a divided World.* Retrieved September 29, 2008, from the United Nations Development Program's Website http://www2.unpan.org/egovkb/ global_reports/08report.htm

United Nations. (2004). *United Nations economic and social commission for Western Asia interim report. Foundations of ICT Indicators Database.* New York: United Nations.

United Nations Department of Economic and Social Affairs (UNDESA). (2003). World Public Sector Report 2003: "e-government at the Crossroads".

United Nations Development Programme. (2004). *Human development report 2004: Cultural liberty in today's diverse world.* New York: United Nations Development Programme.

United Nations. (2002). *Benchmarking e-government: A global perspective—Assessing the progress of the UN member states.* 81pp. Retrieved September 29, 2008, from the United Nations, Division for Public Economics and Public Administration & American Society for Public Administration Website http://unpan1.un.org/intradoc/ groups/ public/documents/un/unpan021547.pdf

United Nations. (2003). *UN Global E-government Survey 2003.* Retrieved November 16, 2006 from: http://unpan1.un.org/int radoc/groups/public/documents/un/ unpan019207.pdf

United Nations. (2004). *UN Global E-government Readiness Report 2004: Towards Access for Opportunity.* Retrieved November 16, 2006 from: http://unpan1.un.org/ intradoc/ groups/public/documents/un/unpan019207.pdf

United Nations. (2005). *Global E-government Readiness Report 2005: From E-government to E-Inclusion.* Retrieved November 16, 2006 from: http://unpan1.un.org/ intradoc/ groups/ public/documents/un/unpan021888.pdf

United Nations. (2008). *E-Government Survey 2008: from e-Government to Connected Governance.* United Nations, NY. Retrieved from http://unpan1.un.org/intradoc/groups/ public/documents/un/unpan028607.pdf

United Nations. (2008). *UN e-government survey 2008: From e-government to Connected governance.* Retrieved May 30, 2008, from the United Nations, Department of Economic and Social Affairs, Division for Public Administration and Development Management Website: http:// unpan1.un.org/intradoc/group s/public/documents/un/ unpan028607.pdf

United Nations. (2010). *E-Government Survey 2010: Leveraging E-Government at a time of Financial and Economic Crisis.* United Nations, NY. Retrieved from http://unpan1.un.org/intradoc/groups/ public/documents/ un/unpan038851.pdf

United Nations. 2001. *E-Commerce and Development Report 2001.* Prepared by United Nations Conference on Trade and Development. New York & Geneva: United Nations. [online edition available at http://www.unctad. org/en/docs/ecdr01ove.en.pdf]

United States Census Bureau. Census 2000. "QT-PL. Race, Hispanic or Latino, and Age: 2000. Data Set: Census 2000 Redistricting Data (Public Law 94-171) Summary File Geographic Area: National City city, California." Available from http://factfinder.census.gov/ servlet/QTTable?_bm=y&-geo_id=16000US0650398&-qr_name=DEC_2000_PL_U_QTPL&-ds_name= D&-_lang=en&-redoLog=false, accessed March 2, 2006.

United States Department of Labor. The Family and Medical Leave Act of 1993, 29 CFR 825.300, 2006. Available from http://www.dol.gov/dol/allcfr/ESA/T itle_29/Part_825/29CFR825.300.htm accessed March 24, 2006.

United States Office of Management and Budget (USOMB). (2005), E-Gov: Powering America's Future With Technology. Retrieved October 5, 2005 from http://www.whitehouse.gov/omb /egov/index.html.

URT-MoCT. (2003). *National Information and Communications Technologies Policy*. Dar es Salaam: Ministry of Communications and Transport, The United Republic of Tanzania.

Uslaner, E. M. (2002). *The Moral Foundations of Trust*. Cambridge: Cambridge University Press. doi:10.1017/CBO9780511614934

van Dijk, J. (2004). *The deepening divide: Inequality in the information society*. London: Sage.

van Dijk, J. (2006). *The network society* (2nd ed.). London: Sage.

van Dijk, J., & Hacker, K. (2000). Summary. In Hacker, K., & van Dijk, J. (Eds.), *Digital democracy: Issues of theory and practice*. London: Sage Publications.

van Dijk, J. (1996). Models of democracy-- behind the design and use of new media in politics. *Javnost/The Public, 3*, 43-56.

van Dijk, J. (2002). A framework for Digital Divide research. *The Electronic Journal of Communication, 12*(1 & 2). Retrieved July 31, 2005 from http://www.cios.org/getfile/vandijk_v12n102

Van Parijs, P. (2004). *Cultural Diversity versus Economic Solidarity*. Brussels: De Boeck Universite.

Van Slyke, C., Belanger, F., & Comunale, C. L. (2004). Factors Influencing the Adoption of Web-Based Shopping: The Impact of Trust. *The Data Base for Advances in Information Systems, 35*(2), 32–49.

Vanston, L., Hodges, R., & Savage, J. (2004). *Forecasts for higher bandwidth broadband services*. Retrieved August 21, 2005 from http://www.tfi.com/pubs/r/r02004_broadband.html

Vassilakis, C., Lepouras, G., Fraser, J., Haston, S., & Georgiadis, P. (2005). Barriers to electronic service development. *E Service Journal, 4*(1), 41–64. doi:10.2979/ESJ.2005.4.1.41

Vassilakis, C., Lepouras, G., Fraser, J., & Georgiadis, P. (2005). Barriers To Electronic Service Development. *e-Service Journal, 4*(1), 41-63.

Vassiliev, A. (2000). *The history of Saudi Arabia*. New York University press.NY,10003.

Venkatesh, V., & Davis, F. D. (2000). A Theoretical extension of the technology acceptance model: four longitudinal field studies. *Management Science, 45*(2), 186–204. doi:10.1287/mnsc.46.2.186.11926

Venkatesh, V., & Morris, M. G. (2000). Why don't men ever stop to ask for directions? Gender, social influence, and their role in technology acceptance and user behavior. *Management Information Systems Quarterly, 24*, 115–139. doi:10.2307/3250981

Venkatesh, V., & Davis, F. D. (1996). A model of the antecedents of perceived ease of use: Development and test. *Decision Sciences, 27*, 451–481. doi:10.1111/j.1540-5915.1996.tb01822.x

Venkatesh, V., & Davis, F. D. (2000). A theoretical extension of the technology acceptance model: Four longitudinal field studies. *Management Science, 46*(2), 186–205. doi:10.1287/mnsc.46.2.186.11926

Venkatesh, V., Morris, M., Davis, G., & Davis, F. (2003). User Acceptance of Information Technology: Toward a Unified View. *Management Information Systems Quarterly, 27*(3), 425–478.

Venkatesh, V., & Davis, F. D. (2000). Theoretical extension of the technology acceptance model: Four longitudinal field studies. *Management Science, 46*(2), 186–204. doi:10.1287/mnsc.46.2.186.11926

Verdegem, P., & Verleye, G. (2009). User-Centered E-Government in Practice: A Comprehensive Model for Measuring User Satisfaction. *Government Information Quarterly, 26*(3), 487–497. doi:10.1016/j.giq.2009.03.005

Viswanath, V., Micha, M., Grdon, D., & Davis, F. (2003). User Acceptance of Information Technology: Toward a Unified View. *MIS Quarterly, 27*(3).

Volken, T. (2002). "Elements of Trust: The cultural dimensions of internet diffusion revisited", *Electronic Journal of Sociology*, Available online http://epe.lac-bac.gc.ca/100/201/300/ejofs ociology/2005/01/volken.html, accessed March 10, 2007.

Waller, P., Livesey, P., & Edin, K. (2001). e-Government in the Service of Democracy. *ICA Information, 74*.

Walsham, G. (1995). Interpretative case in IS research: nature and method. *European Journal of Information Systems, 4*, 74–81. doi:10.1057/ejis.1995.9

Wang, R. Y., & Strong, D. M. (1996). Beyond accuracy: What data quality means to data consumers. *Journal of Management Information Systems, 12*(4), 5–34.

Wang, Y. S., Wang, Y. M., Lin, H. H., & Tang, T. I. (2003). Determinants of user acceptance of internet banking: an empirical study. *International Journal of Service Industry Management, 14*(5), 501–519. doi:10.1108/09564230310500192

Wang, Y., & Emurian, H. (2005). Trust in E-Commerce: Consideration of Interface Design Factors. *Journal of Electronic Commerce in Organizations, 3*(4), 42–60.

Wang, Y., Lin, H., & Luarn, P. (2006). Predicting consumer intention to use mobile service. *Information Systems Journal, 16*(2), 157–179. doi:10.1111/j.1365-2575.2006.00213.x

Wang, Y. S. (2003). The adoption of electronic tax filing systems: an empirical study. *Government Information Quarterly, 20*(4), 333–352. doi:10.1016/j.giq.2003.08.005

Wangpipatwong, S., Chutimaskul, W., & Papasratorn, B. (2005). Factors influencing the use of eGovernment Web sites: Information quality and system quality approach. *International Journal of the Computer, the Internet and Management, 13*(SP3), 14.1–14.7.

Warkentin, M., & Gefen, D. (2002). Encouraging citizen adoption of e-government by building trust. *Electronic Markets, 12*(3), 157–162. doi:10.1080/101967802320245929

Warren, B. (2001). *Uma força Irresistível*. HSM Management.

Warrington, T. B., Abgrab, N. J., & Caldwell, H. M. (2000). Building trust to develop competitive advantage in e-business relationships. *Competitive Review, 10*(2), 160–168.

Warshaw. (1980). A New Model for Predicting Behavioral Intentions: An Alternative to Fishbein. Journal of Marketing Research, 17 (2), 153-172.

Wassenaar, A. (2000). *E-Governmental Value Chain Models*. DEXA (pp. 289–293). IEEE Press.

Wastell, D., Kawalek, P., Langmead-Jones, P., & Ormerod, R. (2004). Information systems and partnership in multi-agency networks: an action research project in crime reduction. *Information and Organization, 14*, 189–210. doi:10.1016/j.infoandorg.2004.01.001

Watson, R. T., & Mundy, B. (2001). A Strategic Perspective of Electronic Democracy. *Communications of the ACM, 44*(1), 27–30. doi:10.1145/357489.357499

Weber, Y., & Schweiger, D. (1992). Top management culture conflict in mergers and acquisitions: a lesson from anthropology. *The International Journal of Conflict Management, 3*(4), 285–302. doi:10.1108/eb022716

Weerakkody, V., & Dhillon, G. (2008). Moving from E-Government to T-Government: A Study of Process Reengineering Challenges in a UK Local Authority Context. *International Journal of Electronic Government Research, 4*(4), 1–16.

Weerakkody, V., Janssen, M., & Hjort-Madsen, K. (2007). Realising Integrated E-Government Services: A European Perspective. *Journal of Cases in Electronic Commerce, 3*(2), 14–38.

Weerakkody, V., Dwivedi, Y.K., Brooks, L., Williams, M. & Mwange, A. (2007). E-government implementation in Zambia: contributing factors. *Electronic Government, an International Journal, 4(4),* 484-508.

Weick, K. E. (1995). *Sensemaking in organisations.* London: SAGE Publications.

Weick, K. E. (2001). *Making Sense of the Organization.* Malden, MA: Blackwell publishing

Weill, P., & Broadbent, M. (1998). *Leveraging the New Infrastructure – How Market Leaders Capitalize on Information Technology.* Boston: Harvard Business School Press.

Weill, P., & Ross, W. J. (2004). *IT Governance: How Top Performers Manage IT Decision Rights for Superior Results.* Boston: Harvard Business Press.

Weinstein, N. D. (1980a). Optimistic Bias about personal risks. *Science, 246,* 1232–1233. doi:10.1126/science.2686031

Weinstein, N. D. (1980b). Unrealistic optimism about future life events. *Journal of Personality and Social Psychology, 39,* 806–820. doi:10.1037/0022-3514.39.5.806

Welch, E. W., Hinnant, C. C., & Moon, M. J. (2005). Linking Citizen Satisfaction with E-Government and Trust in Government. *Journal of Public Administration: Research and Theory, 15*(3), 371–391. doi:10.1093/jopart/mui021

Welch, E. W., Hinnant, C. C., & Moon, M. J. (2005). Linking Citizen Satisfaction with E-Government and Trust in Government. *Journal of Public Administration: Research and Theory, 15*(3), 371–391. doi:10.1093/jopart/mui021

Wellman, B., & Haythornwaite, C. (2002). The Internet in Everyday Life: An Introduction. In Wellman, B., & Haythornwaite, C. (Eds.), *The Internet in everyday life* (pp. 3–41). Oxford, U.K.: Blackwell Publishing. doi:10.1002/9780470774298

Werth, D. (2005). *E-Government Interoperability,* In Khosrow-Pour, M. (Ed.) *Encyclopedia of Information Science and Technology,* I-V., 985-989, Hershey, PA: Idea Group Inc.

West, D. M. (2004). E-Government and the Transformation of Service Delivery and Citizen Attitudes. *Public Administration Review, 64*(1), 15–27. doi:10.1111/j.1540-6210.2004.00343.x

West, D. M. (2005). *Digital Government: Technology and Public Sector Performance.* Princeton, NJ: Princeton University Press.

West, D. M. (2004). E-Government and the Transformation of Service Delivery and Citizen Attitudes. *Public Administration Review, 64*(1), 15–27. doi:10.1111/j.1540-6210.2004.00343.x

West, D. M. (2000). "Assessing E-government: The Internet, Democracy, and Service Delivery by State and Federal Governments", Available online,http://www.insidepolitics.org/ egovtreport00.html, accessed March 2, 2007.

West, D. M. (2003). "Global E-government, 2003". Available onlinehttp://www.insidepolitics.org/egovt03int.pdf, accessed March 3, 2007. White, H., "A Heteroskedasticity—Consistent Covariance Matrix Estimator and a Direct Test for Heteroskedasticity," *Econometrica,* 48(4): 817–838.

Whyte, G., & Bytheway, A. (1996). Factors Affecting Information Systems' Success. *International Journal of Service Industry Management, 7*(1), 74–93. doi:10.1108/09564239610109429

Wikipedia, (2006). Paralanguage. Available from http://en.wikipedia.org/wiki/P aralanguage#Linguistics accessed September 19, 2006.

Wilkin, C., & Castleman, T. (2003). Development of an instrument to evaluate the quality of delivered information systems. *Proceedings of the 36th Hawaii International Conference on System Sciences.* Retrieved December 11, 2007 from: http:// csdl2.computer.org/comp/proce edings/hicss/2003/1874/08/187480244b.pdf

Williamson, O. E. (1975). *Market and Hierarchies, Analysis and Antitrust Implications. A study in the economics of internal organization.* New York: Macmillan.

Wimmer, M. A. (2004). A European perspective towards online one-stop government: the e-Government project. *Electronic Commerce Research and Applications, 1*(1), 92–103. doi:10.1016/S1567-4223(02)00008-X

Wimmer, M. A. (2002). Integrated service modeling for online one-stop Government. *EM - Electronic Markets, special issue on e-Government, 12*(3), 1-8.

Wimmer, M., & Traunmuller, R. (2000). Trends in e- government: managing distributed knowledge. *Proceedings of the 11th International Workshop on Database and Expert Systems Applications.*

Wimmer, M., Liehmann, M., & Martin, B. (2006). Offene Standards und abgestimmte Spezifikationen - das osterreichische Interoperabilittskonzept. In *Proceedings of MKWI.*

World Economic Forum. (WEF). (2007). *The Networked Readiness Index,* available online, http://www.weforum.org/pd f/gitr/rankings2007.pdf, accessed March 2, 2007.

World Fact Book. (2005). Available online at http://www.cia.gov/cia/publ ications/factbook/index.html. Accessed in 6/10/2005.

World Values Study Group. (1994). *World Values Survey, 1984-93* [Computer File]. ICPSR version. Ann Arbor MI: Institute for social Research [producer], 1994. Ann Arbor, MI: Inter-university Consortium for Political and Social Research [distributor].

Wu, Y. L., Tao, Y. H., & Yang, P. C. (2007). Using UTAUT to explore the behavior of 3G mobile communication users. IEEE. 199-203.

Yamane, T. (1973). *Statistics: An Introductory Analysis* (3rd ed.). New York, NY: Harper & Row.

Yang, K., (2003). Neoinstitutionalism and E-government: Beyond Jane Fountain in *Social Science Computer review.* 4(21), p.432-442

Yankelovich, D. (1991). *Coming to judgment: Making democracy work in a complex world.* Syracuse, NY: Syracuse University Press.

Yates (2005). Emerging technology scan: smart cards. Available at <http://www.emory.edu/smartcard.htm> (Accessed December 12, 2006).

Yeow, P. H. P., Loo, W. H., & Chong, S. C. (2007). User acceptance of multipurpose smart identity card in a developing country. *Journal of Urban Technology, 14*(1), 23–50. doi:10.1080/10630730701259862

Yeow, P. H. P., & Miller, F. (2005). The Attitude of Malaysians towards MyKad. In N. Kulathuramaiyer, A. W. Teo, Y. C. Wang & C.E. Tan (Eds.), Proceedings of the 4th International Conference on Information Technology in Asia 2005 (CITA'05), Hilton, Kuching, 12-15 Dec 2005 (pp.39-44). Kuching, Malaysia: Faculty of Computer Science and Information Technology, Universiti Malaysia Sarawak.

Yesser (2006). *E-government project:-website of Saudi Arabian e-government.* Available online at:-http://www.yesser.gov.sa /english/default.asp.

Yildiz, M., (2007). E-government research: Reviewing the literature, limitations and ways forward in *Government Information Quarterly,* 24(3), p. 646-665

Yin, R. K. (1994). *Case Study Research Design and Methods.* London: Sage.

Yin, R. K. (1994). *Case Study Research - Design And Methods* (2nd ed.). London: Sage Publications.

Yin, R. K. (2003). *Case study research: design and methods* (3rd ed.). Thousand Oaks, CA: Sage Publications.

Yong, J. S. L., & Koon, L. H. (2003). E-government: Enabling public sector reform. In Yong, J. S. L. (Ed.), *Enabling Public Service Innovation in the 21ˢᵗ Century E-Government in Asia* (pp. 3–21). Singapore: Times Editions.

Yong, J. S. L., & Leong, J. L. K. (2003). Digital 21 and Hong Kong's advancement in E-government. In Yong, J. S. L. (Ed.), *Enabling Public Service Innovation in the 21ˢᵗ Century E-Government in Asia* (pp. 97–116). Singapore: Times Editions.

Yuan, Y., Zhang, J., & Zheng, W. (2004). Can e-government help China meet the challenges of joining the World Trade Organization? *Electronic Government, 1*(1), 77–91. doi:10.1504/EG.2004.004138

Yusuf, T. O., & Babalola, A. R. (2009). Control of insurance fraud in Nigeria: an exploratory study. *Journal of Financial Crime, 16*(4), 418–435. doi:10.1108/13590790910993744

Zak, P. J., & Knack, S. (2001). Trust and Growth. *The Economic Journal, 111*(April), 295–321. doi:10.1111/1468-0297.00609

Zakareya, E., & Irani, Z. (2005). E-government adoption: Architecture and barriers. *Business Process Management Journal, 11*(5), 589–611. doi:10.1108/14637150510619902

Zeithaml, V. A., Berry, L. L., & Parasuraman, A. (1996). The behavioral consequences of service quality. *Journal of Marketing, 60*(2), 31–36. doi:10.2307/1251929

Zhiyuan, F. (2002). E-Government in Digital Era: Concepts, Practice and Development. *International Journal Of The Computer. The Internet and Management, 10*(2), 1–22.

Ziemann, J. (2010). *Architecture of Interoperable Information Systems – An Enterprise Model-based Approach for Describing and Enacting Collaborative Business Processes.* Berlin: Logos Verlag.

Ziemann, J., Kahl, T., & Matheis, T. (2007). *Cross-organizational Processes in Public Administrations: Conceptual modeling and implementation with Web Service Protocols. 8.* Karlsruhe: Internationale Tagung Wirtschaftsinformatik.

Zinnbauer, D. (2001). Internet, Civil Society and Global Governance: The Neglected Political Dimension of the Digital Divide. *Information & Security., 7*, 45–64.

Zucker, L. G. (1986). Production of trust: Institutional sources of economic structure. *Research in Organizational Behavior, 8*(1), 53–111.

About the Contributors

Vishanth Weerakkody is a member of faculty in the Business School at Brunel University (UK). VW was previously a faculty member in the Department of IS and Computing at Brunel University and he has held various IT positions in multinational organizations, including IBM UK. VW is a member of the British Computer Society, Chartered IT professional and a Fellow of the UK Higher Education Academy. He is the current Editor-in-Chief of the International Journal of Electronic Government Research.

* * *

Hana Al-Nuaim is an assistant professor in Computer Science and the Dean of King Abdulaziz University (Women's Campuses), Jeddah, Saudi Arabia. She received her Ms, DSc in CS from George Washington University, USA, and her Bachelours in CS from the University of Texas at Austin. She was a former Department Chair and as Vice Dean of Distance Learning launched the first e-learning program in the kingdom for women. She has extensive faculty training background and referred papers for publications. Her publications and primary research interests are in HCI, usability, multimedia, e-learning, e- government and knowledge cities and has been involved in many web-based research projects.

Shafi Al-Shafi is a PhD researcher at the School of Information Systems, Computing and Mathematics, Brunel University, UK. He obtained his Master in Business Administration MBA (Information Technology) from the Faculty of Engineering and Computing, Coventry University in the UK. His research interests include e-government adoption and diffusion, impact of integration technologies, strategic management and e-commerce.

Kenneth Allendoerfer is an engineering research psychologist with the Federal Aviation Administration, where he has been developing and evaluating air traffic control systems for more than 10 years. His research interests include designing user interfaces for high-stakes environments, information visualization, usable information security, and incorporating user-centered methods into formal software development methodologies. Mr. Allendoerfer holds degrees in psychology from Carleton College and the State University of New York at Buffalo, and is currently a doctoral candidate at Drexel University.

James (Jim) R. Beatty, PhD, is a Professor of Information and Decision Systems, an Adjunct Professor of Management, and coordinator of Quantitative Methods in the College of Business Administration at San Diego State University. He serves on the faculty for the Executive MBA program and was one of the founding coordinators for SDSU's Institute for Quality and Productivity. He has served on the

National Board of Examiners for the Malcolm Baldrige National Quality Program nine times and has led site visits and assisted organizations through Baldrige self-assessment. He was named Professor of the Year for SDSU in 2001. He has over 100 publications, textbooks, and papers. He is a Certified Compensation Professional (CCP) with WorldatWork and a Senior Professional in Human Resources (SPHR) with the American Society for Personnel Administration. He earned a Ph.D. with a double major in statistics and psychology and completed postdoctoral training in multivariate statistics.

John J. Burbridge, Jr. is a Professor of Business Administration and the former Dean of the Love School of Business at Elon University. Prior to Elon, Dr. Burbridge was Associate Dean of the Joseph A. Sellinger School of Business at Loyola College in Maryland from 1992 to 1996. Dr. Burbridge received his B.S., M.S., and Ph.D. in industrial engineering from Lehigh University in Bethlehem, Pennsylvania. He has made numerous presentations and authored refereed journal articles in journals such as Journal of International Business Studies, Competitiveness Review, Journal of Global Information Management, Journal of Applied Business and Economics, and others.

Lemuria Carter is an assistant professor at North Carolina Agricultural & Technical State University. Her primary research interests include e-government diffusion, internet voting, Web-based trust and risk perceptions. She has published in several top-tier journals including, Journal of Strategic Information Systems, Information Systems Journal, and Communications of the ACM. She also serves as an e-government mini-track co-chair at Americas Conference on Information Systems (AMCIS) and the Hawaii International Conference on Systems Sciences (HICSS).

Wichian Chutimaskul received his Ph.D. from the University of Sheffield, U.K. in 1994. He is a member of the International Federation for Information Processing Technical Committee on Information System for Public Administration (IFIP 8.5). He is now working as the Senior Associate Dean for Academic Affairs at the School of Information Technology, King Mongkut's University of Technology Thonburi, Thailand. He is also a member of the University Academic Broad, and the chairperson for the M.Sc. and Ph.D. programs in information technology. His international involvement also includes Thai-Japanese ICT scientific exchange, Thai-Austria Scientific exchange, and Thai University Administrators' Shadowing Program in Canada. His current research is in the areas of e-Government and software development methodology.

Jayoti Das is an Associate Professor of Economics at the Love School of Business at Elon University. She holds a Masters degree and a Ph.D. from University of Cincinnati. Her research interests are international trade and finance, industrial organization, global business and development issues. She has published in journals such as the Journal of International Business Studies, Journal of International Trade and Economic Development, Competitiveness Review, Journal of Applied Business and Economics, International Business and Economic Research Journal, Journal of Global Information Management, Journal of Global Business, Journal of Contemporary Business Issues, Journal of Teaching in International Business, and others.

Cassandra E. DiRienzo is an Assistant Professor of Economics at the Love School of Business at Elon University. She holds a M.E. and a Ph.D. from North Carolina State University. Her research interests are econometrics and statistical forecasting as well as economic and business applications of nonparametric, spatial, and multivariate statistics. She has published in the Journal of International Business Studies, Challenge, Competitiveness Review, Journal of Applied Business and Economics, and others.

Ferne Friedman-Berg is a Federal Aviation Administration engineering research psychologist. She has conducted research on air traffic control systems, tools, and websites for over six years. Her research interests include performing human-in-the-loop simulations and analyzing user interaction patterns to better understand expert cognition and to aid in the design of more effective and more usable air traffic control tools. She is also interested in using eye-movement metrics to measure workload and as an index of interest in display design components. Dr. Friedman-Berg received her B.A. in psychology from Temple University and her M.S. and Ph.D. in cognitive psychology from the University of Massachusetts, Amherst.

Katarina Lindblad Gidlund is an Associate Professor in Social Informatics, Mid Sweden University, Sweden, where she is the leader of the CITIZYS Research Group, a multidisciplinary research group focusing on IT and sustainable development. Her research interest is directed towards accountability in IT innovation and design processes in order to enhance quality. She is the co-coordinator of the national eGovernment Research Network in Sweden, editor of the International Journal of Public Information Systems and part of the programme committee of a number of international conferences. She is also on the board for the National Association for eCompetence (REK) and a member of the jury for Guldlänken, a national award for the best public e-service in Sweden.

Petter Gottschalk is Professor in the Department of Leadership and Organization at the Norwegian School of Management in Norway. He teaches knowledge management and information systems. He received his MBA in Germany (Technical University of Berlin), MSc in the US (Dartmouth and MIT), and DBA in the UK (Henley Management College). He has been the CEO of ABB Datacables, Norwegian Computing Center, and Norwegian Information Technology Inc.

Kenneth L. Hacker (Ph.D., University of Oregon) is Professor of Communication Studies at New Mexico State University. He co-edited the book (with Jan van Dijk) Digital Democracy: Issues of Theory and Practice (Sage, 2000) and is currently editing (Jan van Dijk, co-editor) a second volume on digital democracy (Digital Democracy in a Network Society, Hampton Press).

Marijn Janssen is an Associate Professor within the Information and Communication Technology section and Director of the interdisciplinary SEPAM Master programme of the Faculty of Technology, Policy and Management at Delft University of Technology. He is also in charge of the "IT and business architecture" of the Toptech executive master. He has been a consultant for the Ministry of Justice and received a Ph.D. in information systems (2001). He serves on several editorial boards (including International Journal of E-Government Research, Government Information Quarterly and Information Systems Frontiers), has many research projects and is involved in the organization of a number of conferences. His research interests are in the field of e-government, design science, orchestration and shared services. He was ranked as one of the leading e-government researchers and published over 160 refereed publications. More information: www.tbm.tudelft.nl/marijnj.

Murray E. Jennex is an associate professor at San Diego State University, editor in chief International Journal of Knowledge Management, co-editor in chief International Journal of Information Systems for Crisis Response and Management, editor in chief of Idea Group Publishing's Knowledge Management book series, and president of the Foundation for Knowledge Management (LLC). Dr. Jennex specializes in knowledge management, system analysis and design, IS security, e-commerce, and organizational effectiveness. Dr. Jennex serves as the Knowledge Management Systems Track co-chair at the Hawaii International Conference on System Sciences. He is the author of over 100 journal articles, book chapters, and conference proceedings on knowledge management, end user computing, international information systems, organizational memory systems, ecommerce, security, and software outsourcing. He holds a B.A. in chemistry and physics from William Jewell College, an M.B.A. and an M.S. in software engineering from National University, an M.S. in telecommunications management and a Ph.D. in information systems from the Claremont Graduate University. Dr. Jennex is also a registered professional mechanical engineer in the state of California and a Certified Information Systems Security Professional (CISSP).

Janet Kaaya is a research scholar in the Department of Information Studies, Graduate School of Education and Information Studies, University California Los Angeles (UCLA). One of her primary research interests is using institutional theories to examine the adoption of ICT-related applications, focusing on e-government implementation – especially in developing countries. Other areas include studying national and international information equity issues; the influence of information and associated communications technologies (ICTs) on institutional or social change; language barriers and Internet access; and exploring and assessing web-based information resources for research and pedagogical uses. She received her PhD in information studies from UCLA and was until recently a CLIR postdoctoral fellow at Charles E. Young Research Library, UCLA. She previously worked as a field researcher and an information specialist in the national agricultural research system of Tanzania.

Muhammad Kamal is a Research Fellow at the Brunel Business School, Brunel University in the UK. He received his PhD from Brunel University in the area of Enterprise Application Integration (EAI) Adoption in the in the Local Government Authorities (LGAs), in addition, he holds two MSc's – Distributed Computing Systems (Greenwich University, London) and Computer Sciences (Punjab Institute of Computer Science, Lahore, Pakistan). His current research interest includes investigating and evaluating factors influencing the decision making process for technology adoption in the local government domain, electronic service delivery and transformation in the public sector and integration of Health and Social Care services. He specialises in the use of data analysis methods such as Discrete Choice Analysis and Analytical Hierarchy Process. In his recent appointment, Dr. Kamal worked as a Post-Doctoral Research Assistant on a European funded research project (Ref: proposal n° 2006/VP021/30137) namely – the REFOCUS (senioR Employees training on inFOrmation and CommUnication technologieS) Project. Prior to that, Dr. Kamal acted as a Business Analyst in a research project that focused on the integration of Customer Relationship Management (CRM) applications in the London Borough of Havering (co-funded by ORACLE UK). Currently, he is an active editorial review board member for International Journal of Electronic Government Research (IJEGR) and he has published in several leading journals and international conferences. Dr. Kamal is a member of the Association of Information Systems (AIS) and Information Systems Evaluation & Integration Network (ISeng) at Brunel Business School.

Dionysis Kefallinos is a network engineer for the General Secretariat of Information Systems of Greece. He holds a Dipl.-Ing. degree in Electrical and Computer Engineering from the National Technical University of Athens, Department of Electrical and Computer Engineering and he is a doctoral candidate in the same university.

Bram Klievink is a researcher with the ICT group of the Faculty of Technology, Policy and Management at Delft University of Technology. He holds a MSc degree in political science as well as a degree in business information systems. He is currently working on a PhD research, focused on coordination mechanisms for public-private service networks. For this research, he developed a role-play game to investigate and facilitate coordination arrangements in multi-actor service delivery. Furthermore, he participated in research projects on integrated, demand-driven e-government and on multi-channel management in government. He is also a board member of the Dutch Alliance for Vital Governance, a strategic research alliance for collaboration and knowledge transfer between government organizations and research institutes in the Netherlands

Maria A. Lambrou is an assistant professor at the University of the Aegean, Business School, Department of Shipping, Trade and Transport. She received her Ph.D. from the National Technical University of Athens (NTUA), Department of Electrical and Computer Engineering – Computer Science Division, in the area of Service Engineering (1998) and she holds an M.A. in Management studies and Industrial Relations from Brunel University (1993).

Karl Löfgren, PhD, associate professor in public administration at the Department of Society and Globalisation, Roskilde University and member of the Centre for Democratic Networks. He is mainly studying issues around ICT-policy with special reference to electronic government and democracy, and is currently involved in research on network governance and electronic government. Recent publications in the area are: Hall, P. & Löfgren, K. (2005) The Rise and Decline of a Visionary Policy: Swedish ICT-policy in Retrospect, Information Polity, 9(3-4): 149 – 165, Löfgren, K (2007) The Governance of E-government: A Governance Perspective on the Swedish E-government Strategy, Public Policy and Administration, 22(3): 335-352.

Peter Loos is director of the Institute for Information Systems (IWi) at the German Research Center for Artificial Intelligence (DFKI) and head of the chair for Business Administration and Information Systems at Saarland University. His research activities include business process management, information modelling, enterprise systems, software development as well as implementation of information systems. He graduated in business administration and information systems at Saarland University and wrote his PhD thesis on the issue of data modelling in manufacturing systems. During his earlier career Prof. Loos had been chair of information systems & management at University of Mainz (2002-2005), chair of information systems & management at Chemnitz University of Technology (1998-2002) as well as deputy chair at University of Muenster. Furthermore, he had worked for 6 years as manager of the software development department at the software and consulting company IDS Scheer. He has written several books, contributed to 30 books and published more than 100 papers in journals and proceedings.

W.H. Loo is a Lecturer in Stanford College, Melaka, Malaysia. She is also a Master of Philosophy student in the Faculty of Business and Law, Multimedia University, Malaysia. She has a first class honour Bachelor of Business Administration degree from Multimedia University.

Fabio Perez Marzullo is currently a Doctorate student in the Database Laboratory of the Federal University of Rio de Janeiro's Department of Computer Science and Engineering. He received his M.Sc. degree in Computer Science and Engineering from UFRJ in 2006. He has been a member of research staff in the Brazilian Ministry of Defense, Brazilian Navy and COPPE/UFRJ since 2007. Author of the book: SOA na Prática: inovando seu negócio por meio de soluções orientadas a serviços, and his research interests are Services Oriented Architectures, Model-Driven and Domain-Driven Developments, Database, Project Management, Component Based Development, Component Architecture, and Frameworks Design, Components for software development automation and IT Governance.

Shana M. Mason (M.A., New Mexico State University) is a Communication Studies instructor at the Dona Ana Branch of New Mexico State University. Her work has appeared in IT & Society and Ethics and Information Technology. Her research interests include CMC and political participation.

Thomas Matheis studied at Saarland University and holds a Diploma in Computer Science (Dipl.-Inform.). Since 2005 he works as researcher and PhD student at the Institute for Information Systems (IWi) at the German Research Center for Artificial Intelligence (DFKI). He is involved in various national and international research and consulting projects at the DFKI (e.g. R4eGov IP). His main research interests are collaborative business process management, collaborative performance measurement, business integration and eGovernment. In this area he published various articles in journals and proceedings. Besides his project work he has been a lecturer at the University of Leipzig in the field of business process modelling and information systems.

Eric L. Morgan (Ph.D. University of Massachusetts) is an Associate Professor of Communication Studies at New Mexico State University. His research interests include cultural/ intercultural, international, and environmental communication.

Borworn Papasratorn is an associate professor of information technology at King Mongkut's University of Technology Thonburi (KMUTT), Thailand. He is currently a Dean of the School of Information Technology at KMUTT. Professionally he is executive director of Thailand Telecommunication Research and Industrial Development Institute. He is a member of the IEEE Computer and Communication Society. He received his D.Eng. (Electrical Engineering) from Chulalongkorn University. His research interests include e-learning, computer graphics, performance analysis, and networking.

Shantanu Pai works as a Human Factors Specialist with Engility Corporation and has been supporting human factors research endeavors at the Federal Aviation Administration Research Development and Human Factors lab for the past seven years. He has a Masters degree in human factors from the State University of New York at Buffalo and is currently a doctoral candidate in Drexel University's College of Information Science and Technology.

Christopher G. Reddick is an Associate Professor and Chair of the Department of Public Administration at the University of Texas at San Antonio, USA. Dr. Reddick's research and teaching interests is in e-government. Some of his publications can be found in Government Information Quarterly, Electronic Government, and the International Journal of Electronic Government Research. Dr. Reddick recently edited the book entitled Handbook of Research on Strategies for Local E-Government Adoption and Implementation: Comparative Studies.

Ludwig Christian Schaupp is an assistant professor at West Virginia University. His primary research interests include website success metrics, e-government adoption and diffusion, and REA. He has published in several top-tier journals including Communications of the ACM, Journal of Computer Information Systems, and Information Systems Frontiers. He has also presented at several international conferences including Americas Conference on Information Systems (AMCIS) as well as the Hawaii International Conference on Systems Sciences (HICSS).

Roy H. Segovia was born and raised in Dallas, Texas. He received his BSEE from Rice University in Houston, Texas. He lived more than 20 years in the southern California region, mostly in San Diego where he worked for a number of technology companies. He also became involved in the Latino communities of the border region of San Diego and Tijuana. In 2002 he started in the MBA program at San Diego State University with a focus on entrepreneurship and bilingual communications on the world wid web. He completed the MBA program and his research on "paralingual" websites for government in August of 2006. He is currently living in the Silicon Valley area of California and is working on various technology projects, including further use of paralingual pages for government agencies.

Daniel M. Schmidt studied informatics with focus on business informatics and graduated at Koblenz University. He started his academic career at the Research Group Modelling and Simulation of Prof. Dr. Klaus G. Troitzsch at Koblenz. His primary working field was research on supply chain management in the automotive industry. Since 2006, Daniel M. Schmidt is member of Prof. Dr. Maria A. Wimmer's Research Group eGovernment at Koblenz University. He does research on the challenges of interoperability in public administration with scope placed on cross-organizational design and implementation of business processes, business process patterns and document engineering for information exchange. Furthermore, he works on application of Service Oriented Architectures and semantic web technologies to ensure cross-organisational and cross-border interoperability in eGovernment. He is engaged in EC-co-funded research and development projects, in German national eGovernment projects, participating in CEN BII, member of IEEE and the eGov expert group of GI e.V.

Jano Moreira de Souza graduated in Mechanical Engineering from Universidade Federal do Rio de Janeiro (1974), Master of Computer and Systems Engineering, Federal University of Rio de Janeiro (1978) and Ph.D. in Information Systems - University of East Anglia (1986). Currently a professor at the Federal University of Rio de Janeiro. He has experience in computer science, with emphasis on the following topics: Database, Knowledge Management, Support Systems Trading, Autonomic Computing, and Computer Supported Cooperative Work.

Katarina Giritli Nygren is a senior lecturer in Sociology, Department of Social sciences, Mid Sweden University, Sweden. Her research interests are connected by a main theoretical interest on how to understand the relation between social reproduction and social change from a critical perspective. Her doctoral thesis focused on the organizational consequences and social meaning of e-government in a local government setting. The aim was to study e-government as an administrative reform, from the global creation of a superstandard to the local practices of everyday work. She has also written about the co-construction of ICT, gender and skills. In her most recent research project she focuses on place as a site where processes of social reproduction and social change take place in terms of inclusion and exclusion concerning gender, ethnicity and class.

Eva Sørensen, PhD, professor in public administration and democracy at the Department of Society and Globalisation, Roskilde University, vice-director of the Centre for Democratic networks governance, and director of a large research project on collaborative innovation in the public sector. She has published extensively on new forms of governance and their impact on democracy and effective problem solving. Recent publications are: Sørensen, E. & P. Triantafillou (2009) The Politics of Self-governance. London: Ashgate. and Sørensen, E. & J. Torfing, (2009) 'Enhancing Effective and Democratic Network Governance through Metagovernance' Public Administration, 87(2): 234-258.

Efstathios D. Sykas is a professor in Communications Engineering, at the School of Electrical and Computer Engineering of the National Technical University of Athens. He received Dipl.-Ing. and Dr.-Ing. degrees in Electrical Engineering, both from the National Technical University of Athens.

Sivaporn Wangpipatwong received her Ph.D. in information technology from King Mongkut's University of Technology Thonburi, Thailand. She is a former chairperson of the Computer Science Department, School of Science and Technology, Bangkok University, Thailand. She is now working as an assistant professor at the School of Science and Technology, Bangkok University. Her current research interests include e-government, e-learning, and the diffusion of information technology in developing countries.

Maria A. Wimmer is Professor for eGovernment at the University of Koblenz. She graduated in computer Sciences (Linz, AT), and first worked in the area of designing safety critical systems for train traffic control in Italy. Her doctoral thesis suggests holistic design methods for this area. In 1999, she returned to the University of Linz and in 2004-2005, Maria A. Wimmer was with the Federal Chancellery in Austria. Since 2005 Maria A. Wimmer is with Koblenz University. She researches and teaches ICT in the public sector since about ten years, thereby specifically addressing among others: strategic modernization issues, process management, interoperability, systems architectures, ontology and semantic web usage, eParticipation. She is engaged in a number of EC-co-funded research and development projects, is co-organizer of scientific conferences, chair of IFIP WG 8.5, member of ACM, IEEE, and published more than 140 articles in eGovernment.

Paul H.P. Yeow is a Senior Lecturer in the Faculty of Business and Law, Multimedia University, Malaysia. He graduated from the National University of Singapore, Multimedia University and University of Science Malaysia with Bachelor of Computer Science, Master of Engineering, and Doctor of Philosophy in Management. Over the past 10 years, he has been giving lectures on Management Information Systems and Electronic Commerce to undergraduate and postgraduate students. He has published about 50 international refereed journal and conference papers in areas of Information Systems, Usability, Human Factors, Quality and Knowledge Management. He is the programme leader of Service Embodied Technology and Application research program. He is a member of the editorial board in Journal of Urban Technology and International Journal of Applied Decision Science. Previously, he was the Head of Information Technology in United Industry Chemicals, Singapore and the Quality and Training Manager of Flextronics Technology Private Ltd., Malaysia.

Jörg Ziemann has studied at the University of Göttingen, the University of California (Irvine, USA) and the University of Hamburg, were he received a Diploma in Business Computer Science (Dipl.-Wirt.-Inf.). In 2004 he joined the Institute for Information Systems (IWi) at the German Research Center for Artificial Intelligence (DFKI) where he worked on various EU projects in the area of cross-organizational business processes, including Interop NOE, ATHENA IP and R4eGov IP. His research interests comprise the fields of collaborative business process management, enterprise-model driven development of service-oriented architectures and eGovernment. In this area he published various articles in journals and proceedings. Currently, Jörg is finalizing his PhD-thesis.

Index

Breinigsville, PA USA
18 December 2010
251717BV00005B/3/P